High-Pressure Shock Compression of Condensed Matter

Springer
New York
Berlin
Heidelberg
Barcelona
Budapest
Hong Kong
London
Milan
Paris
Santa Clara
Singapore
Tokyo

High-Pressure Shock Compression of Condensed Matter

J. Asay and *M. Shahinpoor* (Eds.): High-Pressure Shock Compression of Solids
A.A. Batsanov: Effects of Explosion on Materials: Modification and Synthesis Under High-Pressure Shock Compression
R. Chéret: Detonation of Condensed Explosives
L. Davison, D. Grady, and *M. Shahinpoor* (Eds.): High-Pressure Shock Compression of Solids II
R. Graham: Solids Under High-Pressure Shock Compression
M. Sućeska: Test Methods for Explosives

Lee Davison Dennis E. Grady
Mohsen Shahinpoor
Editors

High-Pressure Shock Compression of Solids II

Dynamic Fracture and Fragmentation

With 527 Illustrations

Springer

Lee Davison
Engineering Science Center
Sandia National Laboratories
Albuquerque, NM 87185
USA

Dennis E. Grady
Experimental Impact Physics Dept.
Sandia National Laboratories
Albuquerque, NM 87185
USA

Mohsen Shahinpoor
University of New Mexico
Department of Mechanical Engineering
Albuquerque, NM 87131
USA

Editor-in-Chief:
Robert A. Graham
Advanced Materials Physics and Device Department, 1153
Sandia National Laboratories
Albuquerque, NM 87185-1421, USA

Library of Congress Cataloging in Publication Data
High pressure shock compression of solids II: dynamic fracture and fragmentation / Lee Davison, D.E. Grady, M. Shahinpoor, editors.
p. cm.—(High pressure shock compression of condensed matter)
Includes bibliographical references and index.
ISBN-13: 978-1-4612-7501-5 e-ISBN-13: 978-1-4612-2320-7
DOI: 10.1007/978-1-4612-2320-7
1. Materials at high pressures. 2. Shock (Mechanics) 3. Materials—Compression testing. 4. Fracture mechanics. I. Davison, L.W. (Lee W.) II. Grady, D.E. III. Shahinpoor, Mohsen. IV. Series.
TA417.7.C65H55 1995
620.1'1242—dc20 94-37843

Printed on acid-free paper.

Softcover reprint of the hardcover 1st edition 1996

Production managed by Terry Kornak, manufacturing supervised by Jeffrey Taub.
Typeset by Asco Trade Typesetting Ltd., Hong Kong.

9 8 7 6 5 4 3 2 1

Preface

This volume concerns the fracture and fragmentation of solid materials that occurs when they are subjected to extremes of stress applied at the highest possible rates.

The plan for the volume is to address experimental, theoretical, and computational aspects of high-rate dynamic fracture and fragmentation, with emphasis on recent work. We begin with several chapters in which the emphasis falls on experimental methods and observations. These chapters address both macroscopic responses and the microscopic cause of these responses. This is followed by several chapters emphasizing modeling—the physical explanation and mathematical representation of the observations. Some of the models are deterministic, while others focus on the stochastic aspects of the observations. Often, the overall objective of investigation of dynamic fracture and fragmentation phenomena is provision of a means for predicting the entire course of an event that begins with a stimulus such as an impact and proceeds through a complicated deformation and fracture process that results in disintegration of the body and formation of a rapidly expanding cloud of debris fragments. Analysis of this event usually involves development of a continuum theory and computer code that captures the experimental observations by incorporating models of the important phenomena into a comprehensive description of the deformation and fracture process. It is to this task that the work of the last few chapters is devoted. The authors whose work comprises this volume have been selected on the basis of their long experience with the subject matter or because they bring new insight and/or methods to the investigation. In all cases, they are among the foremost experts in their field.

The most extreme loads are those that are imposed by high-velocity impact, contact with a detonating explosive charge, or sudden deposition of thermal energy. The material response to these stimuli is formation and propagation of a shock wave, a disturbance that provides access to the highest obtainable rates of deformation. The nature of these loading processes is such that the load is applied only briefly, so the compression produced by the shock is quickly relieved by a following decompression wave.

As a result, the loading event is communicated into the interior of the material by a compression pulse propagating at approximately the soundspeed in the medium. As is well known, a propagating compression pulse reflects from a stress-free surface as a tensile pulse. In the absence of fracture, this tensile pulse has magnitude and duration similar to those of the incident compression pulse. Under typical experimental conditions, the peak compressive stress produced by the loading event falls in the range of 0.1 to 100 GPa (1 GPa = 10 kbar = 145,038 psi) and the duration of the compression pulse ranges from a few nanoseconds to a few microseconds. Because of the very large amplitude of these pulses, the tensile stresses that arise upon reflection can easily exceed the strength of even the strongest materials. The rate at which this tension develops is dependent on both the loading conditions and the properties of the material, but it is usually higher than tensile loading rates obtainable by other means. The brief duration of load application in a typical experiment precludes growth of a single, extended, crack. Rather, damage occurs as an aggregation of smaller, independently formed cracks or voids. These small defects may coalesce into one or more large cracks that are the macroscopic manifestation of the event.

Fractures produced by reflection of strong compressive pulses, now called *spall* fractures, were discussed at least as early as 1912, and became a subject of intensive investigation beginning with the war period of the 1940s. In even the earliest investigations, it was recognized that spall fracture was an evolutionary process in which the material proceeded from an undamaged state through various levels of damage to complete separation of the material body into several, or even very many, discrete parts. For many years, and for a variety of reasons, investigation of this process proceeded independently of investigations of other fracture phenomena. However, this research is now beginning to draw on information obtained in the course of more conventional investigations of fracture phenomena, and the lines of investigation are converging in a subject area called *continuum damage mechanics* that is based on explicit recognition of both the distributed nature of the damage and its continuous evolution in time.

Research in dynamic fracture and fragmentation was originally motivated by military applications, and these applications continue to provide much of the support for this work. Nevertheless, the results have been applied to problems of explosive metalworking, rock blasting, design of shields to protect satellites from impact by space debris, astrophysical events, and even to understanding the mechanism by which impacting sand particles damage automotive finishes. Investigation of fracture and fragmentation is a natural extension of long-standing concern for material deformation, a topic that has broadened from concentration on small elastic deformations to consideration of large inelastic deformations and, inevitably, to material failure. The novel viewpoint and experimental methods brought to this subject by scientists studying spallation have had, and continue to have, a profound influence on development of the field generally.

Among the topics not covered in this volume are the growth and coalescence of individual cracks and voids. These matters are of great importance in engineering and serve as a basis for understanding all fracture phenomena, but form a subject much too large to delve into in this volume. It is possible to gain insight into fracture and fragmentation processes through molecular–dynamic simulations, but this topic was not addressed. All of the work discussed in this volume concerns fracture as a single event, so consideration of complicated comminution processes is omitted. Most regrettably, little material on applications has been included.

This is the second volume on this topic. The first volume* is of more general and tutorial nature, and the reader is invited to consult it for background material and further references.

Lee Davison
D. E. Grady
Mohsen Shahinpoor

---

* *High-Pressure Shock Compression of Solids*, edited by James R. Asay and Mohsen Shahinpoor, Springer-Verlag, 1993

Contents

Contributors

J.A. Ang
Sandia National Laboratories, Albuquerque, NM 87185, USA.
S.J. Bless
Institute for Advanced Technology, University of Texas at Austin, Austin, TX 78759, USA.
D.R. Curran
SRI International, Menlo Park, CA 94025, USA.
R.D. Dick
Department of Mechanical Engineering, University of Maryland, College Park, MD 20742, USA.
J.K. Dienes
Theoretical Division, Los Alamos National Laboratory, Los Alamos NM 87545, USA.
J. Eftis
Department of Civil, Mechanical, and Environmental Engineering, The George Washington University, Washington, D.C. 20052, USA.
R. Englman
Soreq NRC, Yavne 70600, Israel.
W.L. Fourney
Department of Mechanical Engineering, University of Maryland, College Park, MD 20742, USA.
I. Gilath
Department of Applied Physics and Mathematics, Soreq NRC, Yavne 70600, Israel.
D.E. Grady
Sandia National Laboratories, Albuquerque, NM 87185, USA.
B.D. Hansche
Sandia National Laboratories, Albuquerque, NM 87185, USA.
G.I. Kanel'
High Energy Density Research Center, Russian Academy of Sciences, Moscow 127412, Russia.

M.E. Kipp
Sandia National Laboratories, Albuquerque, NM 87185, USA.
M.W. Lewis
Advanced Engineering Technology Group, Los Alamos National Laboratory, Los Alamos, NM 87545, USA.
Liu Cangli
Southwest Institute of Fluid Physics, Chengdu, Sichuan 610003, China.
M.-A. Meyers
Department of Applied Mechanics and Engineering Sciences, The University of California, San Diego, CA 92093, USA.
A.J. Piekutowski
University of Dayton Research Institute, Dayton, OH 45469, USA.
A.M. Rajendran
U.S. Army Materials Technology Laboratory, Watertown, MA 02172, USA.
S.V. Razorenov
High Energy Density Research Center, Russian Academy of Sciences, Moscow 127412, Russia
H.W. Schreyer
Department of Mechanical Engineering, The University of New Mexico, Albuquerque, NM 87131, USA.
L. Seaman
SRI International, Menlo Park, CA 94025, USA.
Sun Chengwei
Southwest Institute of Fluid Physics, Chengdu, Sichuan 610003, China.
D.L. Tonks
Applied Theoretical Division, Los Alamos National Laboratory, Los Alamos, NM 87545, USA.
A.V. Utkin
High Energy Density Research Center, Russian Academy of Sciences, Moscow 127412, Russia.
Wang Yanping
Southwest Institute of Fluid Physics, Chengdu, Sichuan 610003, China.
Zhuang Shiming
Southwest Institute of Fluid Physics, Chengdu, Sichuan 610003, China.
A.K. Zurek
Los Alamos National Laboratory, Los Alamos, NM 87545, USA.

CHAPTER 1

Spallation in Solids Under Shock-Wave Loading: Analysis of Dynamic Flow, Methodology of Measurements, and Constitutive Factors

G.I. Kanel', S.V. Razorenov, and A.V. Utkin

1.1. Introduction

Spallation occurs when two rarefaction waves interact in a body and produce enough tension to fracture it. The principal content of investigations of spall phenomena that have been conducted are: (i) metallographic examination of the spall zone in recovered samples to obtain the correlation of the degree of damage with various features of the stress history [1–3] and (ii) instrumental measurement of resistance to dynamic fracture [4–6]. In the former case the result is information about the ability of the material to withstand shock loading, about the mechanism of nucleation and development of micro-cracks or pores, and statistical description of these processes. Dynamic measurements (free-surface velocity profiles or pressure on the interface with a soft material) during the shock-wave loading give the most correct data about stresses present during the spall process. The theoretical background of these measurements is discussed in Section 1.2 of this chapter.

The damage rate, which is approximately equal to the product of concentration of damage nucleation sites and the average growth rate of damage at the sites, cannot be arbitrarily large. Under conditions of rapid dynamic loading, fracture is an essentially rate-dependent process. As a result, the spall strength value is not a material constant. It is more correct to speak about the resistance to fracture as a function of strain rate and other parameters of state. Since the fracture evolves over a period of time comparable with the duration of a shock wave experiment, we have to identify the stage of fracture which corresponds to the fracturing stress as this stress is determined by dynamic measurements. The acoustic analysis of spallation in relaxing media presented in Section 1.3 shows a correlation among the initial damage rate, the unloading rate in the incident shock pulse, and the free-surface velocity profile that is measured. Using this analysis, the initial damage rate can be estimated directly from the free-surface velocity profile.

The typical behavior of materials of different classes—metals and alloys, brittle single crystals and glasses, and elastomers—is described in Section 1.4. A series of additional factors can influence the fracture process that occurs under shock-wave loading. Extremely high plastic strain rates are produced in the shock front. The shock wave process produces an increase in the temperature of the material. The influence on the resistance to spall fracture of the peak intensity of the shock load, the duration of application of this load, and the orientation of the stress field was investigated in the experiments to be discussed. These results, along with the relation between dynamic bulk tensile strengths of polycrystalline metals and single crystals, permit making some conclusions about the constitutive factors influencing spall fracture.

A spall strength value only defines conditions for damage nucleation; it does not exhaustively characterize the material response. It is necessary to expend additional work to complete the process of separation of the body into parts. Section 1.5 contains a short discussion of the work of dynamic fracture and of the role of edge effects in defining the possibility of complete fracture and separation of the spall element from the body.

1.2. Background: Wave Interactions at Spalling and the Methodology of Spall Strength Measurement

Measurements of spall strength are based on analysis of the one-dimensional motion of compressible, continuous, condensed media following reflection of a shock pulse by the free surface of the body. This analysis is made by the method of characteristics. Characteristics are trajectories in a time–distance plane with slopes $dx/dt = c$ (C_+ characteristics) or $dx/dt = -c$ (C_- characteristics), where c is the sound velocity. Thus, characteristics describe propagation of weak perturbations. Pressure, p, and particle velocity, u, values along the characteristics are connected through the Riemann invariants

$$u = u_0 - \int_{p_0}^{p} \frac{dp}{\rho c} \qquad \text{along } C_+$$

and

$$u = u_0 + \int_{p_0}^{p} \frac{dp}{\rho c} \qquad \text{along } C_-,$$

where u_0 and p_0 are the integration constants, ρ is a density, and c is a sound velocity.

Figure 1.1 shows time–distance, (t–x), and pressure–particle velocity, (p–u), diagrams that illustrate the dynamics of reflection of a triangular shock pulse when it encounters a free surface of a body. In the t–x diagram, the shock front trajectory is described by curve OO'. A family of C_+ characteristics represents the unloading wave overtaking the shock front. When the shock front reaches the free surface, the velocity of the latter undergoes a jump from zero up to $u_0 = 2u_s$, where u_s is the particle velocity behind the

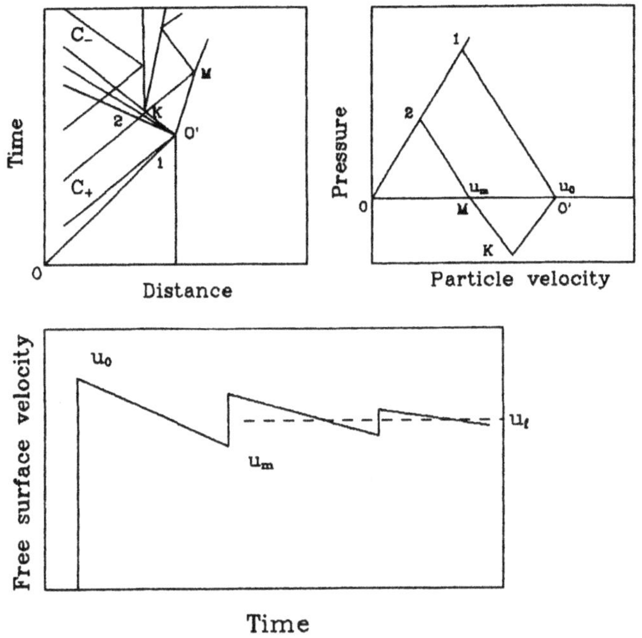

Figure 1.1. Wave interactions producing, and resulting from, spalling. Points identified in the pressure-particle velocity diagram correspond to similarly identified points in the time-distance diagram and in the free-surface velocity profile.

shock front. The following unloading decreases the free-surface velocity. Reflection of the shock wave by the free surface produces a centered rarefaction wave which is described by a fan of C_- characteristics. The state of particles must satisfy conditions on both the C_+ and the C_- characteristics and is determined in the p–u diagram by intersection of Riemann trajectories describing states of matter along the C_- and C_+ characteristics which pass through a given particle at some given time.

The maximum tensile stress is reached at each particle as it is traversed by the terminal characteristic of the centered rarefaction wave. So, the peak tensile stress in a spall plane just before the fracture corresponds to the intersection of the trajectories $O'K$ and $2K$ in the p–u plane of Fig. 1.1. Line $O'K$ describes the change of state along the tail C_--characteristic of the centered rarefaction wave; $2K$ is the trajectory of the change of state along the last of the C_+-characteristics of the incident wave crossing the spall plane before the fracture.

Fracture of material allows the tensile stress to decrease rapidly to zero. As a result, a compression wave appears in the stretched material adjacent to the spall plane. This wave propagates to the free rear surface and forms a so-called spall pulse in the free-surface velocity profile, $u_{fs}(t)$. Subsequent reverberations of the spall pulse between the free surface and spall plane are accompanied by damped oscillations of $u_{fs}(t)$.

The peak free-surface velocity, u_0, and the free-surface velocity, u_m, just before the spall pulse, are determined directly from the free-surface velocity profile. The tensile stress value just before spalling is then determined by intersection of Riemann trajectories passing through point $(p = 0, u = u_0)$ for C_- and $(p = 0, u = u_m)$ for C_+. Within the acoustic approach, the linear approximation

$$\sigma^* = \tfrac{1}{2}\rho_0 c_0 \Delta u_{fs} \tag{1.1}$$

is used [4], where $\Delta u_{fs} = u_0 - u_m$ is the so-called "velocity pullback."

Nonlinearity of material compressibility has to be taken into account in the case of high tensile stresses. This can be done by extrapolation of the material isentrope in the p–u plane to the negative pressure region. In reality, this correction does not exceed 10% in practical cases.

Dynamic measurements of the spall strength are based on Eq. 1.1. Plane shock waves for spall strength measurements are created by impacting the sample to be studied with a flyer plate or by detonating an explosive plane wave generator in contact with it. Measurements in the range of shortest load durations are performed with laser or particle beams as shock-wave generators. Capacitance gauges [6, 7] and laser velocimeters of VISAR [8] or ORVIS [9] types are used for monitoring free-surface velocity profiles.

Figure 1.2 shows examples of measurements of free-surface velocity profiles for the titanium alloy BT6 (Ti-6%Al-4%V) [10]. One can see that the free-surface velocity profile practically replicates the form of the compression pulse in the sample if the load intensity is small (profile 1). The elastic-plastic compression wave and the following complete unloading are recorded. A small velocity hysteresis is explained by hysteresis of the cycle of elastic-plastic deformation.

Tensile stresses developed in the body after reflection of the compression pulse by the free surface increase with growth of shock intensity. When the peak tensile stress reaches the fracturing magnitude, the nucleation and growth of cracks is initiated. Then, tensile stresses relax to zero with development of the fracture. As a result, a compressive disturbance called a "spall pulse" appears on the free-surface velocity profile. Thereafter wave reverberation is observed within the scab between the free surface and the damage zone. The period of velocity oscillation is a measure of the thickness of the scab. As discussed above, the velocity pullback Δu_{fs}, is a measure of incipient fracture strength of the material. Experiments show that increasing the shock amplitude does not influence the magnitude of Δu_{fs}.

For most solids, the free-surface velocity profiles exhibit elastic-plastic properties. There are several sound velocities in the elastic-plastic body. In the case of one-dimensional processes, weak perturbations propagate with longitudinal sound velocity, c_l, if the deformation is elastic and with bulk sound velocity, $c_b < c_l$, in the plastic deformation region. We have to decide

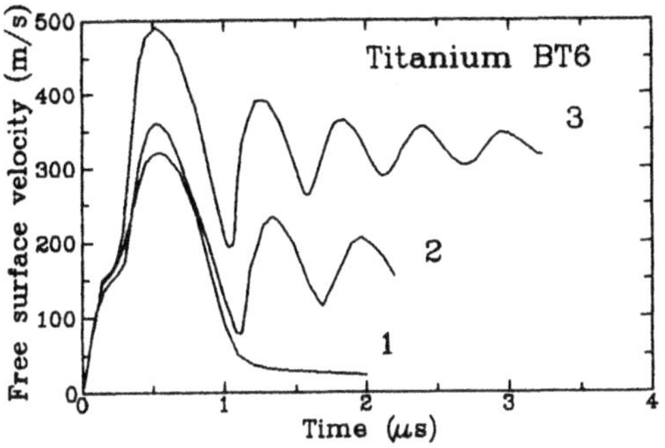

Figure 1.2. Free-surface velocity profiles for samples of Ti-6%Al-4%V alloy 10–12 mm thick resulting from impacts of aluminum plates 2 mm thick [10]. Impactor velocities are 445 m/s (profile 1), 600 m/s (2), and 700 m/s (3).

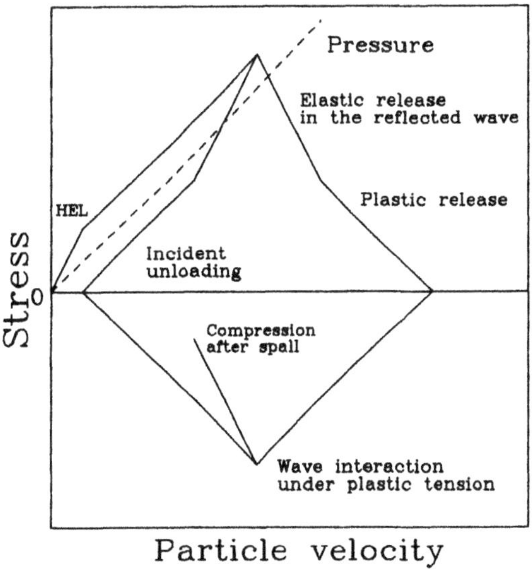

Figure 1.3. States of an elastic-plastic material at reflection of a compression pulse by the free surface.

which sound velocity should be used in Eq. 1.1 to calculate the tensile stress at a spall plane.

Figure 1.3 shows the diagram of states of axial stress and particle velocity (σ_x, u) for wave interaction when a plane square compressive pulse is reflected by a free surface of the elastoplastic body. The process of uniaxial compression is elastic until the stress reaches the Hugoniot elastic limit (HEL). The slope of the initial elastic part of Hugoniot below the HEL in

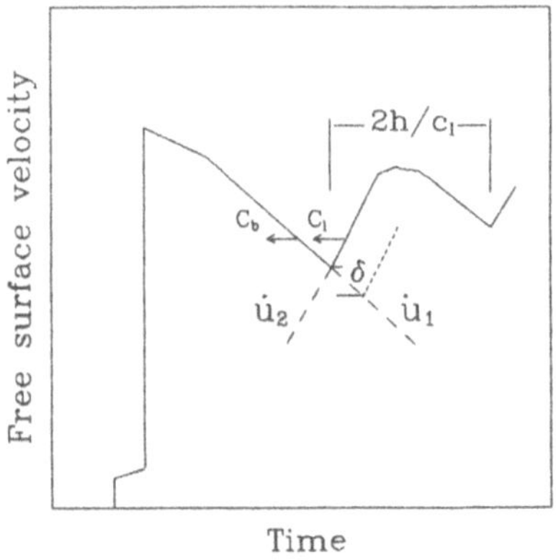

Figure 1.4. Distortion of the wave profile resulting from the different propagation velocity of the unloading wave and the spall pulse front in an elastic-plastic body.

these coordinates is $dp/du = \rho c_l$. The slope in a plastic deformation region above the HEL is equal to ρc_b. Unloading of shock-compressed matter is initially elastic in both the incident and the reflected waves. The elastic part of unloading has magnitude equal to two HEL. After that, all expansion processes occur in the plastic region.

Thus, if the shock-wave amplitude exceeds the HEL at least two times, tension produced under interaction of rarefaction waves takes place in the plastic deformation region. This means that Riemann's trajectories in the tension region have slopes defined by the bulk compressibility and we have to use the bulk sound velocity for calculation of stress using Eq. 1.1. On the other hand, the spall pulse is a compression wave which propagates through the extended material and, therefore, must have an elastic precursor. Thus, the spall pulse front propagates with the longitudinal sound velocity whereas the rarefaction plastic wave ahead of it propagates with the bulk sound velocity (Fig. 1.4). As a result, the compression wave, which appears due to stress relaxation at fracture, overtakes the release part of the incident shock pulse and the measured velocity, u_m, exceeds the value which we should use in Eq. 1.1 for calculation of stress.

The corrected expression for calculation of tensile stress before spalling is

$$\sigma^* = \tfrac{1}{2}\rho_0 c_b(\Delta u_{fs} + \delta).$$

where δ is a correction for the profile distortion due to the elastic-plastic properties of the material. Taking into account the velocity gradient ahead of the spall pulse, $\dot{u}_1$, and the gradient in its front, $\dot{u}_2$, the correction magnitude is estimated as

$$\delta = \left(\frac{h}{c_b} - \frac{h}{c_l}\right)\frac{|\dot{u}_1 \dot{u}_2|}{|\dot{u}_1| + \dot{u}_2},$$

where h is the thickness of spall plate.

Obviously, any estimate of similar corrections can only be approximate. It is impossible to be sure that the extrapolation used for $u_{fs}(t)$ is accurate. Actually, an undistorted profile can contain a smooth minimum near the

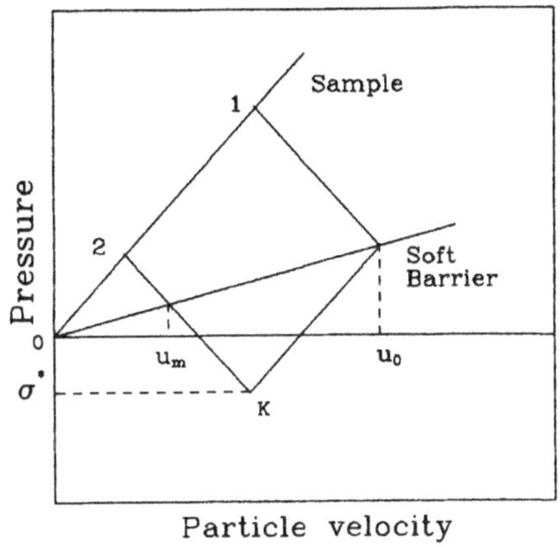

Figure 1.5. States of matter produced when a compression wave is reflected at an interface with a soft barrier.

measured value of u_m as well as a sharp minimum near its corrected significance with the same effect on the real free-surface velocity profile. This means that experiments must be designed to provide the smallest correction magnitude. It can be shown that the correction value is ~10% of Δu_{fs} in the case of an incident load pulse of triangular form and reaches 50% or more in the case of a pulse of rectangular form.

In typical experiments, the shock load is created in samples by impact of a flyer plate. In this case, pressure and particle velocity profiles keep an approximately rectangular form until the distance of propagation reaches ~5 thicknesses of the impactor. After that, the unloading wave overtakes the shock front and the load pulse becomes triangular. Thus, the optimal ratio of a sample thickness to the impactor thickness must exceed 5 to have the smallest correction and smallest possible error of the spall strength value.

Tensile stresses are created inside of a body due not only to reflection of compression pulse from the free surface but also due to reflection from an interface with a softer material. Measurements of spall strength can be based on the pressure or particle velocity profiles on this boundary. The p–u diagram for this case is shown in Fig. 1.5. It is clear that replacement of the free surface by a soft barrier leads to an increase in the distance between the spall plane and the plane at which measurements are made. As a result, the possible distortion of the wave profile also increases. An additional source of error in this case is in the equations of state used for the sample and barrier materials.

1.3. Acoustic Analysis of Flow During Continuous Spall Fracture

In the case of fracture resulting from shock-wave loading, the fracture time can be comparable with a total load duration. This means that, for many applications, dynamic fracture has to be considered as a continuous process with some kinetics of development of damage. In general, the events which take place inside of the body affect the wave profile structure and, thus, results of dynamic measurements contain information about the kinetics of damage evolution. In the case of instantaneous fracture, the spall pulse should have a very sharp front. Increase of damage evolution time should cause a decrease in the slope of the spall pulse front. The question is: What is the quantitative relation between the damage rate and the wave profile structure? One can hope that analysis of flow associated with stress relaxation during spalling will provide information about kinetics of fracture in the sense of the dependence of damage rate on the stress and the degree of damage.

Let us consider the evolution of a triangular compression pulse after its reflection from the free surface in the context of an acoustic analysis [11]. The damage is assumed to be initiated by tensile stresses exceeding the critical value, $\sigma^* = -P_k$, and is described by the specific volume of voids, V_v. The total volume is equal to the sum of V_v and the volume of undamaged (solid) material: $V(P, V_v) = V_v + V_s(P)$. Let the damage kinetics depend only on the specific volume of pores and be a power function of V_v. The form of the kinetic relation is chosen to be convenient for the analysis. In general, of course, the damage rate depends on the acting stress, the temperature, and the degree to which the damage has evolved. We are interested in the initial stage of spallation and, therefore, the collapse of voids under compression will not be considered. In the context of this model, the behavior of the material is described by the following set of equations in Lagrangian coordinates

$$
\begin{aligned}
&\frac{\partial V}{\partial t} - \frac{1}{\rho}\frac{\partial u}{\partial h} = 0,\\
&\frac{\partial u}{\partial t} + \frac{1}{\rho}\frac{\partial P}{\partial h} = 0,\\
&\frac{\partial V_v}{\partial t} = \frac{(\rho V_v)^\alpha}{\rho \tau_\mu},\\
&P = \rho^2 c^2\left(\frac{1}{\rho} - V + V_v\right),
\end{aligned}
\tag{1.2}
$$

where t and h are the time and the Lagrangian coordinate, ρ and c are the initial density and the sound velocity, τ_μ, is the characteristic damage time, and $\alpha < 1$ is a constant. The pressure is defined by the equation of state through the specific volume of the solid material.

Figure 1.6 shows a flow field in the t–h plane. The free surface has a coordinate, $h = 0$. The region 1 is free from the interaction of the incident load pulse with the reflected rarefaction wave. The dependencies of the particle velocity and pressure on time and coordinate in this region correspond to a triangular compression pulse,

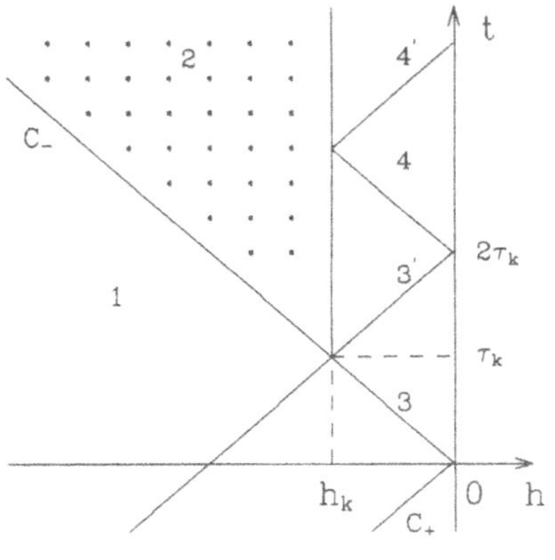

Figure 1.6. The field of flow at spalling. Explanations are in the text.

$$u(h, t) = u_0 - k(ct - h)$$
$$P(h, t) = \rho c u(h, t), \qquad (1.3)$$

where u_0 is the peak particle velocity, k = constant, and $\tau_0 = -h_0/c$ defines the length, $2h_0$, of the pressure pulse through the relation $u_0 = -2kh_0 = 2k_c\tau_0$.

Tensile stresses arise in region 3 and, at the point $h = h_k = P_k/(2\rho kc)$, $t = \tau_k = -h_k/c$, the pressure reaches the critical magnitude, $P_k = -\sigma^*$. The damage occurs in region 2 and the flow is determined here by solution of system (1.2) with boundary conditions at $h = h_k$ and as $h \to -\infty$ and initial conditions along the C_- characteristic through the J_+ Riemann invariant.

To obtain a solution of Eq. (1.2), let us eliminate terms V_v and V and change independent variable t for $T = t + (h/c)$. Then, applying the Laplace transformation, we obtain a system of ordinary differential equations. The solution which satisfies the initial conditions and does not grow exponentially when $h \to -\infty$ is

$$\hat{P}(h, s) = \frac{2k\rho c}{s}\left[h - h_0 - \frac{c}{2s}(1 - \exp(-2s(h - h_0)/c))\right]\theta(h - h_0) - \frac{\rho c}{2}F(s)\left(h - \frac{c}{2s}\right) + a$$

$$\hat{u}(h, s) = \frac{2k}{s}\left[h - h_0 - \frac{c}{2s}(1 - \exp(-2s(h - h_0)/c))\right]\theta(h - h_0) + \frac{1}{2}F(s)\left(h + \frac{c}{2s}\right) - \frac{a}{\rho c}. \qquad (1.4)$$

Here s is the Laplace variable, $\hat{P}$ and $\hat{u}$ are the Laplace images of pressure and particle velocity, $F(s)$ is the Laplace image of damage rate, $\rho \dot{V}_v$, and $\theta(x)$ is the Heaviside unit function. The constant a is defined in such a manner as to keep the J_- Riemann invariant at $h = h_k$. For example, within the time interval $0 \le T \le 2\tau_k$, a is given by

$$a = \frac{k\rho c}{s}\left(2h_0 + \frac{c}{s}\right) + \frac{\rho c}{2}F(s)h_k. \tag{1.5}$$

Equations (1.4) and (1.5) give the solution of Eqs. (1.2) in Laplace images in region 2 of the flow field (Fig. 1.6) for $0 < T \le 2\tau_k$. Some results can be obtained from Eq. (1.4). Let's find the pressure history along the right side of the C_- characteristic for $h > h_0$ using the known property of the Laplace transformation [12]: $\lim_{s\to\infty} sG(s) = G(+0)$:

$$P = 2k\rho ch - \frac{\rho c}{2}(h - h_k)L, \tag{1.6}$$

where $L = 0$ when $0 < \alpha < 1$ and $L = 1/\tau_\mu$ for $\alpha = 0$. It follows from this equation that the pressure corresponds to the case of undamaged material if the initial damage rate is equal to zero. When $\alpha = 0$, we have another situation. After the beginning of damage evolution at the point (h_k, τ_k), the pressure continues to decrease if $\tau_\mu > 1/(4k)$; otherwise it begins to increase if $\tau_\mu < 1/(4k)$. When $\tau_\mu = 1/(4k)$, the pressure is constant and equal to P_k.

The free-surface velocity can be found from Eqs. (1.4). Taking the conservation of the J_+ Riemann invariant along the C_+ characteristic in regions 3′ and 4 (Fig. 1.6) into account, and using the known formulas of inverse Laplace transformation [12], we obtain the following expression for the free-surface velocity within the time interval $2\tau_k \le t < 4\tau_k$:

$$\frac{u(0, t)}{2u_0} = 1 - \frac{t}{2\tau_0} + \frac{c}{4u_0}\left[(1 - \alpha)\frac{t - 2\tau_k}{\tau_\mu}\right]^{1/(1-\alpha)}. \tag{1.7}$$

Let's analyze this solution for various values of α.

The Case $\alpha = 0$: In this case, the damage rate is constant. Figure 1.7 shows the free-surface velocity profiles plotted by means of Eq. (1.7) for different τ_μ. As one can see, there is a critical value of relaxation time equal to $1/(4k)$ (curve 2 in Fig. 1.7). In this case the velocity is constant after the moment of time $t = 2\,\tau_k$ when information about the beginning of damage reaches the free surface. When $\tau_\mu < 1/(4k)$ (curve 1), the damage is displayed as spall pulse in the free-surface velocity profile. If $\tau_\mu > 1/(4k)$, the velocity continues to decrease during the fracture after $t = 2\,\tau_k$. Introducing the damage rate $\dot{V}_v = 1/\rho\tau_\mu$ and the expansion rate in the unloading wave of the incident pulse, the result obtained can be formulated as: The spall pulse on the free-surface velocity profile is formed only if the initial damage rate is more than four times as great as the expansion rate in the unloading wave of incident pulse. The slope of spall pulse front is equal to

$$\frac{d}{dt}\left(\frac{u(0, t)}{2u_0}\right) = \frac{1}{8\tau_0}(\dot{V}_v/\dot{V} - 4), \qquad t > 2\tau_k. \tag{1.8}$$

Using Eq. (1.8), the initial magnitude of damage rate can be estimated from experimental free-surface velocity profiles.

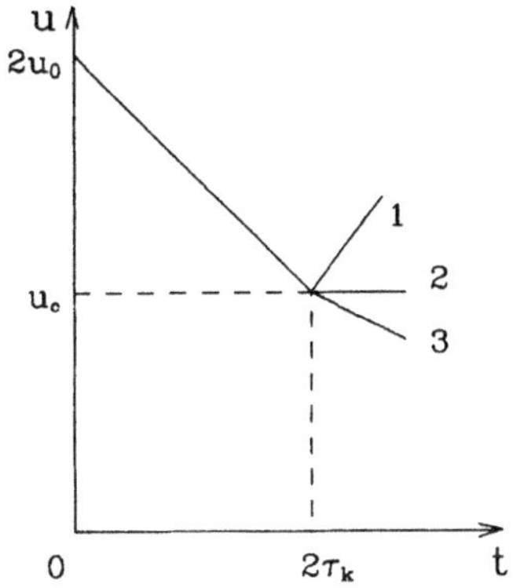

Figure 1.7. Free-surface velocity profiles at various constant damage rates. The damage rate in case 2 is four times greater than the expansion rate in the incident unloading wave.

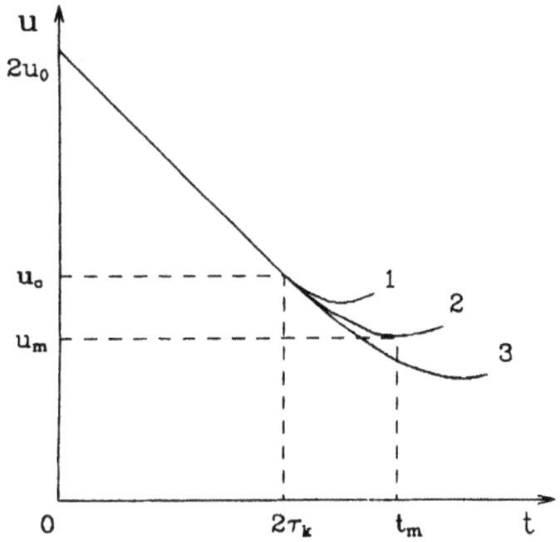

Figure 1.8. Free-surface velocity profiles for damage evolution at accelerating rates.

The Case $\alpha > 0$: In this case the damage evolves at an accelerating rate, beginning at an initial rate of zero. Figure 1.8 shows the profiles of free-surface velocity for this case. Curves 1, 2, and 3 correspond to increasing α or τ_μ. Unlike the case of constant damage rate, the derivative of free-surface velocity in this case is continuous at point $t = 2\,\tau_k$, and a minimum is reached at $t = t_m > 2\,\tau_k$, where

$$t_m = 2\tau_k + \frac{\tau_\mu}{1-\alpha}(4k\tau_\mu)^{(1/\alpha)-1}. \tag{1.9}$$

The corresponding velocity magnitude, u_m, is derived from Eq. (1.7). In practice, the spall strength is determined through the velocity pullback $\Delta u_{fs} = 2u_0 - u_m$. Obviously, $\Delta u_{fs} = -2P_k/\rho c$ in the case of $\alpha = 0$. In the general case, the velocity pullback also depends on the damage kinetics and the expansion rate in the incident pulse through the equation

$$\Delta u_{fs} = -\frac{2p_k}{\rho c} + \frac{\rho c^2 \alpha}{2\rho c(1-\alpha)}(4\tau_\mu \rho \dot{V})^{1/\alpha}. \tag{1.10}$$

Initially, the negative pressure reaches the value $p^* = -\rho c \Delta u_{fs}/2$ in the plane with coordinate $h^* = -\sigma^*/(2k\rho c) < h_k$. Let us estimate the damage rate value which gives the minimum in a free-surface velocity profile. In terms of the model of kinetics of damage that has been used, the damage rate is largest in the plane with coordinate h_k where fracture first began. Going back from the free surface to this plane and from the moment of time, t_m, when velocity minimum occurs to the moment $t = t_m - \tau_k$, we find the damage rate to be equal to $\dot{V}_v = 4k/\rho = 4\dot{V}$. This result coincides with that for constant damage rate. The minimum in the free-surface velocity profile and, consequently, the beginning of spall pulse formation, is observed when the damage rate in the spall plane is equal to the expansion rate in the unloading part of the incident pulse multiplied by four.

It can also be mentioned that, in the case of $\alpha > 0$, the apparent delay of fracture is observed in the free-surface velocity profile as a difference in duration of the first and the following velocity oscillations.

Thus, the analysis shows the relation between the structure of the free-surface velocity profile at spalling and the rate of damage. It follows from the expression obtained that the minimum in the free-surface velocity profile is formed when $\dot{V}_v \geq 4\dot{V}$. The rise time of the spall pulse front is related to the damage rate in the body. This means that not only can the spall strength of the material be found directly from the free-surface velocity profile, but also the kinetics of damage evolution can be obtained.

1.4. Spallation in Materials of Different Classes

1.4.1. Metals and Alloys

Metals are the materials most thoroughly investigated in research on spall phenomena. Figure 1.9 shows results of experiments with the strong Fe-Cr-Ni-Mo steel, 35X3HM [13]. Free-surface velocity profiles shown as solid lines were obtained under loading in the rolling direction, while profiles shown as dashed lines were obtained under loading in the lateral direction. The profiles shown display the influence of load direction on dynamic strength. The resistance to dynamic fracture is 4–4.4 GPa in the case of loading in the rolling direction and 10–15% less in the case of loading in the lateral direction. More rapid decay of the velocity oscillations in the latter case is correlated with a more highly developed fracture surface.

Fracture of this steel is relatively brittle in nature and has been completed by the end of the first half-period of velocity oscillation. A protracted, slowly evolving spall fracture was observed for more viscous materials such as stainless steel [14] (Fig. 1.10). In this case deceleration of the spall plate continued for a long time after the spall pulse appearance. Similar behavior was also observed for other materials when the incident pulse duration was small enough.

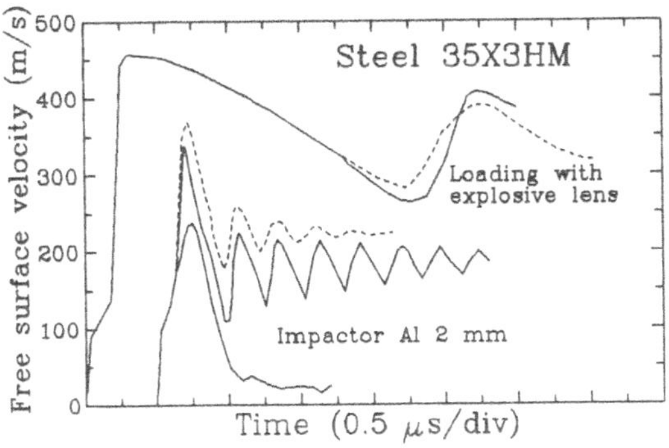

Figure 1.9. Results of spall strength measurements for a high-strength structural steel. Solid lines show free-surface velocity profiles under loading in the rolling direction, dashed lines are from experiments involving loading in the lateral direction [13].

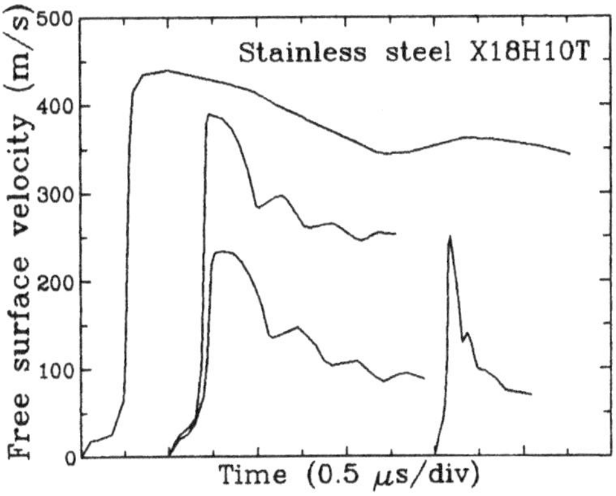

Figure 1.10. Spallation in a viscous stainless steel [14].

Figure 1.11 shows the measured spall strength of copper [15–17] as a function of strain rate in the unloading part of incident shock pulse. The strain rate was calculated as

$$\dot{\varepsilon} = \dot{V}/V_0 = -\dot{u}_1/2c_b.$$

The resistance to spall fracture rises with increasing characteristic expansion rate and can be represented as a power function of the latter. Results of measurements for copper presented in Fig. 1.11 are described by the relation

$$\sigma^* = 0.15(\dot{V}/V_0)^{0.2} \qquad \text{(GPa)}.$$

The damage rate is approximately equal to the product of the concentration of damage nucleation sites times their average growth rate and cannot be arbitrarily large. As a consequence, significant overstressing of the material is possible if the load is applied at a high rate. On the other hand, it is known

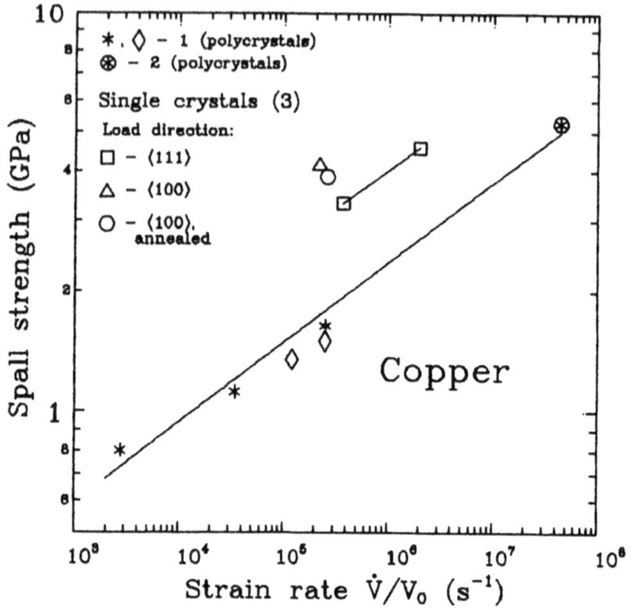

Figure 1.11. Spall strength of commercial-grade copper and copper single crystals as a function of the strain rate in the incident unloading wave (1): [15, 16], (2): [17], (3): [16].

that several thresholds of spalling (incipient fracture, intermediate stage, and the main crack formation) can be observed in samples after testing with load pulses of various durations [1]. The existence of several thresholds is a consequence of evolution of the fracture process in time. The question arises: Which stage of fracture is related to the spall strength determined from the free-surface velocity profile?

The rate of damage that can be observed in the wave profiles is correlated with the decompression rate in the incident load pulse. Actually, the spall pulse can be seen on the free-surface velocity profile if the fracture is fast enough. A slower fracture process leads to an increased rise time of the spall-pulse front. At some damage rate, we can obtain a spall pulse with zero gradient in its front. In this regime, the compression wave produced by stress relaxation during the fracture process compensates for the rarefaction in the incident release wave. Obviously, the suitable damage rate will be different for different load parameters.

The situation with wave compensation is realized at the point of minimum of the $u_{fs}(t)$ profile ahead of the spall pulse. The acoustic analysis presented in Sec. 1.3 shows that a minimum in the free-surface velocity profile appears when the rate of damage is four times that of decompression in the incident load pulse. In other words, each spall stress value corresponds to some specific damage rate.

So, the measured dependence, $\sigma^*(\dot{\varepsilon})$, reflects the damage-rate dependence on the applied tensile stress. Using the acoustic analysis and experimental data covering a wide range of strain rates, we can conclude that, in the case of copper, the damage rate depends on the tensile stress as

$$\dot{V}_v \approx 4V_0(\sigma/0.15)^5,$$

where stress is measured in GPa.

Comparison of the duration of the first velocity pulse and the periods of following velocity oscillations (after beginning of fracture) on the $u_{fs}(t)$ profile can give the delay time for achieving this damage rate. The period of oscillation has to be shorter than the first velocity pulse if any delay exists. Correcting for elastic-plastic properties of the material is simple enough in this case. Similar analysis for metals shows that the damage rate discussed usually corresponds (practically without any delay) to the moment when the tensile stress in the spall plane reaches σ^*. This means that the relation $\dot{V}_v(\sigma)$ obtained describes the initial damage rate.

Comparison of spall strength measured instrumentally and thresholds of spallation measured through the metallographic examination of recovered samples shows that the spall strength values obtained agree with the incipient fracture threshold or are less than this threshold. Two other thresholds—intermediate and complete fracture—occur at higher shock intensities. This does not mean that development of fracture occurs under higher tensile stresses because stress relaxation limits their growth with increasing shock-wave intensity.

It can be pointed out that irregularities observed on the fracture surface become finer with decreases in the load-pulse duration accompanied with an increase in the overstress in the sample. Smaller and more numerous damage nucleation sites are activated as the peak stress increases during propagation of the tensile wave into the sample. This leads to an increase of the initial damage rate. Obviously, the growth of coarse defects initiated near the body surface at lower tensile stresses is stopped by unloading due to the fast fracture in deeper layers of the sample. This correlates with the strong dependence of the initial damage rate on applied stress.

Under conditions of shock-wave testing, materials undergo fast compression and heating before the beginning of tension and fracture. The question is: In what measure do these factors influence the resistance to spalling? Experiments were carried out [18] with titanium (Fig. 1.12) and aluminum alloys in the peak shock pressure range 2–90 GPa. The temperature of samples before the spall fracture reached 1100 K as a result of irreversible heating in the shock front, and the total deformation was up to 60%. Nevertheless, measurements have shown that fracturing stresses are practically independent of the peak shock pressure and, consequently, of the shock-

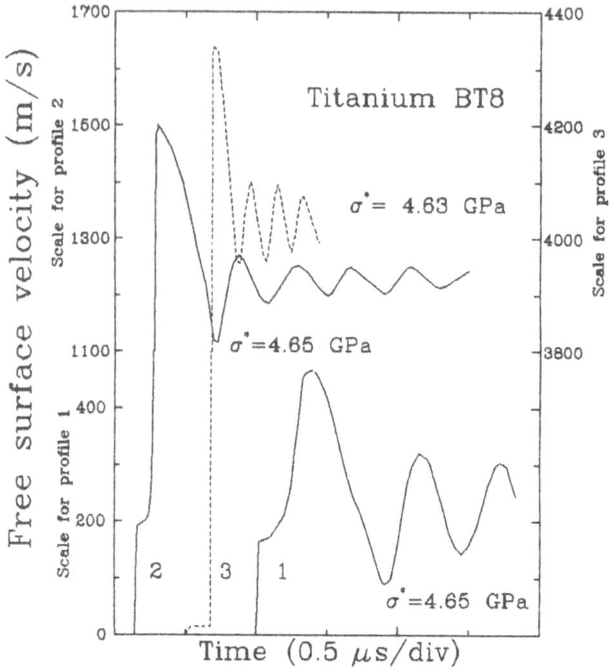

Figure 1.12. Results of the spall strength measurements for BT8 titanium alloy under various impact velocities [18]. Impact is by a 2 mm-thick aluminum plate with velocity 0.66 km/s (profile 1), 1.9 km/s (profile 2), and 5.3 km/s (profile 3). The weak precursor of profile 3 is the result of a shock wave forming in air ahead of the impactor.

induced temperature rise and preceding deformation. Reduction of the effect of temperature on resistance to high-rate fracture can be explained in terms of a change of the plastic deformation mechanism in the vicinity of discontinuities. It is possible that transfer from thermally activated to athermal overbarrier dislocation slip occurs under large overstressing in the material [19].

On the other hand, these experiments, as well as experiments involving variation of the load direction, show that, at the spall conditions, damage nucleation sites are relatively coarse pre-existing defects such as inclusions, micropores, and grain boundaries. Since deformation in the shock front and shock-induced changes in the material microstructure do not influence the resistance to spalling, it is natural to assume that higher stresses are necessary for damage nucleation on a dislocation level. Study of spall phenomena in single crystals gives information about conditions of damage nucleation on a structure level close to that of the ideal crystal structure.

Figure 1.13 shows results of measurements of free-surface velocity profiles for copper single crystals and commercial-grade copper [16]. Differences in the response of single crystals and polycrystalline material are immediately

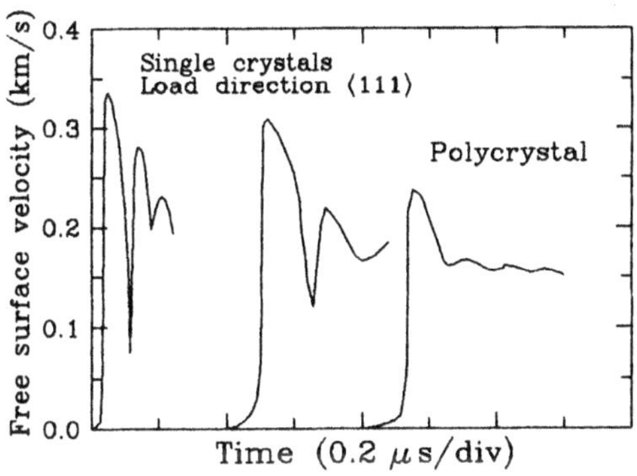

Figure 1.13. Free-surface velocity profiles at spalling in copper single crystals and polycrystalline copper [16].

apparent from comparison of these profiles. The spall strength of single crystals is approximately three times larger than the strength of polycrystalline copper (see Fig. 1.11). In the case of single crystals, the spall pulse is more clearly pronounced. It has a steeper front and higher amplitude than in the case of polycrystalline copper. It is possible to conclude that fracture of single crystals is more brittle in the sense that there is a more pronounced threshold and more rapid evolution of the damage.

Experiments with textured steel samples, and polycrystalline and monocrystalline copper samples, show that the dynamic strength is a structure-sensitive parameter. The dynamic strength of copper single crystals is about 20% of the ultimate theoretical strength and three to four times higher than for polycrystalline samples. This difference is undoubtedly associated with the much lower defect concentration in the monocrystalline material. Damage nucleation at a dislocation level can take place only in the case of single crystals. Independence of the spall strength on the peak shock pressure indicates that microdefects induced by high-rate plastic strain do not influence the strength of commercial-grade alloys. Damage nucleation sites in this case are relatively coarse defects pre-existing in the source material which are activated at lower stresses and lead the process.

1.4.2. Brittle Single Crystals and Glasses

Realization of ultimate tensile strength in the practically undistorted initial structure is possible for single crystals with high Hugoniot elastic limits. This is because generation of large tensile stresses without any plastic deformation is possible below the HEL. In this case, fracture is initiated at a structure level near that of the ideal crystal lattice.

Figure 1.14 shows free-surface velocity profiles measured for X-cut quartz samples [20]. Shock-pulse amplitudes near the rear surface were 2.8 GPa (curve a), 4.6 GPa (curve b), and 5 GPa (curve c). According to [21], the HEL of X-cut quartz is 6 GPa.

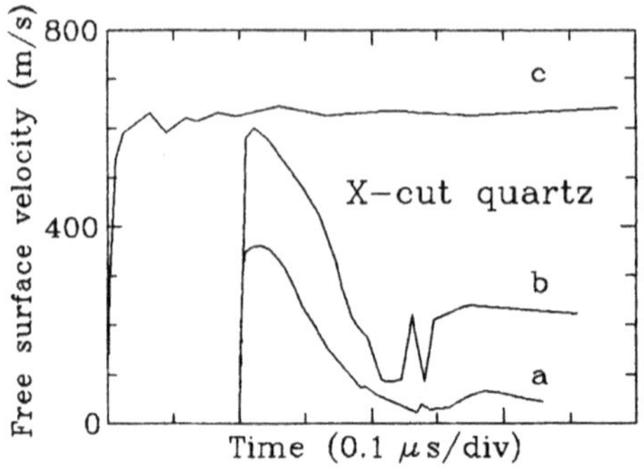

Figure 1.14. Spallation in X-cut quartz impacted at stresses below and above the Hugoniot elastic limit [20].

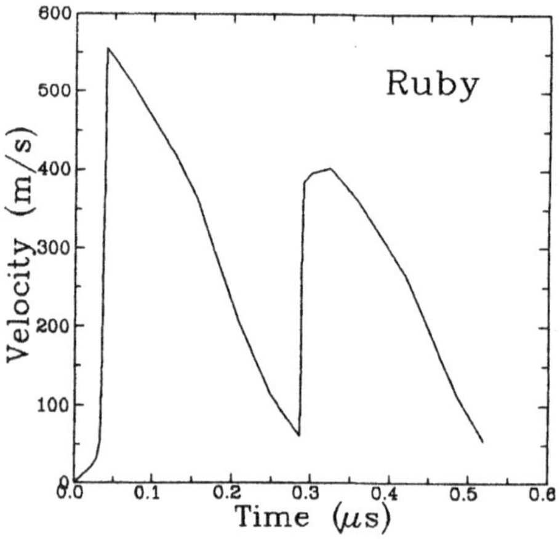

Figure 1.15. Free-surface velocity profile at spallation in ruby. The small disturbance in front of the shock is the consequence of an elastic precursor in the aluminum base plate placed ahead of the ruby sample.

Free-surface velocity profile of curve a in Fig. 1.14 replicates the form of the compression pulse inside the sample and does not display any symptoms of spallation. Increasing the pulse amplitude to 4.6 GPa, corresponding to the case of curve b, causes spalling and the appearance of a spall pulse. A further increase in pulse amplitude (curve c) modifies the free-surface velocity profile dramatically. The unloading part of the pulse does not pass to the surface, which means that the tensile strength is practically zero in this case.

So, the measurements have shown that the dynamic tensile strength of X-cut quartz reaches 4 GPa for shock loading below the HEL and falls to zero for loading close to the HEL. Obviously, cracking of brittle single crystals under compression is a reason for the tensile strength loss near the HEL.

Analogous experiments were carried out with ruby [22]. Measurements were made at shock-wave intensities lower than the Hugoniot elastic limit,

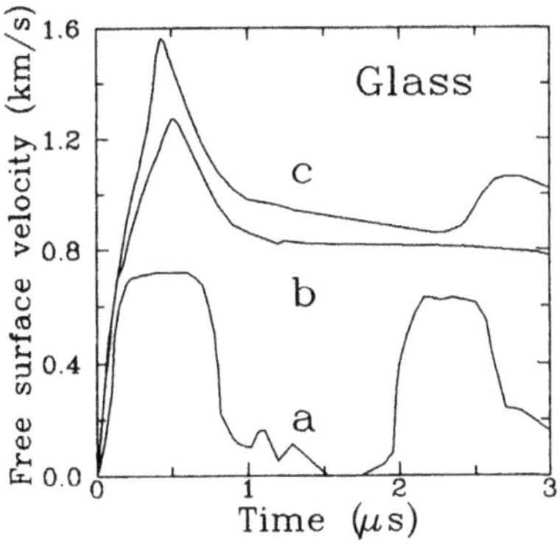

Figure 1.16. Free-surface velocity profiles for glass samples [24]. In cases b and c impactors were backed with paraffin.

which is 14–20 GPa for alumina single crystals [23]. Figure 1.15 shows the result of an experiment at a shock intensity of 13.5 GPa. The spall pulse in this case has a drastically steep front (the rise time is less than 10 nanoseconds). This testifies to the occurrence of a fast destruction process in the material. The measured spall strength value for ruby reaches 10 GPa.

In contrast to the behavior of single crystals, the spall strength of glasses is not sensitive to shock intensity at levels near the HEL. Figure 1.16 shows free-surface velocity profiles for K19 glass [24]. Spall was not observed in these shots, which means that the spall strength of the glass exceeds 4.2 GPa both below the HEL and above it. It is known that an irreversible densification occurs in glasses compressed beyond the elastic limit in both static and dynamic conditions [25]. The process of densification can play the role of a plastic deformation mechanism for glasses [26], which preserves them against cracking under compression.

The spall strengths of homogeneous brittle materials are significantly greater than the strengths of ceramics and minerals, which do not exceed 0.3 to 0.5 GPa.

1.4.3. Dynamic Tension of Rubber

It is known that elastomers are characterized by an ability to undergo large reversible deformations. Study of spall phenomena in rubber [27] has shown that the behavior of this material at fracture is different from that of other solids, including polymers.

Experimental results for rubber are presented in Fig. 1.17. The particle velocity profile 1 in Fig. 1.17 was obtained in a shot with an ethanol window, while profile 2 was obtained in a shot with a hexane window. The dashed line is the free-surface velocity profile calculated from results of window tests with the assumption that rubber is not damaged in the negative pressure region. If the material does not possess any remarkable resistance to tension, its surface velocity behind a shock front should be constant. The measured

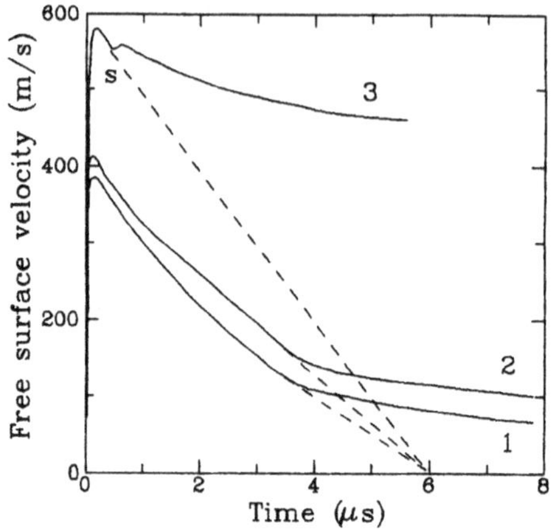

Figure 1.17. Results of spall strength measurements for rubber [27]. Profiles 1 and 2 were obtained with water and ethanol windows; 3 is the free-surface velocity profile.

free-surface velocity profile (curve 3) takes some intermediate position between these two extreme cases. A small initial part of the measured free-surface velocity profile corresponds to the incident loading pulse. A weak spall pulse is registered at point S. Then, a slow, protracted deceleration of the surface velocity is observed. Visual examination of the recovered samples does not reveal any remarkable damage. The measured value of rubber spall strength is 27 ± 3 MPa.

It is known that void formation precedes rupture of elastomers. These voids nucleate under much lower stress than is required to break the sample. Origination of microdiscontinuities is not yet a destructive event itself. Thus, in tests with triaxial tension of natural vulcanizates [28], voids formed at stresses of 1–3 MPa with insignificant deformation. Then, after this stage, samples underwent further deformation, reaching several hundred percent. Therefore, this permits us to suppose that the spall strength value obtained characterizes the nucleation of microdiscontinuities in the elastomer but does not characterize its failure.

1.5. Fracture Work and Edge Effects

A spall strength value only defines conditions for damage nucleation; it does not exhaustively characterize the material response. Development and completion of the spall process occur at reduced stress, but require consumption of additional energy for the growth of flaws and plastic deformation of the material around them. When the initial load pulse is short, fracture, once started, may not proceed to complete separation of the body into distinct parts.

An energy criterion [29–31] determines the possibility of complete rupture through comparison of the work of fracture and the amount of energy stored in the body. The work of fracture is the energy expended per unit cross-sectional area of the body in the course of fragmentation. In reality the dissipation of energy due to the fracture takes place in some layer of finite thickness and, therefore, the work of fracture, generally speaking, rises with increasing size of the failed zone.

Estimation of the energy dissipated in the course of a fracture process can be based on either of two kinds of experiments. The first kind of experiment determines the critical impact velocity which produces spall for given thicknesses of impactor and sample. Using the critical impact velocity thus determined, and parameters of the experiment, the energy dissipated is determined through consideration of the balance of energy and momentum as

$$E_d = \rho h_i v^2 (1 - h_i/h_t)/2,$$

where v is the impact velocity, and h_i and h_t are thicknesses of the impactor and target, respectively. The dissipated energy thus calculated is an upper bound for the magnitude of the work of fracture. The second kind of experiment involves measurement of a free-surface velocity profile. The loss of kinetic energy of the spall plate as it decelerates during the spall process can be inferred from this profile and can be used for estimation of the work of dynamic fracture [30].

The *fracture stress* and *work of fracture* describe the strength properties of materials when subjected to one-dimensional dynamic tension. However, these two parameters are not enough to predict the occurrence of complete separation of scab or to estimate its velocity after the separation.

Figure 1.18 shows the evolution of spalling in a limited area of a plane body [32]. The incident shock is attenuated not only by the axial unloading

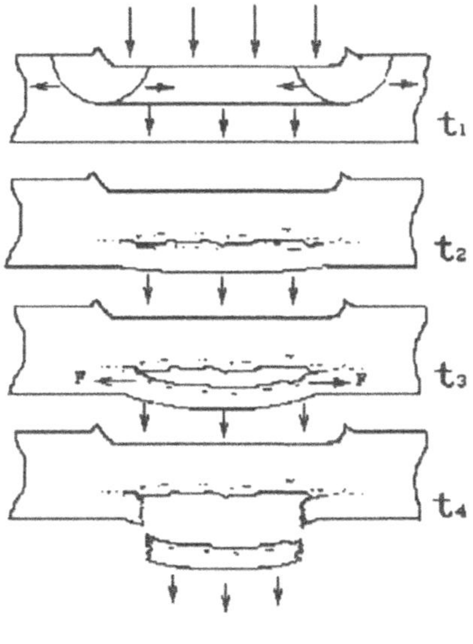

Figure 1.18. Spallation within a limited area of the sample.

waves but also by the lateral release. At reflection of this attenuated load pulse from the free surface, the tensile stresses and energy stored, which are sufficient for the complete fracture, are realized only near the axis. Thus, the fracture of the body by the plane wave is limited to some inner region. The subsequent evolution of the process is determined by the kinetic energy stored in the scabbed element.

The inertial motion of the scabbed layer is decelerated by bonding forces in the periphery of the scab. The scab motion can even be stopped by the edge effects. Additional work is necessary for plastic deformation and fracture along the edge of a spall element. The work of edge deformation and fracture is proportional to the length of the spall element perimeter, whereas the value of energy stored in the element is proportional to its area. The ratio of these two values increases with reduction of the spall element radius. In the case of small radius, development of spalling can be stopped at some intermediate stage and the spall element remains connected to the main body. Investigation of spalling with spall elements of different radii allows us to determine a critical stored energy value for complete separation of the scab. This variation of the spall element radius can be arranged through varying of the impact area radius or by placement of a limiting ring on the rear surface of the sample [32].

1.6. Conclusion

Numerous investigations of spallation have been performed for various practical applications as well as for basic research. Unfortunately, too many contradictory results of spall strength measurement have been obtained. This is natural, because it is impossible to perform any direct measurements of tensile stresses inside of the body during the spall process. All known methods of spall strength measurement are indirect. Because of this, a correct choice of experimental method is very important.

Dynamic measurements made during shock-wave loading give the most correct data about the fracturing stress values under spalling. The theoretical background of these measurements is simple enough, and sources of possible errors and ways to correct them are quite clear. As shown here, free-surface velocity profiles contain information not only about the spall strength of materials but also about stress relaxation during fracture and the kinetics of damage evolution.

Spall experiments that have been performed show the broad capability of dynamic strength measurements. This method was applied to ductile and brittle materials, metals, inorganic single crystals and glasses, and elastomers in a wide range of load intensities and durations. One can hope that this method will find its own place in the physics of strength.

Experiments with plane shock waves provide a unique base of information about strength of solids under small one-dimensional strain and stress states close to three-dimensional tension. Neither the surface of the body nor isolated coarse defects contribute to the main development of the spall fracture. The small duration of stress application makes possible creation of large overstresses in the material near the minimum of the potential curve, $p(V)$, and the study of conditions of elementary fracture events on a structure level close to that of the ideal crystal structure. One can say that shock-wave testing permits measuring the fundamental strength properties of matter. Studying the resistance to fracture in nanosecond and subnanosecond ranges of load duration; direct observations of fracture dynamics in transparent matter; investigations of the dynamic fracture of composite materials and reinforced concretes, biological tissues, and other complicated materials can be most interesting in the near future.

1.7. References

1. Smith, J.H. Three low-pressure spall thresholds in copper, *ASTM Spec. Techn. Publ.* 1962, No. 336 (1962), pp. 264–281.
2. Meyers, M.A. and Aimone, C.T. Dynamic fracture (spalling) of metals, *Prog. Mat. Sci.* **28**, 1–96 (1983).
3. Curran, D.R., Seaman, L., and Shockey, D.A. Dynamic failure of solids, *Phys. Rep.*, **147**, 253–388 (1987).
4. Novikov, S.A., Divnov, I.I., and Ivanov, A.G. The study of fracture of steel, aluminum and copper under explosive loading, *Phys. Metals Metallogr. (USSR)* **21**, 608–615 (1966).
5. Davison, L. and Graham, R.A. Shock compression of solids, *Phys. Rep.* **55**, 255–380 (1979).
6. Bushman, A.V., Kanel, G.I., Ni, A.L., and Fortov, V.E. Intense Dynamic Loading of Condensed Matter. Taylor & Francis, 1993.
7. Ivanov, A.G. and Novikov, S.A. The method of capacitor gauge for registration of momentary velocity of moving surface, *Apparat. Exp. Tech. (USSR)* **7**, 135–138 (1963).
8. Asay, J.R. and Barker, L.M. Interferometric measurement of shock-induced internal particle velocity and spatial variations of particle velocity, *J. Appl. Phys.* **45**, 2540–2546 (1974).
9. Bloomquist, D.D. and Sheffield, S.A. Optically recording interferometer for velocity measurements with subnanosecond resolution, *J. Appl. Phys.* **54**, 1717–1722 (1983).
10. Kanel, G.I. and Petrova, E.N. The strength of titanium BT6 at shock-wave loading. In: II Workshop on Detonation. Chernogolovka, 1981, pp. 136–142.
11. Utkin, A.V. Influence of the initial damage rate on the forming spall pulse, *J. Appl. Mech. Techn. Phys. (USSR)*, **34**, 140–146 (1993).
12. Handbook of Mathematical Functions. Edited by M. Abramowitz and I.A. Stegun. National Bureau of Standards, *Appl. Math. ser.* **55** (1964).
13. Glusman, V.D., Kanel, G.I., et al. The resistance to deformation and fracture of steel 35X3NM in the shock loading conditions, *Prob. Strength (USSR)* No. 8, pp. 52–57 (1985).

14. Kanel, G.I. The resistance of metals to spall fracture, *Phys. Combust. Expl. (USSR)* No. 3, pp. 77–84 (1982).
15. Kanel, G.I., Razorenov, S.V., and Fortov, V.E. The spall strength of metals in a wide range of shock load durations, *Acad. Sci. USSR Doklady (USSR)* **275**, 369–371 (1984).
16. Razorenov, S.V. and Kanel, G.I. The strength of copper single crystals and the factors Governing metal fracture in uniaxial dynamic stretching, *Phys. Met. Metallogr.* **74**, 526–530 (1992).
17. Paisley, D.L., Warnes, R.H., and Kopp, R.A. Laser-driven flat plate impacts to 100 GPa with sub-nanosecond pulse duration and resolution for material property studied, In: Shock Compression of Condensed Matter, Ed.: S.C. Schmidt, R.D. Dick, J.W. Forbes, and D.G. Tasker, Elsevier Scientific Publishers 1992, pp. 825–828.
18. Kanel, G.I., Razorenov, S.V., and Fortov, V.E. The spall strength of metals in a wide range of shock load amplitudes, *Acad. Sci. USSR Doklady (USSR)* **294** 350–352 (1987).
19. Merzhievsky, L.A. and Titov, V.M. The criterion of metal lasting in microsecond range. *Acad. Sci. USSR Doklady (USSR)*, **286**, 109–112 (1986).
20. Kanel, G.I., Razorenov, S.V., and Fortov, V.E. The failure waves and spallation in homogeneous brittle materials. In: "Shock Compression of Condensed Matter," Ed.: S.C. Schmidt, R.D. Dick, J.W. Forbes and D.G. Tasker, Elsevier Scientific Publishers 1992, pp. 451–454.
21. Graham, R.A. Shock-wave compression of X-cut quartz as determined by electrical response measurements, *J. Phys. Chem. Solids*, **35**, 355–372 (1974).
22. Razorenov, S.V., Kanel, G.I., and Yalovets, T.N. Dynamic strength of ruby, *J. Chem. Phys. (USSR)* in press.
23. Graham, R.A. and Brooks, W.P. Shock-wave compression of sapphire from 15 to 420 kbar. The effects of large anisotropic compression, *J. Phys. Chem. Solids* **32**, 2311–2330 (1971).
24. Razorenov, S.V., Kanel, G.I., Fortov, V.E., and Abasehov, M.M. The fracture of glass under high-pressure impulsive loading, *High Pressure Res.* **6**, 225–232 (1991).
25. Sugiure, S.H., Kondo, K., and Sawaoka, A. Dynamic response of fused quartz in the permanent densification region, *J. Appl. Phys.* **52**, 3375–3382 (1981).
26. Ernsberger, F.M. Role of densification in deformation of glasses under point loading, *J. Am. Ceram. Soc.* **51**, 545–547 (1968).
27. Kalmikov, Y.B., Kanel, G.I., et al, The behaviour of rubber in shock and rarefaction waves, *J. Appl. Mechan. Techn. Physics (USSR)* No. 1, pp. 126–130 (1990).
28. Gent, A.N. and Lindley, P.B. Internal rupture of bounded rubber cylinders in tension, *Proc. R. Soc. (Lond.), Ser. A,* **249** (1959).
29. Ivanov, A.G. Spallation in the quasiacoustic approach, *Phys. Combust. Expl. (USSR)* No. 3, pp. 475–480 (1975).
30. Kanel, G.I. The work of spall fracture, *Phys. Combust. Expl. (USSR)* No. 4, pp. 84–88 (1982).
31. Grady, D.E. The spall strength of condensed matter, *J. Mech. Phys. Solids, 36*, 353–384 (1988).
32. Razorenov, S.V. and Kanel G.I. The study of edge effects under spallation, *Prob. Strength (USSR)* No. 9, pp. 52–55 (1991).

CHAPTER 2

Microstructural Aspects of Dynamic Failure

Anna K. Zurek and Marc André Meyers

2.1. Introduction

The term dynamic failure is used to distinguish special characteristics of failure of materials subjected to dynamic loading. The major feature that differentiates dynamic failure from quasi-static behavior is the presence of stress waves. These waves arise due to the applied load or due to the stresses released from a crack tip at fracture. Several phenomena are particularly relevant to dynamic failure, including:

1. Rapid crack propagation. The velocity of the crack can approach the shock wave velocity; the Rayleigh velocity is widely accepted as the limiting velocity, but it is rarely achieved in real materials, where the maximum values are closer to 1000 m/s [Ravi-Chandar and Knauss (1984a,b); Zehnder and Rosakis (1990)].
2. Fast nucleation, growth, and coalescence of voids. The rapid rate of loading makes the independent nucleation and growth of microvoids possible.
3. Shear band formation: localization of plastic deformation in a narrow region takes place when thermal softening is more pronounced than strain and strain-rate hardening combined. The adiabaticity, or quasi-adiabaticity of the process due to high-strain-rate deformation enhances the propensity for this response. Shear bands often lead to failure by separation of the two sides of material along the band.

Dynamic failure is of considerable importance in several practical applications including crack arrest in engineering structures, fracture plane control in blasting, fragmentation in mining, meteorite impact of space structures, planetary impact or crash worthiness of vehicles and structures. Finally, the military applications of importance are the fragmentation of bombs and shells, the breakup of projectiles, and penetration and fracture of armor.

A good example of the different phenomena associated with the dynamic response of a material can be illustrated by the impact of a projectile against a target. Figure 2.1 captures several important dynamic events associated with such an impact. At the impact surface, as well as throughout the target,

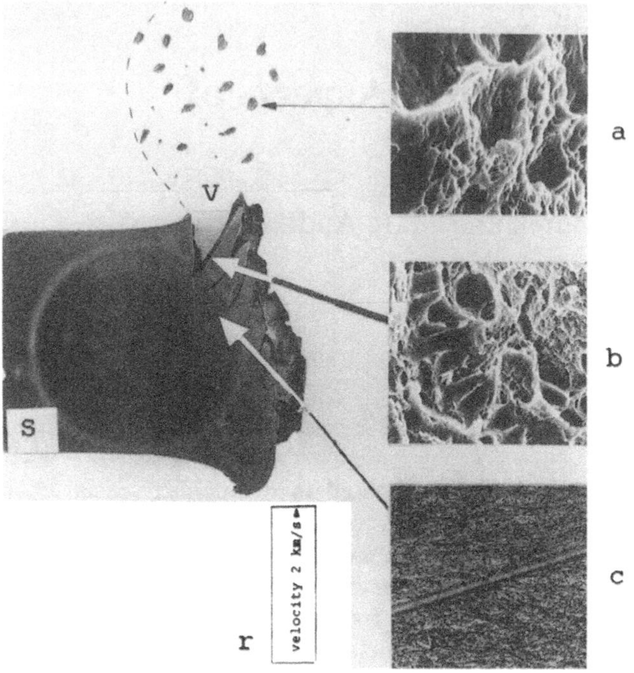

Figure 2.1. A penetration experiment showing important dynamic events. Rolled homogeneous armor steel s, penetrated with a rod r (schematically represented). Opposite the impact surface is a large cavity created by ejection of a cloud of debris. An example of an actual fracture surface of one debris particle is shown in insert a. Insert b shows the fracture surface of the spall cavity. The micrograph of an adiabatic shear band is shown in insert c. [From Zurek (1989)].

one usually observes dynamic deformation events such as shock-wave strengthening in the form of a significant increase in the dislocation density or twinning and other forms of deformation (microbands, slip bands) especially in metals. In other words, the shock wave, if of sufficient amplitude, produces substructural changes in the target. The shock wave can affect the mechanical properties, such as increasing the flow stress, thereby generally decreasing the ductility, and can therefore promote subsequent fracture.

Pressure-induced phase transformations (such as α to ε transition in steels, occurring at about 13 GPa) can also play an important role in changing the response of the material.

In many materials, as the strain rate increases due to the increasing impact velocity, the local temperature increases due to the heat generated from the deformation accumulation; this, in turn, may lead to a localization of the deformation to a narrow portion of the sample. This phenomenon is commonly known as *shear banding*. Shear bands play a very important role in dynamic fracture and fragmentation processes.

Spalling, on the side opposite the impact surface in Fig. 2.1, occurs due to the tensile stresses generated by the interaction of two release waves, one incoming from the free surface of the projectile and the other reflected from the back free surface of the target. If the tensile stress so generated is higher than the dynamic strength of the target, fracture processes in the sample will be initiated. Whether spall is complete or incipient depends on the duration of the pulse as well as the peak shock amplitude. The thickness of the ejected piece from the back of the target will correspond roughly to the thickness/length of a projectile under simple one-dimensional loading. Spalling will produce release waves at the newly created internal surfaces. More complex loading situations can generate internal stresses in the material and promote fragmentation in the spalled portion of the target.

Another type of dynamic fracture (not shown in Figure 2.1) may occur during the explosion of bombs or shells, and this can be represented by the expanding ring technique. In this technique, the hoop strains generated in the explosively expanding ring dynamically fragment the ring. The fracture occurs at the points of weakness in the material corresponding to the internal or external flaws of the ring. The size and the distribution of the fragments created in the expanding ring experiment depend on the flaw strength, the relative position of defects, and release wave propagation velocity as well as the microstructure of the material.

This chapter focuses on the microstructural aspects of dynamic failure. The mechanics of stress-wave propagation, crack, and shear-band formation were intentionally left out and only very schematic physical pictures of the phenomena are presented. Emphasis is placed on the microstructural effects (grain size, dislocation mechanisms, twinning, shock modification, heat treatment effects, and composition effects) and the primary source of information is the work carried out by the authors and co-workers. Therefore, many significant research efforts are not discussed here, for the sake of conciseness. Since another entire section of this book deals with fragmentation, this important topic is left out.

This chapter is divided into three sections, in addition to the introduction: Section 2.2 (Dynamic Fracture), Section 2.3 (Adiabatic Shear Instability), and Section 2.4 (Stress-Wave Effects in Ceramics and Rocks). Each section is divided into subsections dealing with specific methods of approach or materials.

2.2. Dynamic Fracture

2.2.1. General Description of Phenomena

Dynamic (high-velocity) crack propagation has unique aspects that differentiate it from quasi-static fracture. The main differences are:

a. There is a limiting crack velocity, as stated in Section 2.1. For a detailed treatment the reader is referred to Freund (1990). The analyses that led to limiting crack velocity are due to Broberg (1960), Craggs (1960), and

Freund (1972, 1973, 1974). The physical explanation for this effect is simple. Energy has to be delivered to the tip of the crack to drive it. Most of this energy travels along the crack surfaces, as waves [in a way analogous, but opposite to the energy of an earthquake, most of which is transported away by surface (Rayleigh) waves]. When the crack velocity approaches the Rayleigh wave velocity, it becomes more and more difficult to provide energy to its tip, and therefore this velocity reaches a plateau. Zehnder and Rosakis (1990), and Ravi-Chandar and Knauss (1984a,b) measured maximum velocities in real materials (a polymeric resin and a high-strength steel) and found that they were approximately 500 and 1200 m/s, respectively. This is approximately 50% of the Rayleigh speed. In real materials, grain boundaries, texture, differences between grain sizes, and other effects tend to slow the rate of extension of cracks.

b. At a sufficiently high velocity, there is a tendency for cracks to bifurcate, thereby lowering the overall energy of the system. Thus, quasi-static failure tends to produce one large crack while dynamic failure produces many small fragments.
c. The fracture toughness and yield stress of materials are strain-rate dependent.

Failures by crack and void nucleation, growth, and coalescence occurring dynamically are called brittle and ductile fracture, respectively. Dynamic fracture occurs in a variety of loading situations, not all of which involve stress wave propagation. For instance, a statically loaded system can generate a dynamic fracture event because a crack that grows initially at a low velocity is exposed to an increasing stress intensity factor as its length increases. Thus, the driving energy increases. Steel can undergo a ductile-to-brittle transition as the velocity reaches a critical value. Thus, the crack will accelerate even further by the lowering of the toughness: brittle toughness is generally less than ductile toughness.

For research purposes, a variety of experimental configurations are used to produce dynamic fracture and some of these are:

a. spallation experiments
b. expanding ring experiments
c. tensile stress wave propagation through pre-cracked specimens.

Since most of the experimental results described here are based on spallation experiments, this method will be described in detail.

Spallation, as a dynamic fracture process, occurs in a material due to tensile stresses generated by the interaction of two release (rarefaction) waves. Figure 2.2 is used here to help elucidate the spallation process.

The left-hand side of the diagram represents the projectile, while the right-hand side represents the target. At time $t = 0$, the projectile impacts the target. Elastic waves are emitted into the projectile and the target; they are

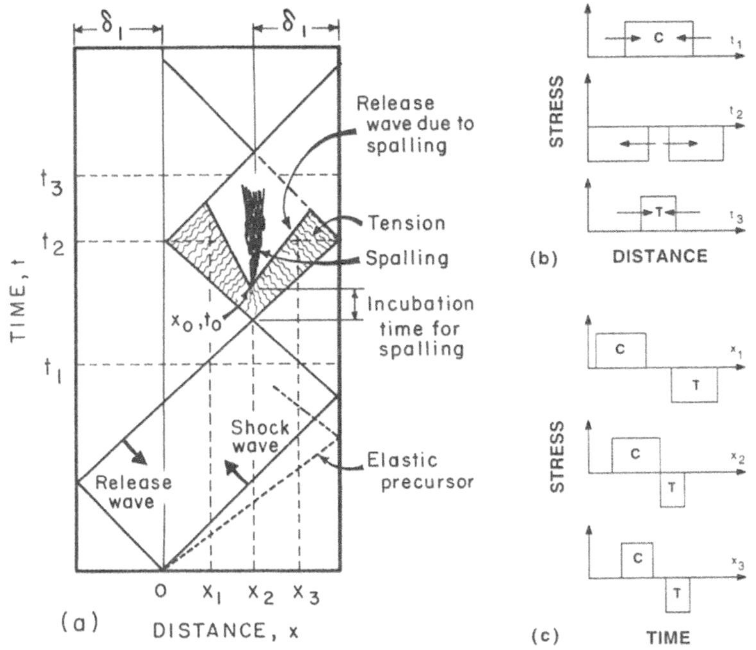

Figure 2.2. Distance–time plot for impact of projectile on target showing spalling induced by tensile stresses.

subsequently followed by a plastic wave. In the plot, the inverse of the slope of the diagonal lines gives the velocities of the waves. It can be readily seen that the elastic precursor has a higher velocity than the plastic shock wave. The elastic and plastic waves encounter the free surfaces of the target and projectile, and they reflect back. The regions in tension and compression are shown. The elastic precursor in this schematic is assumed to be small with respect to the total pulse amplitude. Spalling is dependent on both the amplitude and duration of the tensile pulse. If the tensile pulse is of a sufficient magnitude, spalling will occur. Spalling will produce release waves at the newly created internal surfaces which alter the subsequent configuration of pulses. If the projectile and target are of the same material, the distance of the spall from the back free surface is roughly equal to the thickness of the projectile, δ_1. The travel time of the wave in the flyer plate determines the duration of the compressive wave in the target and this establishes the thickness of the spalled layer.

Stress vs. distance plots at different times and stress vs. time plots at different positions are also shown in Figure 2.2. Multiple spalling is produced when a triangularly shaped pulse is reflected from the back free surface and has an amplitude substantially higher than the critical stress required for fracture. The shape of the pulse determines the distances between the consecutive spall planes. The spalling process repeats itself until the stress value of the wave has been reduced to less than the critical stress for spall. Spalling experiments generate a multitude of microcracks or microvoids which grow independently and coalesce.

2.2.2. Ductile and Brittle Spall Fracture in Iron and Steels

The majority of shock-spall studies have concentrated on investigating the dynamic fracture and shock strengthening of annealed or stress-relieved metals and alloys [Leslie (1981), Murr (1981)]. Few studies have explored the shock response of materials possessing a pre-existing dislocation substructure formed via heat treatment, quasi-static deformation, or prior shock loading [Gray (1993)]. Iron and steels undergo the reversible, α to ε allotropic phase transformation which occurs at approximately 13 GPa, which poses an additional complication in understanding the dynamic behavior in these materials. Above 13 GPa pressure, the spall morphology and the measured spall strength of iron and alloyed steel changes drastically. The reversible phase transformation, α to ε (during loading above the 13.1 GPa) and back to α (during unloading below 9.8 GPa), has been linked to a change in the mode of fracture from brittle (below the 13 GPa applied stress) to ductile (above the 13 GPa applied stress) and the change from rough to smooth topography at this spall surface, respectively [(Erkman (1961), Banks (1968), Barker and Hollenbach (1974), Ivanov and Novikov (1961), Balchan (1963), Meyers et al. (1980), Meyers and Aimone (1983)].

In spall fracture it is difficult to consider the separation of nucleation and propagation events. In the brittle regime, at lower pressures, cleavage is influenced by the local hydrostatic pressure [Teirlinck et al. (1988)]. Thus, in the spall process the fracture occurs under triaxial tension. As the superimposed hydrostatic tensile pressure increases with shock pre-stress, the spall strength of the material decreases.

The question arises as to why there is such a pronounced change in the mode of fracture in iron and steels associated with the phase transformation, as shown in Figure 2.3.

Several studies have investigated this problem. Erkman (1961), Banks (1968), and Barker and Hollenbach (1974) have proposed that formation of a smooth spall is related to the existence of a rarefaction shock and a sudden rise in a tensile stress pulse occurring in a very narrow region. Therefore, the fracture region is very localized [Meyers and Aimone (1983) and Zurek et al. (1992)]. Figure 2.4 shows a smooth spall produced above 13 GPa.

In the case of a spall below the phase transformation pressure (<13 GPa), the tensile stress pulse increases at a slower rate, allowing for the deformation to occur in a wider region and thus leading to the formation of a rough spall surface, as in Figure 2.5.

The spall strength of 4340 steel was found to be independent of pre-shocks

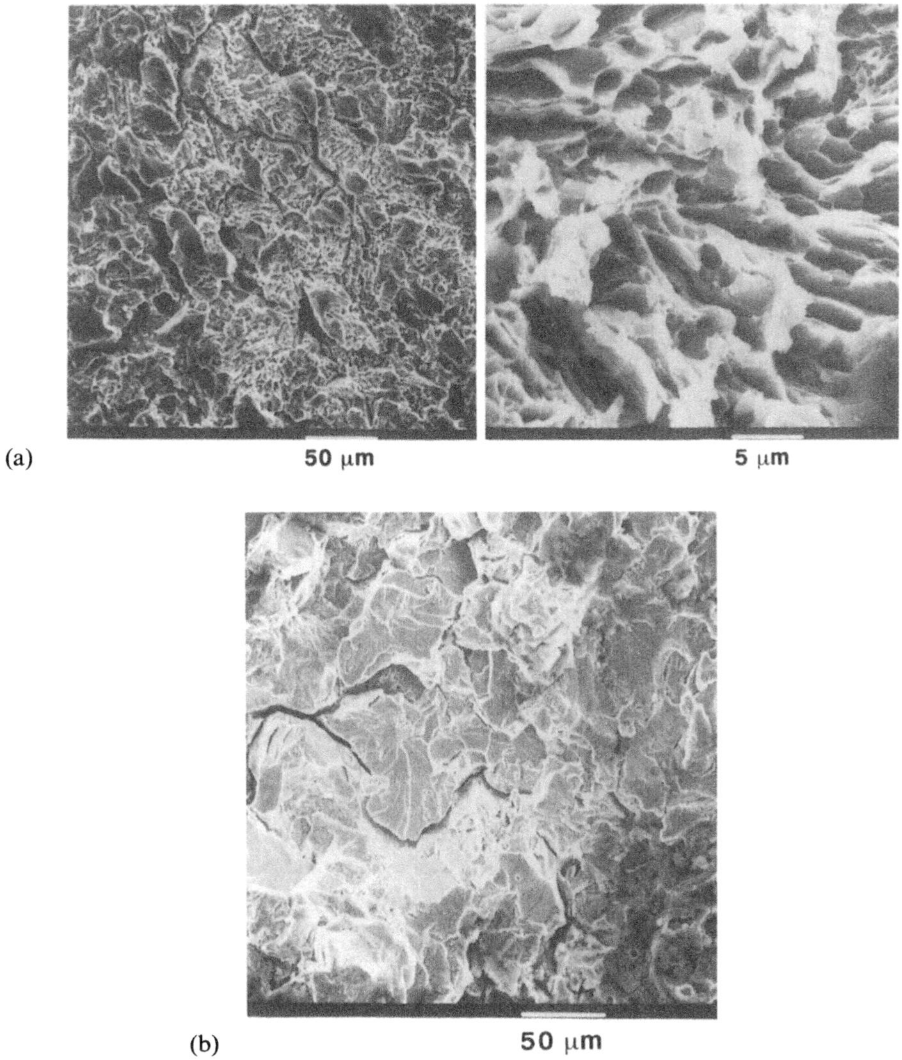

Figure 2.3. Ductile (a), and brittle (b), fracture in 4340 steel. [From Zurek et al. (1990)].

from 10 GPa and 15 GPa in the brittle regime (Zurek et al. (1992)]. This suggests that the fracture process in this material in the brittle regime is dynamic and therefore independent of work hardening and damage induced by such pre-shock. The pre-shock of 15 GPa (above the phase transformation pressure) decreased the spall strength of the 4340 steel in the ductile regime.

An explanation proposed by Zurek et al. (1992) was that the microstruc-

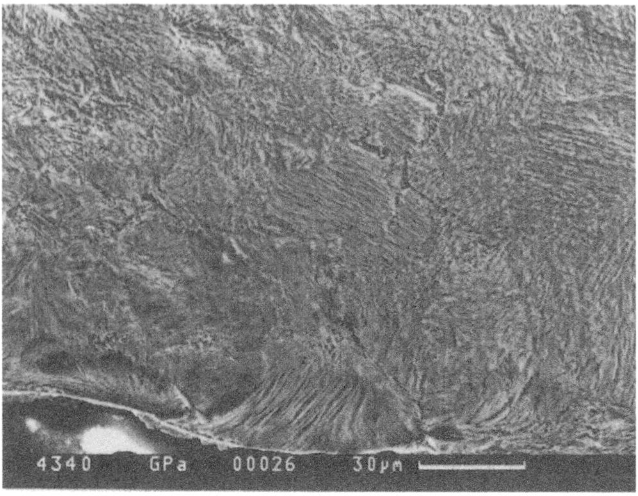

Figure 2.4. Smooth spall in 4340 steel at 15 GPa shown here as an etched cross section, where spall plane is at the bottom of the picture. [From Zurek et al. (1990)].

ture refinement and local deformation due to the phase transformation may create a large number of nucleation sites smaller than the original grains and promote ductile fracture. It is well known [e.g., Meyers and Chawla (1984)] that the ductile-to-brittle fracture transition point can be shifted by grain refinement or by an increase in the strain rate. Also, a sudden local increase in temperature, prompted by the thermodynamics of the phase transformation and combined with a high strain rate localized in a very narrow region, may be an additional reason for the ductile failure mode at a high stress amplitude. Zehnder and Rosakis (1990) used an infrared method to measure the local temperature increase in the vicinity of dynamically propagating cracks in 4340 steel. Their experimental data show that a temperature increase of up to 465°C may occur at the tip of a dynamically propagating crack in 4340 steel, and that the region of intense heating (>100°C) may extend to as much as a third of the active plastic zone size. Others have shown that a temperature increase of 300°C in Armco iron [Curran (1971)], or in low carbon steel [Zurek et al. (1990)], deformed at a strain rate of approximately $10^4\ s^{-1}$, is sufficient to promote ductile fracture even when the spall pressure is far below that necessary for the α to ε phase transformation. However, this temperature is above the dynamic fracture toughness corresponding to the ductile-to-brittle transformation temperature promoting void growth and coalescence. It was proposed by Zurek (1992) that either the substructure modifications due to phase transformation or the local temperature increase may raise the stress for brittle fracture above the peak stress level and promote ductile, smooth spall fracture.

The other observation was that the spall strength in the ductile regime is

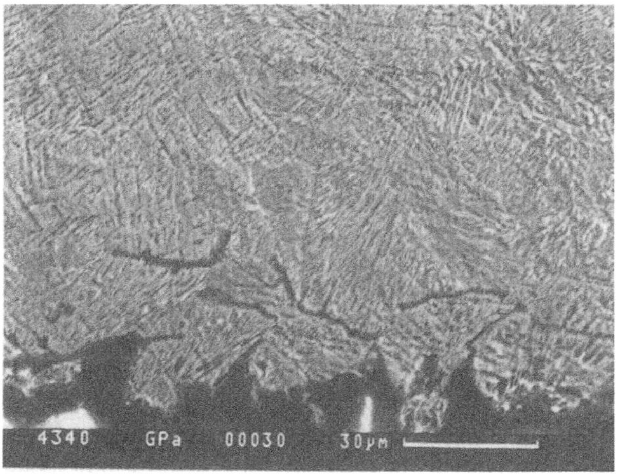

Figure 2.5. Rough spall in 4340 steel at 10 GPa shown here as an etched cross section, where spall plane is at the bottom of the picture. [From Zurek et al. (1990)].

decreased by shock pre-strain. This is consistent with the above hypothesis, because shock pre-straining increases the flow stress and decreases the hardening capacity. These two effects aid the ductile fracture process in spalling and thus lower the spall strength.

Meyers and Guimaraes (1976) found that tensile pulses can generate martensite in austenitic iron–nickel alloys. The effect was used by Thadhani and Meyers (1988) to obtain fundamental kinetic data on martensitic transformation. Dremin et al. (1992) established that the martensitic transformation in the spall region enhances the spall strength of austenitic stainless steel (18/8). This martensitic transformation produces a stress relaxation which results in an anomalous increase in the spall strength of these steels in the 14–18 GPa range.

2.2.3. Spalling in Copper: Microstructural Effects

In this section we will discuss two specific dynamic fracture processes in copper (used here as a model material) and their dependence on the microstructural aspects: (a) tensile fracture under dynamic uniaxial stress conditions (expanding ring experiment); and (b) spalling under dynamic uniaxial strain. Catastrophic failure in an expanding ring is preceded by a large amount of homogeneous plastic strain and little void growth occurs until the latter stages of necking, where void growth and coalescence are responsible for the material fracture. In spalling, void nucleation and growth dominate the tensile plastic flow process from the very beginning and the critical void volume fraction is necessary for ductile spall fracture to occur [Curran, Seaman, and Shockey (1987)]. These observed differences are due to entirely different tensile stress states in the two cases. Under conditions of uniaxial stress (expanding ring test), voids are subject to a nearly uniaxial stress field. Hence, homogeneous plastic strain dominates the flow process in the early stages. It is assumed that the void growth is nonexistent [Johnson (1981)]. Under conditions of dynamic uniaxial strain (spall test) the principal stress components differ by the flow stress, and thus the voids are subject to a nearly isotropic tensile stress field and void growth dominates all stages of the fracture process [Curran, Seaman, and Shockey, (1987), and Johnson (1981)].

The spall strength of copper and ductile fracture morphologies obtained in an expanding ring test and the symmetric spall test were studied by Zurek et al. (1988). In an expanding ring test the grain size in copper samples varied from 10 to 150 μm. The number of fragments generated in the electromagnetically expanded copper rings did not vary strongly with grain size [Gourdin (1987)], but were dependent on the distribution of random physical imperfections which served as nucleation sites for failure. However, the final cross section of the necked region was found to be grain size dependent (Figure 2.6a and b). The smaller grain size material revealed a larger cross sectional area and a larger number of voids per fracture surface area. In the large grain size material, void density was very low and coalescence occurred only in the latter stages of the fracture process.

The homogeneous plastic deformation preceding fracture was not accompanied by significant void growth.

The microstructure scale and purity play an important role in the spall morphology. In contrast to the expanding ring test, void growth and coalescence precedes the fracture process and are necessary elements for fracture in ductile materials [Curran, Seaman, and Shockey (1987)]. Zurek et al. (1988) found that the spall strength of copper can vary depending on the purity and grain size. They measured the lowest spall strength in the large grain ETP (low purity) copper and the highest strength in the small grain size OFHC (high purity) copper.

Figures 2.7a and b show the fracture surfaces obtained in ETP copper of 40 and 200 μm grain sizes. The inclusion particles (copper oxides), serve as void nucleation sites and therefore the spall is void growth controlled. The differences in the spall strength in this material (listed in the figure caption) are the result of the differences in the grain size.

A homogeneously nucleated transgranular spall fracture was found in OFHC small grain size copper (40 μm), as shown in Figure 2.8.

It was suggested [Stevens, Davison, and Warren (1972) and Zurek et al. (1988)], that voids nucleated on the cross slip planes of the edge dislocations, slipping along the $\langle 111 \rangle$ plane of the cube faces. In a large grain size OFHC copper (200 μm), the nucleation was intergranular, as in Figure 2.9.

It was proposed that in the case of a pure material that spall is nucleation controlled and therefore requires more energy and thus the higher spall strength measured. The spall strength differences listed for OFHC copper

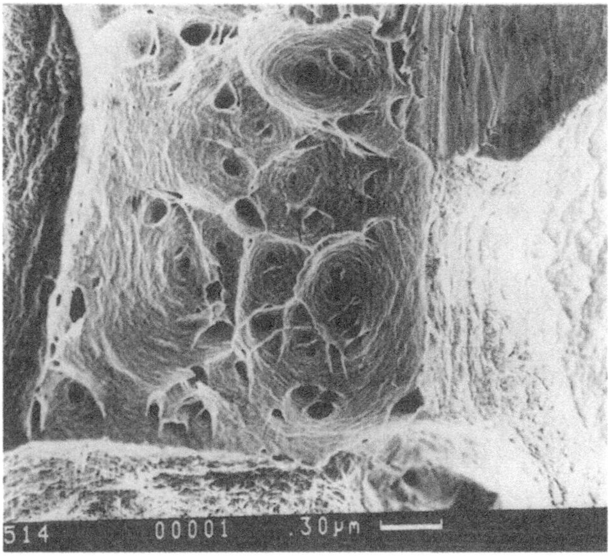

(a)

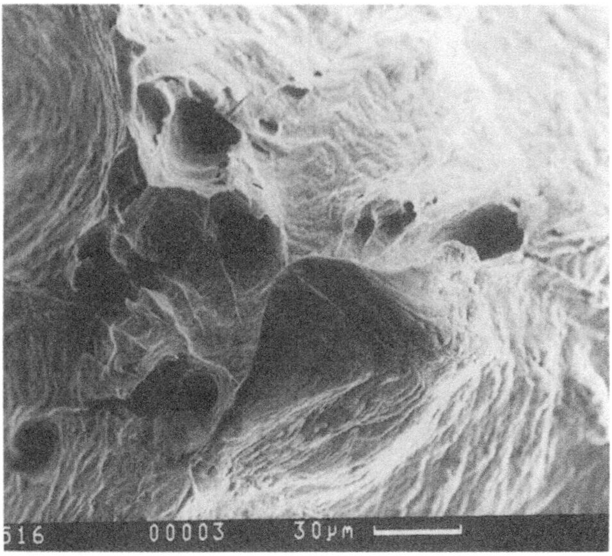

(b)

Figure 2.6. Voids in expanding ring experiment: small grain size (a), large grain size (b). [From Zurek et al. (1988)].

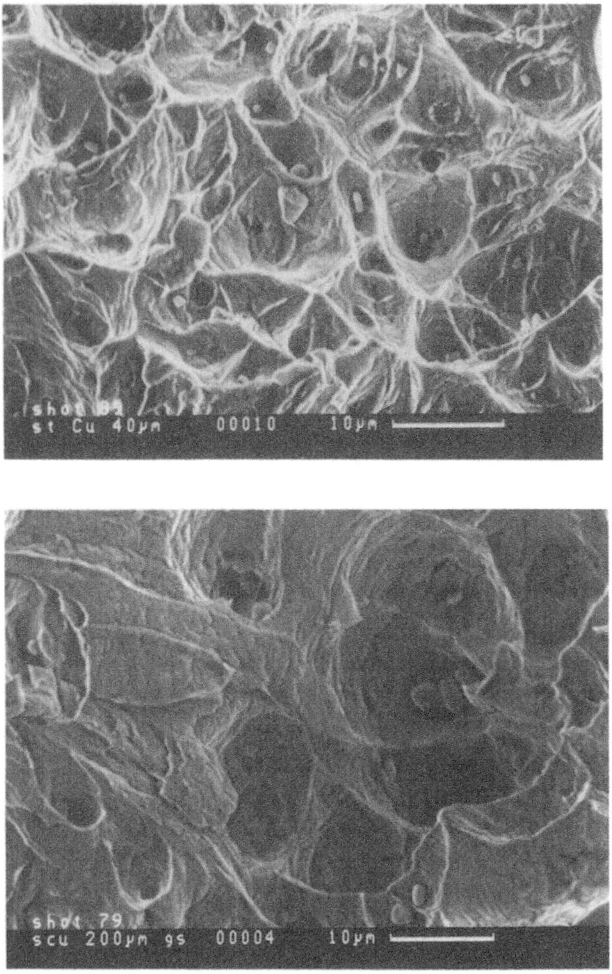

(a)

(b)

Figure 2.7 Fracture surfaces, ETP copper 40 μm. (a) Spall strength −1.5 GPa, and 200 μm. (b) Spall strength −0.4 GPa. [From Zurek et al. (1988)].

are a result of the change in the mode of fracture from intergranular for large grain size material to transgranular voiding in small grained copper.

Christy, Pak, and Meyers (1986) subjected copper with grain size of 20, 90, and 250 μm to spalling and determined the void volume fraction and initiation sites. The metallurgical condition (grain size, purity, and pre-deformation) was found to have remarkable effect on the spall strength. The relative spall strength of the specimens did not correlate with their hardnesses. Whereas the 250 and 90 μm specimens exhibited void nucleation at the grain boundaries, and as a consequence a ductile intergranular fracture, the 20 μm grain size and rolled specimens exhibited void nucleation in the interior of

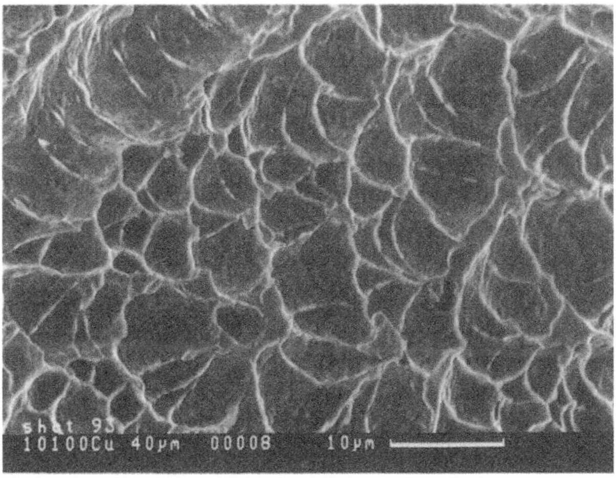

Figure 2.8. Homogeneous nucleation in OFHC copper 40 μm grain size. Spall strength -2.4 GPa. [From Zurek et al. (1988)].

grains. Figure 2.10 shows the differences in void volume fraction observed in the specimens for different metallurgical conditions and at different stress levels. The lower purity (99.5%) copper exhibited a noticeably lower spall strength (more spall damage) than the higher purity (99.99%) copper. The former is the "momentum trap" material marked in Figure 2.10. The explanation for this is simple: the second-phase particles provided easy nucleation sites for voids. The rolled material showed a significantly higher resistance for spalling. Statistical measurements were made and they indicated the following fraction of voids nucleated at grain boundaries:

large grained copper (250 μm)	77%
medium grained copper (90 μm)	65%
small grained copper (20 μm)	30%
rolled copper	49%

The initiation of voids at grain boundaries can be explained by the concentration of impurities at these regions, providing heterogeneous nucleation sites. The larger grain size specimens have a lower interfacial area per unit volume, S_v ($S_v \sim 6/D$; D is a grain diameter). Thus impurities are more severely segregated in these specimens, leading to more pronounced grain boundary failure. Figure 2.11(a) shows a grain-boundary void; markings on the optical micrograph are slip-band traces. Figure 2.11(b) shows a transmission electron micrograph (taken at an accelerating voltage of 1 MeV) of an elongated void. A high concentration of dislocations is noticeable around this void, which has a length of ~ 5 μm. This high dislocation density is the direct consequence of plastic strain necessary to create the void.

Kanel et al. (1992) determined the spall strength of copper mono- and

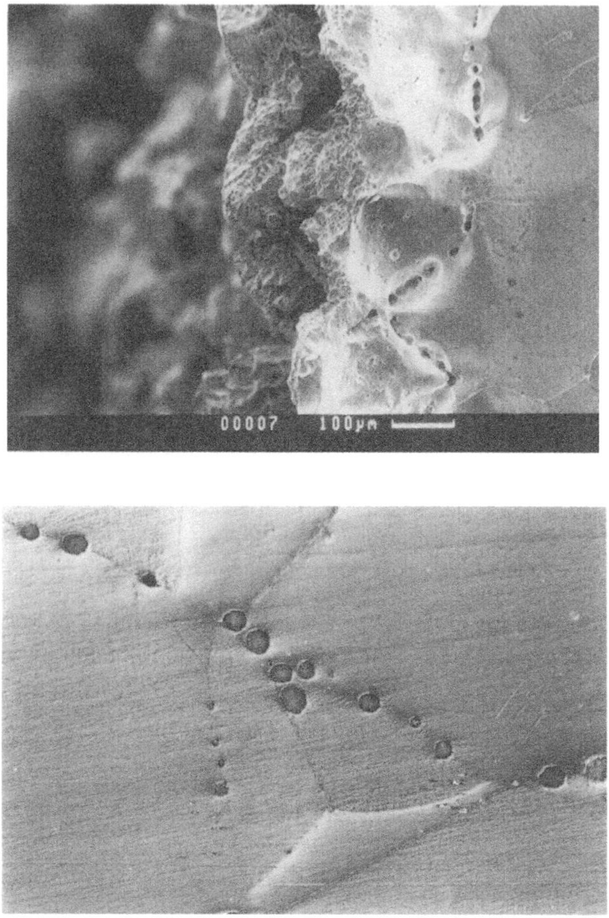

(a)

(b)

5.0 μm

Figure 2.9. Grain boundary nucleation in OFHC copper 200 μm grain size. Spall strength -1.5 GPa. [From Zurek et al. 1988 (a) and Christy, Pak, and Meyers (1982) (b)].

polycrystals and found, surprisingly, that the former had a significantly higher strength than the latter. Polycrystalline copper had a spall strength of ~ 1.36 GPa; on the other hand, monocrystalline copper had a spall strength of ~ 4.6 GPa. This is in contrast with the quasi-static strength of these materials. The explanation for this anomalous behavior lies in the absence of grain-boundary void nucleation sites for the monocrystalline copper.

2.2.4. Dynamic Fracture in Tungsten and Tungsten alloys: Heat Treatment and Alloying Influence

The spall strength of intrinsically brittle tungsten changes dramatically with both heat treatment and alloying. In recent years several investigators have studied the static and dynamic properties of tungsten alloys in sintered and

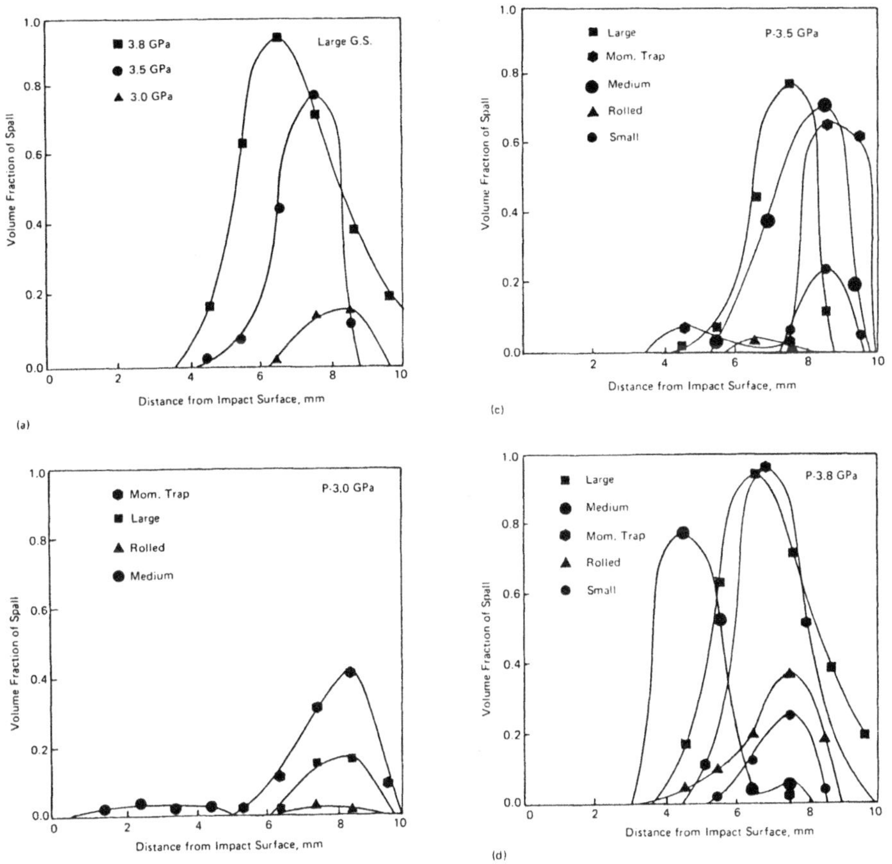

Figure 2.10. Volume fraction of voids formed during spalling experiments in copper (2.8 μs pulse duration); (a) effect of stress on large grain size (200 μm) specimens; (b), (c), and (d) effect of metallurgical condition on spall damage (momentum trap is 99.5% pure copper whereas other copper is 99.9% pure). [From Christy, Pak, and Meyers (1986)].

swaged conditions. Rabin and German (1988) showed that increasing tungsten content in W composites results in strengthening and a drastic decrease in ductility. Bose, Sims, and German (1988) studied the influence of the test temperature and strain rate sensitivity of 90W–7Ni–3Fe alloy and showed that the yield strength increases only slightly with strain rate while the tensile ductility decreases. Meyer et al. (1983) evaluated the dynamic strength and ductility of tungsten alloys in the as-sintered and swaged conditions, showing for each alloy a characteristic increase in strength and a decrease in ductility by increasing the strain rate from 10^{-3} to 5000 s^{-1}. Lankford et al. (1988) proposed a relationship between the strength and the fracture mode of explosively fractured tungsten composite alloys. This relationship indicated a change from cleavage fracture mode in the tungsten particles for

(a)

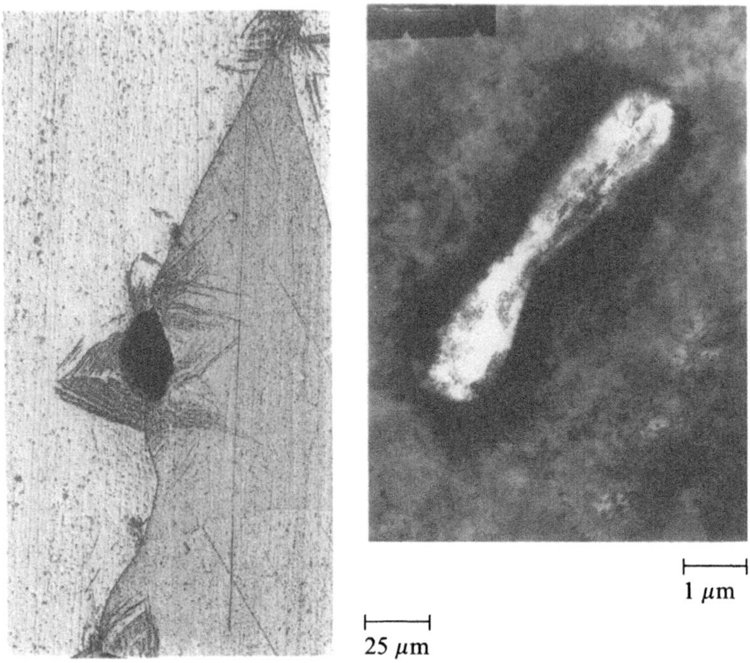

(b)

Figure 2.11. (a) Optical micrograph of intergranular void in copper showing markings due to slip. (b) High-voltage transmission electron microscopy of copper void showing plastic deformation region surrounding void; void is peanut-shaped. [From Christy, Pak, and Meyers (1986)].

Figure 2.12. Intergranular fracture of spalled surface of recrystallized tungsten. [From Zurek and Gray (1991)].

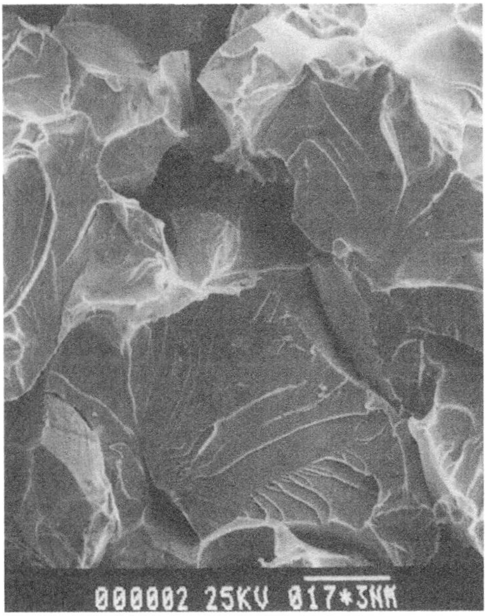

Figure 2.13. Transgranular fracture of spalled surface of cold-rolled tungsten. [From Zurek and Gray (1991)].

low strength tungsten alloys to an interfacial tungsten–tungsten/tungsten–matrix mode of fracture (a combination of W cleavage and interphase fracture) for high strength tungsten alloys.

In the case of recrystallized W, high strength, low work hardening and high sensitivity to impurities on the grain boundaries of this metal account for its extremely poor ductility. The high resistance to plastic flow in tungsten at ambient temperature results in attainment of flow stress levels above the grain boundary fracture stress. This leads to intergranular fracture before significant plastic flow occurs as shown in Figure 2.12, and virtually zero spall strength of this material [Zurek and Gray (1991)].

Addition of mobile dislocations through cold-working and reduction in interstitial segregation to the grain boundaries improves the spall strength. This is reflected by a change in mode of fracture from intergranular to transgranular cleavage (Figure 2.13).

Alloying of the intrinsically brittle tungsten with a total of 10% Ni and Fe by volume (W–Ni–Fe alloy), improves the spall characteristics even further. The significant increase in spall strength of this alloy (spall strength comparable to the spall strength of copper), is believed to reflect the strong influence of the ductile Ni–Fe matrix and W–matrix interfaces rather than the fracture behavior of the tungsten particles (Figure 2.14). As suggested by Krock and Shepard (1963), the ductility of the W–Ni–Fe alloy may be attributed to the crack-arresting effect of the matrix layer which surrounds each tungsten grain. Although for all practical purposes one would not consider this a ductile fracture, one should look at it in contrast with the inter-

Figure 2.14. Ductile fracture of spalled surface of W–Ni–Fe alloy. [From Zurek and Gray (1991)].

granular fracture for the recrystallized tungsten. It should be kept in mind that all the ductile Ni–Fe matrix is concentrated at the boundaries between the individual tungsten grains. What is observed are remnants of a plastically torn ductile ligament at the individual, brittle tungsten grains as observed previously by Lankford (1988). Tungsten grains, fractured transgranularly, are also occasionally observed.

2.2.5. Dynamic Fracture in Uranium Alloy: Cavitation

Cavitation is an unusual phenomenon in spall processes. This is due to the fact that cavitation processes do not occur readily under high hydrostatic pressures. At the spall plane the hydrostatic stress is a major fraction of the spall stress of the material, as shown in Eq. 2.1, Duvall and Fowles (1963).

$$p = \sigma_{\text{spall}} + 2(1 - 2\nu)/3(1 - \nu)\sigma_{\text{HEL}}, \tag{2.1}$$

where p is hydrostatic stress, σ_{spall} is a spall strength, σ_{HEL} is Hugoniot elastic stress, and ν is the Poisson constant.

The fracture surfaces of U-3/4Ti samples spalled at different applied pressures are shown in Figure 2.15.

The fracture mode is predominantly ductile dimples with many large cracks normal to the spall fracture surface. An optical micrograph of the specimen cross section near the fracture surface of the spalled sample is provided in Figure 2.16. It shows major cavitation just below the spall fracture surface [Zurek (1993)].

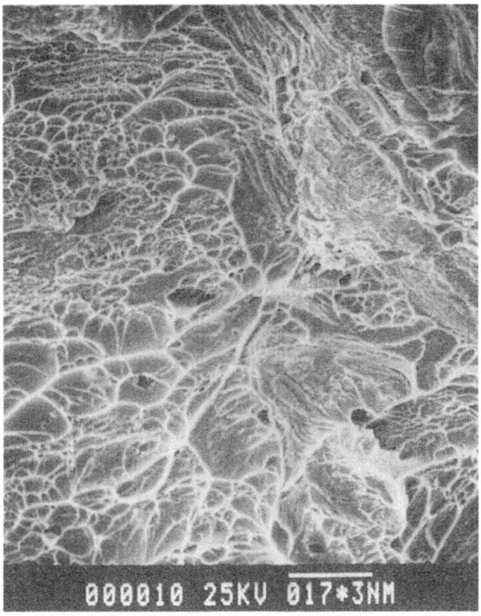

Figure 2.15. Spall fracture surface in U-3/4Ti alloy.

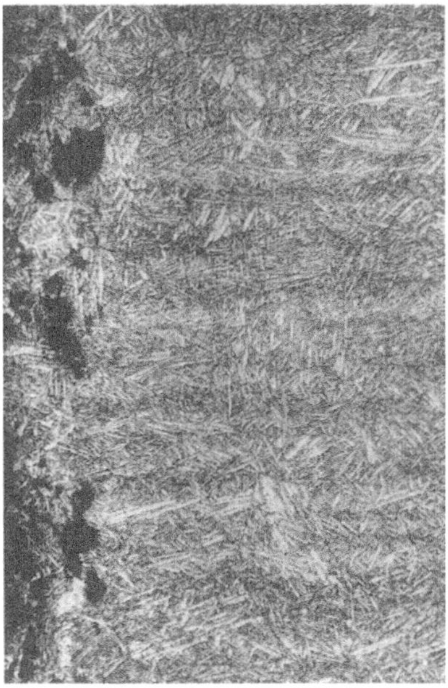

Figure 2.16. Cavitation in subfracture surface of U-3/4Ti alloy.

2.3. Shear Band Instability

2.3.1. General Description of the Phenomenon

Shear bands are macroscopic, non-crystallographic regions where plastic deformation in a material is highly concentrated. The formation of these shear bands is extremely important in dynamic deformation of materials because they are often precursors to fracture.

At low strain rates and moderate levels of strain, slip and twinning are the most common deformation mechanisms in metals and alloys. Both mechanisms are highly correlated with the crystallography of the material. At higher levels of strain and strain rates, deformation instabilities, such as adiabatic shear bands (ASB), may develop. These bands are planar in nature and their formation is related primarily to the specimen geometry, deformation process, and mechanical properties of a material rather than to the local crystallography [Hatherly and Malin (1984)]. Adiabatic shear bands can be formed during compressive or shear loading, and are frequently observed to initiate and propagate cracks or voids followed by fracture and fragmentation of the components. The first reported observation of adiabatic shear bands was made by Zener and Hollomon (1944). They inferred that the observed shear localization was due to the deformation-induced temperature rise at rapid rates of loading. Thermal softening thus overrode strengthening effects due to strain and strain-rate increases.

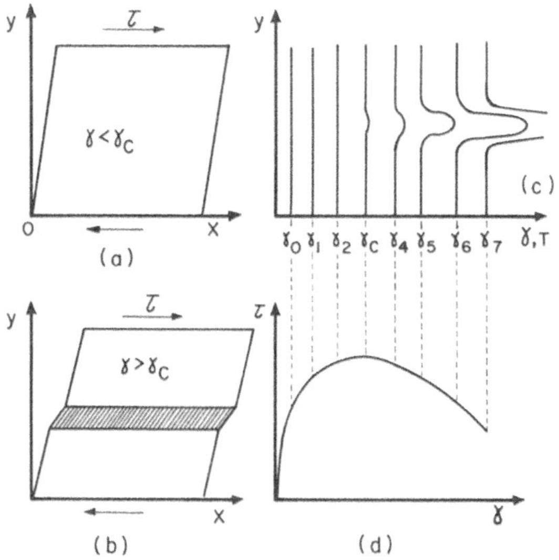

Figure 2.17. Strain evolution and localization beyond γ_c; (a) and (b) homogeneous and localized shear strains; (c) strains and (d) stress–strain curves. [From Meyers (1994)].

The following effects either contribute to or oppose shear-band formation:

opposing factors: strain hardening and strain-rate hardening
favoring factors: geometric softening and thermal softening

Figure 2.17 provides a simple explanation for the mechanics of shear band formation. A parallelepiped is being homogeneously sheared by τ (Figure 2.17a). At a certain strain, γ_c, deformation localizes in a band, as shown in Fig. 2.17(b). Figure 2.17(c) shows the progression of strain (or temperature) as a function of the increasing applied stress τ. The strain is homogeneous throughout the specimen at first: γ_0, γ_1, γ_2. At γ_c a small fluctuation appears, that accentuates itself as the applied stress is increased: γ_4, γ_5, γ_6, γ_7. This localization results from softening dominating hardening. A generic stress-strain curve in Figure 2.17(d) shows that softening dominates beyond γ_c, the strain at which the stress is maximum.

The formation of voids and their coalescence into cracks along shear bands is illustrated in Figure 2.18. These voids (marked by arrows) are produced by tension, due to residual stresses or tensile reflections. Since the flow stress of the material within the band is much lower than in the surrounding

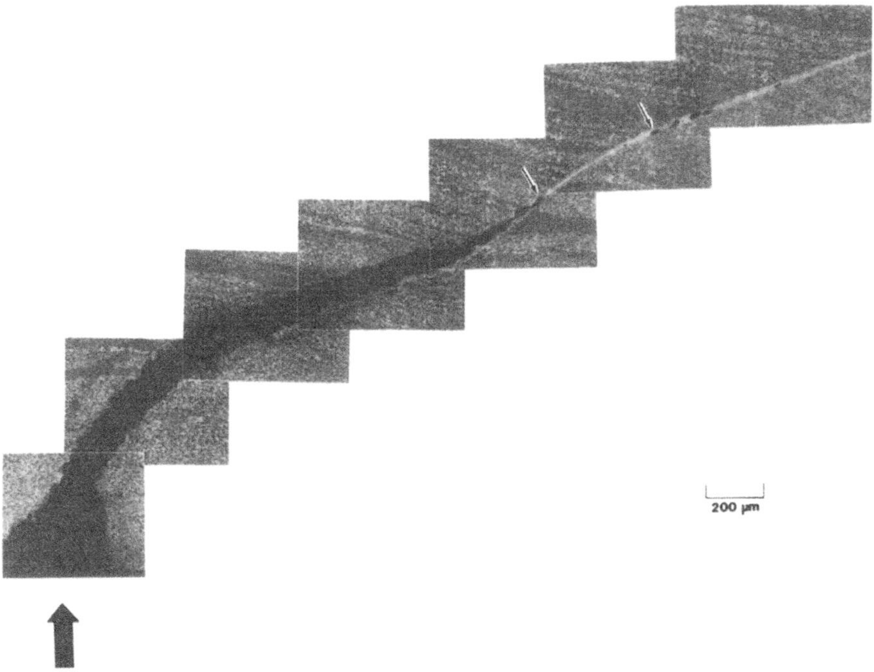

Figure 2.18. Shear-band (top right) and progressive void formation leading to fracture (bottom left) in titanium. [From Grebe et al. (1986)].

material (due to the higher temperature), voids will form there and their diameter is equal to the thickness of the band. When the density of these voids reaches a critical level, they coalesce, forming a crack (lower left-hand side of photomicrograph).

The simplest analytical expression for the formation of a shear band starts with a general constitutive equation represented by:

$$\tau = f(\gamma, \dot{\gamma}, T), \tag{2}$$

where γ, $\dot{\gamma}$, and T are the strain, strain rate, and temperature, respectively. Differentiating:

$$d\tau = \left(\frac{\partial \tau}{\partial \gamma}\right)_{\dot{\gamma},T} d\gamma + \left(\frac{\partial \tau}{\partial \dot{\gamma}}\right)_{\gamma,T} d\dot{\gamma} + \left(\frac{\partial \tau}{\partial T}\right)_{\gamma,\dot{\gamma}} dT \tag{3}$$

$$\frac{d\tau}{d\gamma} = \left(\frac{\partial \tau}{\partial \gamma}\right)_{\dot{\gamma},T} + \left(\frac{\partial \tau}{\partial \dot{\gamma}}\right)_{\gamma,T} \frac{d\dot{\gamma}}{d\gamma} + \left(\frac{\partial \tau}{\partial T}\right)_{\gamma,\dot{\gamma}} \frac{dT}{d\gamma} \tag{4}$$

The condition for instability is:

$$\frac{d\tau}{d\gamma} = 0 \tag{5}$$

If the experiment is conducted at a constant strain rate, $d\dot{\gamma} = 0$, equation 4 is reduced to:

$$\left(\frac{\partial \tau}{\partial \gamma}\right)_T = -\left(\frac{\partial \tau}{\partial T}\right)_\gamma \left(\frac{dT}{d\gamma}\right) \tag{6}$$

A number of theories for the formation of shear bands have been proposed; current formulations use the perturbation method to predict the onset and propagation of shear bands [(Bai (1981), Clifton (1979), Molinari and Clifton (1983), Grady (1980), Wright (1987)]. A perturbation in the strain or temperature "triggers" the localization of deformation at some plastic strain above the maximum of the τ, γ curve ($d\tau/d\gamma = 0$).

The following materials are known to exhibit shear bands: steels: 1040, 1080, 4340, etc. steels, tungsten heavy metal and alloys (W–Cu–Ni), brass, uranium alloys, zirconium alloys, aged aluminum alloys, titanium and its alloys, shock hardened copper, tungsten-titanium alloy, brasses. Most quenched-and-tempered steels possessing a martensitic microstructure exhibit shear bands. The number of alloys exhibiting shear localization under high-strain-rate deformation greatly exceeds the above list, and virtually any material can exhibit shear-band formation if work hardening is appropriately depressed (e.g., by plastic deformation).

Shear bands do not form simultaneously over the entire surface of a discontinuity, but are better visualized as "initiation" and "propagation" events. The schematic depiction in Figure 2.19 shows the two alternative concepts: the left-hand side shows the simultaneous shearing along a surface (or region of defined thickness), whereas the right-hand side shows the initia-

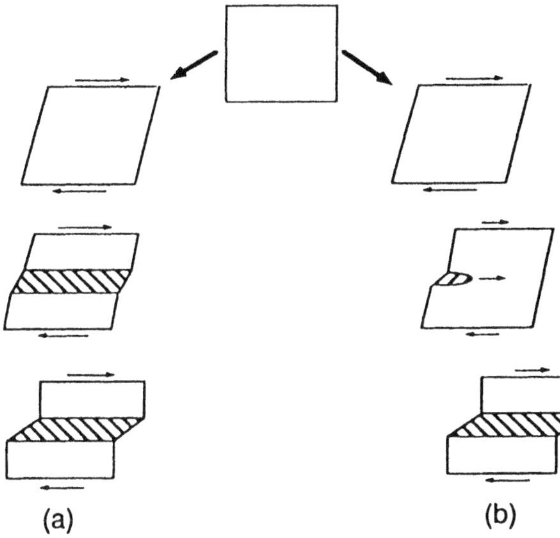

Figure 2.19. Two alternative paths leading to shear-band formation; (a) simultaneous shearing along hatched region; (b) propagation of shear localization by means of advance of shear-band front. [From Meyers (1994)].

tion of a band and its propagation as a shear crack. The latter evolution is thought to represent the actual formation of the shear band. Thus, it is similar to the propagation of a shear crack, leading Grady (1993) to propose a criterion similar to a critical shear-band toughness. Curran and Seaman (1986), Kuriyama and Meyers (1986), and Grady (1993) have modeled the shear band as an advancing region. With this in mind, the initiation of shear-band formation is an important event, and the identification of initiation sites is critical. Figure 2.20 shows a number of possible initiation sites. They have been classified into geometrical and microstructural. External, geometrical sites are easy to envisage, since they are regions where the concentration of strain leads to a temperature rise with thermal softening. The extremities of the softened region are subjected to a stress concentration which, in turn, leads to a stress concentration and the propagation of the band. Internal, microstructural initiation sites operate in the same manner. Several internal sources of strain concentration and/or softening are shown in a schematic fashion. Dislocation pile-ups can, upon their release, pierce through a grain, generating localized heating (Fig. 2.20c). This was initially proposed by Armstrong, Coffey and Elban (1982). If a grain contains a fractured second-phase particle, strain concentration occurs at the extremities of fracture in the matrix, by virtue of sliding of the fractured interfaces (Fig. 2.20d). If a crystal is favorably oriented for rotational softening, its Schmid factor will decrease with plastic strain, decreasing its "effective" flow stress. This leads, in turn, to a concentration of strain in this grain (Figure

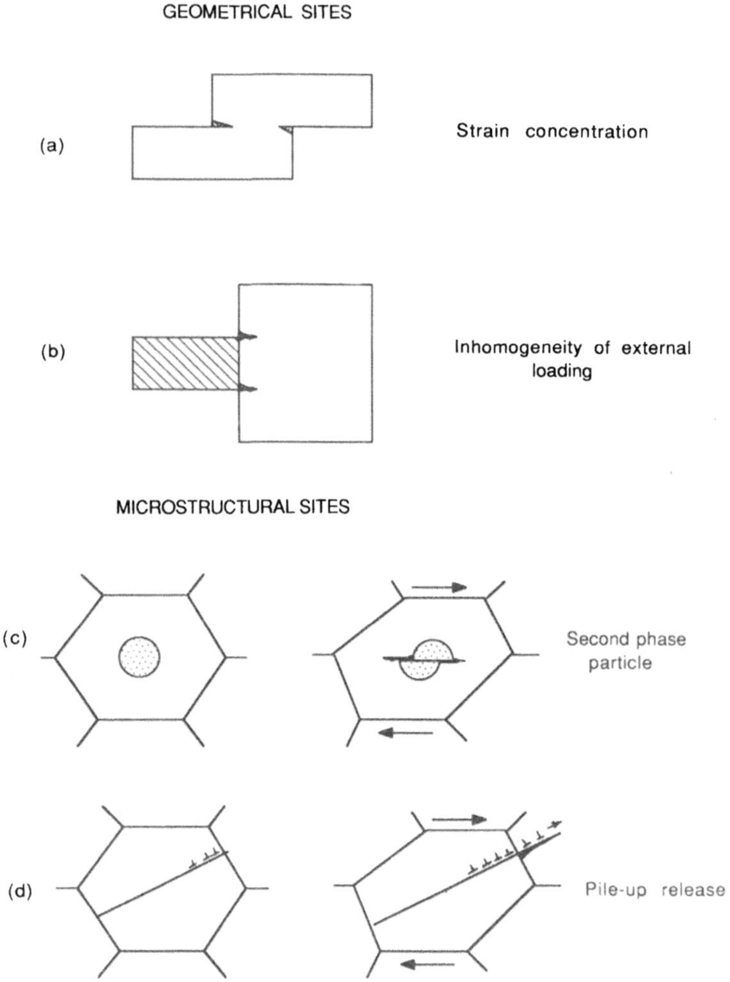

Figure 2.20. Possible initiation sites for shear-band formation; (a), (b) geometrical sites; (c)–(f) microstructural sites. (From Meyers, Subhesh, Kad, and Prasad (1994)).

2.20e). Martensitic lenses or laths are crystallographically oriented within each grain, and interlath or interlens sliding is a favored deformation mechanism because the retained austenite has a lower flow stress than the transformed region. Thus, there will be a tendency for the lenses or laths to orient themselves with the shear direction. This can lead to localized heating, with shear band initiation (Figure 2.20f). A similar situation could arise for deformation twins.

In the subsections that follow we will review the microstructural aspects of shear bands in steels (Section 2.3.2), titanium (Section 2.3.3) and copper (Section 2.3.4).

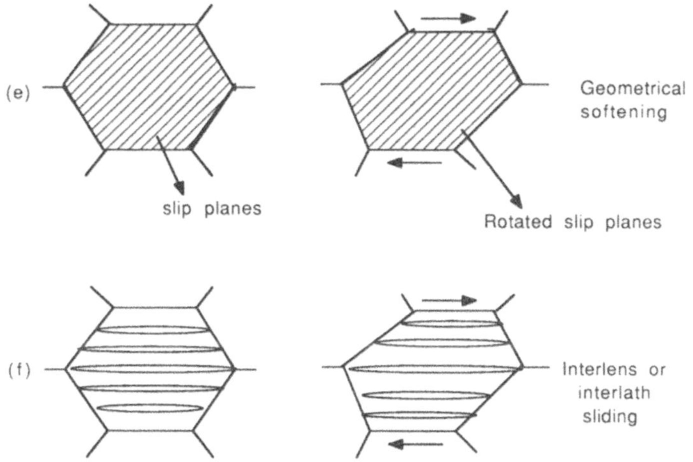

Figure 2.20 (*cont.*)

2.3.2. Shear Bands in Steels

In the case of steels, there is often a "signature" left by the adiabatic shear band in form of a "white etching band" [Rogers (1983)]. Similar etching phenomena were also observed in other materials such as titanium, uranium, etc.. The "white band" is caused by the resolution limit of the optical microscope to resolve the nanometer substructure enhanced by etching. It was believed by early researchers [Hatherly (1984) and Rogers (1983)] that these white etching bands are areas of extremely fine martensite formed during the deformation by locally heating the steel to a temperature above that required for austenitization followed by a rapid quench by the surrounding metal. Definitive proof of this transformation remains to be shown in the literature. Nevertheless, the white etching bands are referred to as "transformed" bands while bands showing high deformation but failing to etch "white" are referred to as deformed shear bands. In addition to the "white" etching signature, these bands are also found to be much harder than the surrounding matrix and are believed to fracture in a brittle manner in steels in the direction normal to the local tensile stress [Rogers (1983)]. Since Zener and Hollomon (1944), there have been numerous works published which have presented experimental observations and models on the subject: Hartley et al. (1987), Asaro and Needleman (1984), Mescall and Rogers (1989), Clifton et al. (1984), Marchand and Duffy (1987), Seaman et al. (1985).

At present, it is generally accepted that adiabatic shear bands occur most readily in materials with a low strain-hardening coefficient, a low strain-rate sensitivity, a low thermal-conductivity and a high thermal softening rate. In addition, shear bands form more readily in high strength materials where the plastic work due to deformation is greater for a given plastic strain increment [Hartley et al. (1987)].

In recent years several investigations have concentrated on the microstructural study of adiabatic shear bands in steels. Glenn and Leslie (1971) studied shear bands produced in ballistic experiments in low carbon martensitic steels. They identified the structure of white etching bands, using a diffraction method, as very rapidly quenched martensite. Wingrove (1971), using transmission electron microscopy, also identified the adiabatic shear band region as atypical looking martensite, with a high density of dislocations and cells. Thorton and Heiser (1971) studied ASB's in explosively-loaded steel cylinders. The X-ray diffraction pattern of the ASB zone, obtained using chromium radiation in a back reflection Laue method, showed it to be a body-centered-tetragonal martensite. Additional analysis using an electron-beam microprobe showed no chemical segregation associated with the ASB. Derep (1987) studied the microstructural transformation induced by adiabatic shearing in French armor steel. Using electron diffraction in the TEM, he characterized the microstructure observed in the center of an ASB as a fine-grain equiaxed delta ferrite structure, containing very narrow laths of martensite. Wittman et al. (1990) and Meyers and Wittman (1990), have presented a study of microstructural analysis of adiabatic shear bands formed in 4340 quenched and tempered steel [Wittman et al. (1990)] and also in a low carbon steel [Meyers and Wittman (1990)]. They reported that adiabatic shear bands formed in explosively deformed 4340 steel consisted of Fe_5C_2 carbides in a martensitic structure similar to the matrix but that the martensite had lost its original shape. The carbides were observed to have segregated on the internal microtwins. In a low carbon steel [Meyers and Wittman (1990)], the adiabatic shear bands were observed only in the quenched and quenched-and-tempered steel subjected to impact in the velocity range of 450 to 1,050 m/s. It has been hypothesized that the ASB are regions of very fine microstructure with a high concentration of supersaturated carbon which would explain a dramatic hardness rise within the adiabatic shear band [Meyers and Wittman (1990)]. However, no actual chemical analysis of the shear band was obtained in this study.

Most recently, a dynamic punch-impact test was designed by Zurek (1993) to produce white-etching adiabatic shear bands in pearlitic and martensitic 4340 steels. The average strain of 0.5 and the associated average strain rate of 18000 s^{-1} were sufficient to induce an adiabatic shear band instability in these steels. Figures 2.21 and 2.22 show results obtained for the martensitic quenched and tempered 4340 steel. The lath structure characteristic of quenching is prominent in the matrix; the corresponding diffraction pattern is also shown. Within the shear band (Figure 2.22) the microstructure is quite different, with small features that could be microcrystals.

These features have sizes between 10 and 30 nm, consistent with the observations of Meunier et al. (1992) and Beatty et al. (1992). Meunier et al. (1992) identified the structure of shear bands produced by ballistic impact

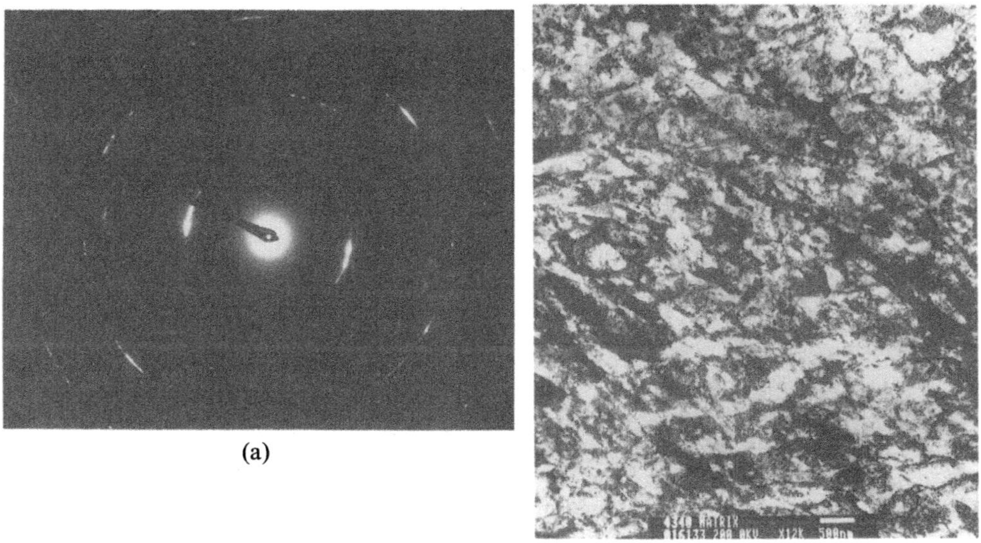

(a)

(b)

Figure 2.21. TEM of 4340 quenched and tempered martensitic matrix and diffraction pattern.

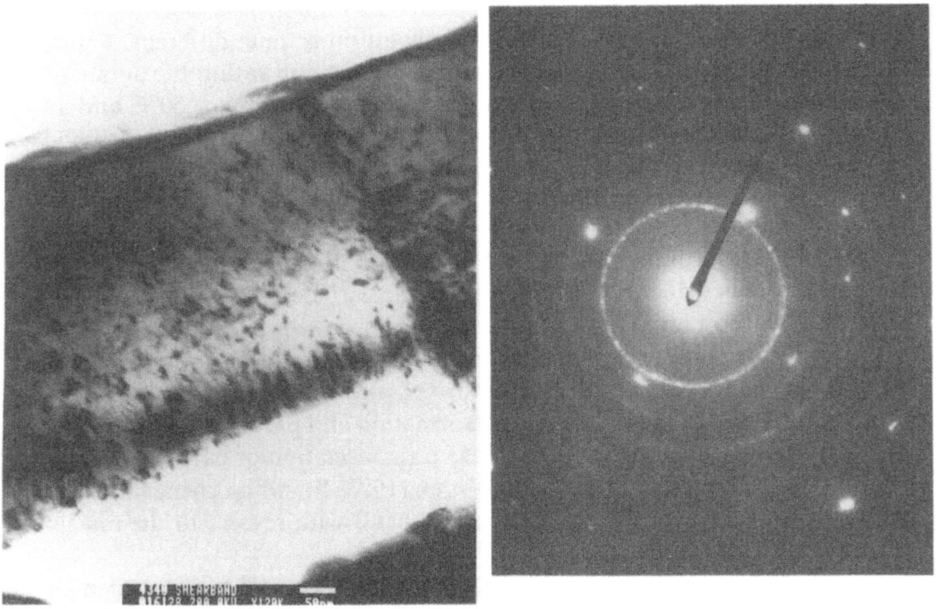

Figure 2.22. TEM of 4340 quenched-and-tempered martensitic shear band and diffraction pattern.

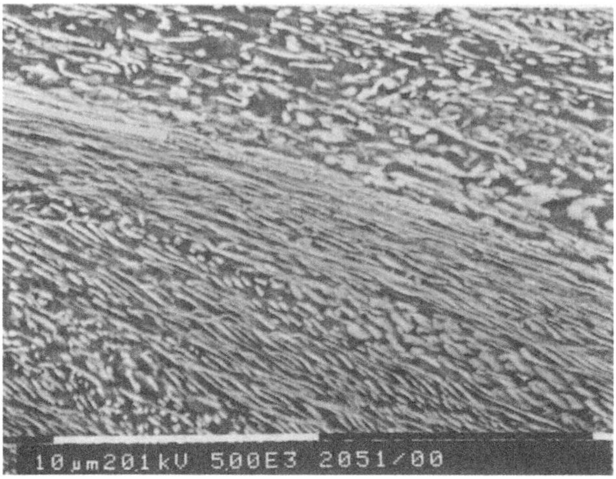

Figure 2.23. Partially spheroidized microstructure inside a shear band in pearlitic 4340 steel. [From Zurek (1993). Submitted for publication.]

and found it consisted of small (<50 nm) grains; carbides were dissolved. Beatty et al. (1992) conducted similar observations on a quenched-and-tempered martensitic 4340 steel. They found grain sizes varying between 8 and 20 nm inside of the shear band.

In pearlitic steels, the microstructural evolution is quite different. Figure 2.23 shows the partial spheroidization of the cementite within the adiabatic shear band. This suggests a local temperature increase above 650°C and an extremely rapid spheroidization. A local shear strain within the shear band of $\gamma = 4$–5 caused the Fe_3C to deform and fragment and at the same time accelerated the kinetics of spheroidization. Uvira et al. (1972), Robbins et al. (1964), Chojnowski and Tegart (1968), Hirano (1963), and Zurek (1993) suggested that during high temperature deformation the lamellar carbides would spheroidize due to both fracture of the carbide plates and to "short-circuiting" carbon and iron diffusion paths associated with the creation of very fine ferrite subgrains. Nanohardness differences found across the adiabatic shear band compared to the deformed and undeformed regions were thought to be due to the refinement, deformation and partial spheroidization of the microstructure, rather than phase transformation or carbon segregation within the adiabatic shear band. Zurek (1993) found no chemical segregation in the region of the adiabatic shear band with respect to the matrix material.

In a shear band, large shear leads to fragmentation and work hardening but the rate of damage accumulation which causes fracture must depend on the level of tensile stress across the band. This supports the observation that ductile fracture was the primary mode of failure at the termination of the shear bands. Finally, the "white" etching bands in steels are beyond the optical microscope resolution limit and are not necessarily an indication of phase transformation.

2.3.3. Shear Bands in Titanium and Titanium Alloys

Titanium shows a great propensity for shear-band formation because of its strain rate sensitivity, low thermal conductivity, and low heat capacity. Grebe, Pak and Meyers (1985), Meyers and Pak (1986), and Meyers et al. (1993), investigated the microstructure of shear bands in titanium and Ti-6Al-4V alloy. Two methods were used for the generation of shear bands: (a)

(a)

(b)

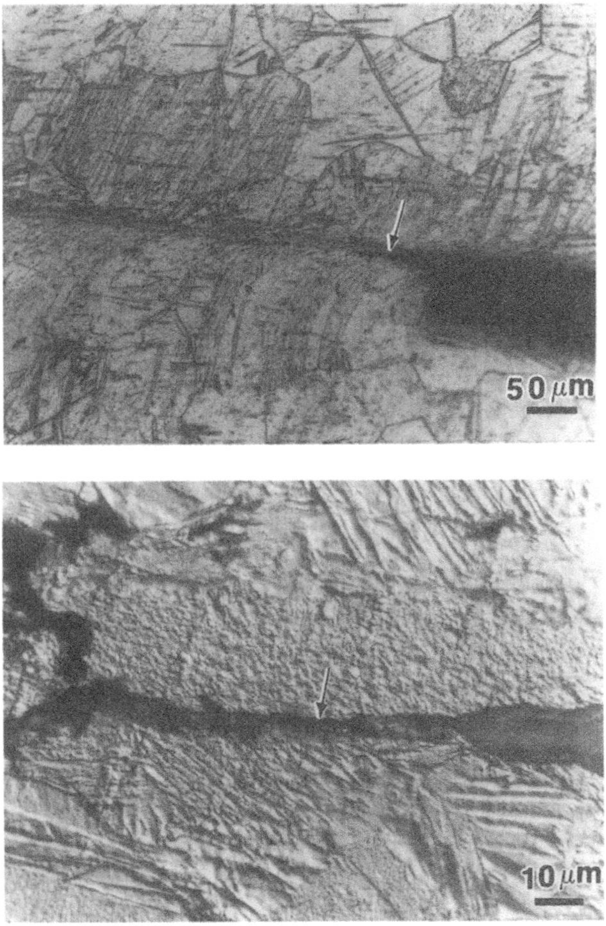

Figure 2.24. Cracks (a) initiated; and (b) propagated along shear band in titanium [From Meyers et al. (1993)].

(a)

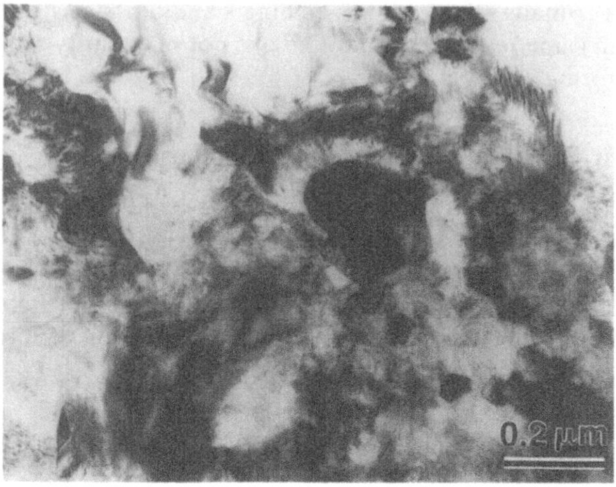

(b)

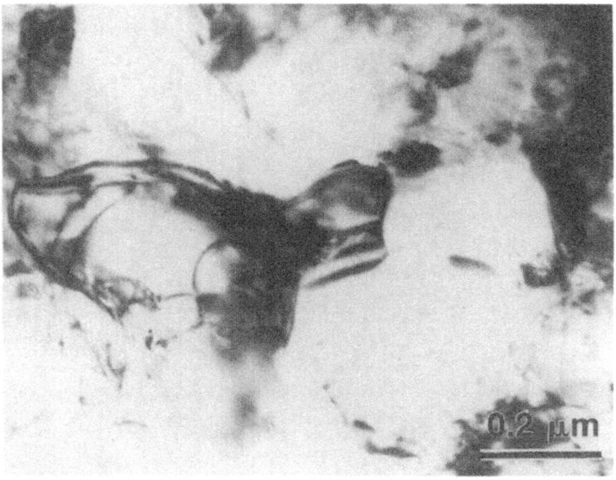

Figure 2.25. Transmission electron micrograph of shear band in titanium. (a) Shear band generated by ballistic impact. [From Grebe et al. (1986).] (b) Shear band generated in Hopkinson bar. [From Meyers et al. (1994).] (c) Diffraction pattern showing microcrystalline structure.

impact at 600 m/s of a cylindrical projectile on a titanium plate; and (b) rapid shearing of a narrow band produced in a hat-shaped specimen placed in a compression Hopkinson bar. These two widely different techniques produced shear bands that were essentially identical. These shear bands varied in width from 2 to 20 μm; their width was found to increase with shear strain. They formed at a critical shear strain of 1.5. These shear bands were inevitably precursors to failure, as shown by the cracks propagating along the shear-band path, Figure 2.24.

(c)
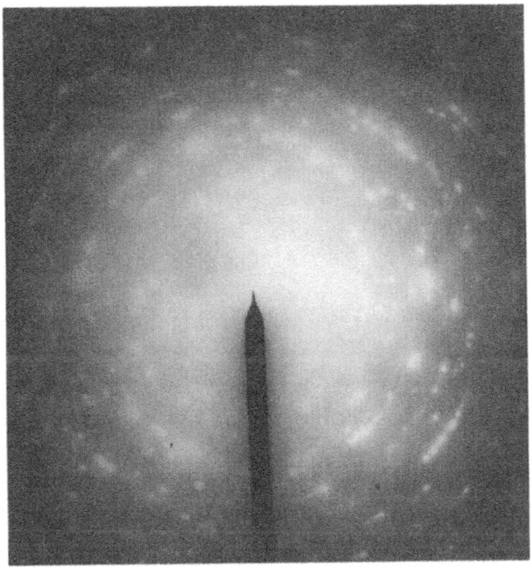

Figure 2.25 (*cont.*)

Figure 2.24(a) shows the onset of fracture, whereas Fig. 2.24(b) shows a fully formed crack. The crack initiated at an external strain concentration (see Figure 2.20(a)). Transmission electron microscopy conducted by Meyers and Pak (1986) revealed that the shear-band interior consisted of very small grains, 0.05 to 0.3 μm in diameter, with a low dislocation density (Figure 2.25(a)). Pak et al. (1986) suggested that this microstructure was formed by a dynamic recrystallization process. The results obtained by Meyers et al. (1994) on shear bands produced under prescribed conditions of strain and strain rate are essentially the same as the earlier study of Meyers and Pak (1986) and Grebe et al. (1983). The shear-band microstructure, produced in Hopkinson bar experiments by Meyers et al. (1994) also consisted of the small grains with a very low dislocation density shown in Figure 2.25(b). The grain boundaries between these grains are well defined and the microstructure between adjacent grains varies between low-angle and high-angle boundaries. The diffraction pattern of Fig. 2.25(c) illustrates the microcrystalline nature of the structure as well as the wide variety of misorientations between micrograins.

2.3.4. Shear Bands in Copper

While annealed copper does not show localization of plastic flow under normal loading situations, the passage of a shock wave can significantly increase the flow stress, and reduce the work hardening rate, rendering it

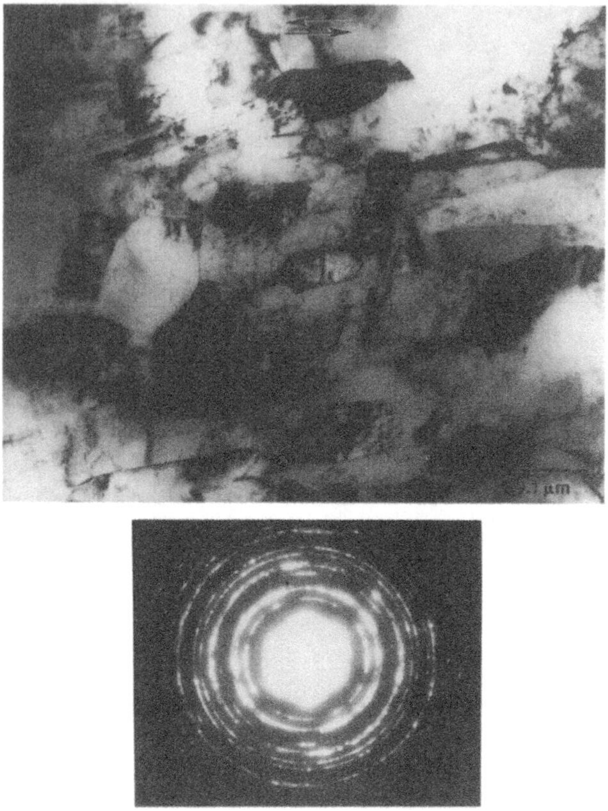

Figure 2.26. Transmission electron micrograph and diffraction pattern of shear band in pre-hardened copper. (From Meyers et al. (1991).]

susceptible to localization. Indeed, Tang et al. (1991) found shear localization regions in recovered explosively forged copper projectiles.

Meyers et al. (1992) and Andrade et al. (1994) subjected copper to shock pre-conditioning at 50 GPa and subsequently studied its response under high strain, high-strain-rate loading. The microstructure within the shear bands (produced in a hat-shaped specimen, using a Hopkinson pressure bar) resembled the ones observed in steel (Section 2.3.2) and titanium (Section 2.3.3). After a critical shear strain $\gamma \sim 2$, the specimen formed a well defined shear band, of a width of $\sim 200\ \mu m$. Figure 2.26 shows a transmission electron micrograph of the shear band and the associated diffraction pattern. The microcrystalline structure within the band is readily evident. Small, equiaxed grains with a low dislocation density are present. The diffraction pattern, consisting of almost continuous rings, shows that the micrograins have a wide range of orientation. This microcrystalline structure is indicative of a dynamically recrystallized structure.

2.4. Stress Wave Effects in Rocks and Ceramics

A shock wave propagating in rocks and ceramics leads to fragmentation rather than shock strengthening. A tensile release wave activates the flaws generated under compression and produces spalling at a low stress level. Models describing the propagation of high amplitude disturbances in rocks and ceramics should incorporate the time dependency of the fragmentation process. Experiments by Aimone, Meyers, and Mojtabai (1984) on a quartz monzonite rock subjected to a uniaxial strain plane compression pulse showed that:

a. The mean fragment size decreases with increasing pulse duration at constant pressure, and with increasing pressure at constant pulse duration. This is shown in Figure 2.27(a) for quartz monzonite copper porphyry subjected to different shock-wave pulses.

b. The internal damage generated in rocks by passage of a shock wave is much larger than the damage simply observed by the fragments. The total surface area of cracks generated internally is higher, by at least one order of magnitude, than the surface area of the observed fragments. This can be verified by converting the internal damage into an "equivalent fragment size." This was accomplished by using the expression $D = 6/S_v$ defined earlier. Figure 2.27(b) shows the measured surface area per unit area (total area of cracks). The "equivalent" particle size is shown in Fig. 2.27(c). The differences in particle size between Figs. 2.27(a) and (c) are of the order of magnitude 10^2. Thus, a great portion of the damage remains hidden inside the fragments.

The time dependent failure of rocks is consistent with the concepts of nucleation, growth, and coalescence. If a description of how a shock wave propagates through a rock is attempted, it has to incorporate these effects. The energy dissipated by the wave in the fragmenting rock or ceramic is associated with the following energy storage mechanisms: generation of cracks, possible phase transformation, twinning, interfacial defects, sliding of the crack surfaces, porosity collapse, and dislocation motion; of these, the energy expenditure associated with pore collapse is the highest dissipative component in rocks.

Spall experiments conducted by Louro and Meyers (1989) on alumina demonstrated that the extent of damage, measured by S_v, the surface area of cracks per unit volume, increased with increasing pulse duration. The results are shown in Figure 2.28(a) and (b).

In Figure 2.28(a) the pressure was varied at a constant pulse duration, for two aluminas: AD 85, and high-purity alumina. The damage in AD 85, an alumina containing considerable porosity and up to 12% of SiO_2, is higher than the high-purity alumina, showing that porosity and grain-boundary phases are a source of damage. The effect of varying the duration of the stress pulse, at a constant stress level, is shown in Figure 2.28(b). The tensile pulse is created by using a capsule with a shock impedance lower than the

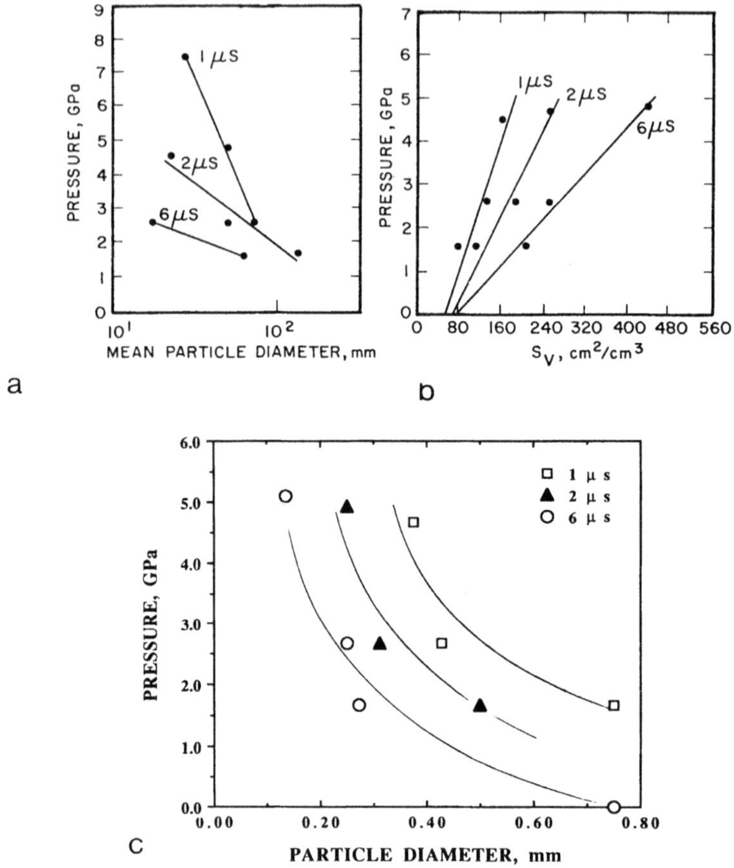

Figure 2.27. Damage in quartz monzonite rock subjected to compressive stress pulses of varying amplitudes and durations; (a) fragment size distribution; (b) and (c) internal damage, expressed as surface area-per unit volume (b) and "effective" particle diameter. [From Aimone et al. (1984).]

alumina, (aluminum). The tensile pulse duration was varied from 0.2 to 1.6 μs, after a constant compressive stress pulse of 4.5 GPa, providing a tensile amplitude of 3.2 GPa.

An important observation made by Louro and Meyers (1989) is that the compressive stress pulse preceding tensile reflection is extremely effective in pre-conditioning the ceramic, by creating cracks that are subsequently activated by tension. Thus, two ceramics subjected to identical tensile pulses can exhibit highly different damage levels. The datum point marked ◪ in Figure 2.29 illustrates this well. Aluminum and epoxy were used as backing materials. Epoxy has a very low shock impedance and the tensile pulse generated, for a specific compression pulse, is much higher than with aluminum backing. For a 4.6 GPa compressive pulse, the tensile pulse has an amplitude of 4.3 GPa. Figure 2.29 shows that the large tensile stress did not cause a signif-

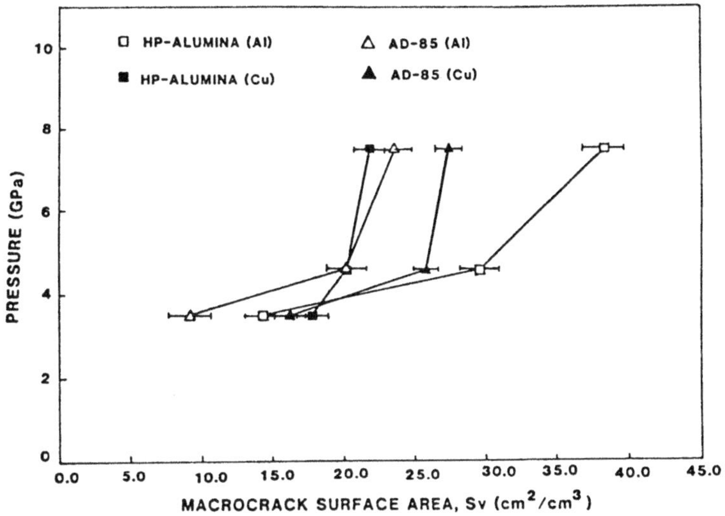

(a)

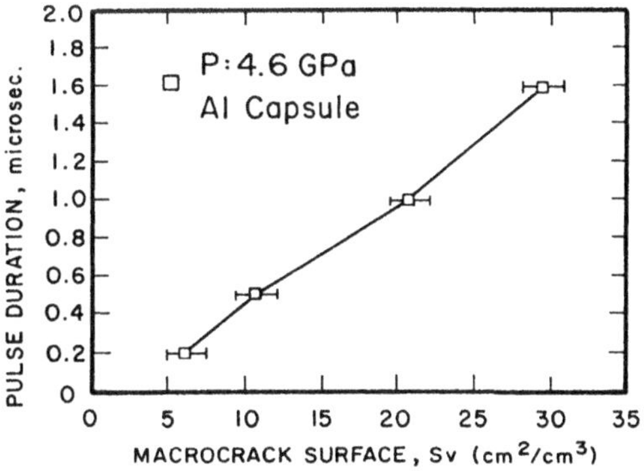

(b)

Figure 2.28. Effect of (a) pressure and (b) pulse duration on the damage produced in alumina. (From Louro and Meyers (1989).]

icant increase in fragmentation. The epoxy-backed alumina (which exhibited a tensile pulse close to the compressive pulse and is marked ◪) is the isolated point in Figure 2.29. All other data points (falling on roughly the same line) correspond to tensile stresses preceded by compressive stresses three times higher in amplitude. The latter pulses were produced in alumina capsules (where the tensile pulse amplitude is ~30% of the compressive pulse). If exposure to compression prior to tension did not have an effect on fragmentation, the point marked ◪ should fall on the same line as the other points, and S_v should be approximately 80 cm^2/cm^3, instead 32 cm^2/cm^3. This result emphasizes the important role played by prior compressive pulse in determining the alumina fragmentation. The effect of grain size on damage was also investigated by Louro and Meyers (1989). The results are shown in Figure 2.30, where two grain sizes, 4 and 24 μm, of the same material, are shown.

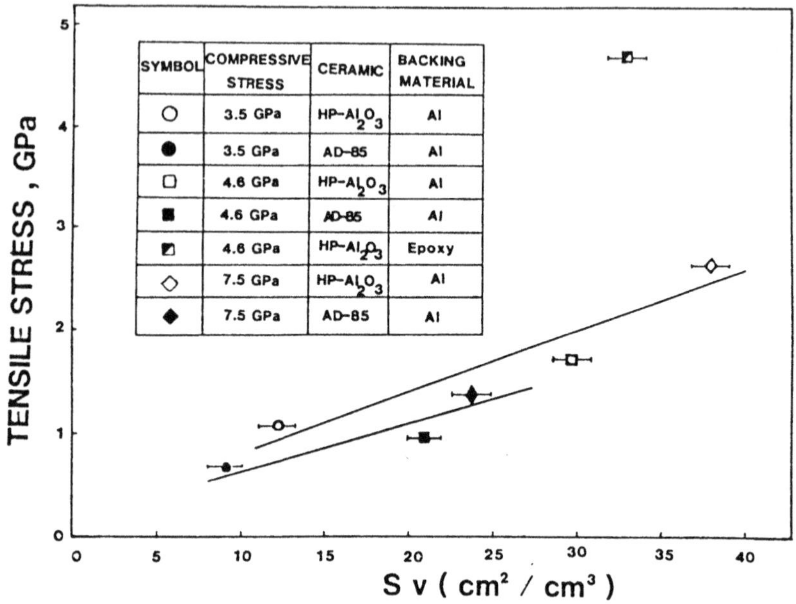

Figure 2.29. Effect of tensile stress amplitude on damage in alumina; notice point ◩ which corresponds to a lower compressive pulse amplitude. [From Louro and Meyers (1988).]

The smaller grain size exhibits greater damage, under the same loading conditions. The explanation for this behavior is that the density of crack nucleation sites is greater for the smaller grain size. The most obvious nucleation sites for flaws are pores and grain boundaries. The mechanisms by which these compressive stresses can generate flaws are shown, in schematic fashion, in Figure 2.31.

The most important mechanisms are:

a. Spherical voids subjected to compression generate tensile stresses. In Figure 2.31(a) we have σ_θ tensile stress. Therefore, cracks are generated. These cracks are parallel to the direction of load application. This problem was first solved by Goodier (1933).

b. For ellipsoidal flaws, the same phenomenon occurs. Indeed, the spherical case (a) is a special case of the elliptical flaw. Brace and Bombolakis (1963) and Horii and Nemat-Nasser (1986) studied the stresses generated by this configuration. The shear stress due to the applied compressive load, τ, generates tensile stresses at the extremities of the elliptical flaw, eventually cracking them [Figure 2.31(b)].

c. Elastic anisotropy of a polycrystalline ceramic aggregate leads to incompatibility stresses at the boundaries. Figure 2.31(c) shows the "softer" and "stiffer" directions of grains marked by a short (>—<) and long (>——<) line respectively. Plastic deformation can occur under compressive loading. When the stress is released, i.e., after the passage of the stress pulse, localized

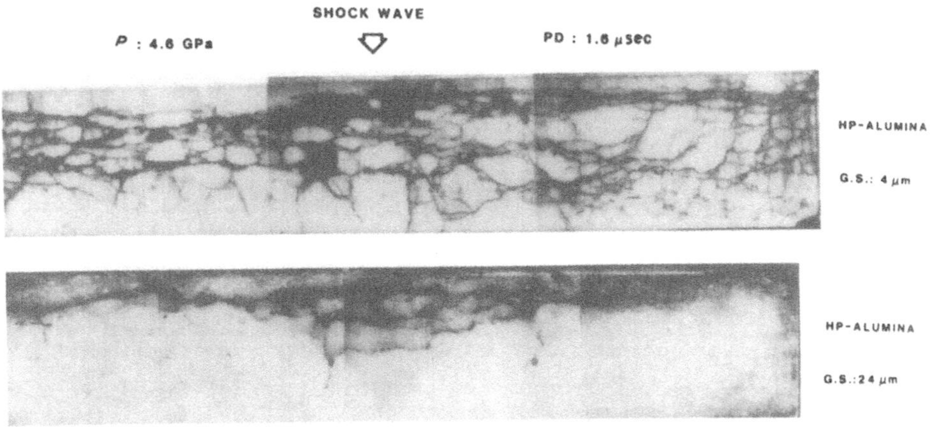

Figure 2.30. Cross-section of alumina disks subjected to compressive stress pulse of amplitude 4.6 GPa and duration 1.6 μs; (a) 4 μm grain size; (b) 24 μm grain size. [From Louro and Meyers (1989).]

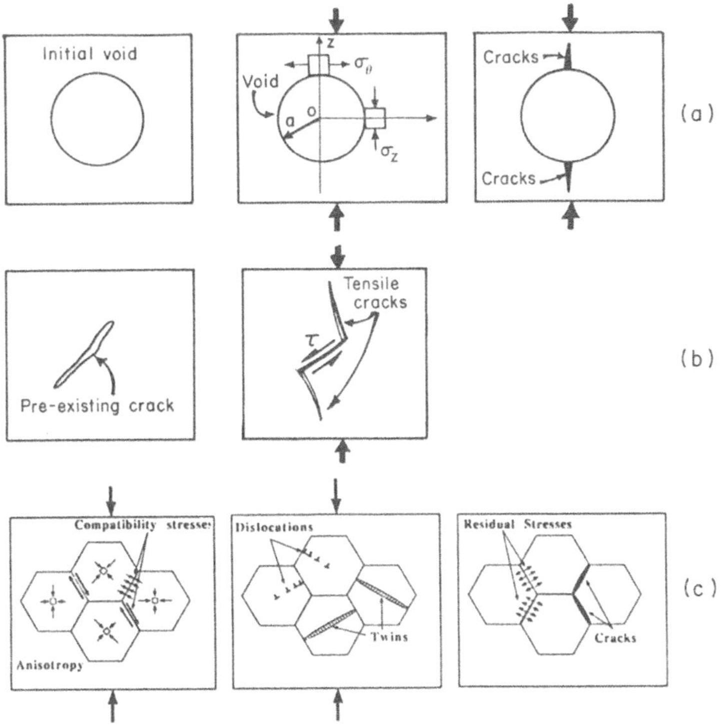

Figure 2.31. Schematic representation of different mechanisms of crack nucleation in ceramics subjected to compressive stress pulses; (a) nucleation at spherical void; (b) nucleation at elliptical flaw; (c) nucleation due to elastic anisotropy of grains. [From Meyers (1994).]

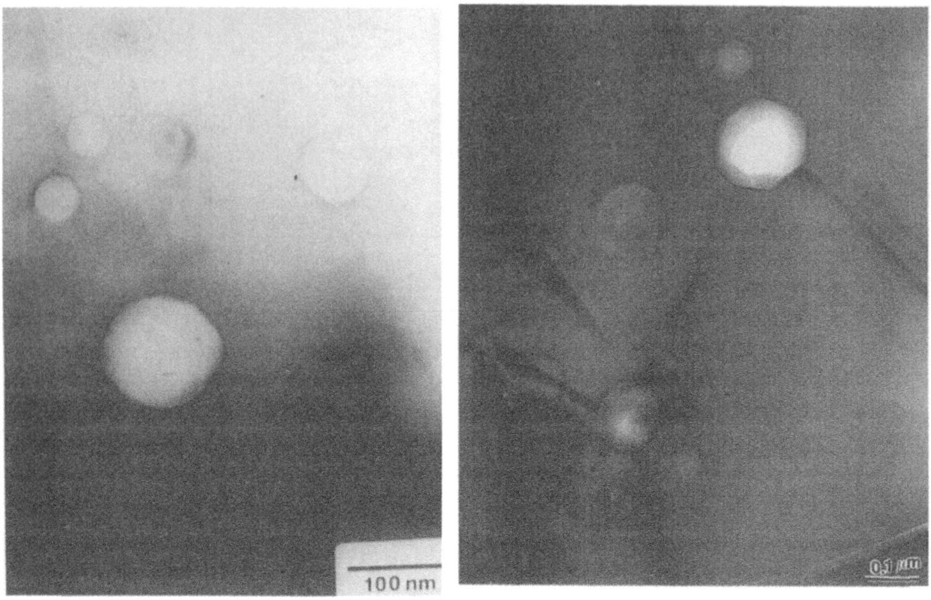

(a) (b)

Figure 2.32. (a) Voids in alumina; (b) cracking at voids after passage of stress wave. [From Louro and Meyers (1989).]

tension regions are created by the residual stresses due to plastic deformation. This can open up cracks. These are not strictly compression cracks but rather "post-compression cracks." Lankford (1977) describes some of these concepts.

Transmission electron microscopy observations can reveal the onset of damage in ceramics subjected to compressive and tensile pulses. Voids in alumina, prior to stress-wave loading, are shown in Figure 2.32(a). The activation of cracks at existing voids is shown in Figure 2.32(b). Cracks, emanating from these voids, are seen very clearly. These cracks have only grown a few micrometers. These cracks can be generated by the mechanism shown in Figure 2.31(a).

Subsequent tensile loading will activate these cracks, if the stress intensity factor and time are sufficient. In Figure 2.33(a), dislocations generated by stress-wave loading (below the material HEL), are shown. The incompatibility stresses can be responsible for the generation of dislocations at the grain boundaries. Alumina, having a hexagonal structure, exhibits an anisotropy of elastic properties. Therefore, stresses build up at the grain-boundary regions, leading to cracks, when unloading occurs. Grain-boundary cracking is shown in Figure 2.33(b). The cracked boundaries are marked by arrows.

The micrographs (TEM) of Figures 2.32 and 2.33 illustrate some of the damage mechanisms in ceramics. In actual ceramics, the presence of grain-boundary phases, of a glassy nature, is often observed. These grain-

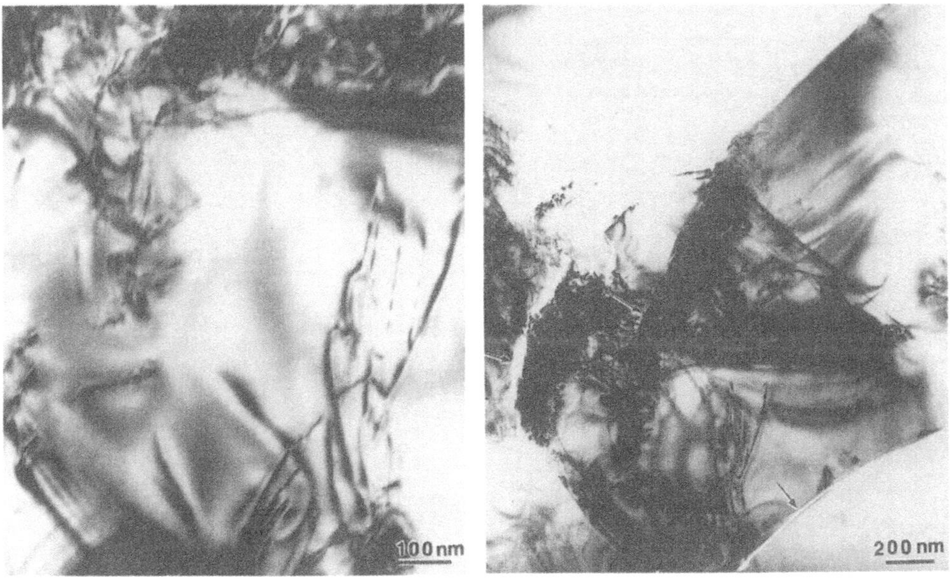

(a) (b)

Figure 2.33. (a) Dislocations in alumina after passage of stress waves, (b) cracks m alumina (marked by arrows). [From Louro and Meyers (1989).]

boundary phases are the result of additives that are part of commercial ceramics to help the sintering process. These glassy phases have a different shock impedance than the surrounding grains and contribute to failure. The process of fragmentation of ceramics under compressive/tensile pulses will not be discussed here. Figure 2.34 shows, in a simplified manner, how pre-existing flaws (a) can generate cracks; (b), under compression; and how these cracks can grow under tension; (c) fragmenting the ceramic (d). A more detailed treatment with quantitative predictions of fragment size is given by Louro and Meyers (1990). Since this is a dynamic process, the stresses in the regions surrounding cracks are reduced by the free surface generated, and therefore nucleation of cracks in these regions is hampered leading to a fragment size that is a function of both stress amplitude and duration. In Figure 2.34, it is clear that the evolution of damage is determined by the existing flaws or microstructural inhomogeneities in material. These microstructural effects together with the stress history parameters determine the final fragment sizes and size distribution.

It has been conclusively shown by Nesterenko, Meyers, and Chen (1995) that the fragmented (or "comminuted") ceramic will flow by shear localization. This is extremely important for the defeat of ceramic armor, since the ceramic fragmented by the shock wave has to be ejected, for penetration to proceed. Experiments conducted on both granular and fragmented alumina revealed that the plastic deformation proceeds primarily by shear localization, beyond a compressive strain of -0.2. Figure 2.35 shows a section of an

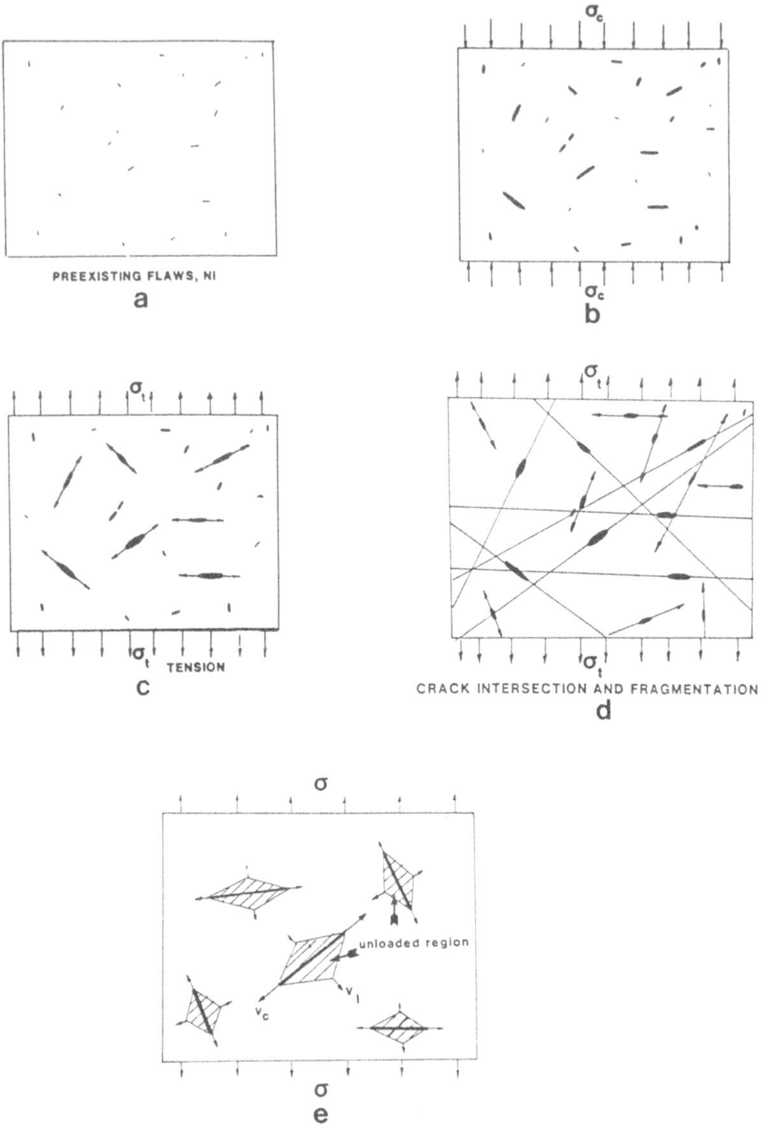

Figure 2.34. Schematic representation of sequence of events leading to fragmentation of brittle material after compression-tension pulse. (a) Initial pre-existing flaws (voids, grain boundary phases, etc.); (b) activation of flaws and formation of cracks by compression; (c) growth of cracks by tension and nucleation of new cracks; (d) intersection of cracks and fragmentation; (e) development of unloaded regions around growing cracks in which no further crack nucleation and growth occurs. [From Louro and Meyers (1990).]

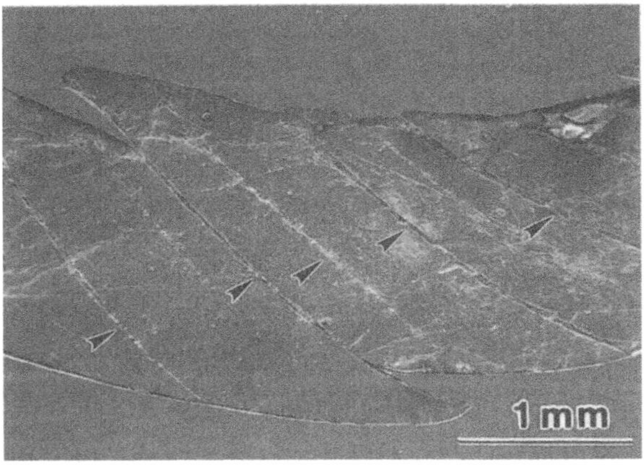

Figure 2.35. Shear localization when densified granular alumina (85% T.D.) is subjected to a compressive tangential strain of ~ -0.3 [From Nesterenko, Meyers, and Chen (1995)].

annular specimen of densified granular alumina after a compressive tangential strain of ~ -0.3. The shear bands are clearly seen (arrows) and have a width of $\sim 10\ \mu m$.

2.5. Summary and Conclusions

The observations reported in this article as well as extensive information in the literature indicate that microstructural aspects are of utmost importance in the dynamic failure of materials.

Dynamic failure was divided into three classes:

1. Dynamic failure by tension; spalling (uniaxial strain state) and uniaxial stress state.
2. Dynamic failure by shear; shear instability is the precursor to this failure mode.
3. Dynamic failure by compression; ceramics, rocks and brittle metals (cast iron, intermetallic compounds) are especially prone to failure under compressive loading.

We summarize below the most important microstructural effects for these three classes of failure.

All microstructural variables play a role in the dynamic failure of materials. The most important variables are:

- grain size
- presence, density, morphology, and size and distribution of second-phase particles
- texture (not mentioned in this chapter)

- impurity atoms and their distribution (e.g., segregation at grain boundaries)
- crystallographic structure (which can be changed by heat treatment, e.g., quenching of steel)
- pre-strain (e.g., annealed vs. shock-hardened material)
- intergranular phases (e.g., glassy phase at grain boundaries of ceramic)
- voids (e.g., porosity in ceramics).

2.5.1. Dynamic Failure by Tension

This failure is determined by the nucleation, growth and coalescence of voids (for ductile failure) or cracks (for brittle failure). The nucleation and growth of different flaws occur independently, in contrast to quasi-static failure, because the rate at which these events occur is such that the stress waves cannot communicate the information to adjoining flaws.

In steels, the transition from ductile to brittle failure is facilitated by a strain-rate increase, because of the strain-rate dependence of the flow stress. The spall failure above 13 GPa has a totally different morphology from the one below 13 GPa, because of the pressure-induced phase transformation.

For copper, grain boundaries and second-phase particles (inclusions) are seen to play a major role in determining the spall strength and fracture morphology. For large-grain-sized specimens, void nucleation takes place at the grain boundaries; although the reasons are not totally understood, grain boundary segregation of impurities (more extensive in large-grain-size specimens, which have a smaller grain boundary area per unit volume) is thought to be the cause. The ratio between nucleation and growth rate determines the size of the dimples and the evolution of damage.

2.5.2. Dynamic Failure by Shear

The occurrence of adiabatic shear bands is determined by the thermomechanical response of the material. The propagation of a shear band occurs by the advance of a shear band tip and the creation, behind it, of a thin region (a few μm) of heated and softened material. This material has a much lower tensile strength than the surrounding matrix, and low tensile stress will induce separation along this region. Upon cooling, the material inside the band often becomes harder and more brittle than the matrix. Therefore, it becomes a favored path for failure; the shear bands can "trigger" both ductile and brittle failure.

All metallurgical parameters which affect the thermomechanical response (and especially, the work hardening, thermal softening, and flow stress) of the material influence its propensity to undergo shear instabilities. These shear bands are the primary path for shear failure. For instance, annealed copper does not undergo shear instabilities under normal loading conditions, whereas it is prone to shear localization after a sufficient hardening. In a similar fashion, pure aluminum does not localize, while aged aluminum forms shear bands. Pearlitic steel localizes, but with much less intensity than quenched-and-tempered steel.

The evolution of the microstructure within the shear bands is determined by the combination of initial microstructure, plastic deformation and heating and may lead, under certain conditions, to dynamic recovery and/or recrystallization; the microcrystalline structures observed in recovered specimens of titanium and copper are supportive of such mechanisms. Ceramics fragmented by compressive and tensile dynamic loading also deform by shear localization; the softening in this case is not, however, a primarily thermal process.

2.5.3. Dynamic Failure by Compression

Whereas ductile metals can undergo large compressive stresses without failure, in brittle materials (ceramics, rocks, intermetallic compounds, and brittle metals such as cast iron) flaws are generated under compressive stresses. It should be clarified that by virtue of microstructural inhomogeneities, the remote compressive loading causes localized regions of tension which on their turn, lead to crack initiation. Thus, microstructural effects are responsible for cracking under compression. Examples of microstructural inhomogeneities that can nucleate cracks under compression are:

- Voids, around which tensile stresses are generated by compression. These voids can have all kinds of shapes, but spheres and ellipsoids are idealized configurations which lend themselves to mathematical analyses predicting localized regions of tension.
- Grain boundaries between grains in materials having elastic anisotropy in such a manner that elastic incompatibility stresses are generated.
- Brittle grain-boundary phases which may fracture under shear resulting from compression. An example is the glassy grain-boundary phase in commercial alumina. Ceramics often contain these phases, which are due to the sintering agents that are added to material to facilitate densification during processing.
- Second phase particles which may have different compressibilities than the matrix, leading to crack nucleation at the interface.
- Destruction of coherency between the matrix and second phases due to differences of elastic properties.

2.6. Acknowledgments

The writing of this chapter was supported by the Joint DOE/DOD Munitions Technology Program (AKZ) and the U.S. Army Research Office University Research Initiative Program (MAM). The research described was supported by the Joint DOE/DOD Munitions Technology Program and by the Eglin Air Force Base (AKZ) and by the U.S. Army Research Office,

National Science Foundation, and the Center for Explosive Technology Research, New Mexico Tech. (MAM).

Anna K. Zurek would like to acknowledge help and collaboration from G.T. Gray III, W. Blumenthal, T. Lowe, W.J. Wright, T. Zocco, C. Trujillo, and M. Lopez. Marc André Meyers would like to acknowledge help and collaboration from S. Christy, A. Grebe, H-r. Pak, J. Isaacs, K.S. Vecchio, B. Kad, U. Andrade, C.T. Aimone, J.C. La Salvia, H.C. Chen, and Y.J. Chen.

2.7. References

Aimone, C.T., Meyers, M.A., and Mojtabai, N. In *Rock Mechanics in Productivity and Design*, (eds. C.H. Darding and M.M. Singh), SME-AIME, 979 (1984).

Andrade, U., Meyers, M.A., Vecchio, K.S., and Chokshi, A.H. *Acta Met. and Mat.*, **42**, 3183 (1994).

Armstrong, R.W., Coffey, C.S., and Elban, W.L. *Acta Met.*, **30**, 2111 (1992).

Asaro, R.J., and Needleman, A. *Scripta Metallurgica*, **18**, 429 (1984).

Bai, Y. In *Shock Waves and High-Strain-Rate Phenomena in Metals: Concepts and Applications*, (M.A. Meyers and L.E. Murr, eds.) Plenum, New York, 227 (1981).

Balchan, A.S. *J. Appl. Phys.* **34**, 241 (1963).

Banks, E.E. *J.I.S.I.*, **206**, 1022 (1968).

Barker, L.M. and Hollenbach, R.E. (1974) *J. Appl. Phys.* **45**, 4872 (1974).

Beatty, J.H., Meyer, L.W., Meyers, M.A., and Nemat-Nasser, S. in *Shock-Wave and High-Strain-Rate Phenomena in Metals*, (M.A. Meyers, L.E. Murr, K.P. Staudhammer eds.), Marcel Dekker, New York, p. 645 (1992).

Bose, A., Sims, D., and German, R.M. *Met. Trans.*, **19A**, 487 (1988).

Brace, W.F., and Bombolakis, E.G. *J. Geophys. Res.*, **68**, 3709 (1963).

Broberg, K.B. *Arch. für Fysik*, **18**, 159 (1960).

Chang, S.N. and Meyers, M.A. *Acta Met*, **36**, 1085 (1988).

Chojnowski, E.A. and McG. Tegart, W.J. *Metals Science Journal*, 1968, **2**, 14 (1968).

Christy, S., Pak, H.-r., and Meyers, M.A. In *Metallurgical Applications of Shock-Wave and High-Strain-Rate Phenomena*, (eds. L.E. Murr, K.P. Staudhammer, and M.A. Meyers), Marcel Dekker, New York, p. 835 (1986).

Clifton, R.J. *Material Response to Ultra High Loading Rates*, Report No. NMAB-356, National Materials Advisory Board, NAS, Washington, Chapter 8, 1979.

Clifton, R.J., Duffy, J., Hartley, K.A., and Shawki, T.G. *Scripta Metallurgica*, **18**, 443 (1984).

Craggs, J.W. *J. Mech. Phys. Sol.*, **8**, 66 (1960).

Curran, D.R. In *Shock waves and Mechanical Properties of Solids*, (J.J. Burke and V. Weiss, eds.), Syracuse University Press, Syracuse New York, p. 121, 1971.

Curran, D.R. and Seaman, L. In *Shock Waves in Condensed Matter* (Y.M. Gupta, ed.), Plenum Press, p. 315, 1986.

Curran, D.R., Seaman, L., and Shockey, D.A. *Physics Reports*, **147**, 253 (1987).

Derep, J.L. *Acta Metall.*, **35**, No. 6, 1245 (1987).

Dremin, A.N., Molodets, A.M., Melkumov, A.I., and Kolesnikov, A.V. In *Shock-Wave and High-Strain-Rate Phenomena in Materials*, (eds. M.A. Meyers, L.E. Murr and K.P. Staudhammer), Marcel Decker, New York, 751 (1992).

Duvall, G.E. and Fowles, G.R. In *High Pressure Physics and Chemistry*, (edited by R.S. Bradley), Academic Press, New York, **2**, 209 (1963).

Erkman, J.O. *J. Appl. Phys.*, **31**, 939 (1961).

Freund, L.B. *J. Mech. Phys. Sol.*, **20**, 129 and 141 (1972).

Freund, L.B. *J. Mech. Phys. Sol.*, **21**, 47 (1973).

Freund, L.B. *Int. J. Engn. Sci.*, **12**, 179 (1974).

Freund, L.B. *Dynamic Fracture Mechanics*, Cambridge University Press, Cambridge, England (1990).

Glenn, R.C. and Leslie, W.C. *Met. Trans.*, **2**, 2945 (1971).

Goodier, N. *J. Appl. Mech.*, **1**, 39 (1933).

Gourdin, W.H. In *Proc. of Intl. Conf. on Impact Loading and Dynamic behavior of Materials*, Bremen, FRG (1987).

Grady, D.E. *J. Geophys. Res.*, **82-B2**, 913 (1980).

Grady, D.E. *Mechanics of Materials.* In press.

Gray, G.T. In *High-Pressure Shock Compression of Solids*, (J.R. Asay and M. Shahinpoor eds.) Springer-Verlag, New York. In press.

Grebe, H.A., Pak, H.-r., and Meyers, M.A. *Met. Trans.*, **16A**, 761 (1985).

Hartley, K.A., Duffy, J., and Hawley, R.H. *Mechan. Phys. Solids*, **35**, **No. 3**, 283 (1987).

Hatherly, M. and Malin, A.S. *Scipta Metallurgica*, **18**, 449 (1984).

Hirano, K-I. *Trans. AIME*, **227**, 950 (1963).

Horii, H. and Nemat-Nasser, S. *Trans. Roy. Soc. Lond.* **319**, 337 (1986).

Ivanov, A.G. and Novikov, S.A. *J. Exp. Theor. Phys. (USSR)*, **40**, 1880 (1961).

Johnson, J.N. *J. Appl. Phys.* **52**(4), 2812 (1981).

Kanel, G.I., Rasorenov, S.V., and Fortov, V.E. In *Shock Wave and High-Strain-Rate Phenomena in Materials*, (eds. M.A. Meyers, L.E. Murr, and K.P. Staudhammer), Marcel Dekker, New York 775 (1992).

Krock, R.H. and Shepard, L.A. *Trans. Metal. Soc. AIME.* **227**, 1127 (1963).

Kuriyama, S., and Meyers, M.A. *Met. Trans. A*, **17A**, 443 (1986).

Lankford, J. *J. Mater. Sci.*, **12**, 791 (1977).

Lankford, J., Anderson, C.E. Jr., and Bodner, S.R. *J. Mat. Sci. Lett.* **7**, 1355 (1988).

Leslie, W.C. In *Metallurgical Effects at High Strain Rates* (R.W. Rohde, B.M. Butcher, J.R. Holland and C.H. Karnes, eds.) Plenum Press, New York, 1981.

Louro, L.H.L. and Meyers, M.A. *J. Matls. Sci.*, **24**, 2516 (1989).

Louro, L.H.L. and Meyers, M.A. In *Shock Compression of Condensed Matter—1989*, eds. S.C. Schmidt, J.H. Johnson, and L.W. Davison. Elsevier, New York, 465, 1990.

Marchand, A. and Duffy, J. An Experimental Study of the Formation Process of Adiabatic Shear Bands in a Structural Steel," Brown University Report, April, 1987.

Mescall, J.F. and Rogers, H. MLT TR 89-104, U.S. Army Materials Technology Laboratory Report, 1989.

Meunier, Y., Roux, R., and Moureaud, J. In *Shock-Wave and High-Strain-Rate Phenomena in Metals* (M.A. Meyers, L.E. Murr, K.P. Staudhammer eds.), Marcel Dekker, New York, p. 637, 1992.

Meyer, L.W., Kunze, H.-D., and Staskewitsch, E. *Proc. Seventh Internl. Ballistic Symposium*, The Hague-Netherlands, p. 289, 1983.

Meyers, M.A. *Dynamic Behavior of Materials*, J. Wiley, New York, 1994.

Meyers, M.A. and Aimone, C.T. *Dynamic Fracture (Spalling) of Metals*, In *Progress*

in Materials Science, J.W. Christian, P. Haasen, T.B. Massalski, eds. Pergamon Press, **28**, 1983.
Meyers, M.A. and Chawla, K.K. *Mechanical Metallurgy. Principals and Applications*, Prentice-Hall, Englewood Cliffs, New Jersey, p. 510, 1984.
Meyers, M.A. and Guimaraes, J.R.C. *Mat. Sci. Eng.*, **24**, 289 (1971).
Meyers, M.A., Meyer, L.W., Vecchio, K.S., and Andrade, U. *J. Physique IV, Coll. C1*, **1**, C3-11 (1991).
Meyers, M.A. and Pak, H.-r. *Acta Met.*, **34**, 2493 (1986).
Meyers, M.A., Sarzeto, C., and Hsu, C.-Y. *Met. Trans.*, **11A**, 1737 (1980).
Meyers, M.A., Subhash, G., Kad, B., and Prasad, L. *Mechanics of Materials*, **17**, 175 (1994).
Meyers, M.A. and Wittman, C.L. *Met. Trans.*, **21A**, 3153 (1990).
Molinari, A. and Clifton, R.J. *C.R. Acad. Sci. Paris*, 296 (1983).
Murr, L.E. In *Shock Waves and High-Strain-Rate Phenomena in Metals*, M.A. Meyers and L.E. Murr, eds.), Plenum Press, New York, p. 607, 1981.
Nesterenko, V.F., Meyers, M.A., and Cheng, H.C., "Shear Localization in High-Strain-Rate Deformation of Granular Alumina", submitted for publication (1995).
Rabin, H.B. and German, R.M. *Met. Trans.*, **19A**, 1523 (1988).
Ravi-Chandar K. and Knauss W.G. *Intl. J. Fracture*, **25**, 247 (1984a).
Ravi-Chandar K. and Knauss W.G. *Intl. J. Fracture*, **26**, 65 (1984b).
Robbins, J.L., Shepard, O.C., and Sherby, O.D. *J. Iron Steel Institute*, **202**, 804 (1964).
Rogers, H.C. *Materials Behavior Under High Stress and Ultra-High Loading Rates*, J. Mescall and V. Weiss eds., Plenum Press, New York and London, p. 101, 1983.
Seaman, L., Curran, D.R., Erlich, D.C., Cooper, T., and Dullum, O. *Journal De Physique*, **46**, 273 (1985).
Stevens, A.L., Davison, L., and Warren, W.E. *J. Appl. Phys.*, **43**, 4992 (1972).
Tang, N.Y., Niessen, P., Pick, R.J., and Worsick, M.J. *Mat. Sci. and Eng.*, **A131**, 153 (1991).
Teirlinck D., Zok, F., Embury, J.D., and Ashby, M.F. *Acta Met.*, **36**, 1213 (1988).
Thadhani, N.N. and Meyers, M.A. *Acta Met.*, **34**, 1625 (1986).
Thornton, P.A. and Heiser, F.A. *Met. Trans.*, **2**, 1496 (1971).
Uvira, J.L., Clay, D.B., Worthington, P.J., and Embury, J.D. *Canadian Metallurgical Quarterly*, **11**, 439 (1972).
Wingrove, A.L. *J. Aust. Inst. Met.*, **16**, 67 (1971).
Wittman, C.L., Meyers, M.A., and Pak, H-R. *Met. Trans.*, **21A**, 707 (1990).
Wright, W.T. *J. Mech. Phys. Sol.*, **35**, 269 (1987).
Zehnder, A.T. and Rosakis, A.J. *Intl. J. Fracture*, **43**, 271 (1990).
Zener, C. and Hollomon, J.H. *J. Appl. Phys.*, **15**, 22 (1944).
Zurek, A.K. The Study of Dynamic Fracture During Armor-Penetrator Interaction, *Los Alamos Report*, LA-11719-MS (1989).
Zurek, A.K. *Met Trans. A*, **25**, 2483 (1994).
Zurek, A.K. Unpublished research, 1993.
Zurek, A.K., Follansbee, P.S., and Hack, J. *Met. Trans.*, **21A**, 431 (1990).
Zurek, A.K., Frantz, Ch. E., and Gray, G.T. In *Shock-Wave and High-Strain-Rate Phenomena in Metals*, M.A. Meyers, L.E. Murr, K.P. Staudhammer eds., Marcel Dekker, New York, p. 759, 1992.
Zurek, A.K. and Gray, G.T. *J. Physique IV*, **1**, 631 (1991).
Zurek, A.K., Johnson, J.N., and Frantz, Ch. E. *J. Physique*, **49**(3/9), 269 (1988).

CHAPTER 3

Dynamic Fracture in Metals at High Strain Rate

Sun Chengwei, Zhuang Shiming, Wang Yanping, and Liu Cangli

3.1. Introduction

The response of materials to shock loading has attracted much attention for its applications in high velocity impact, astronautics, explosive working, and defense technology. Scientists are interested in the strain-rate dependence of material performance, especially as it affects dynamic strength, damage, and failure. Since high-strain-rate loading can only be applied for a short time, the available means are: by a plane shock wave produced by plate impact, explosive detonation, gas- or electrically-propelled gun projectiles, and laser-beam irradiation. Other experiments involve freely-expanding rings or exploding cylinders and projectile penetration or plugging of a target. Spall experiments carried out with plane shock waves are most appropriate for fundamental research because two dimensional effects are minimized and loading parameters are easily adjusted over a wide range.

The topics emphasized in this chapter are dynamic failure and spalling of metals loaded by the above means. However, it should be noted that important physical phenomena observed in high strain rate experiments, such as effects of the equation of state of the material, phase transition, dissipation, and shock attenuation, must often be considered and incorporated into analyses of the mechanics problem. Accurate and direct measurement of mechanical parameters such as the bulk strain rate, $\dot{\varepsilon}$, and the spall strength, σ_s, of elements inside targets seems to be impossible. Moreover, reported data are inconsistent, with uncertainty factors of 2 to 3 for σ_s and a full decade for $\dot{\varepsilon}$ being tolerated.

The main approaches to describing the dynamic response of materials at high strain rate can be sorted into the following categories: micro-statistical fracture mechanics [1], theoretical constitutive models [2, 3], empirical criteria and constitutive relations [4–6], linear elastic fracture mechanics [7], experiments [8, 9], and numerical simulations [1, 10]. Dynamic fracture and spall of solids have been well reviewed in Refs. [1, 11].

Research on spalling and dynamic fracture at the Southwest Institute of Fluid Physics began in the early 1980s. Many experiments were carried out with explosives, lasers, an electric gun, and powder and light-gas guns. The appearance of microflaws in metals loaded at ultra-high strain rate by high-powered laser irradiation has been first observed and reported in this chapter. A stress wave model for spall has been proposed to predict the strain-rate dependence of spall strength. Finally, numerical simulations using a version of the SSS code into which a theoretical constitutive relation and the NAG model have been incorporated have been shown to successfully reproduce the material response, spall parameters, and the statistical distribution of microflaws.

3.2. Loading Techniques

3.2.1. Explosively-Driven Plate Impact (EDPI)

As shown in Figure 3.1, the velocity of an impactor can be adjusted by choice of its thickness and use of an appropriate buffer. An explosive plane wave lens 10 or 20 cm in diameter provides shock loading with excellent planarity and is capable of driving large, thick impactors.

3.2.2. Electric-Gun-Driven Plate Impact (EGDPI)

The schematic of an EGDPI assembly is shown in Fig. 3.2. A current pulse obtained by discharging a capacitor bank converts the metallic foil to an exploding plasma of high temperature, pressure, and density. This plasma accelerates a 0.05 to 0.1 mm thick Mylar flyer to a velocity ranging from 0.5 to 7 km/s before it impacts the specimen. In order to meet the requirements of dynamic elastic-plastic testing, the flyer is allowed to impact a buffer plate backed by a thicker impactor. When the shock wave emerges from the buffer, the impactor is driven toward the target at a velocity of 10 to 100 m/s. The electric guns using impact of the Mylar flyer and of the secondary impactor, respectively, are denoted as EGDPI-I and EGDPI-II.

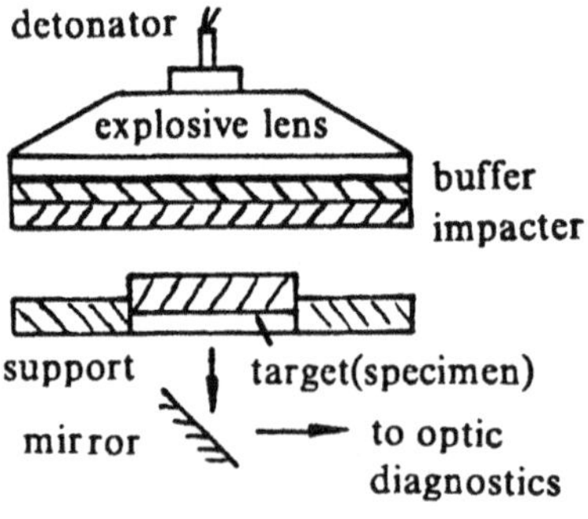

Figure 3.1. EDPI assembly.

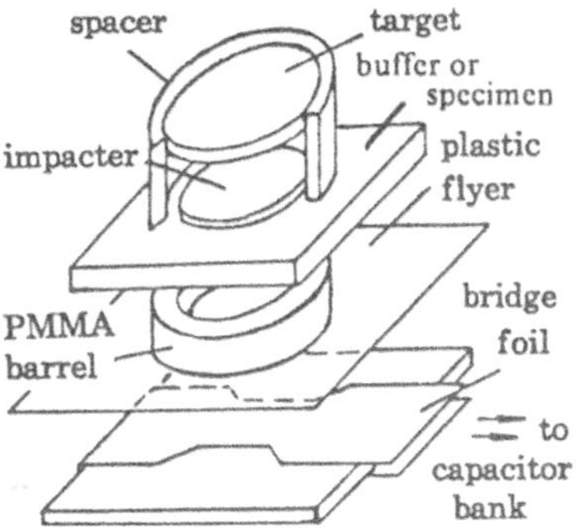

Figure 3.2. EGDPI assembly.

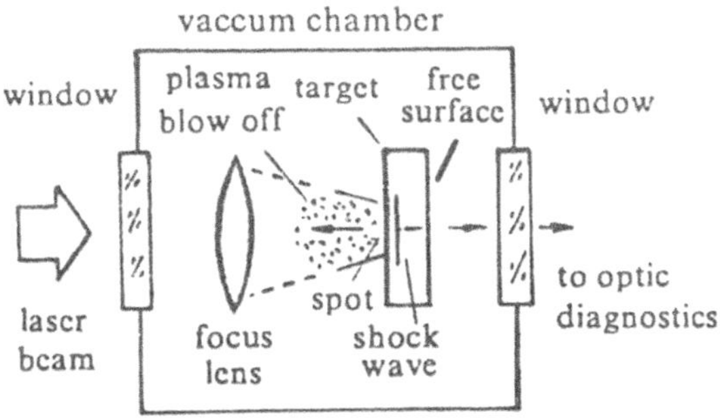

Figure 3.3. Schematic of LDSW experiment.

3.2.3. Laser-Driven Shock Wave (LDSW)

When a target is irradiated by a high-powered laser, the surface layer is vaporized and ionized, with the resulting plasma forming a jet that expands backwards from the target toward the laser. The strong shock resulting from this plasma recoil propagates into the target and induces spall near the rear free surface, as indicated in Fig. 3.3. Since the high-power laser pulse is of very brief duration, for example 100 ps to several ns FWHM, LDSW provides the shortest mechanical loading now obtainable, leading to strain rates of 10^7 to 10^8 s^{-1} and higher. Sophisticated diagnostic equipment and very thin targets (0.1 mm thick) are needed for LDSW experiments.

3.2.4. Projectile Plugging in Target (PPT)

Either elongated projectiles or flat-plate impactors can be launched from gas or powder guns to form an impact crater in a target or even punch through it, i.e. so called plugging. Measurements of shear strain and other parameters in the edge zone of the crater or hole are more difficult than measurements usually made in plate impact experiments.

The velocity of the impactor can be measured with conventional optical and electronic techniques. Piezoresistance and piezoelectric transducers are employed for impact and shock pressure measurements. The pressure gauge used in LDSW experiments must be of at least subnanosecond response [12]. The main measurement in a spall experiment is of free surface velocity history, which can only be accomplished using a high-speed optical device such as a VISAR, Fabry-Perot, or other interferometer. Since there is a strong electromagnetic disturbance in EGDPI and LDSW experiments, scanning electron micrographs of carefully recovered specimens are the main source of data.

3.3. Experiments

3.3.1. Spall in Metals Loaded by EGDPI and EDPI

Zhang [13] conducted spall experiments (EGDPI-I) and numerical simulations on several metals. The results for LY-12 aluminum alloy are listed in Table 3.1. It is concluded that the spall thickness, δ_s, can be reasonably evaluated using a version of the SSS code that incorporates the cumulative damage criterion. The spall threshold of LY-12 aluminum alloy loaded by a stress pulse of 50 to 100 ns duration and at a strain rate of 10^4 to 10^5 s^{-1} is almost the same as that under loading of several microseconds duration. The columns numbered 1, 2, and 3 in Table 3.1 correspond to spall criteria of maximum tensile stress, damage accumulation, and stress gradient, respectively.

Feng and Long [14] have studied the spall threshold of LY-12 aluminum alloy with EGDPI-I loading, as indicated in Table 3.2 and Fig. 3.4, where the unloading bulk strain rate $\dot{\varepsilon}$ is calculated according to Eq. (3.5). The symbols in Table 3.2 are the same as those in Table 3.1, where τ is the calculated duration of the stress pulse using the bulk soundspeed c_0.

The free-surface velocity pullback, Δu_f, and the spall strength, σ_s, of LY-

Table 3.1. EGDPI-I spall experiments and SSS calculations on LY-12 aluminum alloy

Shot No.	Flyer velocity v (km/s)	Mylar flyer thickness δ_f (mm)	Target thickness d (mm)	Impact pressure p (GPa)	Calculated strain rate $\dot{\varepsilon}$ ($10^4 s^{-1}$)	Spall thickness δ_s (mm) cal. 1	cal. 2	cal. 3	Measured
42	0.714	0.10	3	2.11	2.88		no spall		no
43	1.308	0.05	2	2.87	7.84	0.2	0.4	no	0.3 ~ 0.4
35	2.09	0.10	4	5.14	7.02	0.3	0.4	1.23	0.25 ~ 0.45
41	2.09	0.10	6	5.14	7.02	0.4	0.5	/	0.48 ~ 0.52
9	2.17	0.10	7.8	5.39	7.36	0.8	0.6	/	0.45 ~ 0.60
21	2.24	0.05	5	5.61	15.3	0.2	0.3	no	no
25	2.24	0.05	2	5.61	15.3	0.4	0.4	0.2	0.25 ~ 0.35

Table 3.2. Spall threshold of LY-12 aluminum alloy

Target thickness d (mm)	Mylar flyer ($\delta_f = 0.05$ mm, $\tau = 45$ ns)			Mylar flyer ($\delta_f = 0.10$ mm, $\tau = 90$ ns)		
	v (km/s)	p (GPa)	$\dot{\varepsilon}$ (10^5 s^{-1})	v (km/s)	p (GPa)	$\dot{\varepsilon}$ (10^5 s^{-1})
2	2.0 ~ 2.1	4.86 ~ 5.17	1.33 ~ 1.41	/	/	/
3	2.3	5.8	1.58	1.6 ~ 1.8	3.69 ~ 4.27	5.04 ~ 5.83
4	2.5 ~ 2.7	6.43 ~ 7.08	1.76 ~ 1.93	2.1 ~ 2.4	5.17 ~ 6.11	7.06 ~ 8.34
5	3.0 ~ 3.2	8.08 ~ 8.75	2.21 ~ 2.39	2.2 ~ 2.6	5.48 ~ 6.76	7.48 ~ 9.23
9	/	/	/	3.2 ~ 3.5	8.75 ~ 9.78	11.9 ~ 13.4

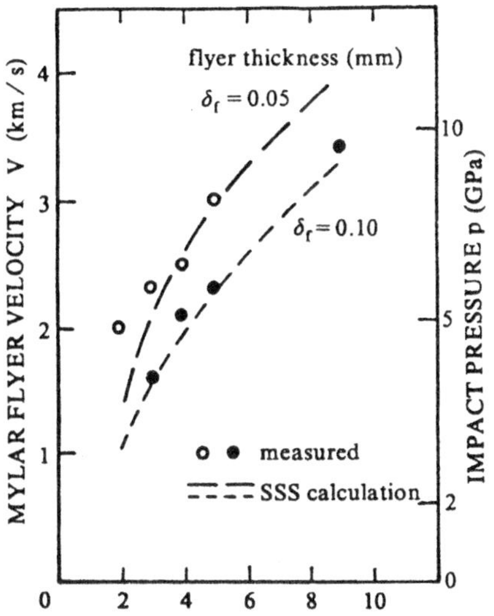

Figure 3.4. Spall threshold of LY-12 Al.

12 have been measured in EDPI experiments with a capacitor transducer [15]. In these experiments, the impactor and target are made of the same material, but have thicknesses of 2 and 6 mm respectively. The results are listed in Table 3.3 and a comparison between measured and SSS-calculated free-surface velocity histories is given in Fig. 3.5.

Two types of titanium-alloy (Ti-6Al-4V) targets were tested with EDPI [16]. In each case, the impactor and target were of the same material and were 6 and 5 cm in diameter and 2 and 6 mm thick, respectively. The first type of target had the so-called Wid structure, with an average grain size from 0.5 to 1 mm. The second type of material had equiaxed α and grain boundary β with grain size from 10 to 20 μm. The free-surface velocity was measured with a VISAR, with the result shown in Fig. 3.6. For the first type target, the spall strength vs. strain rate as evaluated from the velocity pull-

Table 3.3. Measured spall strength of LY-12 aluminum alloy

Shot No.	Impacter velocity v (km/s)	Impact pressure p (GPa)	Strain rate $\dot{\varepsilon}$ (10^4 s^{-1})	Free surface velocity pullback Δu_f (km/s)	Spall strength σ_s (GPa)
4	0.714	5.75	1.00	0.14	1.09
5	0.761	6.16	1.07	0.14	1.09
6	0.787	6.39	1.19	0.17	1.20

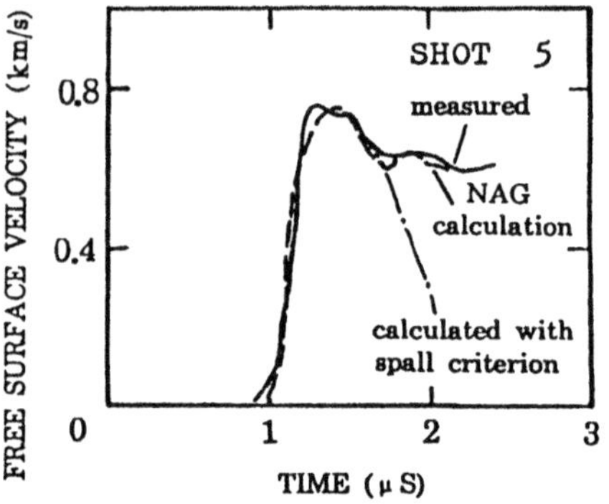

Figure 3.5. Free surface velocity in Shot 5.

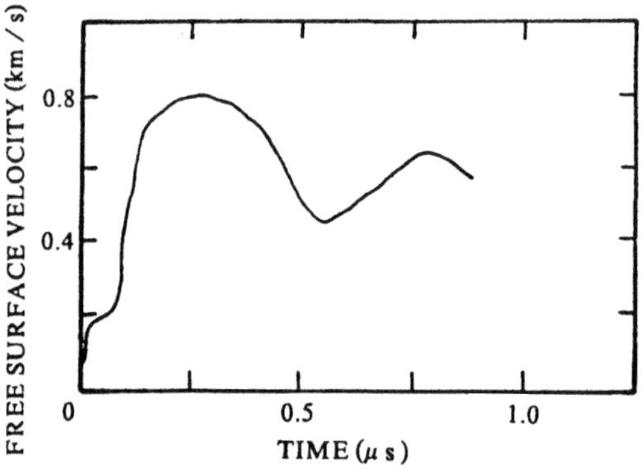

Figure 3.6. Free surface velocity of Ti-6Al-4V.

back is shown in Fig. 3.7. The spall strength of the second type target was found to be 2.2 to 2.6 GPa. These experimental results indicate that the spall strength of Ti-6Al-4V depends on both the strain rate and the microstructure.

Interestingly, strain-concentration bands form in an orientation parallel to the impact surface, as shown in the scanning electron micrograph of Fig. 3.8. There is commonly only one band in coarse-grained specimens (grain size 0.5 to 1.0 mm), but several bands in specimens having fine grain structure. The band shown is of width 20 to 60 μm, within which grain refinement has

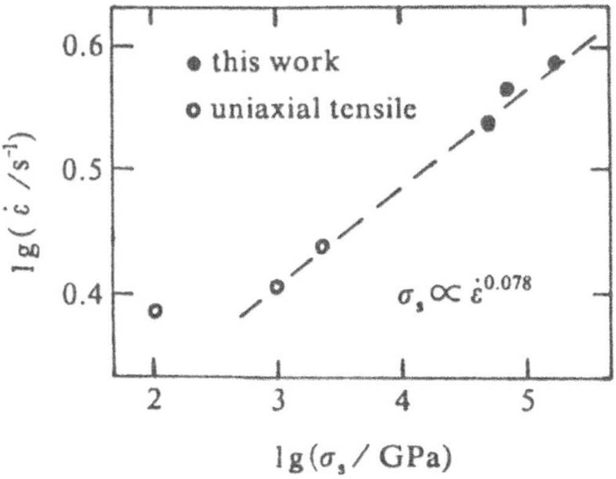

Figure 3.7. Strain rate, $\dot{\varepsilon}$, dependence of spall strength, σ_s, for Ti-6Al-4V.

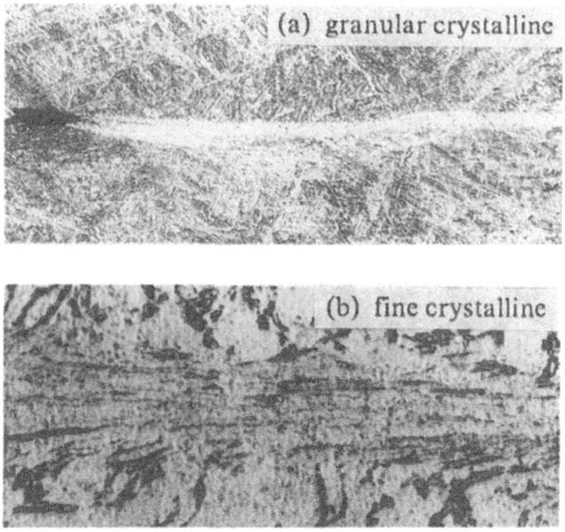

Figure 3.8. Strain bands in Ti-6Al-4V.

occurred. The damage originated from void nucleation and dimples prevail in the spall zone, though there are a few instances of quasi-cleavage fracture. Inclusions and second-phase particles play important roles in dimple formation. From the SEM micrograph of the fracture surface, it can be deduced that the fracture of the titanium alloy (Ti-6Al-4V) is ductile.

3.3.2. Dynamic Failure in Aluminum Loaded by EGDPI

In order to study the micro-flaws formed inside shock-loaded pure aluminum, EGDPI-II experiments were done with 0.7-mm thick impactors at velocities of 144, 267, and 330 m/s. The target thicknesses were 2, 3, and 3 mm, respectively, in shots 1, 2, and 3. The process of nucleation, growth, and coalescence is clearly shown in Figs. 3.9 and 3.20.

Aluminum-lithium alloy, titanium alloy, and pure aluminum at elevated

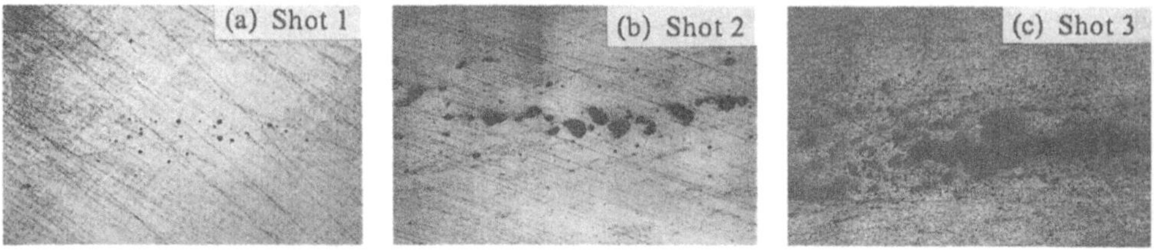

Figure 3.9. Microvoids in aluminum specimens loaded by EGDPI-II.

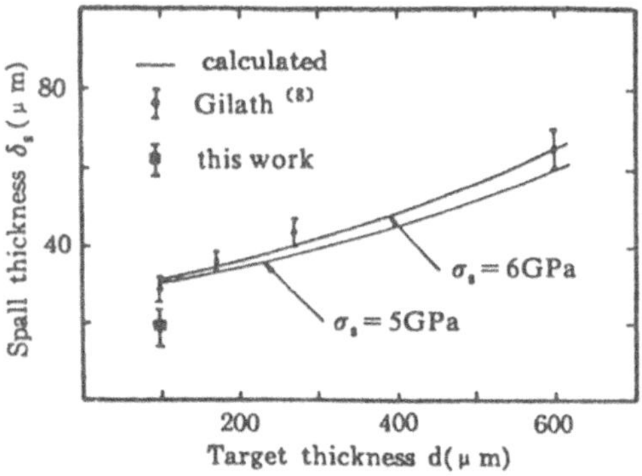

Figure 3.10. Spall thickness, δ_s, *vs* target thickness, *d*, as measured at the Hua Guang facility.

temperatures (<200°C) were also tested by EGDPI-II. It was found that the NAG model was suitable for modeling the observed damage.

3.3.3. Damage and Spall in Metals Loaded by LDSW

The microfracture patterns produced in some metals deformed at ultra-high strain rate (10^7–10^8/s) have been investigated with LDSW loading. Figure 3.10 indicates the experimental data for 0.1-mm thick pure aluminum tested with the Hua Guang Nd : glass laser facility, where the incident laser intensity was 2 to 5 × 10^{11} W/cm^2 and the pulse duration was 4 ns (FWHM). The spall thickness was found to be $\delta_s = 17 \pm 6$ μm and the strain rate was evaluated as 2–6 × 10^7/s.

Figure 3.11 shows an SEM Micrograph of a 0.2-mm thick plate of pure aluminum irradiated by a laser beam of intensity 4.9 × 10^{11} W/cm^2, and a pulse duration of 4 ns FWHM output from the LF-10 laser facility. Experiments were done for target thicknesses ranging from 0.1–0.8 mm and laser intensities varying from 8 × 10^{10}–2 × 10^{12} W/cm^2. The microfracture pattern in a 0.8-mm thick specimen irradiated at 2 × 10^{12} W/cm^2 indicates that the flaws arise from plastic flow and are basically spherical. Some larger voids appear to be somewhat ellipsoidal, as can be seen in Fig. 3.12.

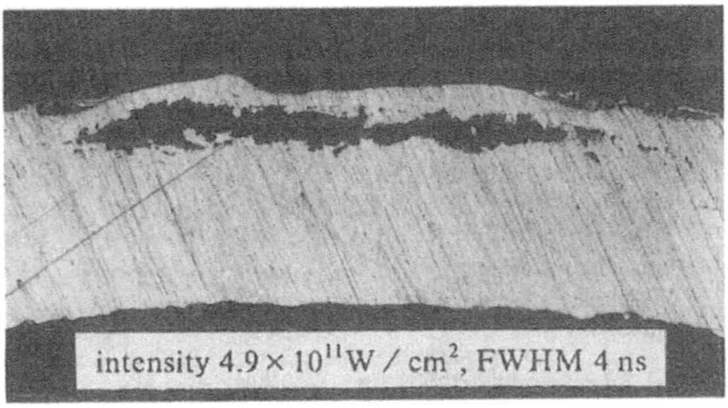

Figure 3.11. Spall in aluminum loaded by the LF-10 facility.

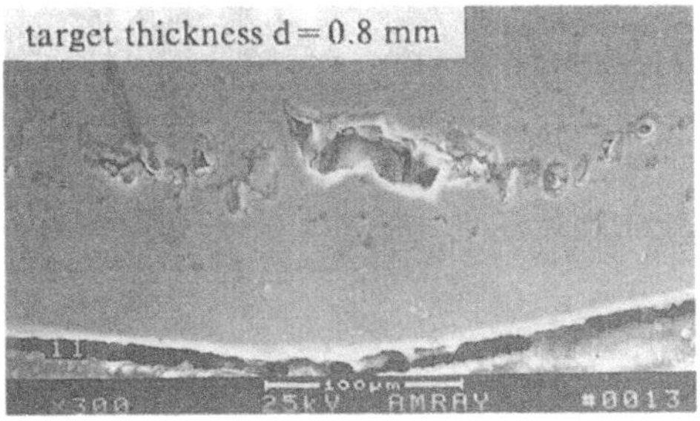

Figure 3.12. Microvoid in aluminum loaded by LDSW at 2×10^{12} W/cm^2.

The measured spall thickness, δ_s, is plotted against target thickness, d, in Fig. 3.13.

More extensive experiments were carried out at the LF-11 ("Xing Guang") facility, where the pulse duration is only 0.75 ns FWHM, the spot diameter is 1.2 mm, and the incident beam intensity ranges from 10^{11} to 10^{13} W/cm^2. Figure 3.14 shows an interesting micrograph of a 0.4-mm thick aluminum target irradiated at 1.4×10^{12} W/cm^2, where the fracture pattern becomes brittle-like cracks rather than typical ductile voids. It seems that the growth of microvoids in the experiment of Fig. 3.14 was limited by the too-brief duration of the loading, and was confined to the plastic flow zone around grain boundaries. The ductile cracks formed by microvoid coalescence extended along grain boundaries, converging to macrocracks. This process occurs only at ultra-high strain rate and has some brittle features such as cleavage and formation of the triple point shown in Fig. 3.14. Consequently, there may be a ductile-brittle transition of the fracture mechanism in this case. The specimen shown in Fig. 3.14 was loaded with a more intense laser beam and at higher strain rate, but did not spall. This implies that the spall strength may increase with strain rate, in contradiction to Grady's prediction.

The typical ductile void pattern appears in a 0.4-mm thick Ti-6Al-4V

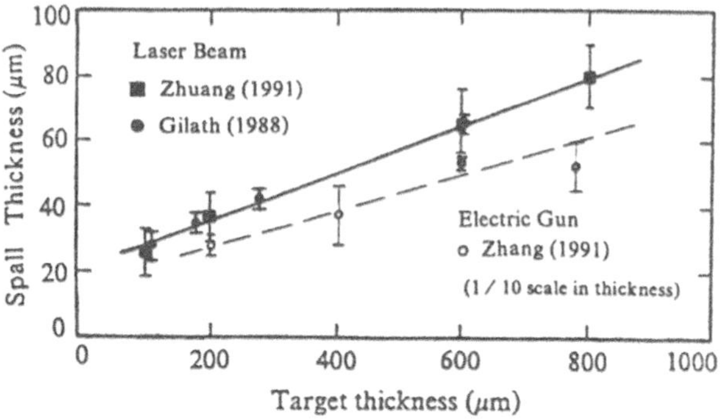

Figure 3.13. Spall thickness, δ_s, *vs* target thickness, *d*, as measured at the LF-10 facility.

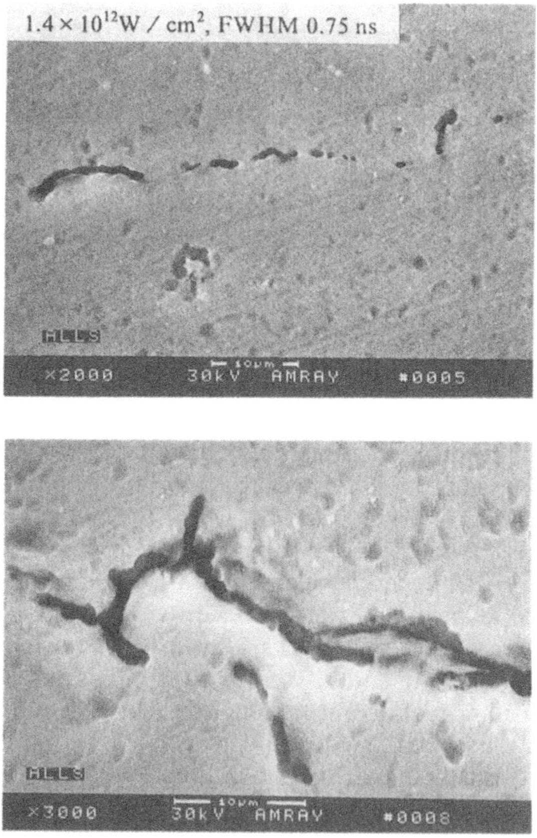

(a)

(b)

Figure 3.14. Brittle-like fracture in pure aluminum loaded by LDSW at ultra high strain rate. Part (a) shows microcracks along grain boundaries and part (b) shows cleavage and a triple point.

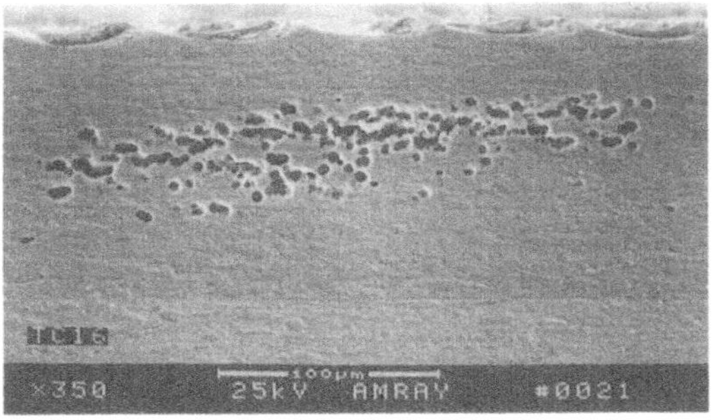

Figure 3.15. Microvoids in Ti-6Al-4V specimen loaded by LDSW at 3.5×10^{12} W/cm^2.

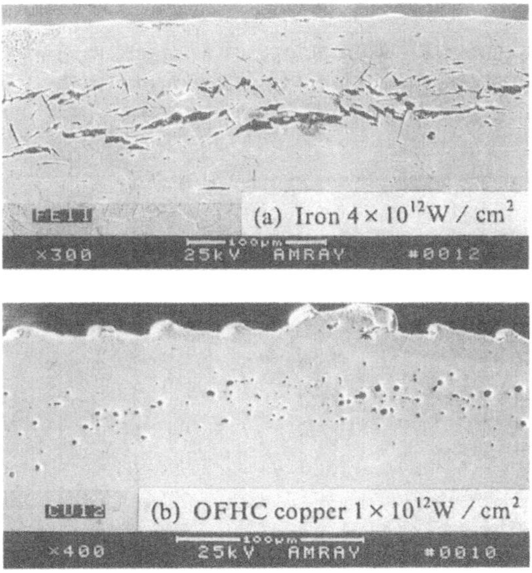

Figure 3.16. Microfracture in metals at ultra-high strain rate.

specimen irradiated by a laser beam of 3.5×10^{12} W/cm^2 (Fig. 3.15), rather than strain-concentration bands observed in EDPI tests at lower strain rate (Fig. 3.8). It could be conjectured that the localization and thermal instability of plastic flow demand a longer incubation period. However, fractures in iron and OFHC copper are insensitive to strain rate, and retain their characteristics at lower strain rate, i.e., brittle cracks and ductile voids, respectively, as shown in Fig. 3.16.

3.3.4. Shear Bands Induced by PPT in Ti-6Al-4V [17]

Tungsten cylindrical projectiles 0.8 cm in diameter were launched by gas or powder gun and penetrated Ti-6Al-4V targets of thickness 0.8–2 cm at velocities of 40–1200 m/s. Figure 3.17 shows a typical deformation morphol-

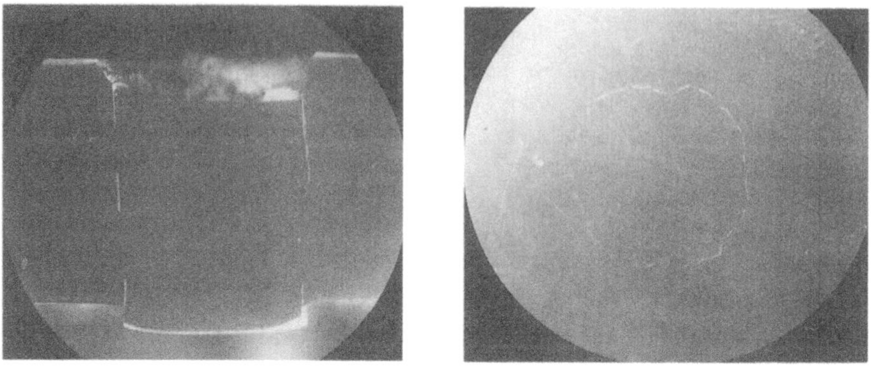

(a) (b)

Figure 3.17. Morphology of a target cross section. The photograph shows views (a) along the axis and (b) perpendicular to the axis.

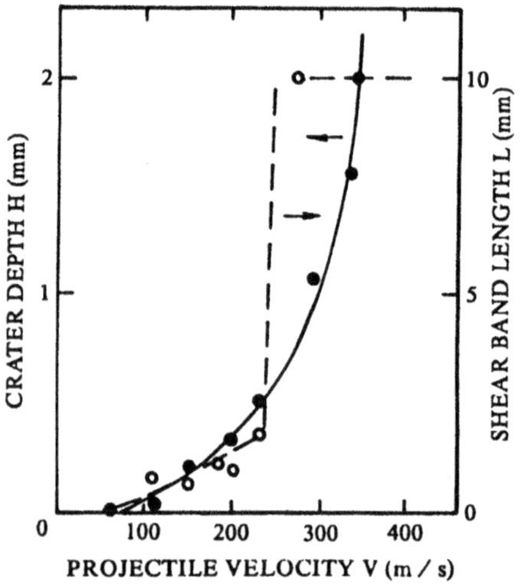

Figure 3.18. Crater depth and shear band length *vs.* projectile velocity as measured in PPT experiments.

ogy. It is seen that the deformation is highly localized and results in simple shear of the target. Figure 3.18 indicates the projectile velocity dependence of the crater depth for the 1 cm thick target. The critical velocity for plugging, u_{cr}, is 240 m/s. The length of shear bands is also shown in the figure; it is found to increase suddenly through the target at u_{cr}. As shown in Fig. 3.19, the damage was mainly localized inside the target-penetrating band and took the form of cavities and cracks. Microcavities prevail near the impact surface; they are spherical when their size is less than the band width and become elliptical when their size is near the band width. The interior surface of the cavities is covered with dimples. Coalescence of these cavities is the major

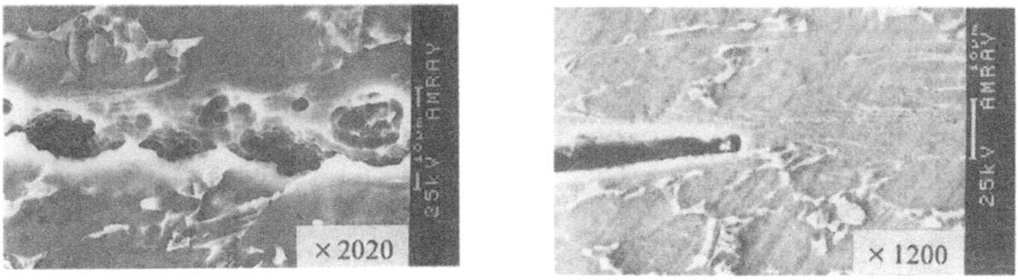

(a) (b)

Figure 3.19. Penetrating shear band feature in Ti-6Al-4V specimen loaded by PPT. The figure shows (a) voids near the impact edge and, (b) cracks at the distant end.

mechanism of damage in the band near the impact surface. Microcracks are dominant in the band near the free surface of the target, with the tip of the cracks being blunt and somewhat like a void.

A strongly luminous ring appears on the specimen's rear free surface during penetration, as was observed by framing photography, demonstrating that the shear band zone had experienced a high temperature. The shear band's propagation velocity was found to be 2.24 km/s, the effective shear strain to be in the range 8–108, and the shear strain rate to be in the range 10^6 to 10^7/s, and the maximum temperature within the bands was determined to be 10^4 K.

3.4. Stress Wave Model for Spall

In 1975, Lawn and Wilshaw [18] proposed a theory to describe the plastic zone at the tip of a ductile fracture, and provided analytic expressions for the fracture-surface energy and a criterion for extension of the crack. In 1988, Grady [7] pursued this work and proposed two conditions for spall, a horizon condition for failure time and a condition for energy balance during fracture. If we denote by E_k and E_e, respectively, the kinetic and elastic potential energy within an element of unit volume, and by U_k and U_y, respectively, the specific fracture energy controlled by the fracture toughness K and the flow stress Y, the general energy balance should be

$$E_k + E_e > U_k + U_y \tag{3.1}$$

Grady pointed out that $E_e \gg E_k$, and assumed that brittle and ductile fracture are characterized by U_k and U_y, respectively, where K and Y might be regarded as constants. He concluded that the spall strength, σ_s, is proportional to the cube root of the bulk strain rate, $\dot{\varepsilon}$, for brittle fracture but is independent of $\dot{\varepsilon}$ for ductile fracture. Furthermore, he believed that there is a brittle-to-ductile transition in the spall behavior, but uncertainty in the microfracture pattern of 6061 aluminum alloy for $\dot{\varepsilon}$ somewhere between

10^5 and 10^6/s. However, his evidence could not be considered sufficient and convincing. In fact, both U_k and U_y contribute to the dynamic fracture process though their ratio may vary with strain rate and other factors. The parameters K and Y are strain-rate sensitive for at least some common materials.

A stress wave model of spall has been suggested [19], in which we assume the stress wave reflection at the rear surface is ideal, i.e., the reflected tensile wave is antisymmetric to the incident compression wave about the surface and that the wave propagates without decay when passing through the zone near the surface. Since the tensile stress in spall experiments at high strain rate is usually much larger than the yield strength of the material, the stress wave discussed here should be considered similar to the shock wave of hydrodynamics.

The first problem we encounter is evaluation of the bulk strain rate, $\dot{\varepsilon}$, for which the accepted approach is incorporation of the measured free-surface velocity pullback into the Riemann relation. Grady and other workers took the analog to plate impact of identical materials,

$$\dot{\varepsilon} = u_f/2\delta_f = 2\sigma_1/\rho_0 c_0^2 \tau, \tag{3.2}$$

where u_f and δ_f are the maximum free surface velocity and impactor thickness, respectively; ρ_0, c_0, and τ are mass density and bulk soundspeed of the materials, and the duration of the rectangular tensile stress pulse, respectively, and σ_1 is the amplitude of the pulse. The right member of Eq. (3.2) is deduced by us from the Riemann relation (or momentum conservation). According to Grady, Eq. (3.2) overestimates $\dot{\varepsilon}$ by perhaps a factor of 2–4.

According to the definition of bulk strain rate, we have

$$\dot{\varepsilon} = -\tau^{-1} \int_0^\tau (\dot{\rho}/\rho)\, dt = \tau^{-1} \ln(\rho_0/\rho_1), \tag{3.3}$$

where the dot denotes the convective derivative with respect to time and the subscript 1 refers to the state of the element during the tensile pulse. The mass and momentum conservation equations give

$$\rho_0 c_0 = \rho_1(c_0 + u_1) \quad \text{and} \quad \sigma_1 = \rho_0 c_0 u_1, \tag{3.4}$$

where u_1 is the particle velocity increment due to the tensile wave. Thus,

$$\dot{\varepsilon} \approx (\rho_0 - \rho_1)/\rho_0 \tau = [\tau(1 + \rho_0 c_0^2/\sigma_1)]^{-1} \approx \sigma_1/\rho_0 c_0^2 \tau, \tag{3.5}$$

where $(\rho_0 - \rho_1)/\rho_0 \ll 1$ and $\sigma_1 \ll \rho_0 \sigma_0$ are assumed. The triangular wave can be regarded as a series of small rectangular waves, the average strain rate of which is

$$\dot{\varepsilon} = \sigma_1/2\rho_0 c_0^2 \tau. \tag{3.6}$$

Better estimates may be obtained from Eqs. (3.5) and (3.6) than from Eq. (3.2).

Using these results, we calculate the tensile wave resulting from interaction of the incident and reflected waves, and employ the spall criteria to deduce threshold conditions for spall [19]. For these calculations only σ_1, τ, target thickness, d, and related material parameters are needed. According to the maximum tensile stress criterion, a rectangular wave cannot induce spall of scaled thickness $\eta = 2\delta_s/c_0\tau < 1$ in targets, but a triangular wave can. This criterion is expressed as spall strength σ_s not less than the material strength σ_b. The stress gradient criterion leads to the same conclusion in the stress wave model. The damage accumulation criterion takes the form

$$\int_0^r (\sigma - \sigma_0)^\alpha \, dt \geq k_{cr}, \tag{3.7}$$

where σ_0 denotes the tensile stress threshold for damage accumulation, k_{cr} is a material constant, and α is usually taken to be 2. It is more consistent with experimental data than the criteria discussed previously, and is essentially an energy condition for spall. From Eq. (3.7), the stress wave model yields the threshold condition for spall in metals as follows:

For rectangular waves,

$$\sigma_s = (\sigma_0 + \sqrt{k_{cr}/\tau}), \qquad \eta \geq 1 \tag{3.8}$$

For triangular waves,

$$\sigma_s = \begin{cases} \sigma_0\left(\dfrac{1-\eta/2}{1-2\eta/3}\right)\left\{1 + \left[1 + \left(\dfrac{1-2\eta/3}{1-\eta/2}\right)\left(\dfrac{k_{cr}}{\sigma_0^2\tau} - 1\right)\right]^{1/2}\right\}, & \eta < 1 \\ \dfrac{\sigma_0}{2}\left\{3 + \left[\dfrac{12k_{cr}}{\sigma_0^2\tau} - 3\right]^{1/2}\right\}, & \eta \geq 1. \end{cases} \tag{3.9}$$

Employing Eqs. (3.5) and (3.6), the strain dependence of spall strength is given as follows:

For rectangular waves,

$$E = (\Sigma - 1)^2/(1 + Z/\Sigma) \approx \Sigma(\Sigma - 1)^2/Z, \qquad \eta \geq 1 \tag{3.10}$$

For triangular waves,

$$E = \begin{cases} \dfrac{\Sigma}{2Z}\left\{1 + \dfrac{(1-\eta/2)^2}{(1-2\eta/3)}\left[\left(\dfrac{1-2\eta/3}{1-\eta/2}\Sigma - 1\right)^2 - 1\right]\right\}, & \eta < 1 \\ \dfrac{\Sigma}{8Z}\left[1 + 3\left(\dfrac{2\Sigma}{3} - 1\right)^2\right], & \eta \geq 1 \end{cases} \tag{3.11}$$

where $\Sigma = \sigma_s/\sigma_0$ and $E = \dot{\varepsilon}k_{cr}/\sigma_0^2$ are scaled spall strength and bulk strain rate, respectively, and $Z = \rho_0 c_0^2/\sigma_0$ is the scaled acoustic impedance of the material. In addition, we get the expression for the threshold spall thickness $\delta_{s,cr}$ in the case of a triangular wave as

$$\delta_{s,cr}/d = c_0\tau\eta/2. \tag{3.12}$$

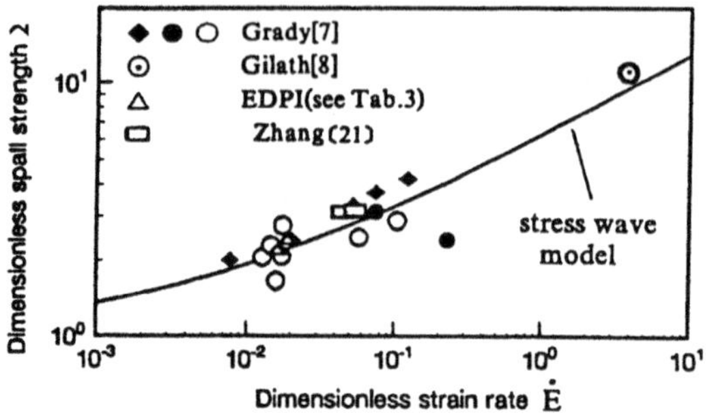

Figure 3.20. Strain rate dependence of spall strength.

In this expression only the right member is a function of Σ, E, and Z. Thus, a relation between Σ and E could be obtained provided the measured data $\delta_{s,cr}$ and τ are available.

Figure 3.20 shows the comparison of Eq. (3.10) with spall data in LDSW experiments. It seems that σ_s increases with $\dot{\varepsilon}$, and to be in contradiction to Grady's conclusion, at least for pure aluminum. Moreover, experimental facts described in Sec. 3.4 support the prediction of the stress wave model. This model has been incorporated into calculations of laser ablation pressure and shock wave attenuation, and successfully used in analyzing laser-induced spall in metals.

3.5. SSS Calculations Incorporating the NAG Model

Based on empirical expressions for the statistical distribution of microflaws, along with assumptions on growth rate and compressibility, Seaman et al. [1] proposed the nucleation and growth (NAG) model for statistics of micro-

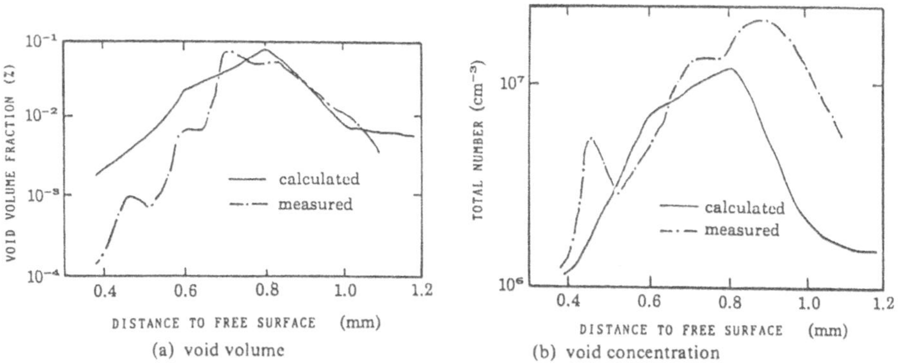

Figure 3.21. Microvoid distributions in aluminum loaded by EGDPI-II for Shot 2. Part (a) shows void volume; and part (b) shows void concentration.

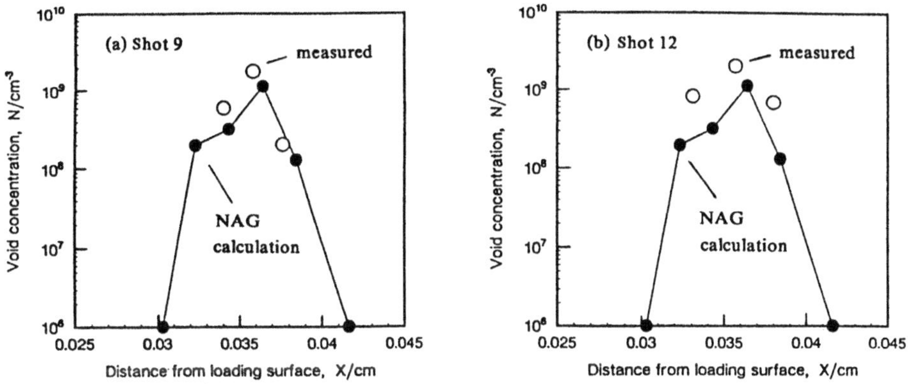

Figure 3.22. Microvoid concentration distribution in copper loaded by LDSW at 1 to 3×10^{12} W/cm^2.

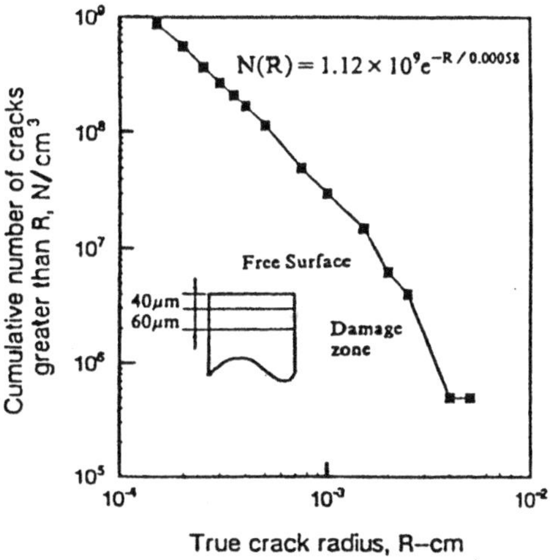

Figure 3.23. Microcrack size distribution in iron loaded by LDSW at 2 to 5×10^{12} W/cm^2.

voids and microcracks. The reactive hydro-elastic-plastic code SSS that has been previously used in detonation and shock wave calculations [20] has now been extended to incorporate a damage and strain-rate dependent constitutive equation [3] and the NAG model. Besides the results shown in Fig. 3.5, comparisons of measured and calculated void distributions in aluminum targets tested by the EGDPI-II method and in OFHC copper specimens tested by the LSDW method are given in Figs. 3.21 and 3.22.

Though the calculation of microcrack distributions in iron specimens tested by the LDSW method is still underway, the preliminary data of Fig. 3.23 indicate that the NAG model is appropriate for this case, too. However, the crack concentration here is 1–2 orders of magnitude higher than that resulting from plate-impact experiments. It is also found that the void concentration in OFHC copper and aluminum specimens tested by the LDSW method is 1–2 orders of magnitude higher and lower, respectively, than observed at lower strain rate, even though these two metals are similar in many respects. We conjecture that the number of potential flaw sites in the NAG model should be treated as a complicated and important variable rather than as a material constant. The nucleation mechanism may be more complicated than previously assumed.

3.6. Acknowledgment

This work was supported partly by the National Natural Science Foundation of China and partly by the National Hightech Program, Laser Technology. The authors are greatly indebted to Feng Shuping, Long Xinping, Yuan Xi, Zhang Zhongzhen, Chen Qiuhua, and Wang Chunyan for their contributions to this work.

3.7. References

1. Curran, D.R., Seaman, L., et al., *Phys. Rep.* **147**(5–6), 253–388 (1987).
2. Gurson, A.L. *J. Eng. Mater. Tech., Trans. ASME*, **99**, 2–15 (1977).
3. Sun Lizhi, Huang Zhuping, The Dynamic Void Growth in Rate Sensitive Plastic Solids. In press.
4. Davison, L. and Stevens, A.L. *J. Appl. Phys.* **43**, 988 (1972).
5. Nemes, J.A., Eftis, J. et al., *J. Appl. Mech.* **57**, 282–291 (1990).
6. Johnson, G.R. and Cook, W.H. *Proc. of 7th Int. Symp. on Ballistics*, 541–547 (1983).
7. Grady, D.E. *J. Mech. Phys. Solids*, **36(3)**, 353–384 (1988).
8. Eliezer, S. and Gilath, I. *J. Appl. Phys.* **67(2)**, 715–724 (1990).
9. Zhuang Shiming, et al., *Proc. of 2nd Int. Symp. on Intense Dynamic Loading and Its Effects*, pp. 612–616 (1992).
10. Larson, A.R., LA-5619-MS, 1974.
11. Meyer, M.A. and Aimone, C.T. *Prog. Mater. Sci.* **28**, 1–96 (1983).
12. Romain, J.P., Bauer, F., et al., *Proc. of 1993 Joint AIRAPT/APS Topical Conf. on High Pressure Sci. & Tech.* (1993).
13. Zhang Zhongzhen, *Chinese J. High Pressure Phys.* **5(2)**, 137–144 (1991).
14. Long Xinping, Feng Shuping, et al., The Spall in Aluminum under Shock Loading by Electric Gun and Its Comparison with Laser Induced Spall. In press.
15. Yuan Xi, Sun Chengwei, et al., *Proc. of 2nd Int. Symp. on Intense Dynamic Loading and Its Effects*, 270–273 (1992).
16. Wang Yanping, Liu Cangli, Deformation Localization Phenomenon in the Spallation of Ti-6Al-4V In: ISID proceeding.
17. Wang Yanping, Liu Cangli, Deformation and Damage Characteristics of Ti-6Al-4V under High Rate of Shear Loading In: ISID proceeding.

18. Lawn, B.R. and Wilshaw, T.R. *Fracture of Brittle Solids*, Cambridge Univ. Press, Cambridge England, 1975.
19. Sun Chengwei, Zhuang Shiming, *Chinese J. High Power Laser Particle Beams*, **2(3)**, 345–352 (1990).
20. Sun Chengwei, *Chinese J. Comput. Phys.* **3(2)**, 142–154 (1986).
21. Zhang Wanjia, et al., *Chinese J. Explosion and Shock Waves*, **1(1)**, 73–75 (1982).

CHAPTER 4

Laser-Induced Spallation and Dynamic Fracture at Ultra High Strain Rate

Irith Gilath

4.1. Introduction

Pulsed high irradiance laser induced shock waves are being used for assessing the dynamic fracture behavior of materials at extreme conditions. Theories [1, 2] have long predicted that very strong shock waves in the kbar-Mbar regime could be produced in small samples with pulsed lasers. Recent experiments have shown that this idea is not only a theoretical possibility, but also a practical way to obtain pressures and strain rates previously unachievable in a laboratory environment.

Absorption of laser radiation by a target material takes place within a very thin layer. The rapid temperature increase in this layer (10^3 eV ns^{-1}) and the associated steep temperature gradient (10^7 eV/cm) induce ablation of material plasma into the evacuated chamber containing the sample. The plasma expansion drives a strong compression shock wave into the cold target material. The shock wave is supported by pressure in the expanding plasma as long as the laser pulse is on. Even before the end of the laser pulse, ablation at the surface drops and a continuous sequence of rarefaction waves propagates into the target, overtaking and attenuating the shock [3–5]. This shock wave is of approximately triangular shape. When the shock reaches the stress-free back surface of the target or a low-impedance interface in a multilayer structure, a tensile wave of increasing negative amplitude is reflected. When this tensile stress becomes larger (more negative) than the dynamic tensile strength of the target, spall occurs (see Fig. 4.1).

When a triangular pressure pulse of length L and a pressure P_0, as shown in Fig. 4.1, reaches the back surface of the target, a constant-spall-stress model predicts formation of spall at a distance Δ into the sample, where

$$\Delta = \frac{L}{2}\frac{P_S}{P_0} \tag{4.1}$$

and P_S is the spall pressure or negative pressure required for spall. There is a minimum P_0 value, $P_0 = P_S$, for spall to occur. For higher values of P_0

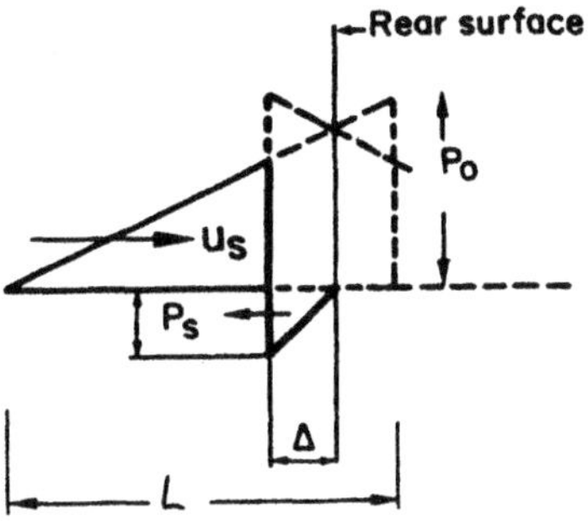

Figure 4.1. Spall caused by a triangular pressure pulse. P_S = spall pressure, P_O = maximum pressure at the back surface, Δ = spall layer thickness, L = pulse length, U_S = shock wave velocity.

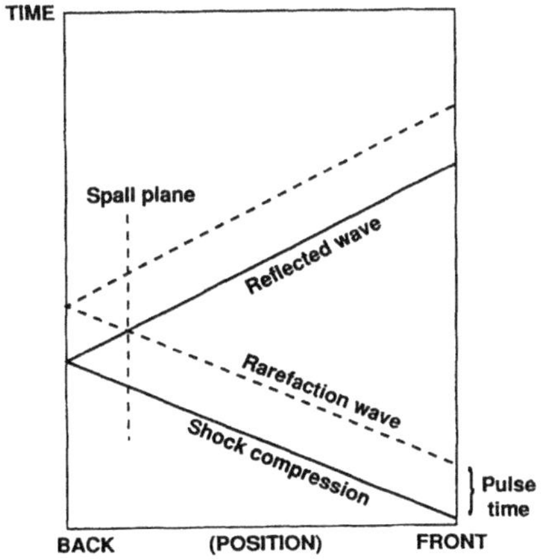

Figure 4.2. Position–time diagram for spall formation in laser experiments.

multiple spall layers can be obtained in brittle materials, while in ductile materials much of the energy is dissipated in plastic deformation.

A simplified model of laser-induced spall is illustrated in the position-time diagram of Fig. 4.2. The solid line in this figure shows the time it takes for the compression wave, i.e., the front of the shock wave, to reach the back surface; the dashed lines show the rarefaction wave. The position-time diagram for laser-induced disturbances is similar to the corresponding diagram for impact-induced disturbances, with the laser pulse duration corresponding to the impact duration. The position of the spall plane depends on the duration of the pulse and the thickness of the target.

Each material is characterized by a threshold energy to produce spall. The threshold energy (or the corresponding stress) for spall is a material constant. However, higher energy input is required on the irradiated (front) side in order to compensate for the energy dissipation through the thickness of the sample by plastic deformation. This subject will be discussed in more detail in the paragraph on shock wave pressure decay.

4.2. Laser-induced Shock Wave Pressure

The ablation-induced shock wave pressure P is directly related to the laser pulse intensity I, scaling approximately as $P \propto I^{3/4}$, and can be calculated by hydrodynamic codes. However as a good approximation, an analytic quantitative estimate of the ablation pressure was given by Key [6],

$$P = 230 A^{7/16} Z^{-9/16} \lambda^{-1/4} \tau^{-1/8} \left(\frac{I}{10^{12}}\right)^{3/4}, \tag{4.2}$$

where the pressure is given in kilobars, A and Z are the atomic number and the ionization number, respectively, of the target material, λ is the laser wavelength in μm, τ is the laser pulse time in nanoseconds, and I is the laser intensity in W/cm^2.

A simplified calculation of the ablation pressure can be made by using the scaling law developed by Trainor and Lee [1]

$$P_{(\mathrm{Mbar})} = 6(I/10^{14})^{0.7} \tag{4.3}$$

or the one developed by Boustie et al. [7],

$$P_{(\mathrm{MPa})} = 12(I/10^{11})^{2/3}. \tag{4.4}$$

Equations (4.2–4.4) are usually applied for plane, one-dimensional (l-D) geometry where the laser spot is large as compared to the target thickness. For small laser spots a correction factor should be included for 2-D effects. This correction in the limit of very small spots (<0.1 mm) reduces the ablation pressure by a constant factor of 1.8 without changing its 1-D intensity dependence [8, 9].

The measurement of very short (1 ns rise time for 3 ns pulse) and ultra-high-pressure pulses is still a challenge. A new piezoelectric polymer (PVDF) shock gauge was developed [10] for pressures up to 20 GPa. The shortest pulse recorded to date with the PVDF gauge is about 8 ns as reported by Paisley [11].

The spall pressure can be determined through measuring the velocity of the free surface of spalled samples by velocity interferometer (VISAR). The spall strength for short duration ($\sim$3 ns) pressure pulses has been recorded for a spall "pull-back" velocity minimum reached in $\sim$1.5 ns [11].

Pressures up to 30–50 kbar, corresponding to lower irradiances, can be measured with quartz gauges. Above this limit this quartz response is non-linear [12].

By placing a transparent plate on the irradiated surface the plasma formed by ablation can be confined, thus increasing the shock pressure. The confinement is limited by the laser induced electric breakdown of the transparent plate: sodium chloride crystal breaks down at an irradiance of 16 GW/cm^2, glass at 30 GW/cm^2 and quartz at 5.4 GW/cm^2 (1 GW = 10^9 W).

4.3. Numerical Simulation of Laser-Induced Spall

A hydrodynamic code, based on MEDUSA code [13], was used to simulate the evolution of the laser-generated plasma and shock wave [14]. The code assumes a plasma of planar symmetry whose spatial expansion is simulated by means of a Lagrangian mesh with each cell having constant mass throughout the computation. The laser energy is absorbed in the low-density coronal plasma, up to the critical surface, by the inverse bremsstrahlung process, with the remaining laser radiation absorbed at the critical surface. The code attributes different temperatures to the electron and the ion subsystems. The laser energy is deposited into the electrons and is transferred to the ions by electron-ion collisions. Classical heat conduction transports energy away from the region of the laser light deposition. A flux limiter inhibits heat transport to values less than the local electron thermal energy. The SESAME tables were used for the equation of state. Variable cell-size target zoning permits fine resolution of regions with sharp density and temperature gradients, particularly near the irradiated surface. Cell size is increased geometrically by about 5% per cell starting with a thinnest cell of about 0.1 μm; 100 cells were used altogether.

The MEDUSA code is restricted to one-dimensional systems, and cannot simulate two-dimensional effects such as lateral shock propagation, production of magnetic fields or the formation of double layers. However, when the focal diameter of the laser is much greater than the foil thickness, these 2-D effects are expected to have only secondary influence. In the original MEDUSA code there was no mechanism of spall or for any effects of negative pressure (stress). A simple spall model was introduced in which the material is spalled at the cell whose (negative) pressure attains a value more negative than the spall pressure. In this simple model the spall pressure is a variable parameter in the range of 10–50 kb. After spalling, the pressure of the cell is kept zero while its mass and energy are conserved.

By using experimentally measured spall thicknesses for specific laser intensities and target thicknesses, the input parameters to the hydrodynamic code were defined [14] so that it could be used to estimate the spall pressure. This simple one-parameter pressure criterion [14] was justified by the good agreement between experiment and simulation.

A typical sequence of profiles of a shock wave propagating [14] in a 600-μm thick aluminum target can be seen in Fig. 4.3. The triangular shock wave decay and widening at later times in the sequence are evident. Position-time plots for shock, rarefaction and reflected waves were computed for several metals. In Fig. 4.4 an example is shown for aluminum foils of 0.1, 0.3, and

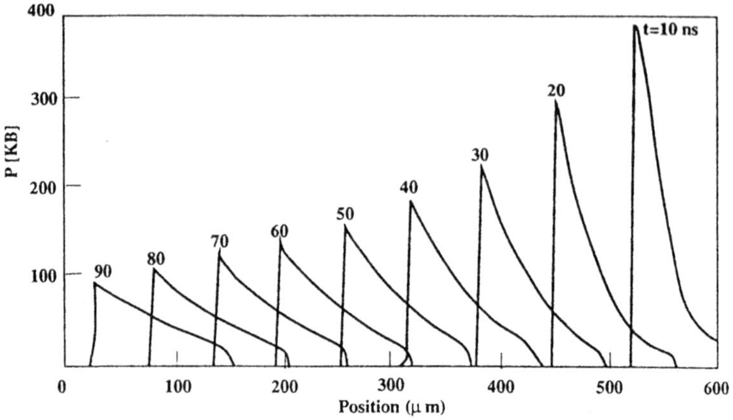

Figure 4.3. Shock wave propagation at different times for a 600 μm Al foil ($I = 9 \times 10^{11}$ W/cm^2) (Ref. 14).

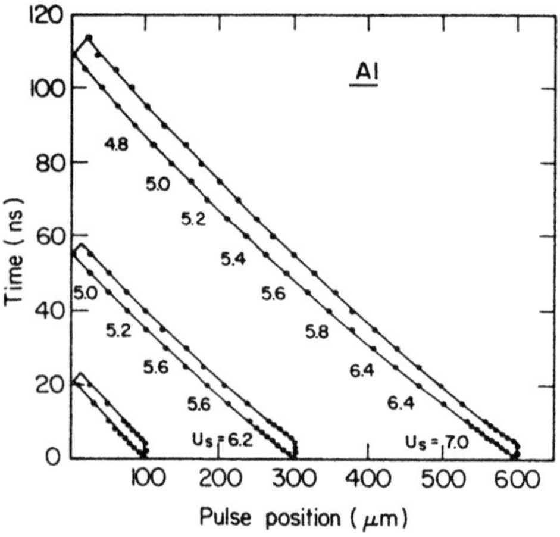

Figure 4.4. Position–time plot for shock, rarefaction, and reflected waves for 0.1, 0.3, and 0.6 mm aluminum foils. Shock velocities are in km/s with ± 0.2 km/s accuracy (Ref. 14).

0.6 mm thickness. The shock wave velocity is pressure dependent, so the lines are not straight as represented in the schematic plot of Fig. 4.2 but are slightly curved.

The strain rate, $\dot{\varepsilon}$, can be calculated from the simulation [14] using the equation

$$\dot{\varepsilon} = \frac{-\dot{\rho}}{\rho} = -\frac{1}{\rho}\frac{\partial \rho}{\partial X} C, \tag{4.5}$$

where $\partial \rho / \partial X$ is the density gradient obtained from the simulations at time and position just before the spall occurs, and C is the speed of sound.

The density gradient is laser-intensity dependent and varied in the range of 5×10^6 to 2×10^7 kg/m^4. Therefore, the strain rate for aluminum and copper spall experiments was in the range of $(1-4) \times 10^7$ s^{-1}.

A two-dimensional (2-D) code was used [15] to describe the spall phenomenon (which is by definition at least a 2-D effect) for Al and Cu targets. Good agreements were obtained between experimental and simulation results for the spall and the spall width.

Another approach to the simulation of laser-induced spall was followed by Cottet et al. [16]. The laser interaction with solid matter was not included in his code and instead the pressure on the irradiated face of the material was calculated using eq. (4.4). The code, named SHYLAC, is a 1-D finite difference program for solving the hydrodynamic equations in Lagrangian form. The code was modified to permit calculation of the spallation process corresponding to the cumulative damage model of Tuler and Butcher [17]. The Tuler and Butcher model was also included in the simulation performed by Larson [2].

4.4. Experimental Method

The laser used [7, 14, 16] for spall experiments is a Nd : glass laser with pulse length of 3–20 nsec, energy up to 100 Joules and wave length of 1.06 μm. The laser system [14] comprises three stages (see Fig. 4.5): the pulse forming and preamplifying stage, the filtering stage and the main amplifying stage. The Gaussian shape of the pulse can be modified by steepening its edges. Using a fast electro-optical shutter, the rise time was reduced to 1 nsec. The amplified beam is focused in the target chamber (TC) through an f/20 lens.

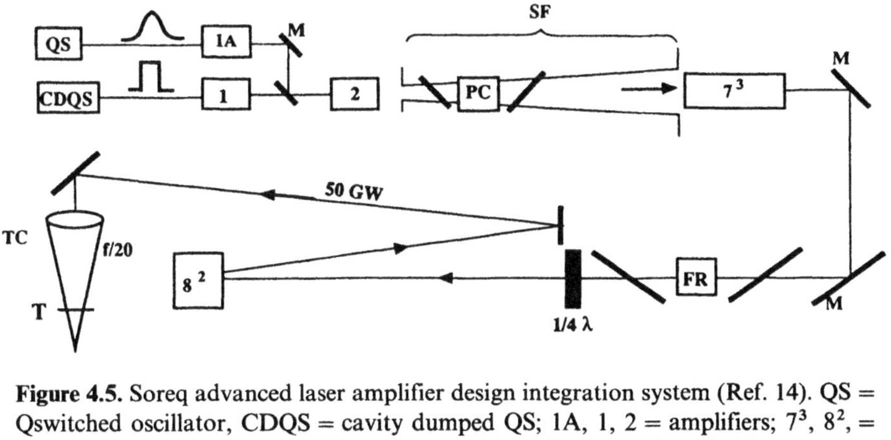

Figure 4.5. Soreq advanced laser amplifier design integration system (Ref. 14). QS = Qswitched oscillator, CDQS = cavity dumped QS; 1A, 1, 2 = amplifiers; 7^3, 8^2, = triple- and double-pass amplifiers, respectively; PC = pockels cell; FR = faraday rotator, TC = target chamber; T = target; SF = spatial filter, M = mirror.

The target chamber is evacuated in order to avoid ionization and breakdown of air due to strong irradiation. The target, T, is placed in the focal spot of the beam. By moving the target out of focus with a micropositioner, the laser spot diameter can be changed between 0.1 and 3 mm. When using a large diameter spot as compared to sample thickness, 1-D conditions prevail (similar to plate impact). When using a small spot, hemispherical shock waves are obtained. For very small spots and thick targets, the conditions are similar to a point explosion.

The experimental work consists of determining the irradiance and energy density required to produce spall from threshold conditions up to sample perforation. The laser energy can be gradually modified at the amplifier stages.

For damage evaluation, metallurgical or scanning electron microscopy are used. For internal damage evaluation, the samples are sectioned, cast in epoxy holders and polished to high quality in order to find damage threshold and damage evolution. The deposited energy E (and energy density) is monitored for each laser shot. The intensity is calculated from the equation

$$I = E/(\pi r^2 \tau), \tag{4.6}$$

where I is in W/cm^2, E is in Joules, r is the laser spot diameter in cm, and τ is the pulse duration in seconds.

4.5. Spall and Dynamic Fracture of Metals

Experimental studies of laser-induced shock wave damage in materials are rather recent [14–16, 18–24]. Several brittle and ductile metals were tested in order to obtain the failure stages at increasing levels of deposited energy density.

The stages of damage development can be seen in Fig. 4.6 representing schematic cross sections through targets for increasing laser intensities. The incipient spall was identified from the level of separate voids for ductile

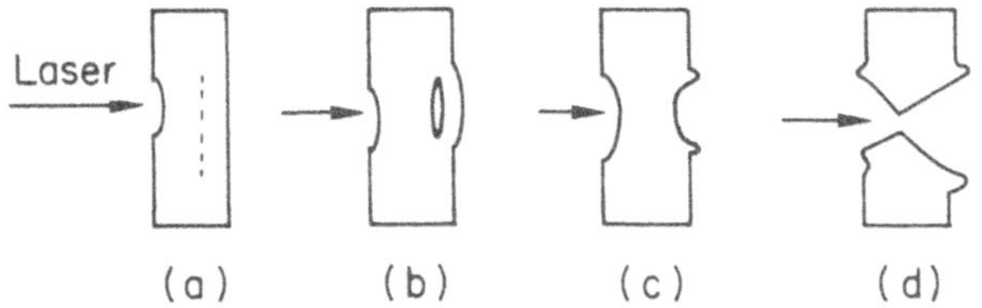

Figure 4.6. Stages of material failure with increasing laser intensities: (a) voids or cracks formation; (b) spall layer formation and plastic deformation; (c) spall layer break away; (d) target perforation.

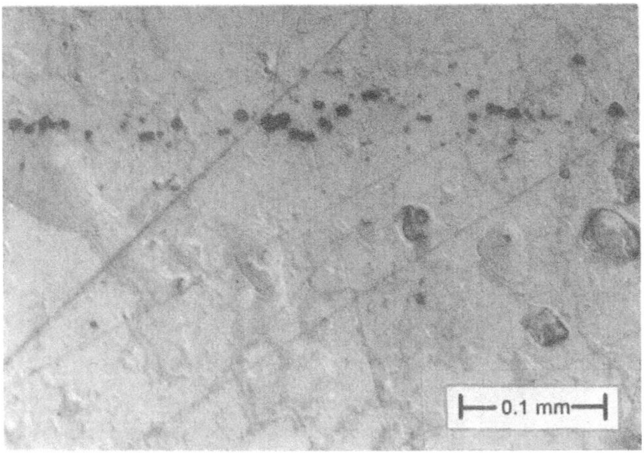

Figure 4.7. Voids at the spall plane in aluminum (cross section × 187).

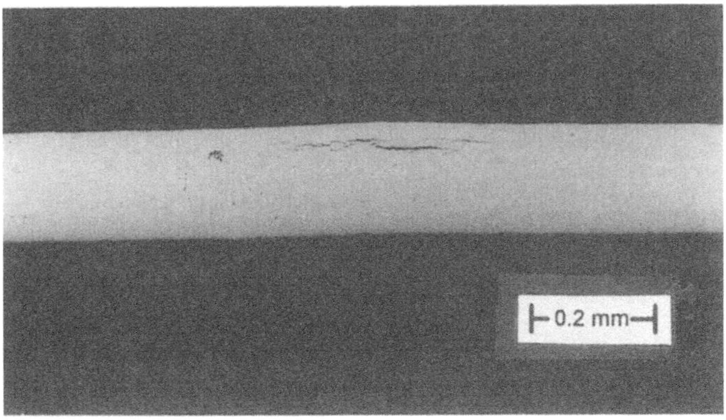

Figure 4.8. Cracks in iron (cross section × 75).

metals (see Fig. 4.7) or cracks for brittle ones (see Fig. 4.8). The incipient damage can be observed only on sectioned samples. For higher energies, the voids or cracks coalesce, resulting in a continuous fracture or spall layer. This stage can be observed as a bulge (swelling) or some cracking on the back surface of the sample. By increasing the deposited energy, the spall layer breaks away revealing the spall plane and, finally, the sample will be perforated. Some examples of laser-induced spall in different metals can be seen in Figs. 4.9 through 4.13.

A cross section through a spalled molybdenum sample can be seen in Fig.

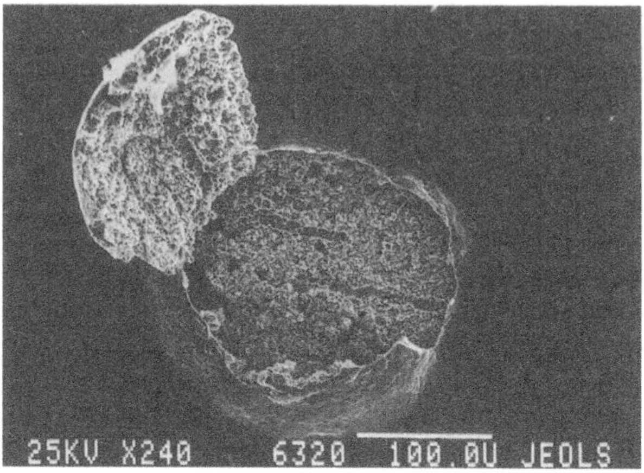

Figure 4.9. Spall layer opening on the back surface of a stainless steel sample (× 168).

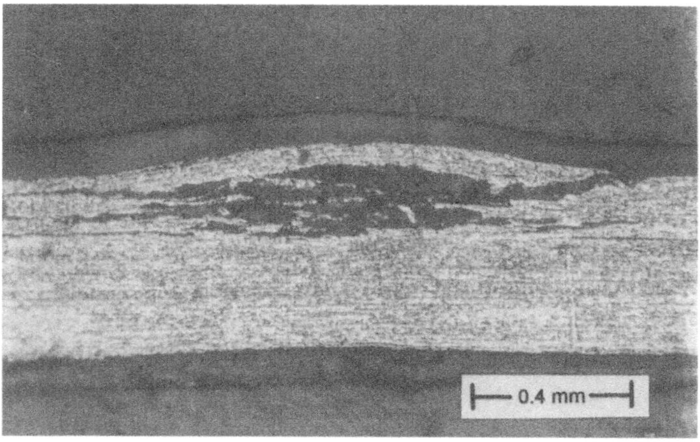

Figure 4.10. Multiple spall layers in molybdenum, (cross section × 53).

4.10. Molybdenum behaves as a ductile metal at quasi-static conditions while for laser-induced high strain rate experiments, brittle failure occurs with several spall layers.

For all samples exhibiting laser-induced spall, the damage is located in a well-defined spall plane. The lateral extent of the damage corresponds to the laser spot size. This phenomenon is a characteristic response to very short and high intensity stress waves with no lateral dispersion effects.

Most experiments were performed in a 1-D geometry using a laser spot

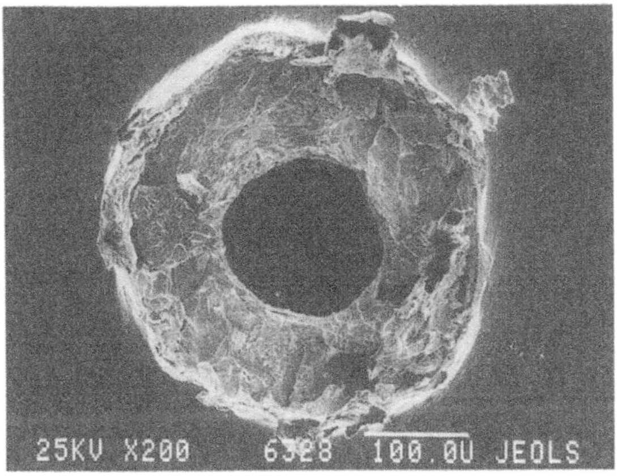

Figure 4.11. Marginal penetration in gold. A spall layer is still visible around the perforation hole.

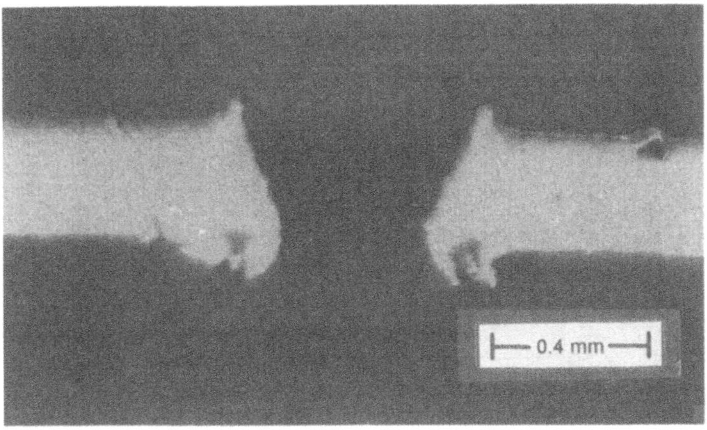

Figure 4.12. Perforated aluminum target. Petalling of the back surface can be observed (cross section × 53).

diameter about three times larger than the sample thickness [23]. For 1-D laser experiments (corresponding to plate impact) very slight, if any, damage can be observed on the front surface. In Fig. 4.14 an example is shown of an event where the laser spot diameter was about seven times greater than sample thickness. The damage on the irradiated side is almost invisible. The back side bulging has the dimension of the laser spot with no lateral dispersion observed.

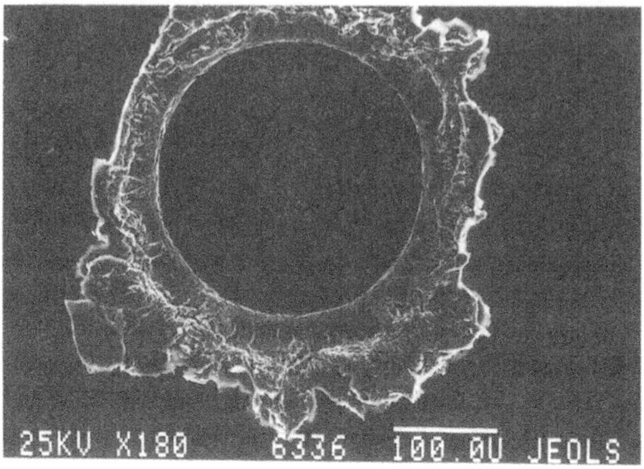

Figure 4.13. Perforated aluminum target. The circular hole is surrounded by petalling (back surface × 126).

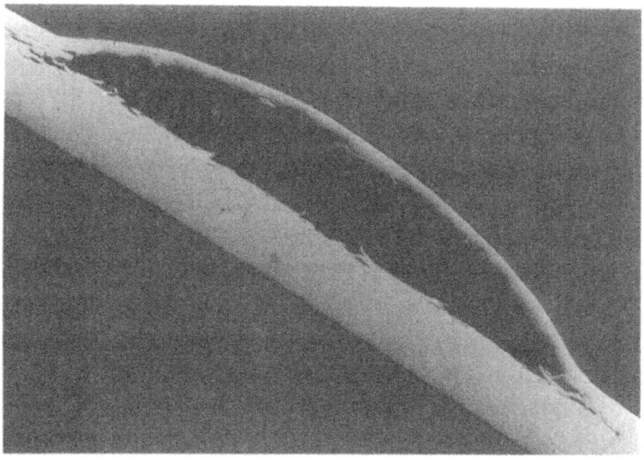

Figure 4.14. Cross section through an iron foil (150 μm thickness) impacted with a 1 mm diameter laser beam.

For hemispherical shock waves, craters typical of those produced by hypervelocity impact can be observed (see Fig. 4.15). In Fig. 4.16 a detail is shown from the laser-induced crater revealing a smooth crater wall and molten splash (solidified) around the crater lips. It is interesting to observe that the spall plane is flat for the hemispherical configuration just as it is for 1-D conditions.

A schematic representation for the planar spall plane formation with hemispherical shock waves is shown in Fig. 4.17 [25].

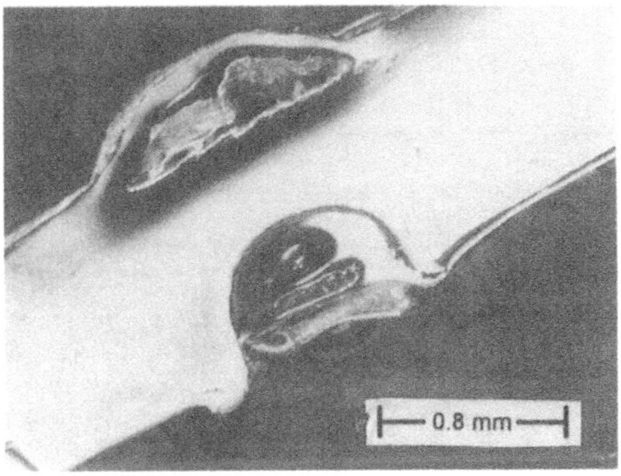

Figure 4.15. Cross section through an aluminum target impacted with hemispherical shock waves (× 26). A typical hypervelocity impact crater can be seen.

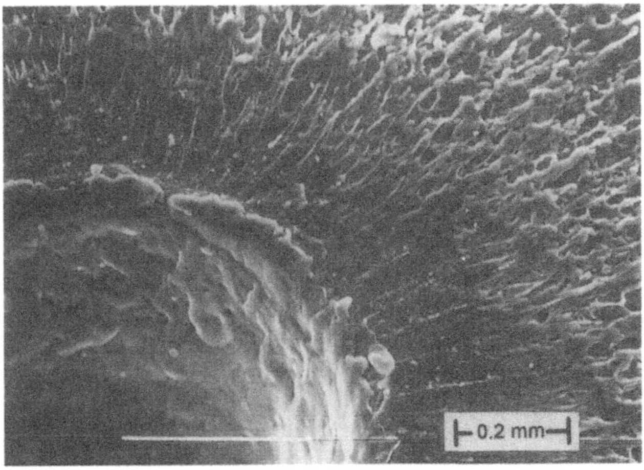

Figure 4.16. Detail from Fig. 4.15 impact crater. The molten aluminum was splashed around crater lips and solidified (× 75).

4.5.1. Spall as a Function of Sample Thickness

Pure aluminum and copper foils in the 0.1–0.6 mm thickness range were irradiated by a pulsed laser and the spall layer thickness was measured [20]. The measurements were performed at threshold spall conditions. The results are presented in Figures 4.18a and b, showing a linear dependence of spall thickness, Δ, with foil thickness, d

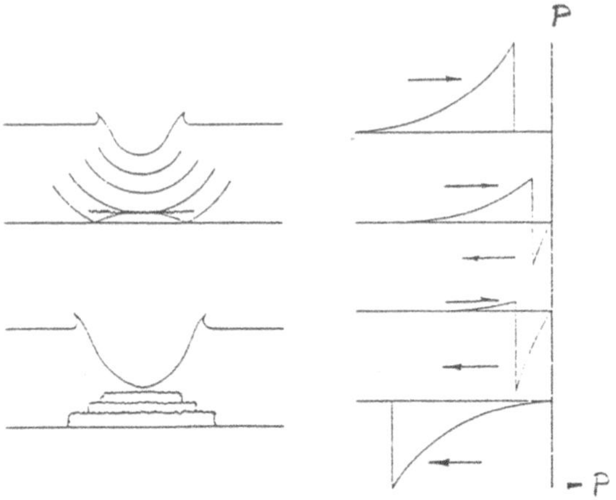

Figure 4.17. Conceptual sketch of a projectile impacting a target of intermediate thickness and a demonstration of tensile wave production on reflection at a free surface.

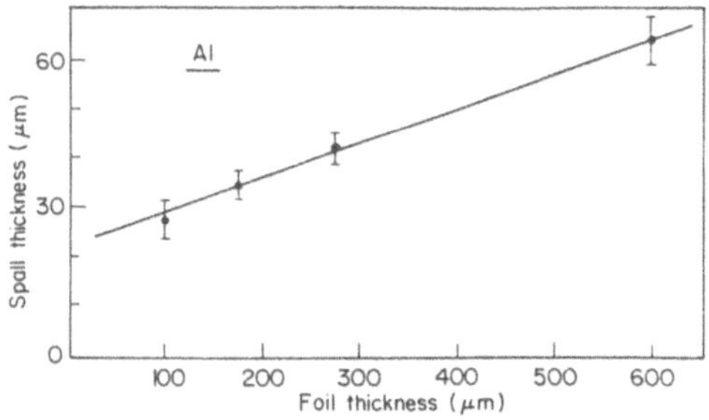

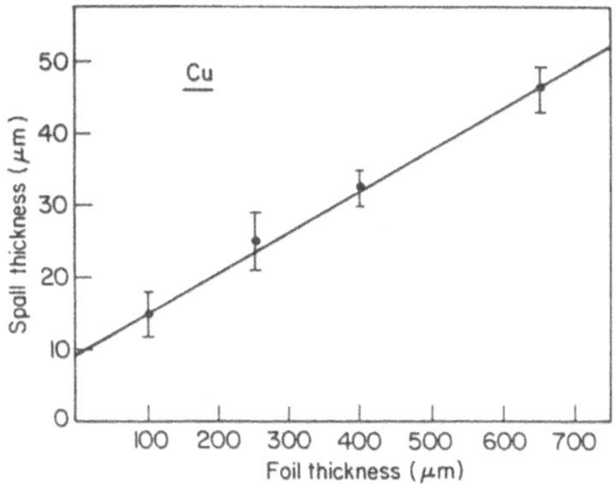

Figure 4.18. Spall thickness as a function of foil thickness in (left) aluminum and (right) copper.

for aluminum: $$\Delta = (0.07 \pm 0.01)d + (22 \pm 4)\mu\text{m} \tag{4.7}$$

for copper: $$\Delta = (0.057 \pm 0.01)d + (9.3 \pm 1.8)\mu\text{m}. \tag{4.8}$$

From geometrical considerations of the position-time plot (see Fig. 4.2), the following expression for the spall thickness as a function of target thickness can be obtained [20]:

$$\Delta = \left(\frac{U_S - C}{U_S + C}\right)d + \frac{\tau U_S C}{U_S + C}. \tag{4.9}$$

The sound speed C, laser pulse duration τ, and the target and spall thickness are known, so the mean value for the shock wave velocity U_S can be calculated.

Although this estimate is based on a very simplified model, the results agree well with those obtained by more time-consuming hydrodynamic codes. By substituting the experimentally measured values for aluminum, the shock wave velocity obtained is $U_S = (6.21 \pm 0.07) \times 10^5$ cm/s. The particle velocity U_p can be calculated from the known equation of state for aluminum:

$$U_S = 5.4 \times 10^5 + 1.34 U_P \text{ cm/s.} \tag{4.10}$$

The strain rate, $\dot{\varepsilon}$, can be estimated from

$$\dot{\varepsilon} = \varepsilon/\tau, \tag{4.11}$$

where

$$\varepsilon = 2U_P/C. \tag{4.12}$$

A value of $\dot{\varepsilon} = 2.9 \times 10^7 \text{ sec}^{-1}$ was obtained for aluminum.

4.5.2. Total Elongation at Fracture at Ultra High Strain Rate

The explosively driven expanding ring test is often employed to measure dynamic elongation of metals. The measurement of dynamic elongation using laser-induced shock waves is suggested as a simple and inexpensive alternative method.

In Fig. 4.19, a photograph of the spalled back surface of an aluminum target is shown. For low laser intensities no damage can be seen on the back

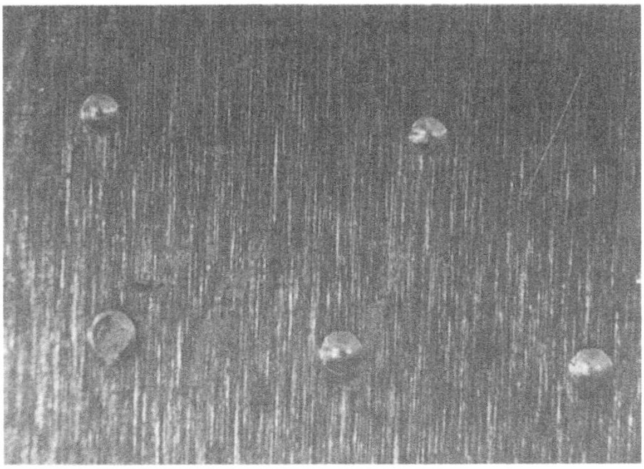

Figure 4.19. Spalled back surface of an aluminum target for increasing laser intensities (energy densities).

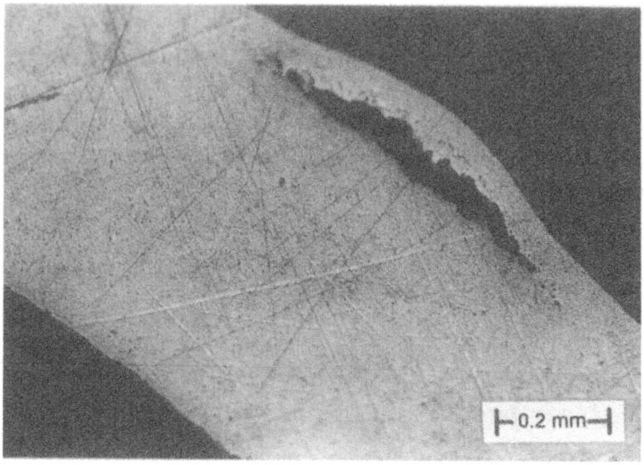

Figure 4.20. Spall layer formation in aluminum, 1.84 kJ/cm^2 ($\times$ 100).

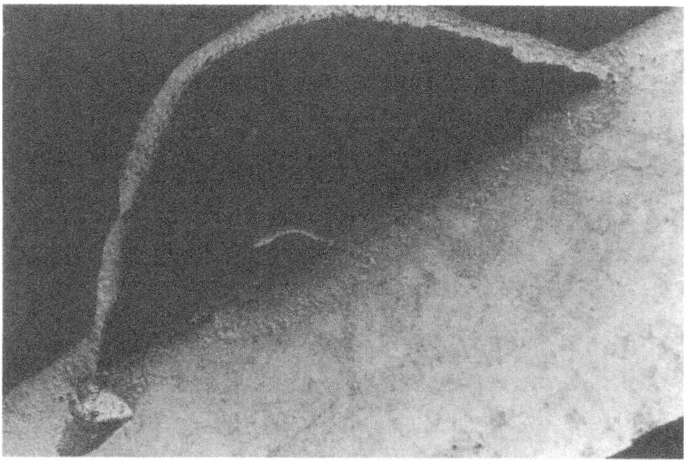

Figure 4.21. Maximum elongation of spalled layer before fracture in aluminum (cross section, 3.25 kJ/cm^2).

surface. The voids, characteristic of threshold spall, are revealed only by sectioning and polishing of the damaged area. For increasing intensities, the bulging of the back surface increases until the spall layer is detached. The spall layer is deformed at very high strain rate, being gradually elongated to failure (see Figs. 4.20–22). By comparing the spall layer length, l, (a segment of the expanding dome) to its initial length, l_0, (see Fig. 4.25) the elongation at fracture at ultra high strain rate can be estimated from the equation [19]

$$\varepsilon = (l - l_0)/l_0. \tag{4.13}$$

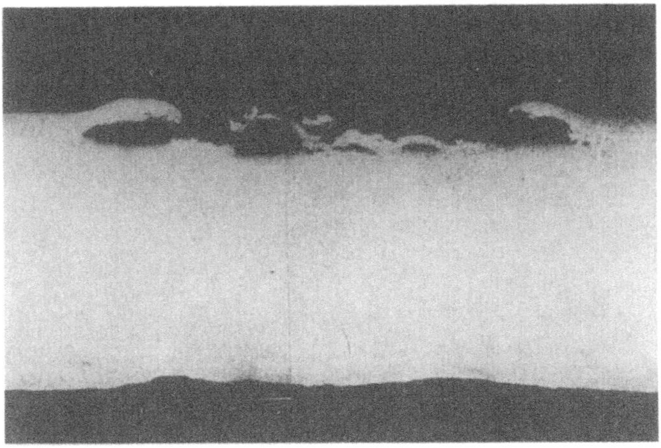

Figure 4.22. Spall layer break away in aluminum corresponding to lower left impact from Fig. 4.19, at 3.8 kJ/cm^2 (cross section).

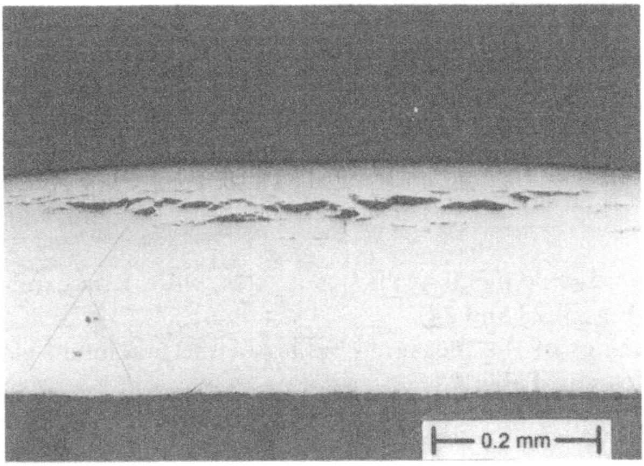

Figure 4.23. Incipient spall in iron at 2.4 kJ/cm^2 (cross section, $\times$ 105).

Table 4.1. Elongation at fracture at ultra-high strain rate

Material	Elongation at Fracture (%)
Aluminum (pure)	33
Aluminum 6061-T6	14
Iron	7.6

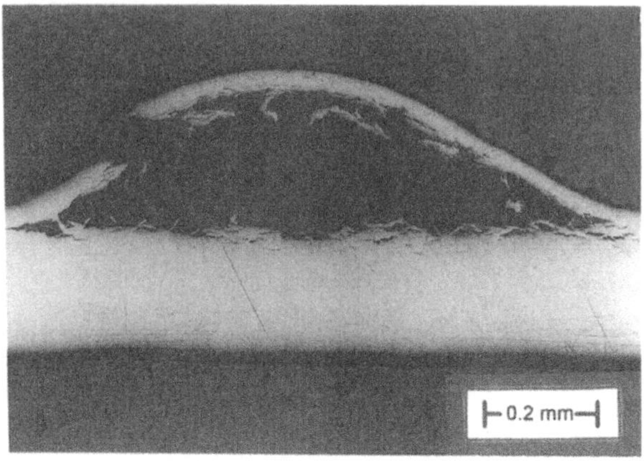

Figure 4.24. Spall layer break-off in iron at 5.3 kJ/cm^2 (cross section, × 75).

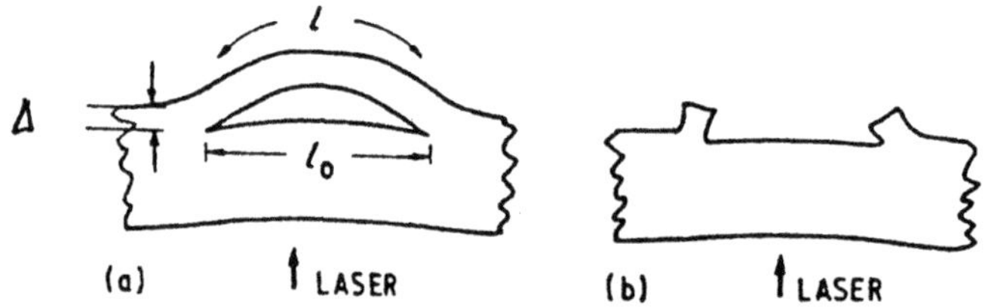

Figure 4.25. (a) Internal fracture, spall; (b) spall layer break away.

Aluminum is a ductile metal at all strain rates, while iron exhibits reduced ductility (see Figs. 4.23 and 24).

A few examples of the measured values of the maximum elongation at fracture are given in Table 4.1.

4.6. Experimental Estimation of the Shock Wave Pressure Decay and Spall Strength in Different Materials

At threshold (incipient) spall, the same spall pressure and the same shock velocity prevail at the spall plane for different target thicknesses. In thicker targets, higher energies (pressures) have to be applied on the front surface to make up for the decay in peak pressure that occurs as the shock travels through the target.

Using targets with decreasing thicknesses and 1-D impact geometry, a linear relationship was found [26] between P_{th} (the ablation pressure necessary to induce threshold spall) and target thickness (see Fig. 4.26). This result suggests that P_{th} represents the spall stress P_S in addition to a value corresponding to pressure decay [26] through the thickness ΔX:

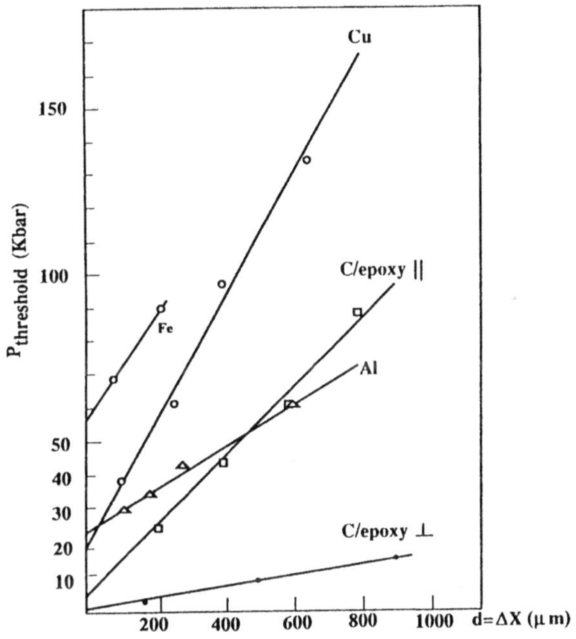

Figure 4.26. Threshold ablation pressure for spall as a function of target thickness for 1-D impact. (is parallel and is perpendicular to fiber direction).

$$P_{th} = P_S + \frac{\partial P}{\partial X}\Delta X + \frac{\partial^2 P}{\partial X^2}\Delta X^2 \cdots. \tag{4.14}$$

For small values of ΔX, Eq. (4.14) can be approximated by the first two terms. Therefore, by extrapolating the experimental values to zero thickness, the limiting value will represent the dynamic spall pressure, while the slope will represent the pressure gradient (attenuation). The results of applying this analysis to several materials are given in Table 4.2.

In Fig. 4.27, the spall stress as a function of strain rate for aluminum is reproduced from Ref. 27. On this plot the spall stress inferred from experi-

Table 4.2. The spall pressure and pressure gradient in different materials.

Material	P_{spall} (kb)	P_{gradient} (kb/mm)
Copper	20	180
Aluminum	25	60
Iron		200
CF/epoxy (perp. to fiber)	0.3	15
CF/epoxy (parallel to fiber)	7	100

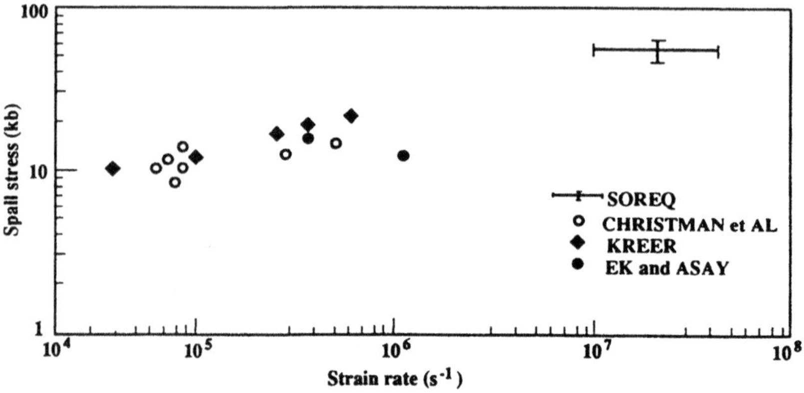

Figure 4.27. Spall strength data for 6061-T6 aluminum vs. strain rate. This is Fig. 9 (p. 371) from Grady's article, Ref. 27, with laser induced spall results for aluminum added at strain rate of (1-4) $\times$ 10^7 sec^{-1}.

Table 4.3. Threshold laser energy for spall (E_L), laser intensity (I_L) and the ablation pressure (P_a) for different target thickness (d).

d (mm)	E_L (J)	I_L (10^{14} W/cm^2)	P_a (Mbar)
0.5	11 ± 2	0.47 ± 0.1	2.4 ± 0.4
0.6	15 ± 2	0.64 ± 0.1	3.1 ± 0.4
1.0	49 ± 5	2.1 ± 0.2	6.9 ± 0.7
1.3	105 ± 7	4.5 ± 0.3	11.7 ± 0.8

ments employing laser-induced shock waves is added at the far right. This is the only experimental value at strain rate of 10^7 s^{-1}.

For hemispherical shock waves the pressure decay is much steeper. A linear relationship between the laser energy E and the cubic target thickness was obtained [24]. The experimental data for aluminum can be fit (see Table 4.3) by the relation

$$E = P_0 d^3 + E_0, \tag{4.15}$$

where $P_0 = 45.3$ J/mm^3, $E_0 = 4.9$ J, and d is the target diameter in millimeters. At the limit of a very small laser spot, the ablation pressure dependence on the laser intensity is the same as in the 1-D case [24] and, therefore,

$$P \sim I^{0.7} \sim E^{0.7} \sim d^{2.1} \tag{4.16}$$

The threshold ablation pressure as a function of target thickness is shown in Fig. 4.28 where:

$$P_{\text{Mbar}} = 6.46 d^{2.1} + 0.54 \qquad (d \text{ in mm}). \tag{4.17}$$

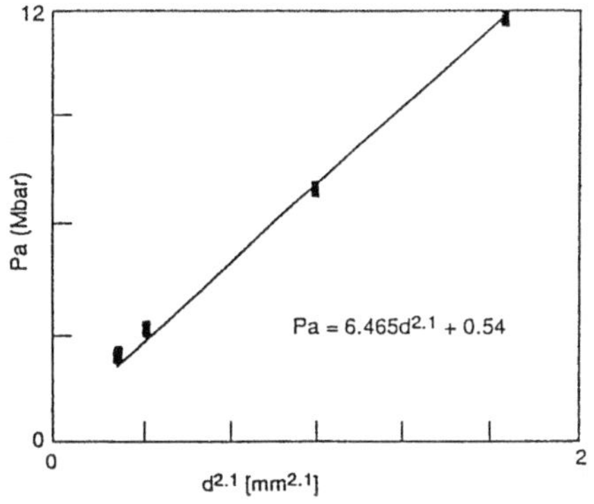

Figure 4.28. Threshold ablation pressure as a function of target thickness for aluminum (hemispherical shock waves).

4.7. Spall Behavior of Composite Materials

A continuous replacement of conventional metals by composite materials is observed in space and other critical applications. Carbon–epoxy composite is widely used for its superior mechanical properties, high ratio of strength to weight, and high chemical resistivity. Carbon–carbon composites provide superior thermal and mechanical properties for temperatures above 2000°C.

Carbon–epoxy unidirectional composites [28] and carbon–carbon composites [29] were tested at ultra high strain rate using laser-induced shock waves. Details of sample preparation and properties can be found in the above references.

In Fig. 4.29, the surface damage of a carbon–epoxy unidirectional composite irradiated by a laser beam propagating perpendicular to the fiber

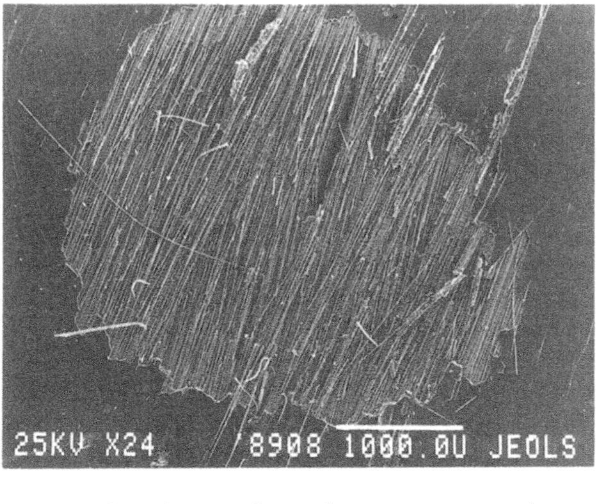

Figure 4.29. Front surface damage for carbon epoxy composite at 1.47 kJ/cm^2, 0.5 mm thickness sample (1-D impact) perpendicular to fiber direction).

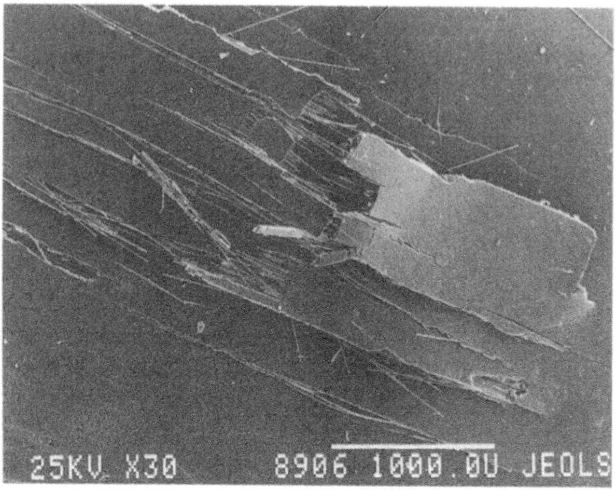

Figure 4.30. Fracture development at back surface of carbon epoxy composite at 0.515 kJ/cm^2.

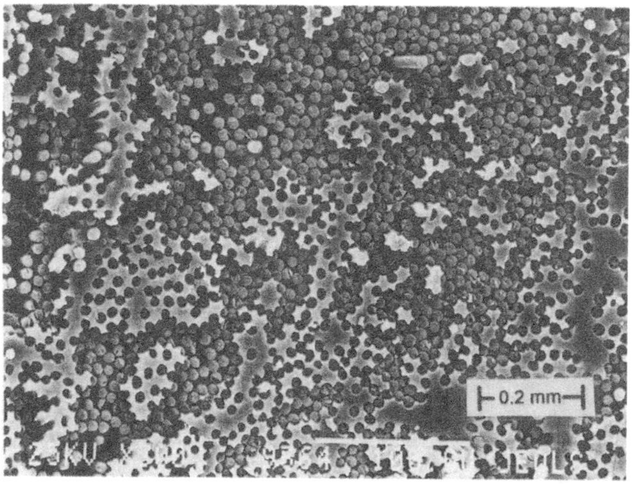

Figure 4.31. Front surface of carbon epoxy composite impacted parallel to fiber direction at 1.32 kJ/cm^2. The partially debounded matrix can be observed ($\times$ 75).

direction can be seen. The damage consists of a shallow crater due to epoxy ablation and breaking of a few fibers. The mild damage (hardly seen without magnification) on the irradiated side for a 1470 J/cm^2 pulse corresponds to severe spall damage at the back of the sample. In Fig. 4.30, the spall damage is shown for an energy density of only 515 J/cm^2. For this low energy density, the surface damage was almost invisible.

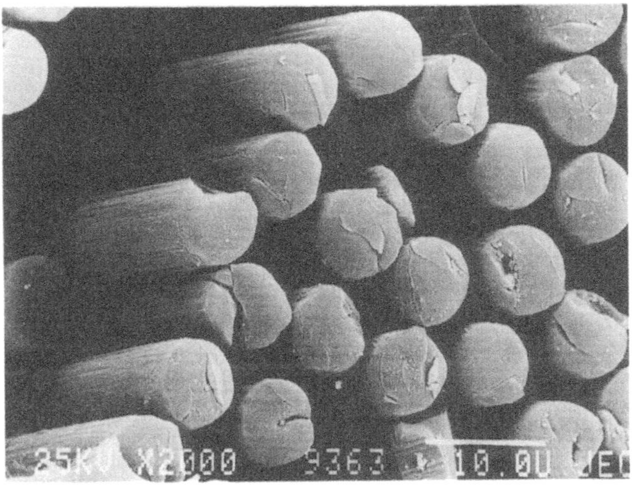

Figure 4.32. Detail from Fig. 4.31. The ablated matrix reveals smooth carbon fiber ends.

Figure 4.33. Fractured carbon fibers as a result of spall, at 132 kJ/cm^2 ($\times$ 1400).

Composites irradiated along fiber direction revealed greater resistance than when irradiated perpendicular to fiber direction. The front surface damage can be seen in Fig. 4.31, with matrix ablation to a depth of approximately 20 μm (as can be seen also in Fig. 4.29). Fig. 4.32 is a detail from Fig. 4.31 showing the smooth ends of the carbon fibers after the epoxy was ablated. In Fig. 4.33, the broken fiber ends due to a spallation can be seen.

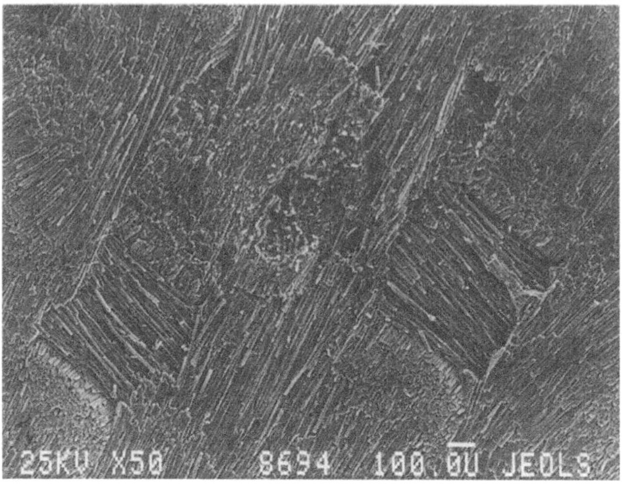

Figure 4.34. Front surface damage of carbon/carbon composite at 2.8 kJ/cm^2.

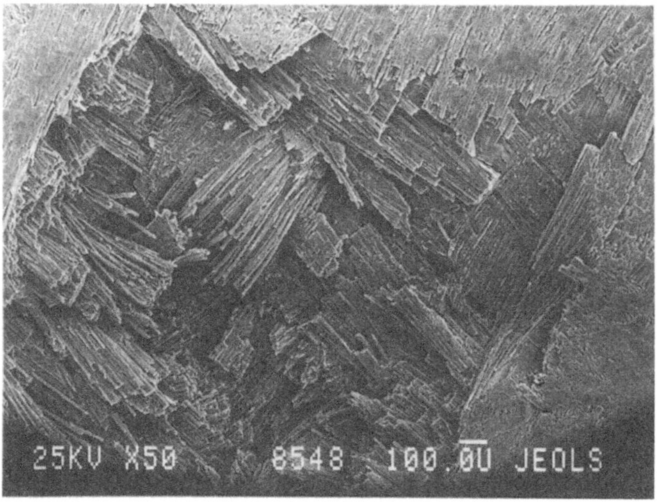

Figure 4.35. Back surface spall damage of carbon/carbon composite at 2.8 kJ/cm^2.

The above experiments were performed on samples of 0.5 mm thickness using a laser spot diameter of 2 mm.

In Fig. 4.26, the threshold ablation pressure for spall as a function of carbon–epoxy composite thickness is shown for irradiation parallel and perpendicular to the fiber direction in the target. The spall pressure and pressure gradient of the composites are given Table 4.2.

It is interesting that incipient damage occurs at a relatively low energy density (195 J/cm^2 for a 0.5-mm thick sample); however, penetration of the sample, unlike in metals, was not achieved at 1500 J/cm^2 for 1-D geometry. Also, samples were not penetrated by a focused beam at 50 kJ/cm^2. No internal delaminations were observed for laser-induced stress waves.

The carbon–carbon composites revealed almost no damage on the front side for high energy density irradiation. In Fig. 4.34, very slight damage can be seen on the front side of a carbon–carbon composite densified in a hot isostatic press (1000 atm and 650°C). The slight damage on the front corresponds to severe spall fracture on the back [29] (see Fig. 4.35). The irradiation resistance of the carbon–carbon composite is dependent on the densification process for the same density and product. Impregnation and densification at high pressure and temperature yielded a product having superior resistance to spallation [29].

4.8. Fracture Modes of Alumina at Hypervelocity Impact Conditions

Ceramics are used for high pressure and high strain rate applications. Alumina and other brittle ceramics are shattered by high-speed impacts in dynamic tests and the recovery of impacted samples is not easy. There is an advantage in using small samples which are easily handled in laser irradiation experiments.

Laser irradiation tests were performed in order to find the fracture patterns produced in alumina by planar and hemispherical shock waves [30].

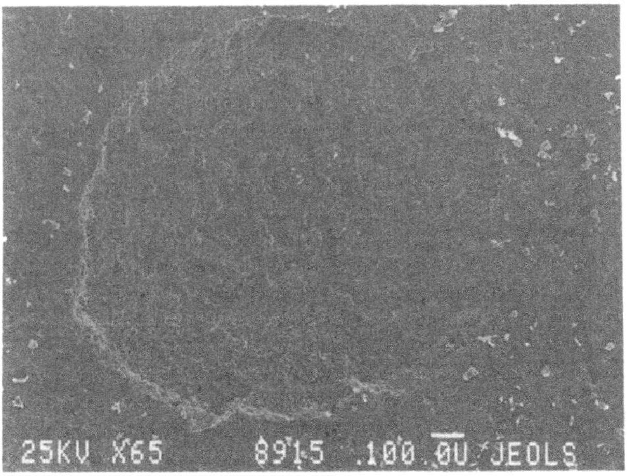

Figure 4.36. Back surface spall in alumina, impacted in 1-D configuration.

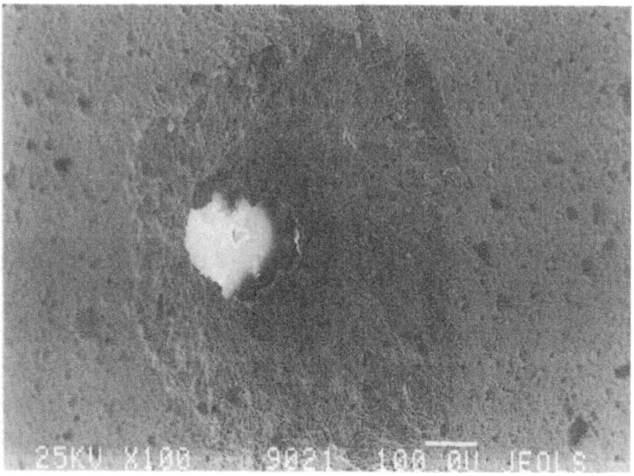

Figure 4.37. Cone cracking spall in alumina as a result of hemisperical shock wave impact.

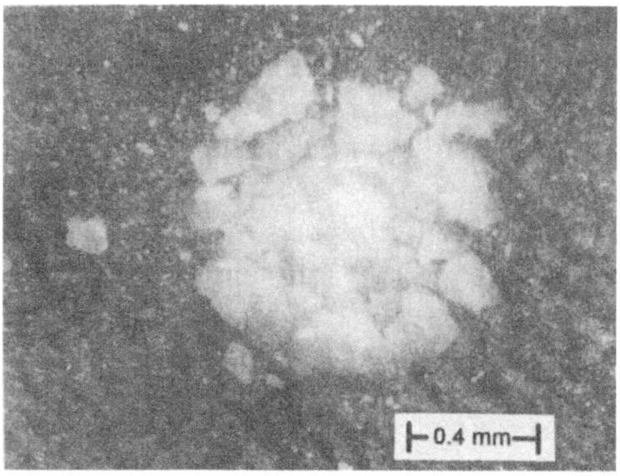

Figure 4.38. Collected debris from cone cracking in alumina (× 38).

For planar shock waves, spall damage was observed on the back (free) surface of the irradiated target in the form of a shallow plane circular crater (see Fig. 4.36). The energy density for spall layer detachment for this brittle material was 500 ± 50 J/cm^2, as compared to 5300 J/cm^2 for iron at similar experimental conditions. On the irradiated side no crater was observed but sintering occurred as a result of some surface melting. The spall diameter (Fig. 4.36) is greater than the irradiated diameter, suggesting dispersion of the shock wave-induced damage when traversing the sample. The angle of divergence measured for damage extension was 58 ± 3°. The angle was calculated by comparing the diameter of the incident laser beam to the spall surface diameter for a definite sample thickness [30]. No radial cracks

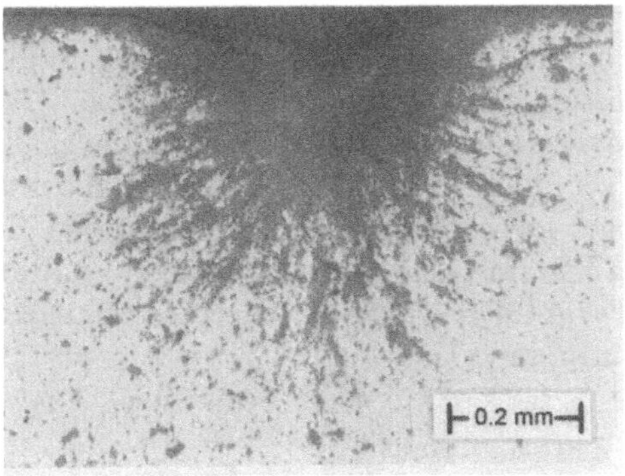

Figure 4.39. Shockwave pattern as revealed by dye absorbtion on a sectioned alumina sample (× 75).

have been observed on the surface up to 2000 J/cm^2. Above this value, the samples were shattered. For hemispherical shock waves, cone cracking (Hertzian) was observed, as shown in the back-surface view of Fig. 4.37. The expelled cone was recovered in the form of fragments of sizes varying from 0.01 to 0.15 mm (see Fig. 4.38). The angle of divergence measured for damage extension was $74 \pm 5°$.

In Fig. 4.39, the shock wave damage pattern in alumina is revealed by dye absorption on a sectioned and polished sample. The damage extended through one-half the sample thickness.

Using laser-induced shock waves, evidence was shown confirming the different fracture patterns in ceramics for plane and hemispherical shock waves.

4.9. Brittle to Ductile Transition in 6061-T6 Aluminum Alloy at Ultra High Strain Rate

The solution of spall and dynamic fracture problems requires full knowledge of the shock wave data, tensile stress history and rarefaction wave interaction.

Grady [27] has developed a theory of the spall process in which a balance between the kinetic and potential energy driving spall fracture and the fracture energy resisting breakup, is achieved within a characteristic time of the event. Brittle spall is characterized by internal crack growth and a corresponding surface energy or fracture toughness. Ductile spall, on the other hand, is characterized by plastic hole growth and the fracture energy is the plastic work dissipated in the hole-growth process. In certain metals, transition from crack-growth, surface-energy dominated spall at lower loading rates to hole-growth, plastic-dissipation dominated spall at higher rates is predicted. This predicted behavior is a consequence of the decreasing spatial scale of the spall event with increasing loading rate. This theory was tested on 6061-T6 aluminum and the transition from brittle to ductile failure was predicted for this material at a strain rate of $4 \times 10^5\ s^{-1}$ [27]. Laser-induced shock wave methods have been used [21] to study the spall behavior of the aluminum alloy at very high strain rates. At threshold conditions, voids are formed, characteristic to the ductile spall failure mode. Thus, the ductile behavior at $10^7\ sec^{-1}$ strain rate was experimentally confirmed. There are a few other alloys (Ti-6A1-4V) which are expected to undergo this unusual transition at high strain rates [27].

4.10. Impact Resistance of Adhesive Joints

A shock wave arriving at the interface between two materials (aluminum and epoxy) splits into transmitted and reflected waves according to the acoustic resistance (impedances) of the materials. According to simulations, only about 40% of the shock pressure will be transmitted to the epoxy at the interface with aluminum [31].

Adhesively bonded joints (representing multilayer structures) were tested for impact resistance [32] using hemispherical shock waves. An epoxy-based structural adhesive (FM-73-American Cyanamid) was used to bond two aluminum layers. The adhesion was enhanced by anodization of the aluminum and by using BR-127 as primer. The early stages of ductile failure in the adhesive were characterized by microvoid formation in the epoxy followed by coalescing to larger voids, delamination and, ultimately, complete failure of the sample. Some typical stages of adhesive joint failure are shown in Figs. 4.40 and 4.41, which represent cross sections through the damaged material.

In Fig. 4.40, a crater is already seen in the aluminum layer. The adhesive is damaged and the second aluminum layer has a plastic deformation (bulg-

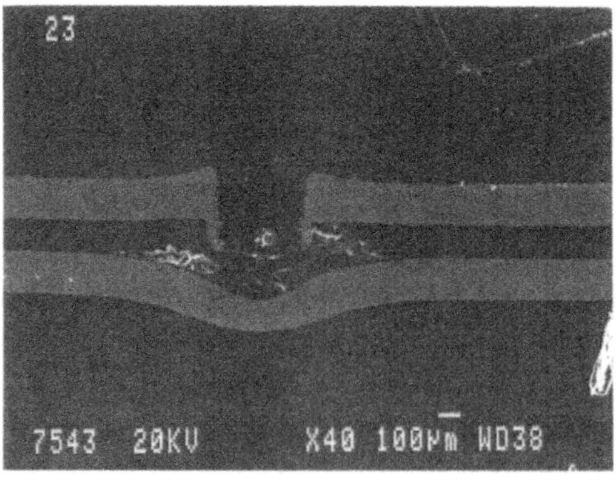

Figure 4.40. Cross section through a damaged adhesive joint.

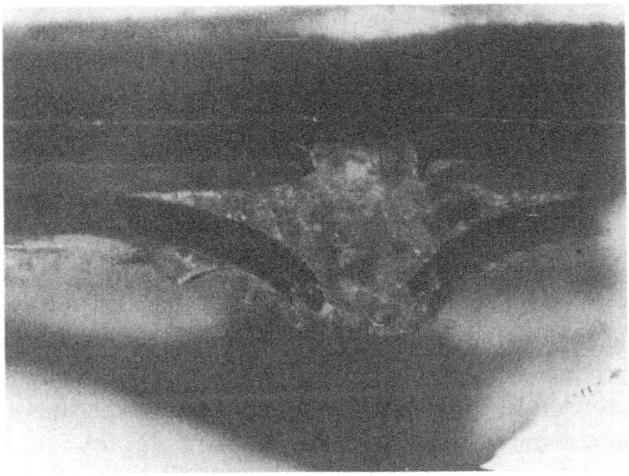

Figure 4.41. Cross section through a perforated and delaminated adhesive joint.

ing) typical of spall development. In Fig. 4.41, the adhesive joint is seen to be completely penetrated. Delamination of the metal layer from the adhesive is greater at the back surface where the stress is maximum [32].

The relative damage as a function of deposited energy can be measured on cross sections of the samples indicating the shock absorption ability of the joint and the weakest link [32]. Shock absorption is greater for toughened adhesives of ductile character than for brittle type where cracks and delaminations can extend to considerable distance, thus ruining the strength of the joint.

4.11. Laser Simulation of Hypervelocity Impacts in Space

In terms of duration and energy transfer, there is a very close correspondence between pulsed laser-induced irradiation and the impact of a hypervelocity microparticle.

Laser simulation of microparticle impacts is based on the following assumptions:

a. The average laser-induced surface ablation pressure is equal to the shock pressure generated by the hypervelocity impactor.
b. The impactor has a diameter equal to the laser spot diameter.
c. The laser-induced impulse is equal to the impactor momentum and the loading time of the target material is equal to the laser pulse duration.

Simulation of hypervelocity micrometeoroid impacts on optical glasses and solar cell cover slides was already reported [33]. The laser simulation method is cheaper and more controllable than other methods of hypervelocity launchers. The achievable pressures in the Mbar range [24] and variable laser spot diameter can extend the results of particle impact simulation to a wider range of conditions than those presented in Ref. 33.

4.12. Summary

Spall resulting from laser-induced shock waves was investigated in several materials and the various stages of material failure at ultra high strain rates were identified. A small quantity of material is needed for testing and the results are reproducible. A linear relation between spall thickness and target thickness for threshold spall formation was found experimentally and explained by a simple model using the time–position plot for laser experiments. Numerical simulations of the experimental conditions were performed including the modeling of laser absorption, shock wave propagation and spall formation.

A new experimental method was developed to estimate the spall stress and the decay of laser-generated strong shock waves. Total elongation at the time of fracture at a very high strain rate was measured for several metals using the phenomenon of spall.

In carbon–epoxy and carbon–carbon composites, severe spall damage often develops on the back face even when damage on the irradiated side is essentially invisible. The shock dispersion ability of ceramics was found using planar and spherical shock waves.

Laser irradiation is a useful technique for selection of adhesives for impact resistance and the detection of the weakest link in a multilayer structure.

Laser simulation of hypervelocity impact on space structures was discussed.

4.13. References

1. Trainor, R.J. and Lee, Y.T. Analytic models for design of laser generated shock-wave experiments *Phys. Fluids* **25**, 1898–1907 (1982).
2. Larson, A.R. Calculation of laser induced spall in aluminum targets, Los Alamos Lab. Report No. LA-5619-MS (1974).
3. Eliezer, S., Loeb, A., Zigler, A., Arad, B., Ludmirsky, A., Gazit, Y., Jackel, S., Salzmann, D., Krumbein, A.D., Borowitz, J.L., Paiss, Y., Gilath, I., Szichman, H., and Givon, M. The evolution of strong shock waves produced by a trapezoidal laser pulse, in *Laser Interaction and Related Plasma Phenomena*, Edited by H. Hora and G. Miley (Plenum, New York), Vol. 7, pp. 279–288 (1986).
4. Loeb, A. and Eliezer, S. An analytical model for creation and decay of strong shock waves caused by a trapezoidal laser pulse, *Phys. Fluids* **28**, 1196–1201 (1985).
5. Arad, B., Borowitz, A., Eliezer, S., Gazit, Y., Gilath, I., Givon, M., Jackel, S.,

Krumbein, A.D., and Szichman, H. Measurement of shock waves generated by a trapezoidal laser pulse, *Plasma Phys. Controlled Fusion* **26**, 845–852 (1984).
6. Key, M.H. The physics of the superdense region. In *Laser–Plasma Interaction*, edited by R.A. Cairns and J.J. Sanderson, Publ. SUSSP, p. 219 (1980).
7. Boustie, M., Cottet, F., and Chauveau, Y. Spalling of aluminum and copper targets by laser shocks p. 741, in *Shock-wave and high strain rate phenomena in materials*, edited by M.A. Meyers, L.E. Murr, and K.P. Standhammer, Marcel Dekker, New York, 1992.
8. Harrach, R.J., Lee, Y.T., Trainor, P.J., Holmes, N.C., Rosen, M.D., Banner, D.L., and Olness, R.J. Livermore National Lab. (U.S.A.) preprint UCRL—86301, 1981.
9. Thompson, P.C. and Roberts, P.D. *Laser Particle Beams*, **2**, 13 (1984).
10. Bauer, F. Metrologie d'ondes de choc a l'aide de polymers piezoelectriques. *J. Physique* **IV**, C3, 427–433 (1991).
11. Paisley, D.L. Laser driven miniature plates for one-dimensional impacts at 0.5–≥6 km/5, p. 1131–1141. In *Shock wave and high strain rate phenomena in materials*, edited by M.A. Mayers, L.E. Murr, and K.P. Standhammer, Marcel Dekker, New York, 1992.
12. Fabbro, R. Physical study of laser produced plasma in confined geometry, *J. Appl. Phys.* **68**, 783–784 (1990).
13. Christiansen, J.P., Ashby, D.E., and Roberts, K.V., *Comp. Phys. Commun.* **7**, 271 (1974).
14. Eliezer, S., Gilath, I., and Bar-Noy, T. Laser induced spall in metals, Experiment and simulation, *J. Appl. Phys.* **67**, 715–724 (1990).
15. Fortov, V.E., Kostin, V.V., and Eliezer, S. Spallation of metals under laser irradiation, *J. Appl. Phys.* **70**, 4524–4531 (1991).
16. Cottet, F., Ng, A., Da Silva, L., and Marty, L. Spallation under the action of a laser induced shock wave, *J. Physique*, **49**, C3, 419–424 (1988).
17. Tuler, F.R., and Butcher, B.M. *Int. J. Fracture Mechan.* **6**, 431 (1968).
18. Gilath, I., Salzman, D., Givon, M., Dariel, M.P., Kornblith, L., and Bar-Noy, T. Spallation as an effect of laser induced shock waves, *J. Mat. Sci.* **23**, 1825–1828 (1988).
19. Gilath, I., Eliezer, S., Dariel, M.P., and Kornblith, L. Total elongation at fracture at ultra-high strain rates, *J. Mat. Sci. Lett.* **7**, 915–917 (1988).
20. Gilath, I., Eliezer, S., Dariel, M.P., Kornblith, L., and Bar-Noy, T. Laser induced spall in aluminum and copper, *J. Physiques* **C3**, **49**, 191–197 (1988).
21. Gilath, I., Eliezer, S., Dariel, M.P., and Kornblith, L. Brittle-to-ductile transition in laser-induced spall at ultra high strain rate in the 6061-T6 aluminum alloy, *Appl. Phys. Lett* **52**, 1207–1209 (1988).
22. Salzmann, D., Gilath, I., Givon, M., and Bar Noy, T. Measurement of the tensile strength of aluminum at a strain rate of 2×10^7 s^{-1}, *J. Phys. D., Appl. Phys.* **22**, 1271–1274 (1989).
23. Salzman, D., Gilath, I., and Arad, B. Experimental measurements of the conditions for the planarity of laser-driven shock waves, *Appl. Phys. Lett.* **52**, 1128–1129 (1988).
24. Gilath, I., Eliezer, S., and Bar-Noy, T. Hemispherical shock wave decay in laser matter interaction. *Laser Particle Beams*. In press.
25. Zukas, A., Nichols, T., Swift, H.F., Gresczuk, L.B., and Curran, D.R. *Impact Dynamics* John Wiley, New York, 1982.

26. Eliezer, S., Gazit, Y., and Gilath, I. Shock wave decay and spall strength in laser-matter interaction, *J. Appl. Phys.* **68**, 356–358 (1990).
27. Grady, D.E. Spall strength of condensed matter. *J. Mech. Phys. Solids* **36**, 353–384 (1988).
28. Gilath, I., Eliezer, S., and Shkolnik, S. Spall behavior of carbon epoxy unidirectional composites as compared to aluminum and iron, *J. Compos. Mat.* **24**, 1138–1151 (1990).
29. Gilath, I., Eliezer, S., and Weisshaus, H. Damage in 2D Carbon–Carbon Composites by short pulsed laser induced shock waves. *J. Reinforced Plastics Composites* **8**, 259–269 (1989).
30. Gilath, I., Eliezer, S., and Gazit, Y. Fracture modes in alumina at hypervelocity impact conditions, *J. Mat. Sci.* **26**, 2023–2025, (1991).
31. Gilath I., Eliezer, S., Gazit, Y., and Barnea, N. Low and high strain rate experiments on model Kevlar-Epoxy microcomposites, *J. Reinforced Plastics Composites* **10**, 84–90 (1991).
32. Gilath, I., Englman, R., Jaeger, Z., Buchman, A., and Dodiuk, H. Impact resistance of adhesive joints using laser induced shock waves, *Int. J. Adhesion Adhesives*. In press.
33. Hill, D.C., Stevenson, T.J., and Zarnechi, J.C. Application of laser simulation of hypervelocity impacts to space particulate damage to metals and optical glasses. In: Proceedings of the fourth Intern Conf. on Spacecraft materials in space environment, Toulouse, France, 1988.

CHAPTER 5

Explosive Fragmentation

W.L. Fourney and R.D. Dick

5.1. Introduction

The ability to predict fragmentation of geological materials due to explosive loading has increased significantly over the past two decades. There are currently a number of hydrodynamic finite difference Lagrangian codes available which do a reasonable job of predicting the fracture and fragmentation that results from explosive loading [1–3]. These codes are particularly useful for estimating the amount of damage that has occurred at a particular time after the explosive has detonated and the resulting shock waves have passed over the material being fragmented. These codes are not of value for later times in the event during which the fragmented material is thrown away from the vicinity of the borehole. There are, however, a number of other codes which use discrete elements that are useful for predicting the final location of the fragmented materials [4–6]. These discrete element codes normally use the fragmentation pattern that results from the stress wave action as the beginning stage for determining the discretation used to predict later particle motion. These elements are normally subjected to a pressure loading which is determined from estimates of the gas generated during the detonation of the explosive.

In this discussion we are interested primarily in the use of the hydrodynamic codes and their prediction of the fracture and fragmentation resulting from the passage of the outward traveling shock (stress) waves and the passage of the reflected waves over the conditioned material. In particular, we will be looking at the types of failure that these codes predict and where these failures occur; and we will be comparing these results with what we have observed in laboratory testing—primarily in tests which were conducted in transparent materials using dynamic photoelasticity to aid in the understanding of the fracture processes.

5.2. Code Results

Although there are some basic differences among the various hydrodynamic codes, most allow for three different types of failure. The first type of failure

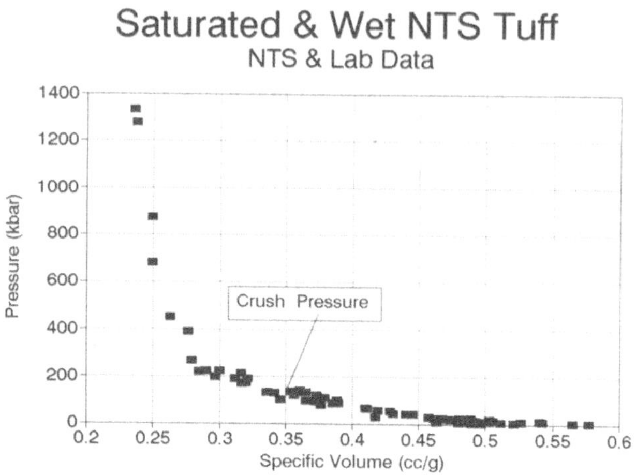

Figure 5.1. Typical pressure–volume data used in P-alpha model for predicting crush.

allowed is crushing of the material and this normally occurs at very high pressures. This is the so called P-alpha model which allows for the air voids to be irreversibly removed within regions when pressures are quite high. Figure 5.1 shows typical pressure–volume data used to estimate how volume changes as a function of pressure. Normally this data is input into the codes by means of a functional relationship. One very important point shown on the curve is the critical crush pressure P_c. At P_c all voids have been irreversibly removed from the material. Figure 5.2 shows output from TOODY which simulates a concentrated energy source at a deeply buried point and shows the region in which the code predicts that crushing failure has occurred. (The area shown in Figure 5.2 actually shows damage due to both crushing and shear failure. At the time shown in the figure, however, there should be very little failure due to shear.)

As implied above, the second mechanism of failure built into the codes is plastic flow due to shear loading (compression). A typical model for this mechanism is given in Figure 5.3 which shows a failure curve (shear stress difference) as a function of mean pressure. This is a Von Mises type criteria and essentially indicates that once the stress difference is greater than the values indicated by the failure curve that plastic flow occurs. Figure 5.4 shows output from the same TOODY simulation as shown in Figure 5.2 and indicates the regions in which damage has occurred. (The damage region shown was determined to be from both crushing and shear failure but most of the additional failure is felt to be due to shear.)

The final failure mechanism built into the codes is failure due to tensile loading. This failure mechanism is triggered whenever the tensile stress exceeds a failure value used as input to the code. The value of tensile failure used is normally quite low when compared to values used in the other two

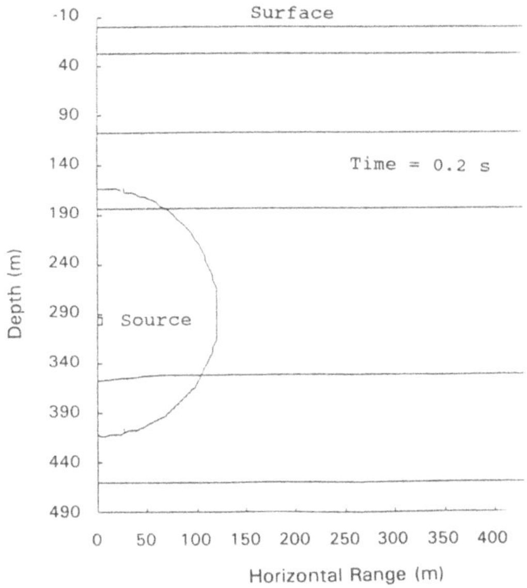

Figure 5.2. TOODY prediction of where fragmentation due to crush is occurring. (Courtesy of Fred App.)

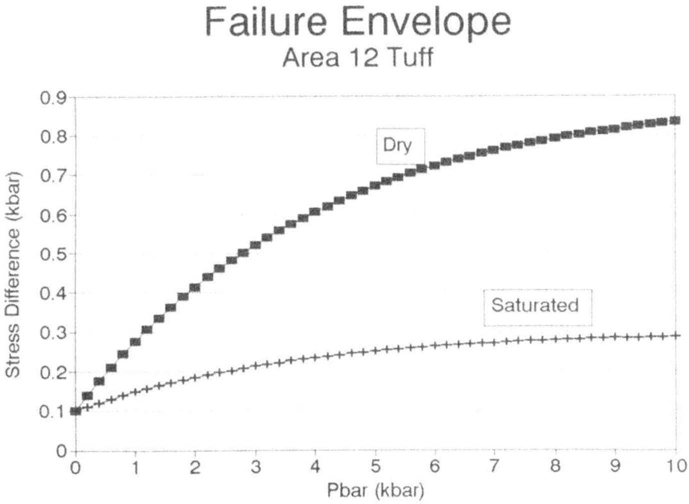

Figure 5.3. Shear strength as a function of pressure used in codes to predict failure by shear.

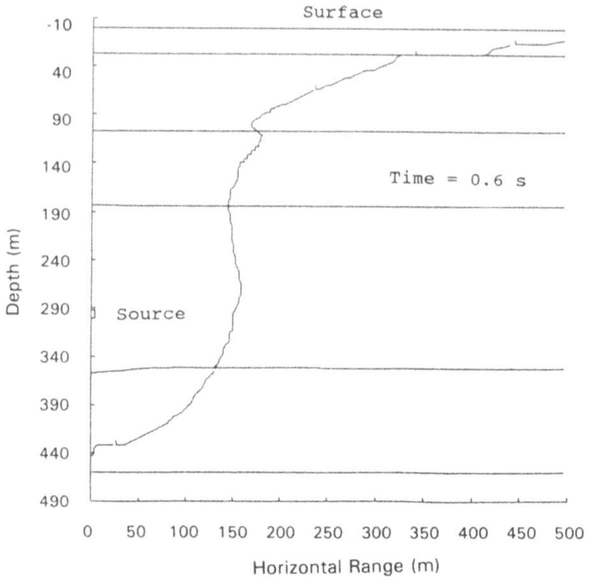

Figure 5.4. TOODY prediction of where fragmentation due to shear is occurring. (Courtesy of Fred App.)

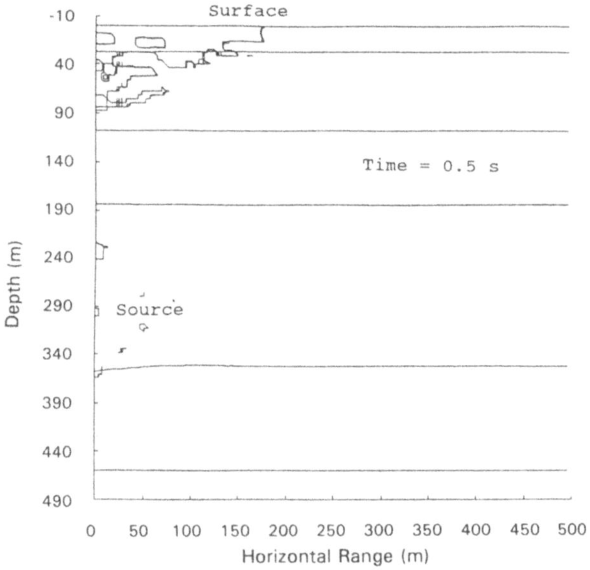

Figure 5.5. TOODY prediction of where fragmentation due to tensile loading is occurring. (Courtesy of Fred App.)

mechanisms and in many cases the presence of any tensile stress is used to signify failure. Figure 5.5 shows the regions which are predicted to fail due to tension for the same TOODY simulation as given before.

With the predictions obtained from the computer codes in mind and an idea of where the various types of failure are predicted to occur we turn our attention to results of model testing in the laboratory and to the various types of failure and where we expect these failures to occur.

5.3. Experimental Results

The experimental results that will be described were obtained using a dynamic polariscope. With this experimental technique transparent birefringent materials are used to make the models which simulate the test geometry of interest. When viewed under polarized light, these materials have the property that the speed of propagation of light through them depends upon the stress state of the material. The polarized light which enters the model is split into two components—one in the direction of the maximum principal stress and the other in the direction of the minimum principal stress. These two components of light are compared with one another after traveling through the model by means of an observed fringe pattern. This fringe pattern is a measure of how much one of the light components is slowed down with respect to the other and hence is a direct measure of the difference in the two principal stresses. These fringe patterns are photographed with a high-speed camera which records the stress information as a function of time. The particular camera used is capable of taking pictures at framing rates up to about a million frames a second and takes 16 frames during the dynamic event. The system is very suitable for recording events which involve explosive loading.

Most of the experimental results presented are for isotropic materials subjected to plane stress loading conditions. Our experience has shown that for the purposes of investigating possible mechanisms of fragmentation these conditions are not overly restrictive. The results obtained must, of course, be investigated in more detail using three dimensional testing of non-isotropic materials but we have found that the basic mechanisms are easily observed in two dimensional model testing using transparent materials.

5.4. State of Stress

When an explosive point source is detonated in the interior of a body a number of different wave types are created which travel outward from the source. Figure 5.6 shows a typical result obtained from testing in a model geometry with an explosive point source. The circular fringes represent the P (primary) wave as it travels from the source. The fringes represent lines along

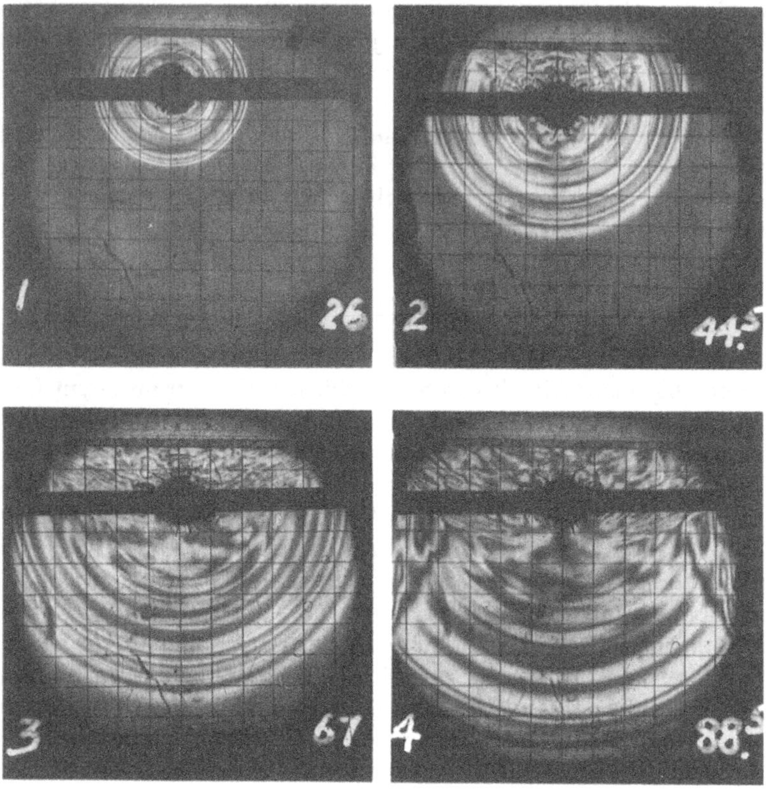

Figure 5.6. Frames from a dynamic test showing outgoing waves.

which the shear stress (difference between the maximum and minimum principal stresses) is a constant. At the fringe which is farthest from the source, the difference between the maximum and minimum principal stress (maximum shear stress) is 6.9 MPa. At the next fringe the maximum shear stress is twice as large and so on as the source is approached and the fringe order number increases. By knowing the fringe order and the model material it is possible to obtain a solution for not only the stress differences but in this case also the individual principal normal stresses. Figure 5.7 shows how stress varies as a function of time for the situation where 200 mg of PETN has been detonated in a PMMA model that is 6.4 mm thick. As seen in the figure, the leading edge of the wave is composed of biaxial compressive stresses, that is, both the radial and the tangential stresses are compressive in the leading part of the wave. Behind this leading edge, the experimental results show that both the radial stress and the tangential stress are tensile. This wave travels out into the media and upon reaching a free surface reflects back into the media as a tensile wave. At points where the incidence with the boundary is non-normal a shear wave is generated as well as the P wave

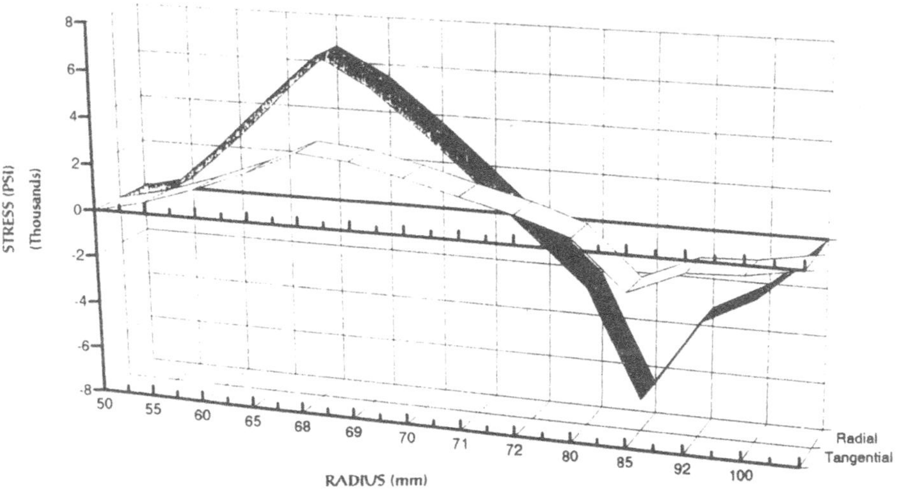

Figure 5.7. Stress state in outward propagating P wave. □, tangential; ■, radial.

being reflected. A second wave is created at the borehole wall as soon as cracking around the borehole begins or distortion of the borehole occurs. This is a shear wave and travels out into the media at a slower speed than the P wave. The outgoing S wave interacts with the inward traveling PP wave and PS wave and upon reaching a free surface also reflects back into the media. At non-normal incidence the shear wave is split into both an inward traveling shear wave and an inward traveling P wave (SP and PS waves). These waves as well as the PP and PS waves continue to reflect back and forth in the media until they eventually decay to zero amplitude. The decay rate is actually quite high and only a few transits between the explosive source and the free surface are needed before the amplitude of the waves becomes quite small.

5.5. Fragmentation Steps

5.5.1. Borehole Crushing

The most difficult mechanism for us to observe experimentally is the creation of the crushed zone in the media around the explosive. The crushed zone occurs very near the borehole and most often we use a smoke deflection or pressure containment device which obscures the crush zone as it is being created. Also, the fact that the transparent material is crushed makes it translucent and prevents observation of the crushed zone as it is being formed. We do observe a crushed zone after the test is over, and two examples are shown in Figure 5.8—one in PMMA and one in Homolite 100. Of

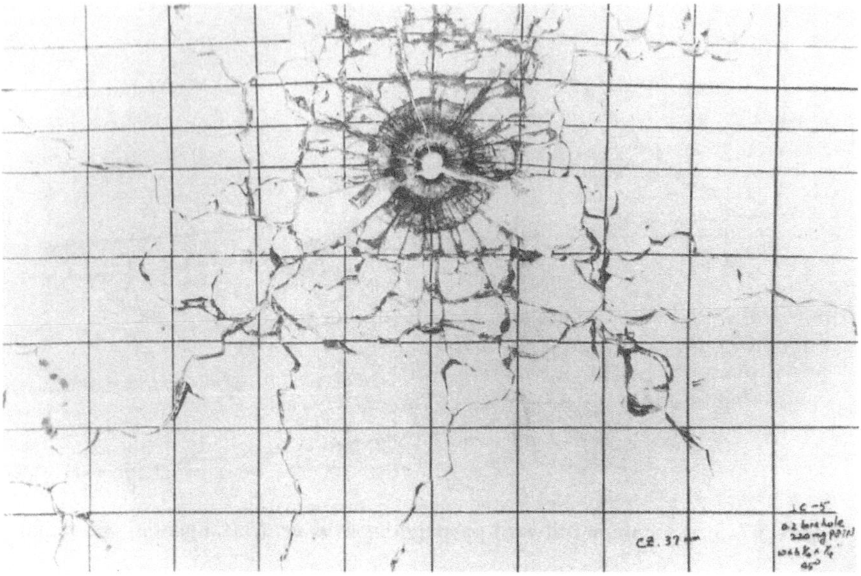

Figure 5.8a. Crushed area in homolite model.

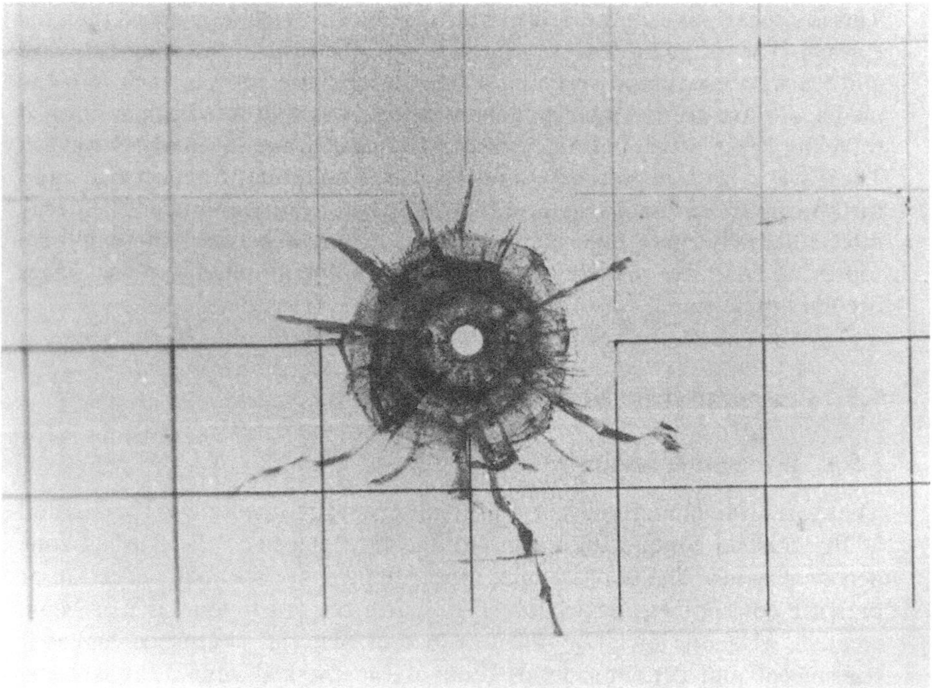

Figure 5.8b. Crushed area in PMMA model.

the two materials the Homolite 100 is more brittle and fragments more easily. The Homolite 100 model had a charge of 220 mg of PETN in a 0.2 mm diameter borehole. The PMMA model was subjected to a loading of 260 mg —also in a 0.2 mm diameter borehole. The Homolite 100 model was 6.4 mm thick while the PMMA model was 8.0 mm thick. Although the amount of cracking outside the crushed zone is much more extensive for the Homolite model the two crushed zones are about the same size. In a study of the effect of charge size on size of the crushed zone [7] we found in two-dimensional model testing that the size of the crushed zone did not depend upon charge size. In fact, there was a fairly constant ratio of crushed zone diameter to original borehole diameter for any given material. For PMMA and Homolite 100 this ratio was found to be between 6.5 and 7.5 with the PMMA crushed zone diameter ratio being slightly larger than for Homolite. These were, however, only two-dimensional tests and it is felt that the results might be significantly different in three-dimensional models where the pressure from the explosive is prevented from venting to atmospheric so quickly. Nonetheless, we do observe significant borehole crushing and the location of this crushing appears to agree well with code predictions.

5.5.2. Radial Cracking

The tangential stress in the trailing portion of the outgoing P wave appears to be very effective in the initiation of radial cracking . As indicated above this stress is tensile (in the trailing portion) and as cracks are initiated the magnitude of the stress decreases as energy is consumed. Very close to the borehole the stress is quite high and a large number of radial cracks are initiated. The existence of this very large number of radial cracks can be seen very clearly in Figure 5.8—especially in the Homolite model. As the distance from the borehole increases, the stress magnitude is greatly decreased and many of the initiated cracks arrest until only a few dominant cracks are propagating. Figure 5.9 shows a frame from the high speed camera taken 85 μs into the event and shows 12 cracks still propagating away from the borehole. These cracks would quickly arrest as the P wave propagates from the borehole were it not for the fact that additional energy is added to these cracks by the shear wave and also by the pressurization of these cracks by the gases created by the detonation of the explosive.

5.5.3. Circumferential Cracks and Branching

The model shown in Figure 5.9 is circular and very limited in extent. As a result, the P wave when interacting with the free surface results in only a PP wave being reflected back into the interior of the model. The fact that the model is small (0.3 m in diameter) will result in the magnitude of the wave at reflection being quite large. As the PP wave propagates back over the crack

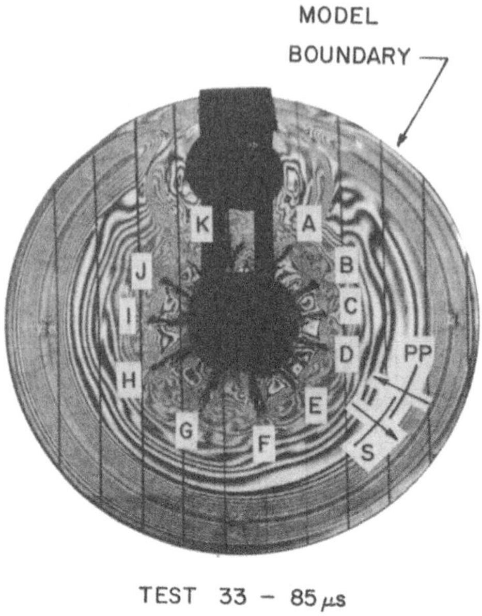

Figure 5.9. Frame from dynamic test showing propagation of dominant radial cracks.

tips, geometric convergence will result and the effects of the reflected waves on the outward propagating cracks will be emphasized. The picture shown in Figure 5.9 was taken 85 microseconds after the detonation of the PETN charge [8]. The blackness in the center of the model is due to a special device that was used to contain the gas pressure in the borehole as long as possible. Notice at 85 μs that the 12 radial cracks are still propagating outward under the influence of the leading edge of a shear wave which was generated when the loading symmetry of the model was destroyed by the initiation of cracks around the borehole. The position of the outward propagating S wave and the returning PP wave are marked on the figure. Some of the cracks shown have branched or are in the process of branching since the energy driving them is enough to drive more than one crack. Crack F, for example, branched quite early and two cracks are propagating outward. Many of the other cracks, B, I, J, D, and A for example, are in the process of branching. Some, such as H show no evidence of branching. Figure 5.10 shows an enlargement of a frame taken slightly later in the same tests—also before the PP wave has reached the crack tips—and shows examples of both successful and unsuccessful branching attempts by the various cracks. Notice from Figure 5.10 that all of the cracks shown are still being driven by a significant energy source. This is evident by the presence of the "butterfly" loops at the tips of the cracks. These loops are an instantaneous measure of the stress intensity K_{Id}, the dynamic crack driving force [9]. Unlike the P wave it is not possible to determine exactly the state of stress in the outgoing S wave. The reason for this arises from an inability to count accurately the fringe order in

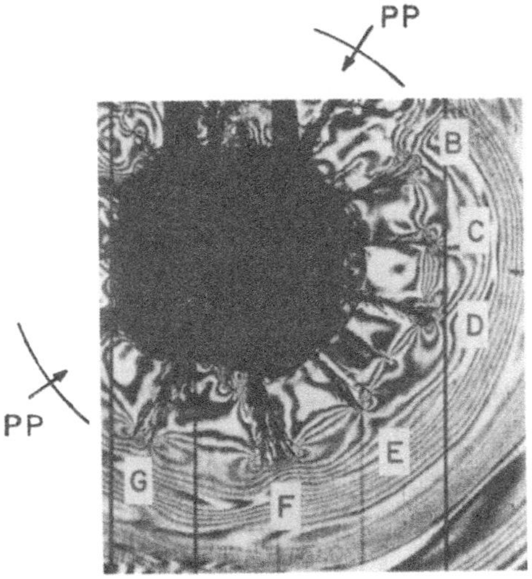

TEST 33 – 104 μs

Figure 5.10. Enlargement of area showing radial propagation.

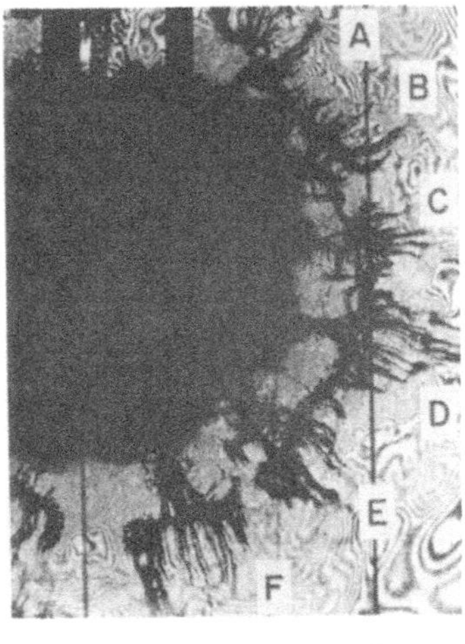

TEST 33 – 153 μs

Figure 5.11. Enlargement showing effects of PP wave passing over tips of outward propagating radial cracks.

the shear wave. It is felt, however that the mechanism driving these radial cracks is tensile Mode I loading.

As the PP wave propagates back over the radial cracks a tremendous amount of energy is added and the direction of the dominate tensile stress changes from the radial to the tangential direction. This results in the radial cracks turning into a circumferential direction and multiple branching taking place. Figure 5.11 shows an enlarged view of the area shown in Figure 5.10

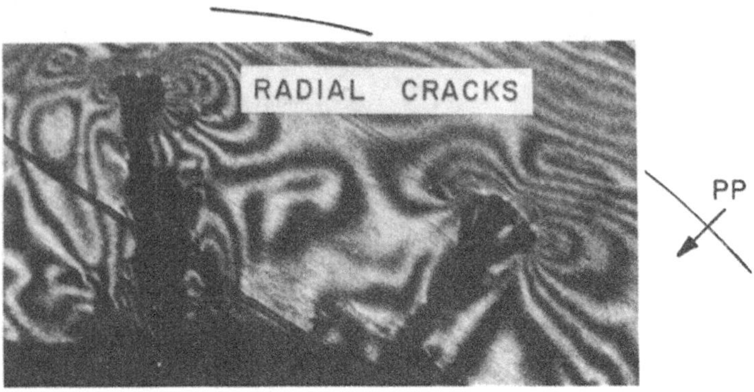

TEST 32 – 101 μs

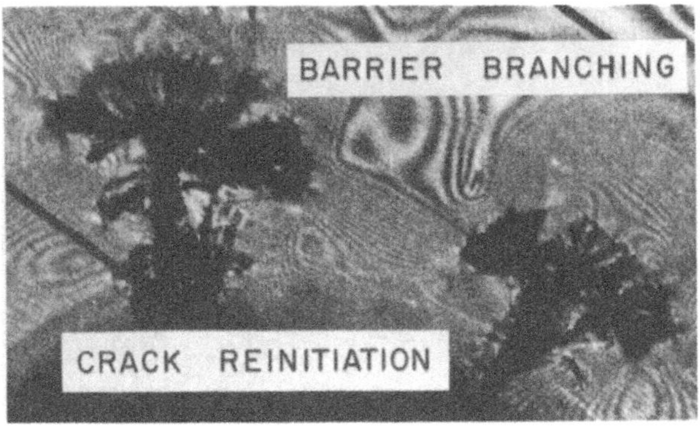

TEST 32 – 119 μs

Figure 5.12. Two frames showing mechanism of barrier branching.

after the leading front of the PP wave has passed the radial cracks. After this wave front passes, the cracks (now there is a very large number of them) turn and propagate in the radial direction. We [10] have termed this mechanism "barrier branching" or "candelabra cracking" since the resulting fracture pattern looks like an ornate candle holder. Figure 5.12 shows an enlarged view of this mechanism from another test. Figure 5.12(a) shows two radial cracks just before the PP wave reaches them. Figure 5.12(b) shows the same cracks 18 microseconds later—just after the PP wave has passed. We call this barrier branching since it appears that the radial cracks hit the barrier of the inward propagating PP wave and turn and branch (multiple times) at the

Figure 5.13a. Postmortem of model with light charge showing lack of barrier branching.

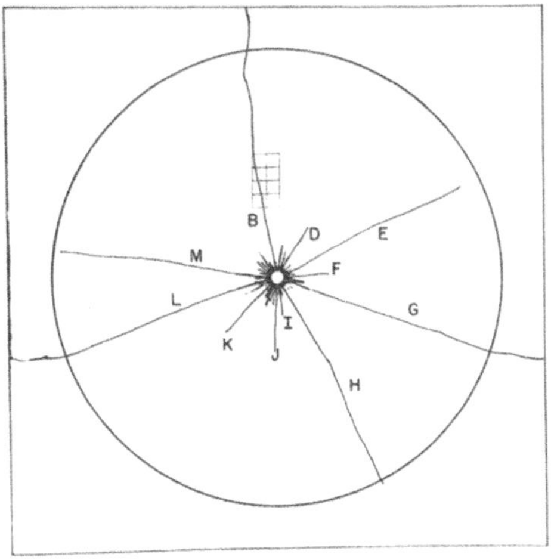

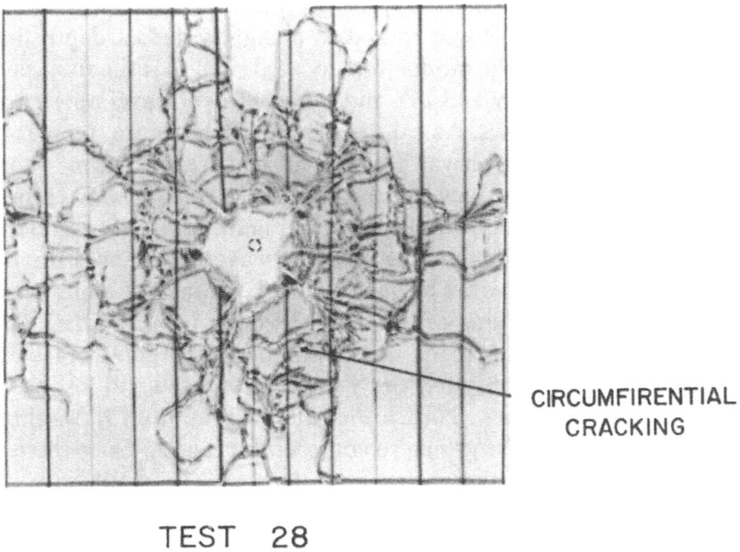

Figure 5.13b. Postmortem of model with heavy charge showing effect of barrier branching on fragmentation.

same instant in time. The barrier then passes over the crack tips and the radial cracks resume their outward propagation.

The branching and especially the barrier branching mechanism can have an enormous effect on the final fragmentation pattern. Figure 5.13 gives a comparison of fragmentation that resulted in two model tests. In one a light charge was used to suppress the branching of cracks. In the second a larger charge was used to enhance branching. It is felt that barrier branching can be very effective even when blasting is done in the absence of a major free face. The reflections in this case occur from joints and from impedance mismatches which are the norm in geologic materials. Since the codes cannot keep track of the enormous number of propagating cracks and since most of the codes do not account for the very complicated geologic structure this mechanism is underemphasized by code predictions.

5.5.4. Flaws and Joints

The lack of geologic structure also affects the ability of the codes to accurately predict fragmentation resulting from the interaction of the stress waves with flaws and joints. Figure 5.14 shows fracture (and fragmentation) that occurs due to stress wave interaction with a small flaw in an otherwise homogeneous material. Actually two flaws are shown in the figure—both were routed into the surface of a sheet of Homolite 100. The grid shown on the specimen is 25.4 mm by 25.4 mm and the location of the borehole is given by the black cross which is located to the right of the area of interest. The lower of the two flaws was routed to a slightly deeper depth than the upper flaw. The edges of the model were located such that the first reflection from the bottom boundary and PP- and PS waves will travel upward and to the left across the two flaws. As can be seen from the figure, the passage of the P wave over the two flaws was insufficient to cause initiations—although the stress state at the lower flaw due to the passage of the P wave was nearly enough to initiate fractures from the flaw. In Figure 5.14(a) the stress state is elevated greatly as the leading edge of the S wave approaches the bottom flaw and, in fact, two cracks are initiated at the bottom flaw—one from each end. The mixed mode loading of the flaw is evident from the fringe pattern in Figure 5.14(a) as well as from an examination of Figure 5.14(b), which shows the bottom flaw 21 μs after the S wave has arrived and just before the PP wave envelopes the flaw. Notice that at this time both flaws that were initiated by the S wave have grown to considerable length. Once these cracks are propagating, they are then acted upon by the passing PP wave causing them to turn and branch—just as the radial cracks did during "barrier" branching. The fracture pattern becomes quite complicated since the location of the lower flaw and the model size results in interactions occurring from the PP wave from the bottom edge of the model and the PP wave from the upper edge. The location of all PP and PS waves are indicated on the figure and show that significant crack growth and branching results due to the passage of the various PP and PS waves after cracks are initiated by the S wave. Since geologic materials contain a very large number of flaws the opportunity for fragmentation due to this mechanism is great.

Figure 5.14. Interaction of stress waves with small flaws. (a) Test 36, 137 μs; (b) Test 36, 155 μs; (c) Test 36, 170 μs

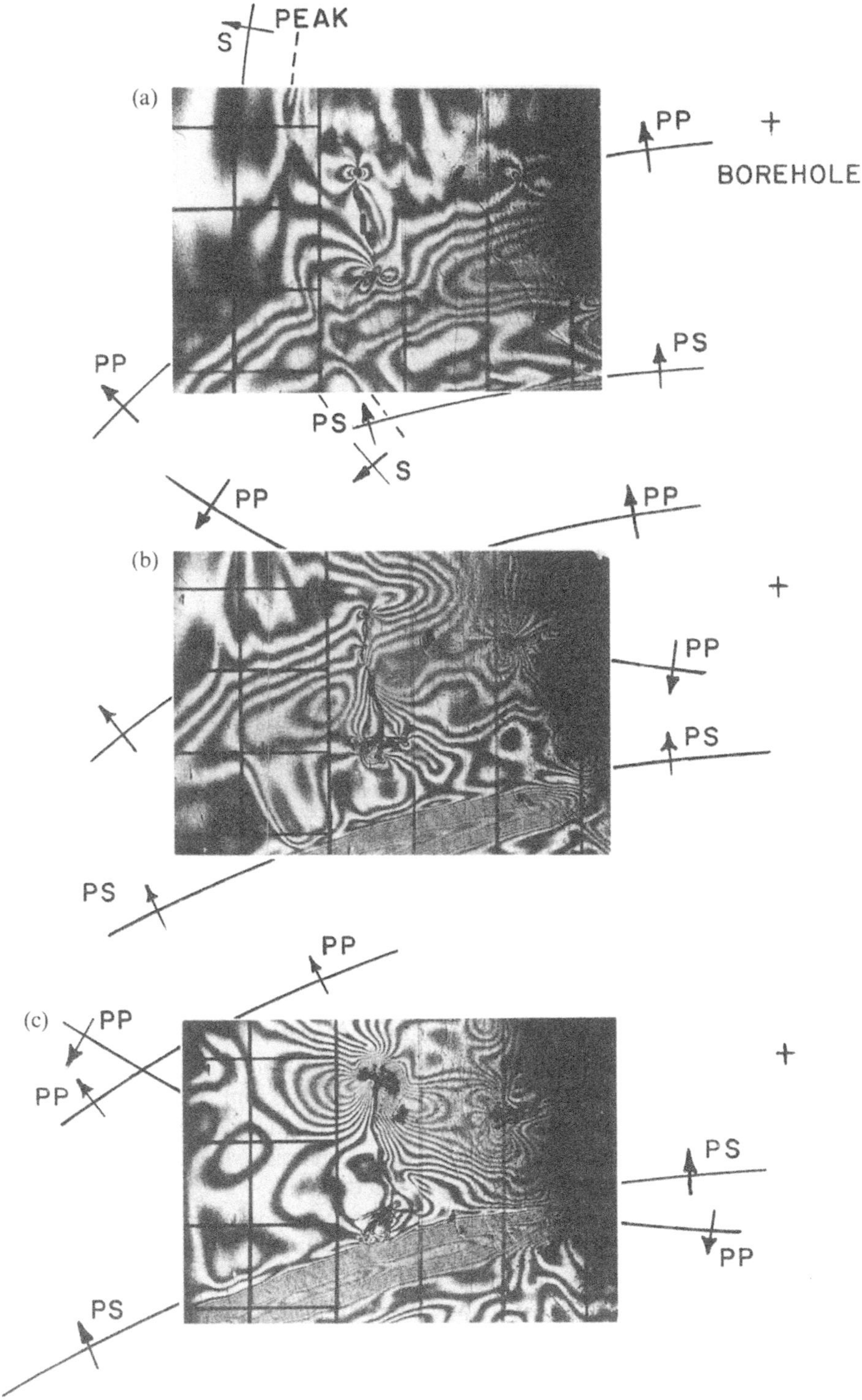
PEAK
S
(a)
PP
BOREHOLE
PS
PP
PS
S
PP
PP
(b)
PP
PS
PS
PP
(c)
PP
PP
PS
PP
PS

Figure 5.15. Bench face showing bedding and joint sets at Pinesburg Maryland Quarry.

Figure 5.15 shows a typical bench face at a quarry in Central Maryland. This bench has been formed as a result of blasting operations that were used to remove limestone for making roadstone. As a result, the joints and bedding planes shown are much more open than they would be in the virgin material. However, Figure 5.15 illustrates that real geologic materials do contain numerous joints and bedding planes. We were interested in knowing if the passage of outgoing stress waves past such joints would aid in the fragmentation process. We constructed models which consisted of strips of Homolite 100 which had their edges routed smooth and had been bonded together with a cyanoacrylate ester. The resulting models were fragile from the standpoint that if bent or dropped they would come apart but yet they were strong enough to stay together until explosive testing was complete. Figure 5.16 shows the result of one of the jointed model tests. The strips shown are 50 mm in height, 300 mm long and 6.35 mm in thickness. The model shown contained two boreholes as this particular test involved an investigation of the effect of time delays between explosive detonations upon the resulting fragmentation. The black shadows shown in the figure are again devices used to retain the gas pressurization for as long a period of time as possible. Figure 5.16(a) shows the model at 34 μs after detonation of the PETN charge. At this time the P wave has traveled about 50 mm and a large number of cracks are visible above the borehole at the first joint. Figure 5.16(b) shows the situation at 62 μs and in addition to the cracks shown

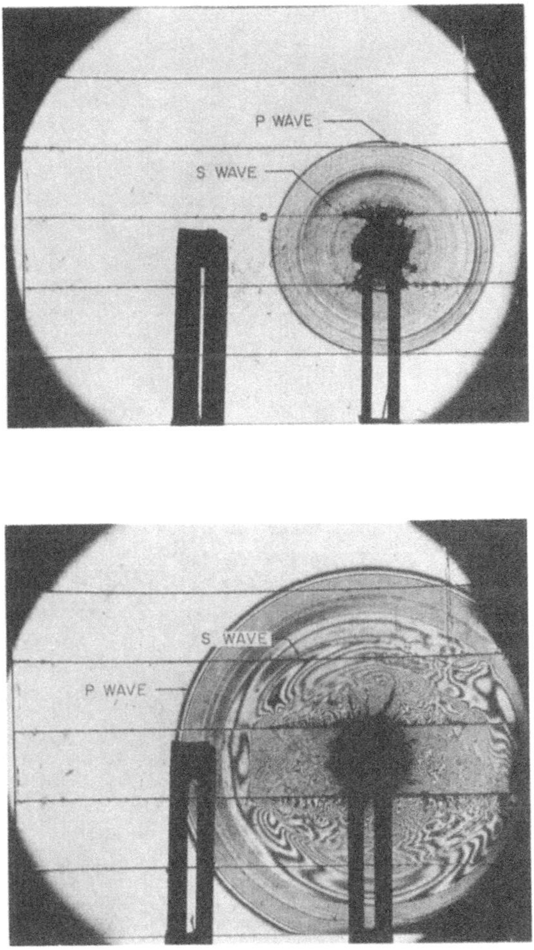

(a)

(b)

Figure 5.16. Interaction of stress waves with large flaws.

in the earlier frame a large number of cracks have been initiated (by the outgoing S wave) all along the joints just above and below the charge. These cracks are propagating away from the layer which contains the charge and are located all along both joint interfaces. Figures 5.16(c) and (d) show the fracture pattern later in the event, at 86 and 116 microseconds after detonation. A very dense system of fractures has been initiated along the two joint interfaces nearest the charge site. These fractures are believed to have been initiated by shear stresses and propagate at an angle of 70 degrees from the interface. These results agree with theoretical predictions for the initiation of fracture from an existing crack (interface) [11] and appear to be a possible

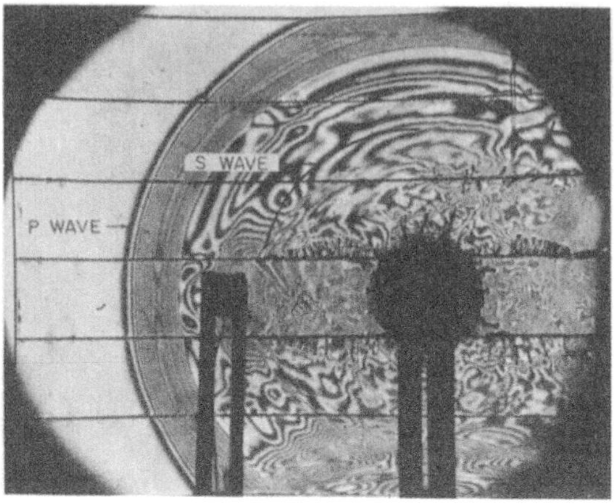

(c)

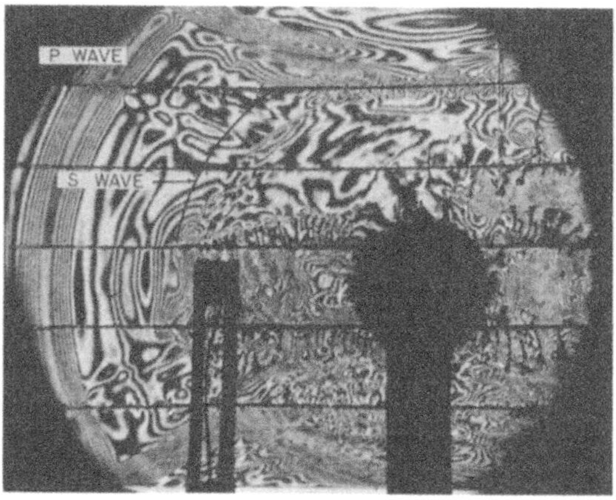

(d)

Figure 5.16 (*cont.*)

source for a major contribution to the resulting fragmentation. We have termed this mechanism of fragmentation "joint initiated fragmentation" [12] and current codes do not utilize this mechanism primarily due to their inability to accurately simulate the complicated geologic materials on a small scale.

5.5.5. Spalling

Spalling occurs when the outgoing compressive P wave interacts with a free surface and produces an inward traveling tensile wave [13]. Depending upon the shape of the stress pulse and the magnitude of the compressive stress at reflection it is possible to have multiple spalling. As the reflection is taking place, failure occurs whenever the resulting tensile stress is greater than the tensile strength of the geologic material. As a new free surface is formed by the tensile failure, reflection occurs at that new surface and the process continues so long as the resulting reflection results in a tensile stress that is greater than the tensile failure stress.

Figure 5.17 shows results from a series of model tests conducted to investigate spalling. In this case the model was 305 × 305 mm and was 6.35 mm thick and was made from Homolite 100. The dynamic polariscope was configured to give isoclinics in addition to isochromatics. That is, we were interested in the direction of the stresses in the outgoing wave as opposed to the stress magnitudes so the fringe patterns will be somewhat different from those seen in earlier pictures. In this series of tests we utilized a device to conduct the smoke and debris away from the field of vision as opposed to trying to contain it—hence the field of view is very clear in all frames presented. The black shadow over the borehole is this smoke deflection device.

Figure 5.17(a) shows the fracture pattern at 33.8 μs after detonation of the explosive. With this smoke deflection device the pattern is extremely clear and the crushed zone as well as the radial crack system that extends beyond the crushed zone is clearly evident. Some attempts at radial crack branching both above and below the deflection device are evident. At the time the photograph was taken, the P wave has reflected from the boundary and is just at the tips of the radial cracks above the borehole. The grid system on the model is every 12.7 mm (vertically) above the borehole and every 25.4 mm below. Notice the dark spots within the first grid from the upper free surface. These are the locations of spall failures which are just being initiated. Figure 5.17(b) shows the same model 6.4 microseconds later. The front of the P wave has now traveled 100 mm and the reflected wave is just passing over the charge location. Notice at this time the well developed spall fractures and the multiple spalling. Figures 5.17(c) and (d) show the further development of the fracture pattern at later times (46.1 and 50.7 microseconds after detonation). Evident from these figures are both the growth of the spall fractures and the development of the fracture pattern due to the barrier branching mechanism. Figure 5.18 shows a fracture pattern obtained from a similar test but very late in the dynamic process. It is evident from this frame (taken at 230 μs) that extensive fragmentation has occurred above the deflection device compared to below the deflection device. Above the charge (on the side towards the free surface) the spalling and circumferential cracking (barrier branching) have combined to result in an extensive system of fractures. Below the device very little circumferential cracking has occurred and no spall is present. Some fracturing is present but no real fragmentation. The codes seem to do a good job with predicting spall failure as is clear from looking again at Figure 5.5 which is the TOODY simulation of tensile failure occurring from a more deeply buried charge.

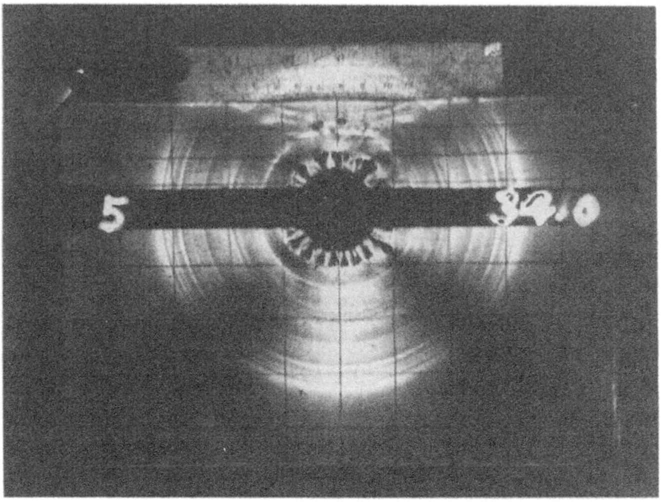

(a)

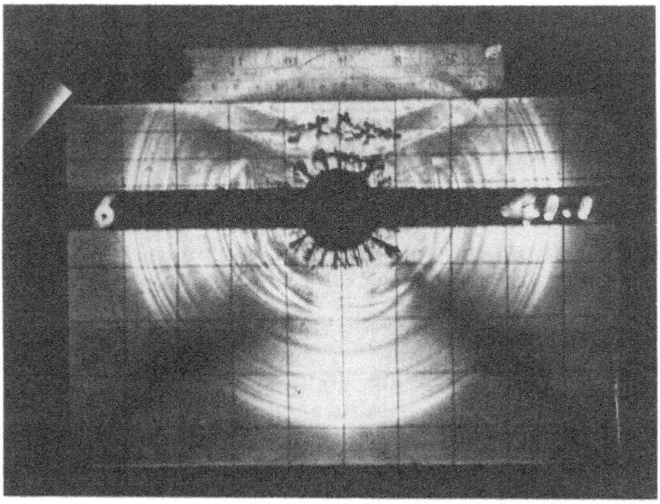

(b)

Figure 5.17. Sequence of photoelastic pictures showing spalling at free surface.

5.6. Comparison to 3D Situations

Figure 5.19 shows particle velocities at several locations as determined with TOODY from a report by App and Brunish [14] for a nuclear event of 10 kt. Also shown on the figure are measured values for the same locations. The accelerometers used to measure the ground motion were located in a vertical hole just above the device. As can be seen from the figure the computations predict quite well the waveforms for the velocities that were measured. The

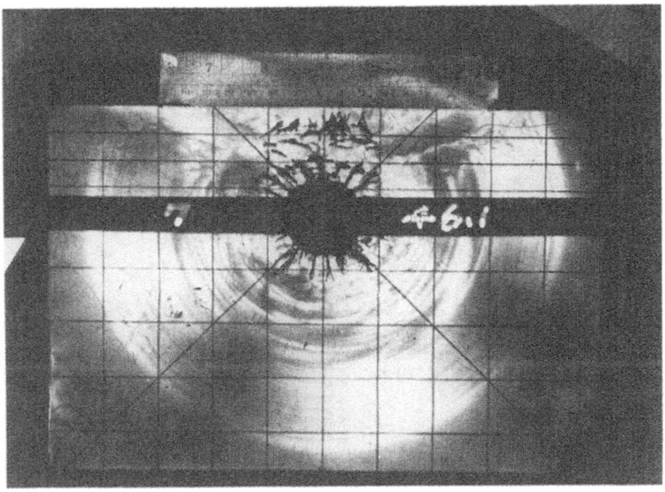

(c)

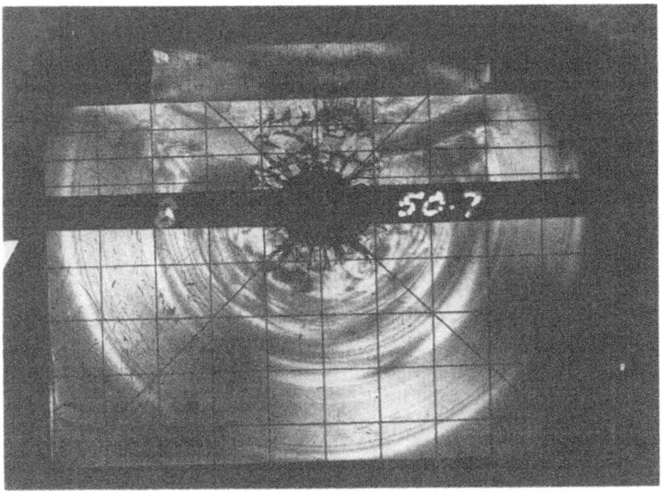

(d)

Figure 5.17 (*cont.*)

event shown was conducted in alluvium which was modelled as a homogeneous material with the exception that sound speeds were varied from layer to layer. Figure 5.20 shows a more global result obtained from TOODY for particle velocities in another nuclear event which was conducted in tuff. In this particular event the source was larger and the test bed was made up of several layers which had considerable differences in material properties from layer to layer. In Figure 5.19 it should be noted that the maximum particle velocity measured was on the order of 0.5 m/s. Since these results are from a

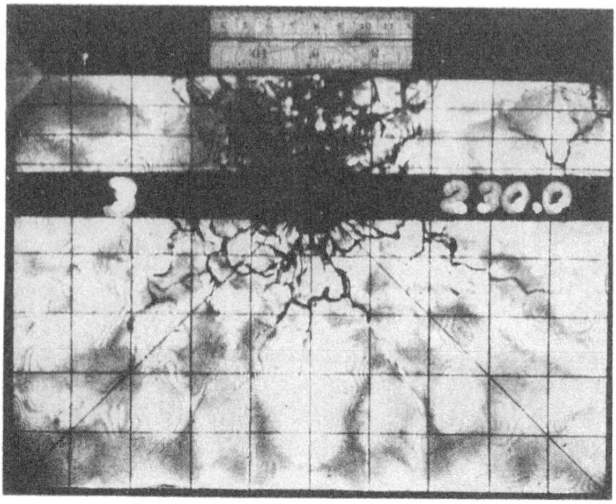

Figure 5.18. Picture taken very late in cratering event showing fragmentation pattern.

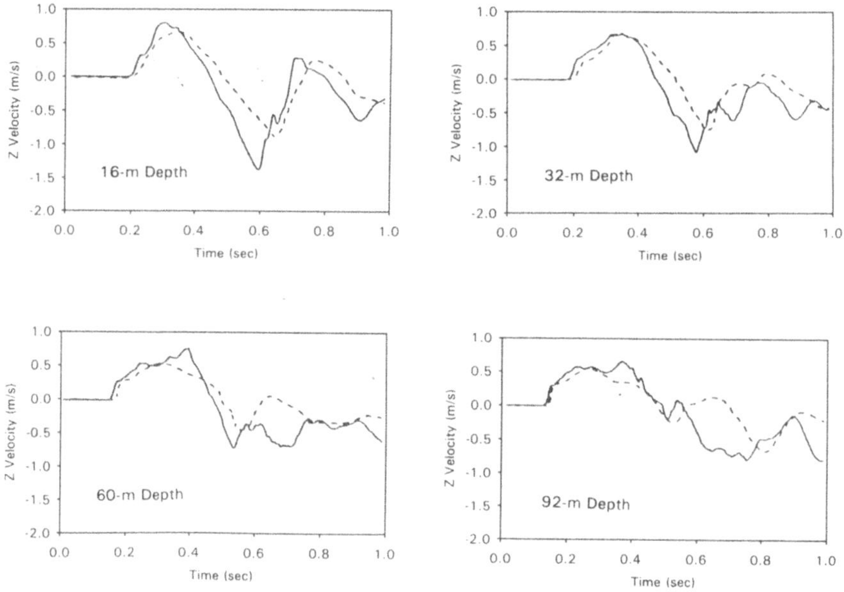

Figure 5.19. Vertical velocities obtained from TOODY calculations for a deeply buried event. (Courtesy of Fred App.) Calculated (- - -) and observed (—)

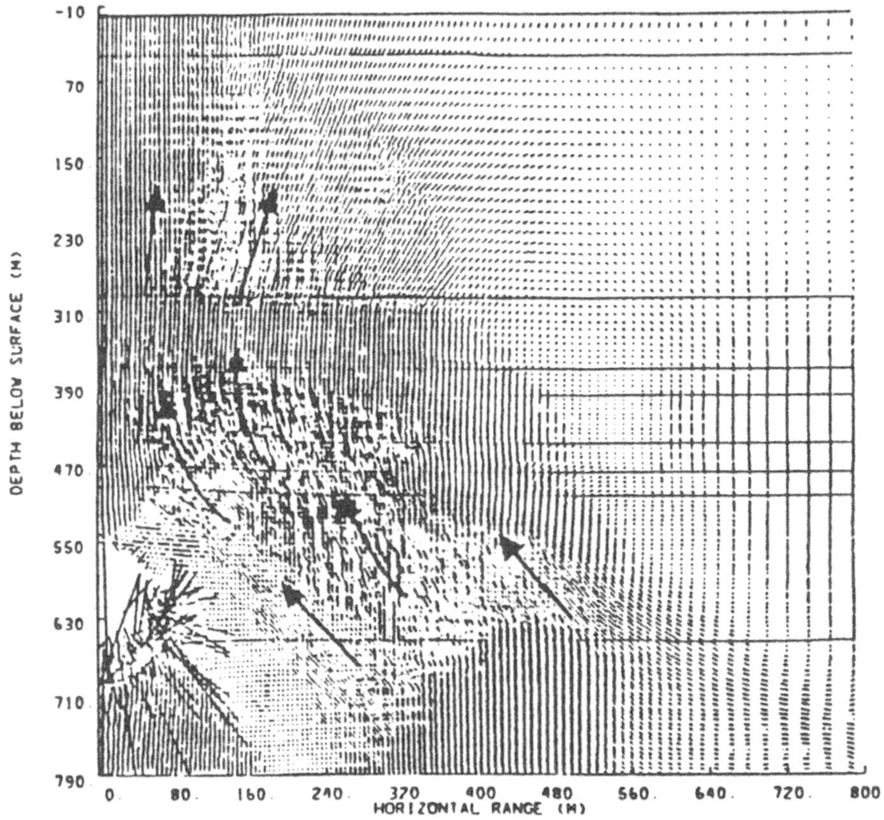

Figure 5.20. Velocity directions calculated with TOODY for a deeply buried event. (Courtesy of Fred App.)

deeply buried event in which containment was successful and for which no cratering occurred these low values seem plausible. The results presented in Figure 5.20 are from an event which was deeply buried and for which no cratering occurred. Notice from Figure 5.20 that material on the same level as the energy source is predicted to move towards the source (inward) and upward. About 230 m above the source the material motion is mainly upward whereas at a location 400 m above the source the motion is upward and outward. In Figure 5.20 the lengths of the heavy arrows also indicate velocity magnitude.

At Maryland we conducted cratering tests in cement-like materials and were successful in measuring particle velocities at various locations from the charge. In particular the models used were 254 mm in diameter, 208 mm in height, and the spherical 1 g charge of PETN was located 57.1 mm from the upper free surface. The material used to construct the models was a very high strength cement with limestone chips as the aggregate. In separate tests elec-

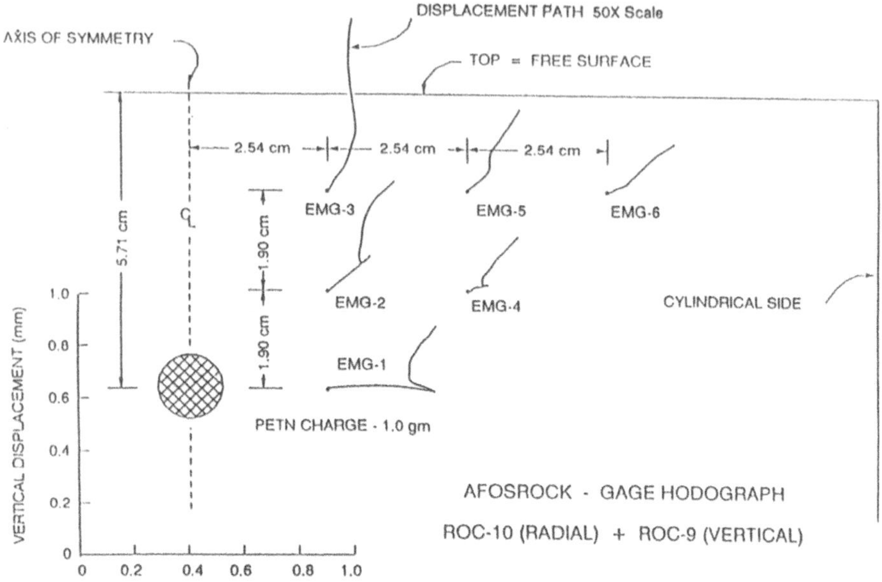

Figure 5.21. Displacements measurements from three-dimensional cratering test in cement-like material.

tromagnetic velocity gages [15] were used to measure horizontal and vertical components of particle velocity at six different locations. The velocities measured at these six locations were integrated to obtain horizontal and vertical displacements of the measurement points. These displacements are presented in Figure 5.21. The displacements shown are those which occur over the first 100 μs after detonation and the curves shown in the figure therefore trace the paths of the six points over that time interval. The lengths of the curves are proportional to the total displacements. The particle located where gauge number three was located is subjected to the maximum displacement and this displacement is primarily upward. A comparison of the displacements given in Figure 5.21 with the velocities given in Figure 5.20 gives a good idea of how the motions differ when cratering occurs as opposed to when it does not.

We wanted to determine if the results of the testing we had conducted in the two-dimensional models predicted the fragmentation that would occur in a three-dimensional situation. Since we did not have a dynamic code at our disposal we decided upon a means to conduct a quasistatic check of our work. We used the results of our two-dimensional model testing to determine the fragmentation state at a given time after explosive detonation and used this configuration in a static finite element code to determine the directions in which particles would be moving. We then compared the results of this quasistatic calculation with the results we had obtained in our three-

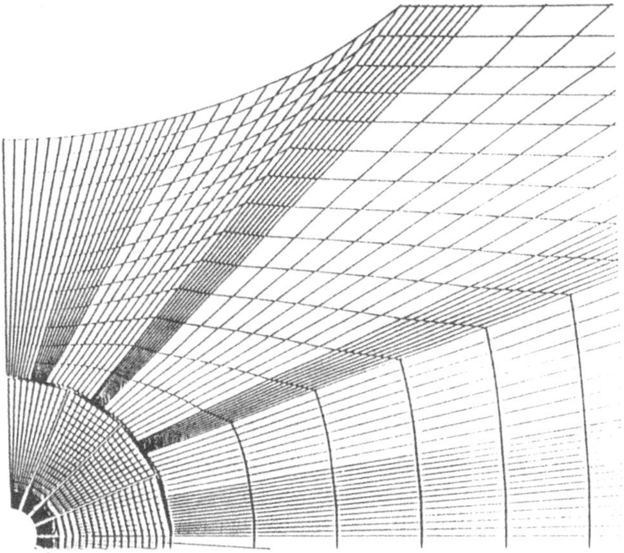

Figure 5.22. Finite element model used in quasi-static calculation of cratering. **a.** Upper right quadrant of finite element model.

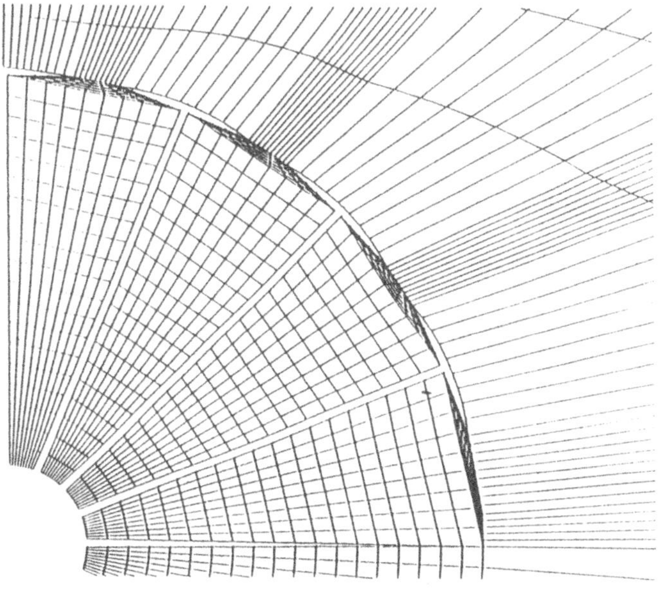

Figure 5.22b. Closeup of borehole region.

dimensional cement testing as described above. For our preconditioned fragmentation state we took the condition of the model as shown in Figure 5.17(d) at 50 μs after detonation. At this time multiple spalling has occurred and on the free surface side of the borehole both radial and circumferential cracking has occurred. On the side of the borehole away from the free surface only radial cracking has taken place. Our finite element model was therefore taken as shown in Figure 5.22(a). The figure represents the upper right hand quadrant of a two-dimensional numerical model. The borehole with radial and circumferential cracks is shown at the lower left hand corner. The curved surface at the top of the figure represents the location of the material broken by the multiple spall. The borehole and all crack surfaces (radial and circumferential) were loaded by a pressure of 110 MPa. This is a pressure that we had measured in earlier model testing at a time of 50 μs after detonation of a similar sized charge. Figure 5.22(b) shows an enlargement of the borehole area giving some details of the radial and circumferential cracks. The particular lengths of these cracks were taken to be the length that we had observed in the model testing.

Figure 5.23 shows a comparison of the normalized displacements calculated by the quasistatic approach with those measured in our three-dimensional test. Since the finite element code was run with material properties which did not exactly agree with the properties of the models, the displacements calculated by the finite element program were adjusted so that the displacement calculated for Gauge 3 was equal to the displacement measured at Gauge 3 in the test. The calculated values are the directions and relative magnitudes computed at the time of 50 μs after detonation. The measured values are for the time period between 0 and 100 μs and are somewhat of an average. Both the preconditioned fragmentation state and the finite element calculation were two-dimensional and were taken to represent a slice taken from the three-dimensional model perpendicular to the borehole. From the figure it appears that the agreement is reasonably good. At gauge locations along the free surface the measured motion is more vertical than the calculated ones but at the other three locations the calculated value represents a good "average" to the measured values. It would be interesting to see if a TOODY calculation would match the three-dimensional model results.

5.7. Conclusions

The finite difference computer codes in use today for predicting fragmentation appear to do a very good job of predicting stresses, pressures, and velocities which occur as a result of explosive detonation. They are less reliable in predicting fracture and fragmentation and are unable to predict final fragment location. With any of the codes it is cumbersome to model the very complicated geological variations found in nature. Logical extension of these codes will require the introduction of these details. For the most part the

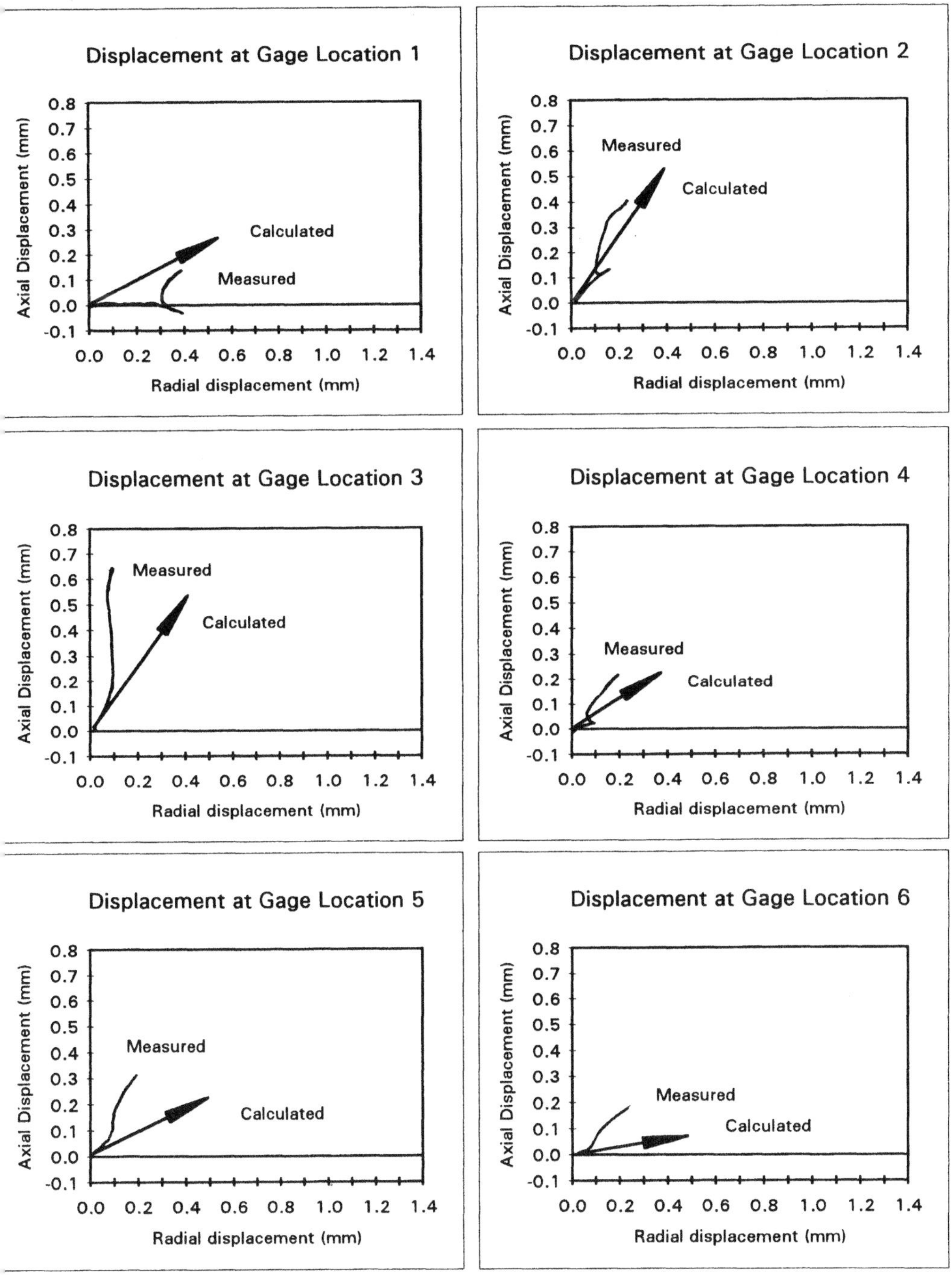

Figure 5.23. Comparison of three-dimensional measurements and results of quasi-static calculation.

codes use static properties rather than more appropriate dynamic ones which are obtained at the strain rates of interest. The codes also use a damage criteria which is used to weaken the material once a given value of stress (or strain) is reached. The problem of the codes being able to predict fracture and fragmentation accurately appears to stem from a proper definition of damage. The damaged areas predicted by the codes are, in fact, damaged but how much damage is necessary for fragmentation to appear is a point for debate. If anything, the codes appear to overpredict the amount of fragmentation. As indicated, this is due to the lack of a true knowledge of what constitutes fragmentation and also the difficulties with accurately reflecting fragmentation mechanisms in the reaction of the codes.

With regard to the laboratory testing, there are several shortcomings which might affect agreement with code results. First, the testing was accomplished using rather small two-dimensional models made of transparent homogeneous and isotropic material. In addition the testing was conducted with no overburden or confining stress being applied to the model. With these limitations in mind, comparisons of damaged areas predicted with TOODY with fragmentation that we have observed in model testing in the laboratory are in better agreement than might be expected.

In the code, the main mechanism is by shear failure whereas in the model testing, it appears that most of the severe fracturing was due to crack curvature and crack branching due to tensile components of stress contained in the passing waves. Nonetheless, the overall picture of damage predicted by the code appears to agree rather well with the observations made in the laboratory.

5.8. Acknowledgments

Some of the results presented were obtained in conjunction with Drs. Donald B. Barker, James W. Dally, and David C. Holloway from the University of Maryland. The authors would like to acknowledge their contribution. We would also like to acknowledge the input to this chapter by Fred App of the Los Alamos National Laboratory. The work described at Maryland was funded by the National Science Foundation under the direction of William Hakala and more recently by the Air Force Office of Scientific Research under the guidance of Dr. Spencer Wu. We gratefully acknowledge the financial support received from those agencies. We also acknowledge the assistance of Mr. Xiang Jun Wang (currently a PhD student at the University of Maryland) in conducting some of the more recent tests.

5.9. References

1. Cherry, J.T. Computer Calculations of Explosion-Produced Craters, *Int. J. of Rock Mechanics and Mining Sci.*, Vol. 4, pp 1–22, 1967.

2. Trent, B.C., Young, C., Barbour, T.G., and Smith, J.W. A Coupled Gas Pressurization Explicit Fracture Model for Oil Shale Fragmentation, Proceeding of 22nd U.S. Symposium on Rock Mechanics, pp 198–204, 1981.
3. App, F.N. and Brunish, W.M. Modelling of the MERLIN Event, Proceedings of the Fifth Symposium on Containment of Underground Nuclear Explosions, CONF-89091163-Vol 2., 1989.
4. Cundall, P.A. A Computer Model for Simulating Progressive, Large Scale Movements in Blocky Rock Systems, Proceedings of Symposium of International Society of Rock Mechanics, 1971.
5. Butkovich, T.R., Walton, O.R., and Heuze, F.E. Insights in Cratering Phenomenology by Discrete Element Modeling, 29th U.S. Symposium on Rock Mechanics, pp 359–368, 1988.
6. Preece, D.S. and Taylor, L.M. Spherical Element Bulking Mechanisms for Modeling Blasting Induced Rock Motion, Third International Symposium on Rock Fragmentation by Blasting, Brisbane, Australia, 1990.
7. Fourney, W.L. and Dick, R.D. Fragmentation Mechanisms in Crater Blasting, International Journal of Rock Mechanics and Mining Sciences, 1993.
8. Barker, D.B., Fourney, W.L., and Dally, J.W. Photoelastic Investigation of Fragmentation Mechanisms, Part I: Borehole Crack Network, National Science Foundation Report, March 1978.
9. Sanford, R.J. and Dally, J.W. A General Method for Determining Mixed Mode Stress Intensity Factors from Isochromatic Fringe Patterns, *Eng. Fracture Mechanics*, 11, pp 621–633, 1979.
10. Barker, D.B. and Fourney, W.L. Photoelastic Investigation of Fragmentation Mechanisms, Part II: Flaw Initiated Network, Report to National Science Foundation, July 1978.
11. Erdogan, F. and Sih, G.C. On the Crack Extension in Plates Under Plane Loading and Transverse Shear, *J. Basic Engineering*, ASME, p. 8514, December 1963.
12. Fourney, W.L., Barker, D.B., and Holloway, D.C. Mechanism of Fragmentation in a Jointed Formation, Report to National Science Foundation, July 1979.
13. Hino, K. Fragmentation of Rock through Blasting, Colorado School of Mines Quarterly Report, pp. 191–207, Vol 51, 1956.
14. App, F.N. and Brunish, W.M. Modelling Surface Motion and Spall at the Nevada Test Site, Los Alamos National Laboratory Report LAUR-92-500, January, 1992.
15. Dick, R.D., Fourney, W.L., Young, C., Wang, X.J., and Wei, Y. Mechanisms of Fracture and Fragmentation by Explosive Loading, Report to AFOSR from University of Maryland, 1992.

CHAPTER 6

Radiographic Studies of Impact Fragmentation

Andrew J. Piekutowski

Notation

c, c_0	Speed of sound in projectile
D	Diameter of projectile
d	Nominal fragment diameter (after Grady, 1982)
d_f	Equivalent diameter of largest fragment
$\tilde{d}_m$	Median Martin's diameter of spall fragments
H	Height of largest fragment
K_c, K_{Ic}	Critical stress intensity factor or fracture toughness of projectile
M_T	Total mass of material in spall shell
$M_{T\,\min}$	Total mass of material in spall shell of test with thinnest bumper in data set
N_T	Total number of fragments in spall shell
R_0	Radius of projectile
s	Nominal fragment diameter (after Grady, 1988)
T	Thickness of largest fragment
t	Bumper thickness
$t_{\min}$	Thickness of thinnest bumper in data set
V	Volume of ellipsoid
V_0	Impact velocity
V_r	Measured radial velocity of spall-shell fragments
W	Width of largest fragment
$\dot{\varepsilon}$	Strain rate, approximated as V_r/R_0
ρ	Density of projectile

6.1. Introduction

Characterization of the fragments and fragment size distributions produced by hypervelocity impact is an area of significant interest. Coincident with this interest is work to incorporate dynamic fracture and fragmentation models in numerical simulations of impact events. As interest in fragmentation has developed, many static and dynamic techniques have been employed to obtain experimental data used to further our understanding of the impact process and to evaluate the models.

Static techniques usually employ a "catcher" material which is placed in the path of the debris and serves to decelerate and capture the fragments. Foams, foam boards, ceiling tiles, and particle boards are examples of some materials that have been used singly or in combination to "catch" debris particles. Fragment size and size-distribution data were obtained when the particles are separated from the "catcher" material and analyzed to determine their sizes and numbers, usually by sieving and weighing, respectively. In other instances, witness plates made of thin sheets or thick plates of material (usually a metal) were placed in the path of the debris. Fragment number, size, and velocity data were inferred from an analysis of the holes left in the thin sheets or from the dimensions of craters produced in the thick plates. While these collection techniques have provided valuable data for fragmentation studies, they do not provide much information regarding the source of the fragments (i.e., projectile or target) unless different materials were used for the projectile and target.

The combination of one or more static techniques with a dynamic technique (e.g., high-speed photography, flash radiography, and, more recently, holography [Ang, 1995]) facilitates identification of some of the fragments. Use of dynamic techniques permits an analysis of fragments during their transport from the impact site to a "catcher" or witness plate, making identification of the source material a trivial matter. Front- and back-lit high-speed photography are the most direct ways of viewing the impact of projectiles with target plates. The light used to expose the film is usually generated by a point source timed to flash when the event occurs. Most hypervelocity impact events generate an extremely intense impact flash. Unless measures are taken to exclude unwanted light from the viewed area, the intensity of the impact flash greatly exceeds the intensity of the external light source, severely overexposing the film during the most critical part of the event. Consequently, flash radiography is frequently used when impact flash, smoke, dust, motion blur, and so forth, will not permit the use of high-speed photography.

This chapter describes the use of flash radiography as part of an analytical procedure to determine the number and size of fragments in debris clouds produced by the impact of aluminum spheres with thin aluminum plates. While the use of flash radiography to observe small aluminum fragments presents certain technical difficulties, the analysis of aluminum-fragment debris clouds is of considerable interest. Aluminum is the material most likely to form pieces of orbital debris, and most commonly used for the projectile and target in studies evaluating the design and performance of spacecraft shield systems. A thorough understanding of debris clouds facilitates the interpretation of the results of these tests and the threat to spacecraft walls posed by the impact of debris.

The following sections describe the experimental and analytical techniques used to obtain fragment size distributions, and present data from several series of tests. This chapter merely presents these data as an illustration of the usefulness of the technique. Rigorous comparisons with existing fragmentation models are not made.

6.2. Experimental Techniques

Background. Flash radiographs are really shadowgraphs. Radiation, emitted when an x-ray source is pulsed, is absorbed by an object between the source and a piece of film; consequently, a shadow of the object is cast on the film. In an ideal situation, the x-ray source is a point source, the object is located near the film, and the film is capable of a complete response to the x-ray pulse. The combination of these conditions will produce a clean, sharp shadow of the object on the film. To analyze the film, a magnification factor resulting from the placement of the x-ray source, object, and film is determined and used to adjust measurements taken from the film.

In practical situations, however, the idealizations described above suffer from the hardships of reality. The manner in which certain attributes of the source and film are treated is very important to the production of useful radiographs. In the following paragraphs, the limitations of the source and film are discussed, and procedures to minimize their effects are described. Of particular interest is the production of radiographs suitable for analysis of small to very small fragments (>0.3 mm diameter).

An x-ray source is not a point source. Rather, it has finite dimensions and emits radiation from an area. The emitting area is a function of the source type and construction and can be as small as 1 mm in diameter. Sources several millimeters in diameter are not uncommon. Because of the "finiteness" of the source, the shadow cast on the film is not sharp. A partially lit region or penumbra surrounds the shadow of the object as illustrated in Figure 6.1. The penumbra hampers accurate measurement of the dimensions of an object in the path of the radiation, particularly for small objects. Steps that can be taken to minimize the effect of the penumbra include: (1) using the smallest diameter source possible, (2) placing the object as close to the film as possible, and (3) maximizing the source-to-film distance.

The relationship of the source, object, and film is crucial to producing sharp shadows in the radiographs. Care must be taken not to place the source too far from the object and/or film. The intensity of the radiation striking the film can be reduced to unacceptably low levels by beam dispersion and absorption of the radiation by the atmosphere and other materials between the source and the film. Objects of interest are usually very energetic and must be kept some distance from the film to protect the film from damage. Reasonably sharp radiographs can be made if the ratio of the source-to-object and object-to-film distance is at least 5 : 1. Use of a larger

Figure 6.1. Penumbra and its relationship to x-ray source, object, and film.

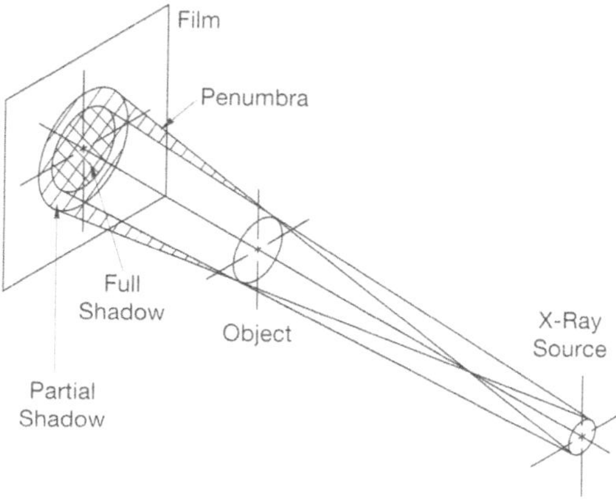

ratio is beneficial in improving detail in the radiographs, but the object must be located within the envelope of allowed source-to-film distances. These distances are best determined by making trial exposures using actual test setups and materials.

Films used for flash radiography must be "fast." Fast films tend to be grainy and not suited to showing great detail. When using "hard" x-rays, most radiation will pass through the film leaving little evidence of its passage. To enhance the response of the film, image-intensifier screens are commonly placed on one or both sides of the film. Radiation strikes the salts in these screens, causing the screens to fluoresce and produce an image on the film. The light emitted by the salts is omnidirectional and tends to diffuse the shadow of the object, further reducing detail in the radiograph. Severe diffusion and blurring of the shadow will occur if the film and screens are not kept in intimate contact during the exposure.

Problems associated with the use of image-intensifier screens can be minimized if a direct-exposure film is used (i.e., a film which responds directly to radiation and does not require screens). However, "soft" x-ray sources are usually required to properly expose this film. Soft x-rays and use of direct-exposure film have several advantages that are beneficial to the study of debris clouds and fragmentation, particularly of aluminum projectiles. Aluminum is fairly transparent to hard x-rays but somewhat opaque to soft x-rays. Thus, use of soft x-rays makes it easier to produce good radiographs of all-aluminum impact tests. Since soft x-rays are easily absorbed by test chamber windows, protective materials placed over the film cassettes, the atmosphere in the target chamber, and so forth, these materials must be of minimal thickness. Again, the best combinations of these materials can be determined by making test exposures using the actual materials and setups.

A shadowgraph and two radiographs (produced using soft x-rays) of 0.79-, 0.90-, 1.00-, 1.14-, and 1.25-mm-diameter, 2017-T4 aluminum spheres are shown at the top of Fig. 6.2 to demonstrate the effect object-to-film distance

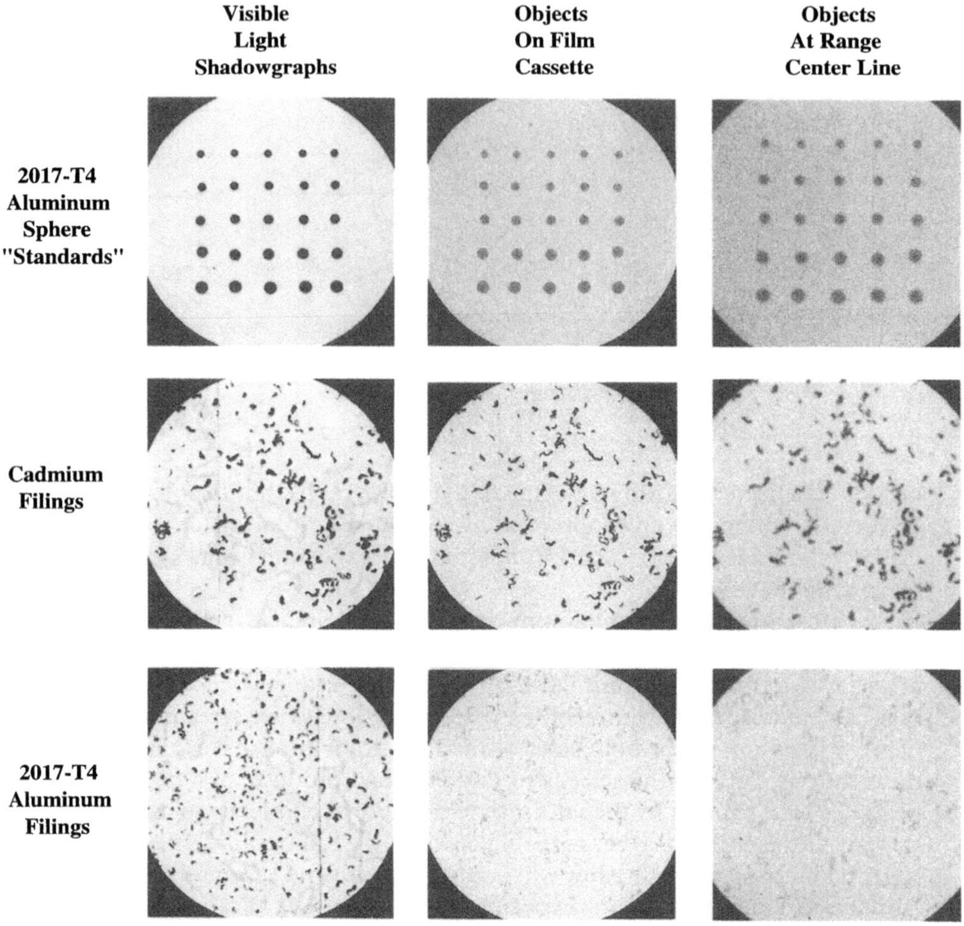

Figure 6.2. Shadowgraphs and radiographs (center and right column) of aluminum sphere "standards," cadmium filings, and aluminum filings. The radiographs demonstrate the effect the penumbra and the absortivity of the target material have on the sharpness and density of the shadows of small fragments.

and penumbra have on the sharpness and density of the shadow of the spheres. The radiographs also demonstrate that the aluminum spheres do not completely absorb the radiation striking them. The "transmissivity" of aluminum is further demonstrated in the radiographs of the cadmium and aluminum filings shown, respectively, at the middle and bottom of Fig. 6.2. The visible light shadowgraphs clearly show that the aluminum filings are larger or of different shapes than they appear in the radiographs. The density of a filing in the radiograph is a function of the presented cross section of the filing (e.g., shadows of filings that are "on edge" are much darker than those that are lying flat). Moving the test objects from the film to the range center line clearly illustrates the effect of the penumbra on the shadows of the smaller particles. Shadows of small cadmium filings at the left center of the viewed area, for example, are only faint smudges when the filings were placed at the range center line.

A final comment regarding use of flash radiography (as well as high-speed photography) has to do with motion blur. The duration of a typical flash x-ray pulse is 25 to 35 nanoseconds. An object travelling at 6 km/s, for example, will move about 0.15 mm during this interval. The motion of the object will produce a shadow that is blurred at the front and rear, with respect to the direction of travel of the fragment. Shadows of fragments travelling faster or slower than 6 km/s will exhibit more or less blur, respectively. Use of an x-ray source with a shorter pulse duration is the only way to reduce the effects of motion blur on the radiographic shadow.

6.2.1. Experimental Design

All tests described in this chapter were performed in the University of Dayton Research Institute (UDRI) Impact Physics Laboratory using a 50/20-mm, two-stage, light-gas gun. Various thicknesses of 6061-T6 aluminum sheet and 9.53-mm-diameter, 2017-T4 aluminum spheres were used as bumpers and projectiles, respectively. Two tests using 12.70-mm-diameter, 1100-O and 2017-T4 aluminum spheres and 6061-T6 aluminum sheets are also described. In all tests, the bumper sheets were positioned normal to the range center line.

Four pairs of fine-source, soft, flash x-rays were used to observe the projectile and debris clouds. The x-ray heads were accurately positioned on the target chamber to provide simultaneous orthogonal views of the debris clouds. A typical test setup is shown in Fig. 6.3. The first pair of x-ray heads was used to view and record the position of the projectile a few microseconds before impact. This view verified projectile integrity and permitted accurate determination of the time after impact for the three views taken of the debris cloud after its formation. Normally, the first two debris-cloud views were taken when the leading edge of the cloud was about 4 cm and 12 cm downrange of the bumper. The two pairs of x-ray heads used to produce these views were positioned directly above and 7.87 cm downrange of the bumper. The fourth pair of x-ray heads was fired when the debris cloud was about 30 cm downrange of the bumper. The delay in firing this pair of heads allowed the cloud to expand and permitted a more detailed examination of the cloud structure. The fourth pair of x-ray heads was positioned 15.8 or 31.5 cm downrange of the front sheet to provide oblique (approximately 13 degrees from normal) or normal views of the cloud, respectively. A quantitative analysis of the radiographs was made to determine axial and radial velocities of a number of characteristic elements of the debris clouds. In addition, the fourth (or late-time) view radiographs were used to determine

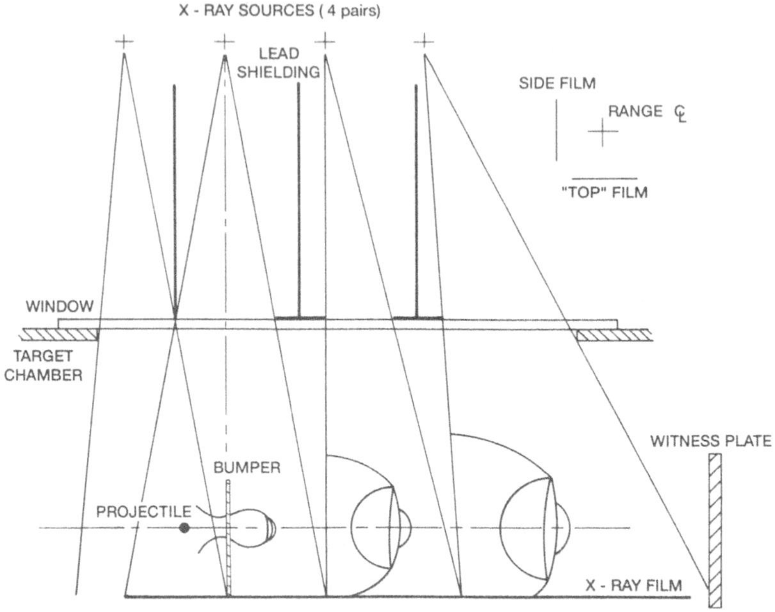

Figure 6.3. Setup used to obtain multiple-exposure, orthogonal-pair, flash radiographs of debris clouds.

the dimensions of the largest fragment in the debris cloud and for the fragment size and distribution analyses that are presented in the following sections.

Field Emission Model 2772, 180 kV dual pulsers and Scandiflash 150 kV remote tubeheads (Model XT 150) with soft x-ray adpators and 1-mm-diameter sources were used to produce the radiographs. The tubeheads were positioned so that the source-to-object (range center line) distance was 64 cm. Kodak Direct Exposure Diagnostic Film (Cat 154 2463) was placed 9.6 cm from the range center line. The source-to-object, object-to-film ratio for this combination is 6.67 : 1.

6.2.2. Debris Cloud Features

Two views of a debris cloud are presented in Fig. 6.4 to illustrate three major features of a debris cloud produced by the impact of an aluminum sphere with a thin aluminum plate. First, an ejecta veil, consisting almost entirely of bumper fragments is ejected from the impact or front side of the bumper. Second, an expanding bubble of bumper debris is formed on the rear side of the bumper. Finally, there is a significant structure of projectile debris located inside and at the front of the external bubble of bumper debris. This internal structure is composed of a front, center, and rear element. For a

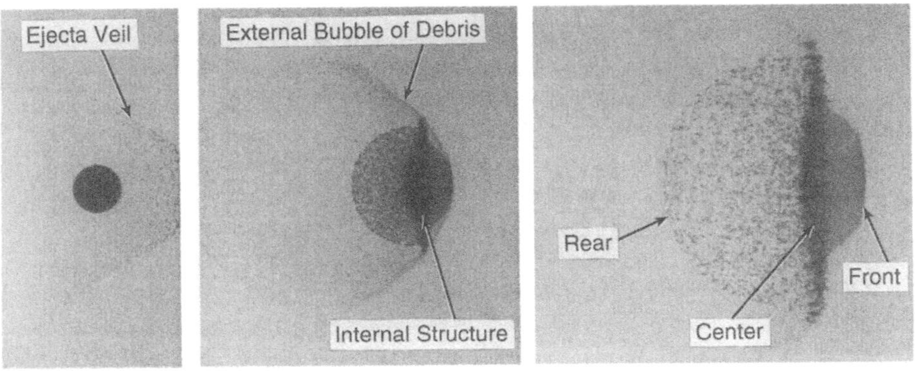

Figure 6.4. Morphological features and elements of a debris cloud. Note that the ejecta veil and projectile are a double exposure in this figure.

6.70 km/s impact, the front element consists of finely divided, molten droplets of bumper and projectile. The disc-like center element is composed of numerous slivers of projectile. This element also contains a single large chunky projectile fragment that is located at the center of the disc and on the debris-cloud center line. A central fragment was observed in all debris clouds where the bumper-thickness-to-projectile-diameter ratio (t/D) was less than 0.2. This fragment was most clearly observed in the late-time view of the debris cloud. Dimensions of the largest fragment, for the various test series, are presented later in this chapter. The rear element of the structure is a hemispherical shell of fragments spalled from the rear surface of the sphere. Results of the analysis of the shell of spall fragments are also presented in this paper. The internal structure of the debris cloud shown in Fig. 6.4 and subsequent figures, is the most significant feature of the debris cloud in terms of potential for rear wall damage.

6.3. Analysis of Film

An enlarged, late-time oblique view of the internal structure of a debris cloud is presented in Fig. 6.5. The view shown in Fig. 6.5 clearly exhibits the features of the internal structure and illustrates the constraints placed on the analysis of the radiographs. The front element of the structure consists of very fine droplets of molten aluminum which appear to be developing in streams. This front element consists of both bumper and projectile material (Piekutowski, 1993); however, this portion of the internal structure was not analyzed. The center element consists of a large number of solid slivers, comma-shaped, and/or chunky pieces of fragmented projectile and a single large fragment of the projectile. The bulk of the post-impact projectile mass is concentrated in the center element. Attempts to determine the size and

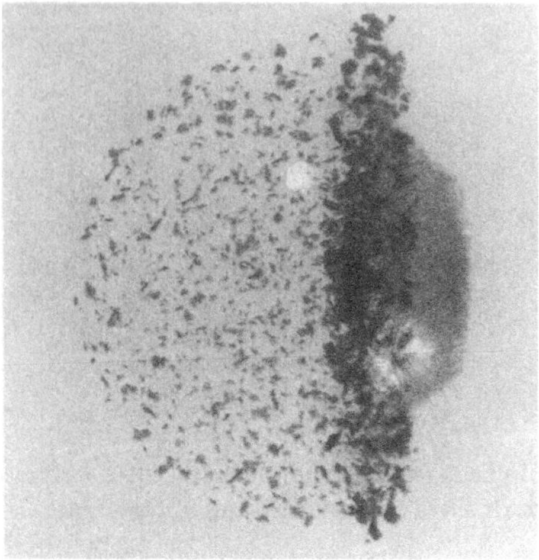

Figure 6.5. Late-time view of debris cloud (46 μs after impact). This view is used in the analysis of the spall-shell fragments and to determine the dimensions of the largest fragment. The white areas are film damage caused by the impact of ejecta from the witness plate used for this test. The projectile was a 12.70-mm-diameter, 2017-T4 aluminum sphere that impacted a 0.59-mm-thick, 6061-T6 aluminum sheet at 6.26 km/s (Shot No. 4-1358).

number of fragments in the center element proved unsuccessful because the fragments were numerous and difficult to separate in the oblique view and impossible to separate in the normal view. The largest fragment could usually be distinguished in all late-time views, and the dimensions of the largest fragment were obtained using this view. The radiograph of the debris cloud was projected on the screen of an optical comparator at a magnification of 10X. Graduated dials were used to measure the movement of a cross hair across the magnified image. Since these large fragments were irregularly shaped, the actual dimensions recorded for the fragment, in effect, were the dimensions of a box that would just enclose the fragment.

Height and one value for the thickness of the largest fragment were obtained from the side x-ray view. Width and a second value for the thickness of the largest fragment were obtained from the top x-ray view. The measured values were then adjusted using the appropriate magnification factors. Every precaution was taken to ensure the measurements were of the largest fragment and not several overlapping fragments. To further ensure that the most reliable measurements were used to determine the volume of the largest fragment, the smaller of the two thickness measurements was used in the computation. When the impact velocity or t/D ratio was low, the length-to-diameter ratio of the largest fragment was near unity. When the impact velocity or t/D ratio was increased, the length-to-diameter ratio of the largest fragment decreased and tended toward values of 0.5 to 0.6. The largest fragment height,

H, width, W, and thickness, T, were used to compute the volume, V, of an ellipsoid according to the following relationship:

$$V = \pi/6\,(\mathrm{HWT}).$$

The diameter of a sphere having the same volume as the ellipsoid was then computed. The diameter of this sphere, d_f, was termed the equivalent diameter of the largest fragment.

The expanding shell of spall fragments forming the rear element is the most noticeable portion of the debris-cloud internal structure in terms of area covered on film. Fragments in this shell tended to be flaky when the t/D ratio and impact velocity were low, but were chunky when the impact velocity was increased. The remainder of this section describes the analysis of the shell of spall fragments in the rear element.

Before proceeding to specific details of the analysis, several comments regarding analysis of the radiographs are in order. Since the shell is three-dimensional, some "overlapping" of fragment shadows is inevitable. During analysis, care was taken to distinguish those fragments that were clearly overlapped and to treat them accordingly. Problems previously illustrated—penumbra and "transmissivity" of the fragments—also contribute to the limitation of this technique. Resolution of the direct exposure film limits the smallest fragments that can be resolved, under ideal conditions, to about 0.2 mm in diameter. A more realistic resolution limit for radiographs of the tests is 0.4 mm. Finally, measurements taken from the films must be adjusted for film magnification. The film magnification factor is the ratio of the source-to-object distance and the source-to-film distance. The measured object is assumed to be in a plane which passes through the range center line normal to the center line of the x-ray beam. However, fragments in the debris cloud are clearly above and below this plane. Appropriate magnification corrections for fragments that are not in the plane through the range center line are difficult to determine. The analysis that follows treats all fragments as though they are in the plane of the range center line. After magnification corrections are applied, fragments above the plane (nearer the source) will appear larger than they actually are. Likewise, fragments below the plane will appear smaller. This distortion of the fragment size distribution cannot be avoided except for those rare instances where some distinguishing feature of a fragment allows its position to be determined in the side and top view.

In principle, analysis of the spall fragments in the radiographs is straightforward—merely count and measure the appropriate dimensions of all fragments in the shell. This kind of an analysis appears to be ideally suited to an automated analysis procedure. The companion-view radiograph of the debris cloud shown in Fig. 6.5 was analyzed using IMAGE ANALYSIS by OLYMPUS CUE-2. In addition to the problems with the general analysis of radiographs, noted in the preceding paragraphs, several additional problems unique to electronic image analysis appeared. After the radiograph was scanned, the contrast of the image was adjusted before the variable-tone image was converted to a black-and-white image. Adjustment of the contrast level caused the apparent size of the fragments to change. Selection of a contrast level, prior to conversion of the variable-tone image, was arbitrary at best. In addition, background noise and graininess of the film produced an extremely large number of single- and double-pixel phantom fragments which had to be eliminated before the actual analysis could be made. Finally, the electronic technique joined or combined fragments that clearly overlapped. Considerable effort was expended to analyze just one radiograph; therefore, an alternate method of analysis was developed.

The alternate method of analysis is manual and uses enlarged prints (~2.5X) of the radiographs. A full analysis (i.e., examining every fragment) would be tedious and was judged not necessary. It was assumed that spall-shell fragments could be characterized using some statistical description of fragment sizes. Accordingly, the procedure illustrated in Fig. 6.6 was developed and used to perform the analysis. The forward boundary of the shell area was determined by drawing a line as close as possible to the rear of the center element. The forward boundary of the sampled area was drawn parallel to the forward shell boundary and through the center of a circle that just enclosed the shell of spall fragments. The sampled area was further divided into circular fields that enclosed approximately 20% of the sampled area. Fragment analysis was limited to those fragments in the circular fields. By the arbitrary convention shown in the upper left of Fig. 6.6, fragments that intersected the lower right boundary of the field were not analyzed. Fragments that intersected the upper left boundary of the field were analyzed. Characteristics of the distribution of fragment sizes in the fields were extended to all fragments in the shell area. The total number of fragments in the shell area, N_T, was estimated by multiplying the number of fragments in the fields by the shell-area-to-total-field-area ratio.

In addition to counting all appropriate fragments in the fields, and for reasons forthcoming, the Martin's statistical diameter of each fragment was measured using a 6X pocket optical comparator. Herdan (1953) defines Martin's statistical diameter as follows: "Martin's 'statistical' diameter is defined as the mean length of a line intercepted by the profile boundary which approximately bisects the area of the profile. The bisecting line is taken parallel to a fixed direction, irrespective of the orientation of each particle. This has the effect of avoiding bias as to the direction in which the profiles coming up for inspection are bisected." A second statistical diameter, Feret's diameter, is shown at the lower left of Fig. 6.6. In discussing the relative precision of various methods of determining the projected diameter of irregularly shaped particles, Herdan states: "Taking the determination of the profile area by planimeter and the corresponding mean projected diameter as a standard, the approximation by using transparent circle comparison had the smallest error, next came Martin's statistical diameter, and then the method using opaque circle comparison. Feret's statistical diameter is not recommended on account of its large errors." He further states that experimental investigations have confirmed that "Martin's diameter is, on the

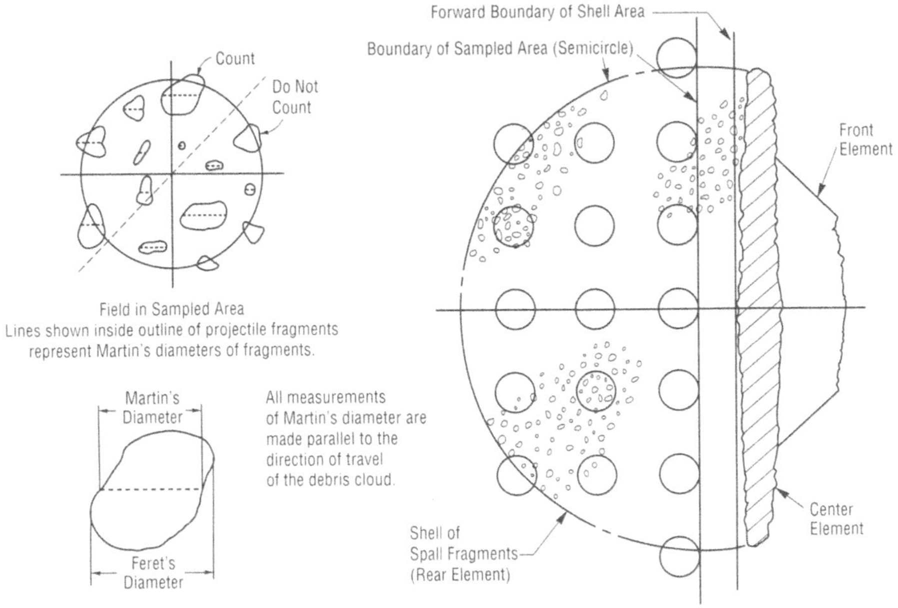

Figure 6.6. Illustration of the sampling procedure used to estimate the number and size of particles in the shell of spall fragments.

whole, less than the mean projected diameter, and Feret's is on the whole, greater."

The radiographs analyzed for use in this chapter were part of several series of test data used to examine debris-cloud formation for t/D ratios ranging from 0.026 to 0.424 (Piekutowski, 1993). Radiographs from tests with lower t/D ratios (thinner bumpers) were more amenable to the analytical procedures just described because the external bubble was much thinner and contributed relatively few fragments to those in the shell area. In addition, the spall fragments for the tests with lower t/D ratios were usually larger and easier to analyze. Numerous, small spall-shell fragments appear in the radiographs of the tests with larger t/D ratios, but attempts to count and measure them were unsuccessful. Data from the tests with the lower t/D ratios were sufficient to establish patterns and trends which could reasonably be extended to tests using higher t/D ratios.

6.4. Results

Radiographs of debris clouds showing the effect of t/D ratio and impact velocity, V_0, on the size, number, and distribution of fragments in the shell of spall fragments are shown in Figs. 6.7 and 6.8. In these figures, the size of the spall-shell fragments and the largest fragment decrease with increasing

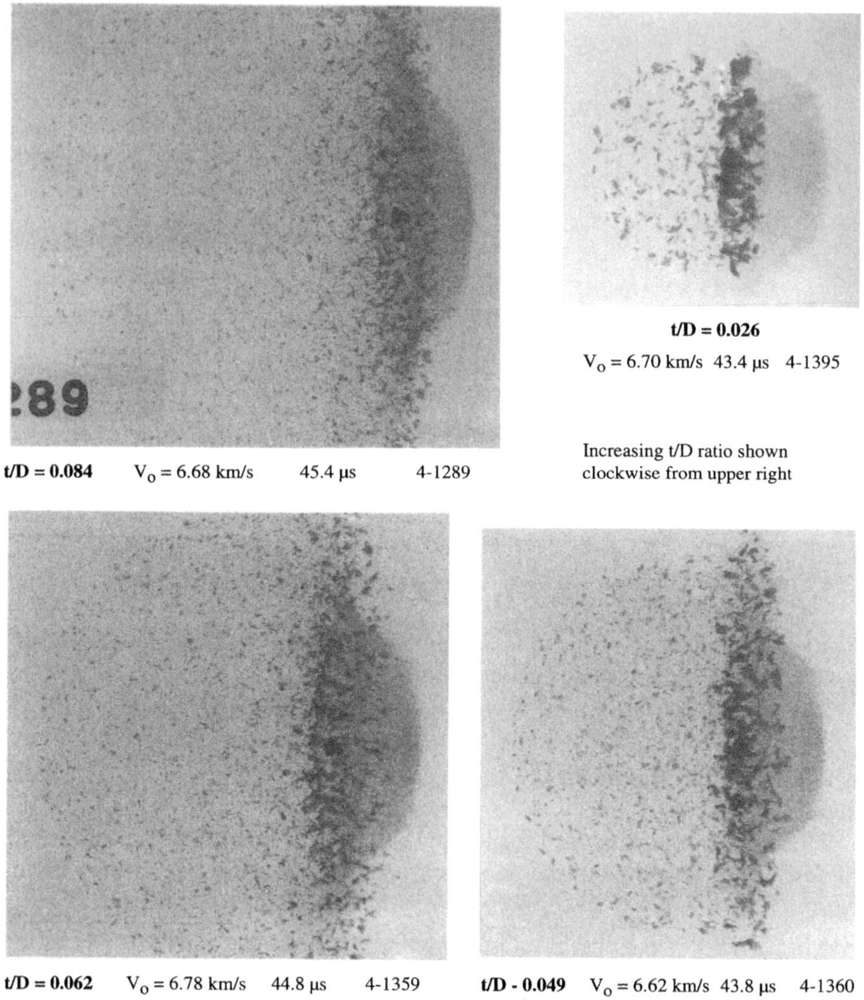

Figure 6.7. Late-time views of debris clouds showing the effect of the t/D ratio on size, number, and distribution of fragments in the spall shell. The projectiles were 9.53-mm-diameter, 2017-T4 aluminum spheres and bumpers were 6061-T6 aluminum sheets. All views are at the same magnification. Time after impact for each view is shown below the radiograph.

t/D ratio or impact velocity. Although not clearly visible in the photographic reproduction of the debris "cloud" from Shot 4-1428 (Fig. 6.8, upper left), the radiograph from this test shows the development of a spalled region inside the projectile.

Radiographs from two tests employing 12.70-mm-diameter aluminum projectiles are shown in Fig. 6.9. Different projectile materials were used for

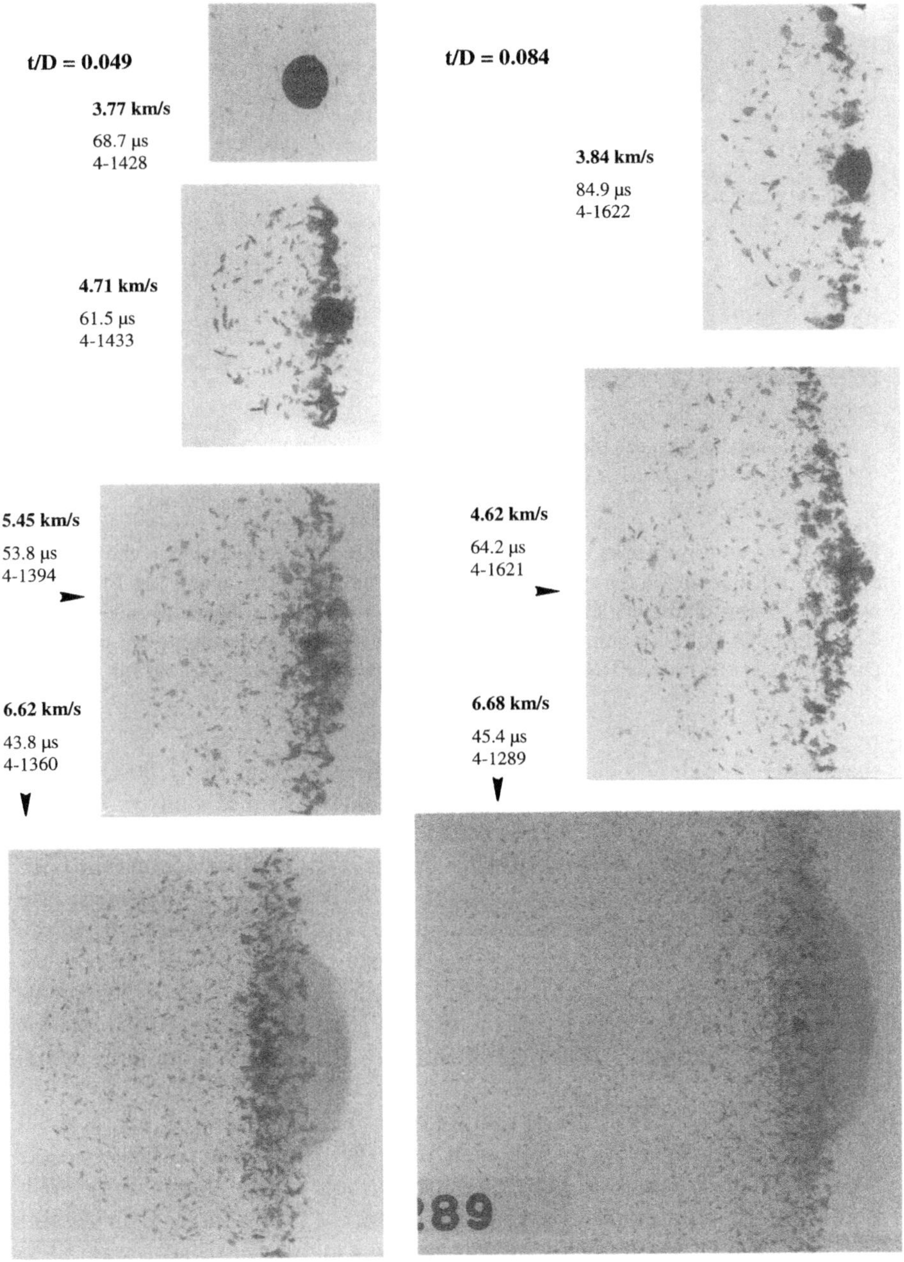

Figure 6.8. Late-time views of debris clouds showing the effect of impact velocity on size, number, and distribution of the fragments in the spall shell for two t/D ratios. The projectiles were 9.53-mm-diameter, 2017-T4 aluminum spheres and bumpers were 6061-T6 aluminum sheets. All views are at the same magnification. Time after impact for each view is shown near each radiograph.

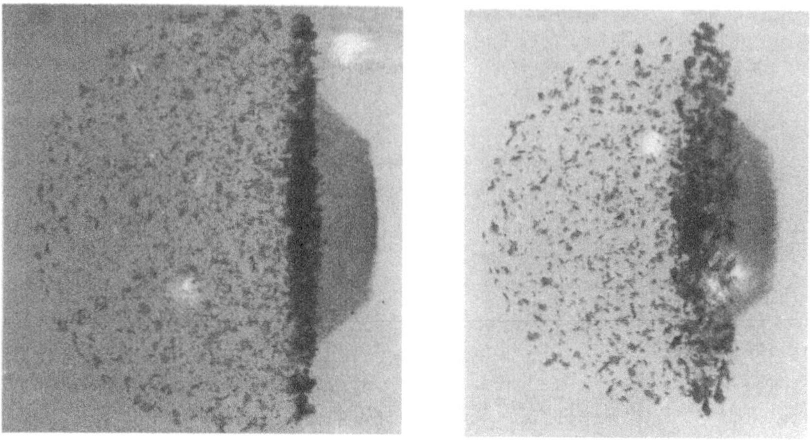

1100-O Aluminum Sphere
6.37 km/s 46.5 μs 4-1601
(Normal View)

2017-T4 Aluminum Sphere
6.26 km/s 46.0 μs 4-1358
(Oblique View)

Figure 6.9. Late-time views of debris clouds showing the differences in the number and size of fragments in the spall shell for projectiles of different materials. Projectiles were 12.70-mm-diameter, 1100-O and 2017-T4 aluminum spheres. Bumpers were 0.59-mm-thick, 6061-T6 aluminum sheets. Views are at the same magnification; other parameters are as noted. Time after impact is shown below each radiograph.

these tests. The debris cloud produced by the impact of the 1100-O aluminum sphere (Shot 4-1601) shows that the cloud consists of many more fragments than are evident in the cloud from the test with a 2017-T4 aluminum sphere (Shot 4-1358). While the front element is essentially the same for both debris clouds, the expansion and growth of the center and rear elements are significantly different. The concentrated mass of projectile fragments in the center element is expanding radially 4% faster for the 2017-T4 aluminum cloud than for the 1100-O aluminum cloud. On the other hand, the normalized radial expansion velocity of the shell of spall fragments is 14% less for the 2017-T4 aluminum debris cloud than for the 1100-O aluminum debris cloud.

Results of the analysis of the radiographs shown in Figs. 6.7, 6.8, and 6.9 are presented in Table 6.1. For each test, the distribution of the particle Martin's diameters was evaluated for log normality using the Shapiro–Wilk goodness-of-fit test at a 0.05 level of significance. The median Martin's diameter, $\tilde{d}_m$, for each test was estimated assuming a log-normal distribution fit to the data (i.e., $\tilde{d}_m = \exp[\overline{\ln d_m}]$). The test for log normality of the distributions was prompted by the following statements by Herdan (1953): "Materials showing a normal distribution of particle size are relatively rare, and are found chiefly among the particulate substances produced by chemical processes like condensation, precipitation." and "If the dispersion is attained by

Table 6.1. Fragment size data
All bumper sheets were 6061-T6 aluminum. All projectiles were 2017-T4 aluminum except for Shot 4-1601. Shot 4-1601 used an 1100-O aluminum projectile.

Shot Number	Impact Velocity, V_0 (km/s)	t/D	Frequency Distribution of Fragments: Mean $\ln d_m$	of Standard Deviation	Log Normality of Distribution	Median Martin's Diameter, $\tilde{d}_m$ (mm)	Number of Fragments, N_T	Diameter of Lgst. Frag., d_f (mm)
			t/D Effects, V_0 = Constant (6.70 km/s), $D = 9.53$ mm					
4-1395	6.70	0.026	−0.206	0.463	Fail to Reject	0.814	240	5.49
4-1360	6.62	0.049	−0.455	0.369	Reject	0.636	897	2.95[a]
4-1359	6.78	0.062	−0.511	0.331	Reject	0.600	1430	2.97
4-1289[b]	6.68	0.084	−0.680	0.279	Fail to Reject	0.506	1995	2.03
4-1283	6.72	0.102	—	—	—	—	3600[c]	1.45
			V_0 Effects, $t/D = 0.049$, $D = 9.53$ mm					
4-1428	3.77	0.049	—	—	—	—	1	~9.53[d]
4-1433	4.71	0.049	−0.214	0.537	Fail to Reject	0.807	178	6.36
4-1394	5.45	0.049	−0.307	0.412	Fail to Reject	0.736	326	4.62
4-1360	6.62	0.049	−0.455	0.369	Reject	0.636	897	2.95[a]
			V_0 Effects, $t/D = 0.084$, $D = 9.53$ mm					
4-1632	3.47	0.084	—	—	—	—	10–20[e]	7.67
4-1631	3.64	0.084	—	—	—	—	40–60[e]	7.51
4-1622	3.84	0.084	−0.160	0.525	Fail to Reject	0.852	136	7.79
4-1621	4.62	0.084	−0.436	0.409	Reject	0.646	414	4.80
4-1289[b]	6.68	0.084	−0.680	0.279	Fail to Reject	0.506	1995	2.03
			Projectile Material Effects, $D = 12.7$ mm					
4-1358	6.26	0.047	−0.316	0.492	Fail to Reject	0.729	764	5.49
4-1601[f]	6.37	0.047	−0.567	0.468	Reject	0.567	1376	3.02

[a] Dimensions available from only one view.
[b] Shadows of fragments faint on film.
[c] Fragments not distinct enough for measurement or accurate count. Value is best estimate.
[d] Projectile did not break up.
[e] Difficult to distinguish between center and rear elements of debris cloud.
[f] Projectile was 1100-O aluminum.

comminution (milling, grinding, crushing), the distribution appears to be governed very often by the log normal law."

6.5. Discussion

Before the test results are presented, several comments regarding the accuracy and/or appropriateness of the analytical procedures used to produce the data presented in Table 6.1 are in order. A comparison of the adjusted diameters of the aluminum sphere "standards" (shown in Fig. 6.2), to their true diameters indicated the overall error of the measurement procedure was about ± 0.05 mm, or the precision of the measurements themselves. It is assumed that all fragment measurements were made with about the same accuracy. As mentioned earlier, the companion view to the radiograph shown in Fig. 6.5 was evaluated using electronic image analysis procedures. When all "corrections" for overlapping fragments were made, the results of the electronic analysis indicated that the shell sample area contained 683 fragments with a mean Feret's diameter of 0.75 mm. The Feret's diameter of each fragment yielded by the electronic analysis was the average of 18 Feret's diameters measured at ten-degree increments for that fragment. The manual procedure described earlier indicated there were 672 fragments with a mean Martin's diameter of 0.73 mm. The "agreement" between the results of the two methods was remarkable.

Finally, the effect of motion blur on the mean Martin's diameter was evaluated for Shot 4-1621 by comparing measurements made as shown in Fig. 6.6 to measurements made perpendicular to the direction shown in Fig. 6.6. The same fragments were measured in both cases. Median Martin's diameters of the fragments in the sampled area, measured parallel and perpendicular to the direction of travel of the debris cloud, were 0.646 mm and 0.619 mm, respectively. Because of a desire to maintain a sense of "length" of the particle and its ability to penetrate a rear wall, all fragment measurements were made as shown in Fig. 6.6. The reader is left to decide whether adjustments to the reported diameters, to compensate for motion blur, are in order.

The development of the shell of spall fragments is shown in the late-time-view radiographs presented in Fig. 6.10. Note that both the t/D ratio and impact velocity are varied for the tests shown in this figure. Although not evident in the photographic reproduction of the radiograph for Shot 4-1428, the post impact views for this test show a narrow region of reduced density just inside the rear surface of the sphere. Apparently, a spall failure developed inside the projectile and formed a thin shell that is loosely attached to the deformed sphere.

In Shot 4-1633, the spall shell has ruptured and formed petals that remain attached to the deformed sphere. (Data presented for this test are limited, in this chapter, to the information shown below the radiograph.) Fragmenta-

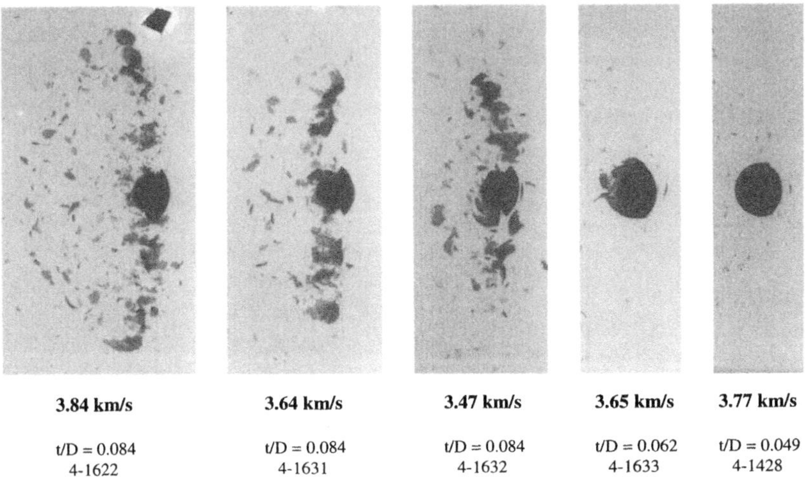

Figure 6.10. Radiographs showing the development of the shell of spall fragments during the transition phase as a function of impact velocity and t/D ratio.

tion of the spall shell is underway but not well developed for Shot 4-1632. A center element is being formed but appears to consist of spall fragments from the rear of the sphere. The spall shell (rear element) is too poorly developed to analyze. Further development of the spall shell has occurred in Shot 4-1632, but again the shell was not sufficiently developed to permit a reliable analysis. The spall shell shown for Shot 4-1622 is fully developed and was analyzed. Spall fragments that separate from the mushroom head appear to do so with a sharp, clean break. See, for example, the spall fragments (petals) shown just below the very large central fragments in Shots 4-1631 and 4-1632.

In all five radiographs shown in Fig. 6.10, formation of a front element has not begun. All projectiles exhibit mushrooming and plastic deformation of the front of the sphere. The impact of the sphere and bumper appears to be elastic for the three tests shown at the right in this figure. A small piece of the bumper sheet is shown, in these views, to rebound from the front of the sphere and move forward along the shot line relative to the deformed sphere. Bumper material from the periphery of the central bumper fragment is also shown to rebound from the sphere, but at various angles to the shot line. In the two tests shown at the left in Fig. 6.10, the plug remains attached to the front of the sphere.

Largest fragment diameter, d_f, and the median Martin's diameter, $\tilde{d}_m$, of the spall fragments are shown in Fig. 6.11 as a function of impact velocity. Data for two series of tests in which the t/D ratio was kept constant are shown with lines in the figure. Fragment diameters for other t/D ratios at an impact velocity of 6.7 km/s are also shown. Fragment size data for particles

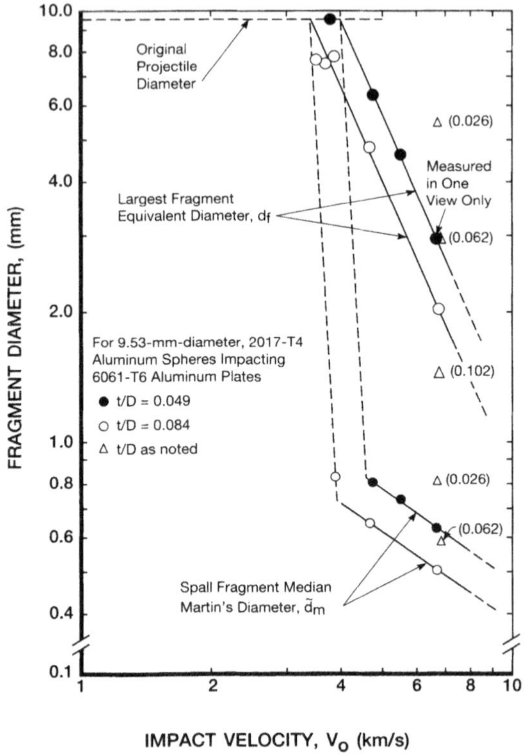

Figure 6.11. Diameter of the largest fragment and median Martin's diameter of spall fragments versus impact velocity for two t/D ratios.

in the center element, if they were available, would lie in the region bounded by the largest-fragment-diameter and the spall-fragment-diameter curves. Clearly, fragment diameter is a function of impact velocity and t/D ratio when all other parameters are held constant.

As shown in Fig. 6.11, a threshold impact velocity must be reached for each t/D ratio before the sphere begins to fragment. Plastic deformation of the front surface of the sphere and development of a spall failure beneath the rear surface of the sphere precedes fragmentation of the projectile (see Shot 4-1428 in Fig. 6.10). The decrease in the equivalent diameter, d_f, of the largest fragment has a power-law dependence on the impact velocity. Changing the t/D ratio does not alter the power-law dependence but does change the size of the largest fragment. Equations of the lines shown in Fig. 6.11 for the largest fragment are:

$$d_f = 204.8\ V_0^{-2.24} \text{ for } t/D = 0.049, \text{ and}$$

$$d_f = 147.1\ V_0^{-2.24} \text{ for } t/D = 0.084.$$

In contrast to the largest-fragment curves, which show a well-behaved departure from the original projectile diameter, the median Martin's diameter of the spall fragments appears to experience a precipitous drop followed by a gradual decrease with increasing impact velocity. The severe transition in spall fragment diameter occurs over a rather narrow range of impact velocities. Insufficient data are available to accurately define the transition phase, but the relationship shown by the pair of dashed lines in Fig. 6.11 appears reasonable.

The lower or threshold velocity for the onset of fragmentation was defined by the intersection of the largest fragment line and the line labeled "Original Projectile Diameter." The upper velocity limit of the transition phase is not as well defined, however. For $t/D = 0.084$, the limit is near the data point at 3.84 km/s. The transition phase seems to occur within a narrow range of impact velocities for each t/D ratio and is complete when the fragmentation threshold velocity is exceeded by 400 to 500 m/s. The decrease in spall-fragment, median-Martin's diameter also showed a power-law dependence on impact velocity. Changing the t/D ratio did not alter the power-law dependence but did change the median Martin's diameter of the fragments. Equations of the lines shown in Fig. 6.11 for the spall-fragment, median-Martin's diameter are:

$$\tilde{d}_m = 2.30\ V_0^{-0.68} \text{ for } t/D = 0.049 \text{ and}$$

$$\tilde{d}_m = 1.85\ V_0^{-0.68} \text{ for } t/D = 0.084.$$

The median Martin's diameters of the spall fragments were compared with diameters computed using the following relationships suggested by Grady (1992):

$$d = (20^{1/2} K_{Ic}/\rho c\dot{\varepsilon})^{2/3} \text{ (from Grady, 1982, Eq. 12) and}$$

$$s = 2(3^{1/2} K_c/\rho c_0 \dot{\varepsilon})^{2/3} \text{ (from Grady, 1988, Eq. 14).}$$

In these relationships, d and s are the nominal fragment diameters, K_c and K_{Ic} are the critical stress intensity factor or fracture toughness ($\sim 3 \times 10^7$ Nm$^{-3/2}$ for aluminum), ρ is the density of the projectile (2.791 g/cm^3), and c and c_0 are the bulk sound speed in the projectile (6420 m/s). The strain rate, $\dot{\varepsilon}$, was approximated by V_r/R_0 where V_r was the radial expansion velocity of the spall fragments measured from the radiographs of the tests and R_0 was the radius of the sphere. Strain rates determined for the tests ranged from 0.9×10^5 s^{-1} for a t/D ratio of 0.026 to 2.6×10^5 s^{-1} for a t/D ratio of 0.084. The nominal fragment diameters computed using these relationships were approximately two times larger than the median Martin's diameters determined from the analysis of the radiographs. The computed diameters may, in fact, be very close to the average diameters of the fragments in the center element of the internal structure.

Largest fragment diameters, normalized by dividing by the projectile diameter, were shown to scale geometrically (Piekutowski, 1992) for projectiles ranging from 6.35 mm to 12.7 mm in diameter, and t/D ratios ranging from 0.026 to 0.163. Data for comparisons at larger t/D ratios were not available. The equivalent diameter of the largest fragment presented in Table 6.1 and Fig. 6.11 for Shot 4-1360 was determined using measurements taken from only one view of the debris cloud. The third dimension required to make the calculations was generated by averaging the two measurements that were available. A comparison of the only data available for scaling of spall fragment data, (with allowance for the difference in impact velocity) Shots 4-1360 and 4-1358, indicates that spall-fragment diameters do not scale geometrically. Formation of spall fragments is a shock-related process, and is sensitive to rate effects and other material properties that do not scale. The largest fragment, on the other hand, appears to originate from near the center of the sphere and is a part of the sphere that remains intact after all processes that work to reduce the size of the sphere are complete; thus it is not a direct result of-shock related processes.

The number of spall fragments, N_T, is shown as a function of impact

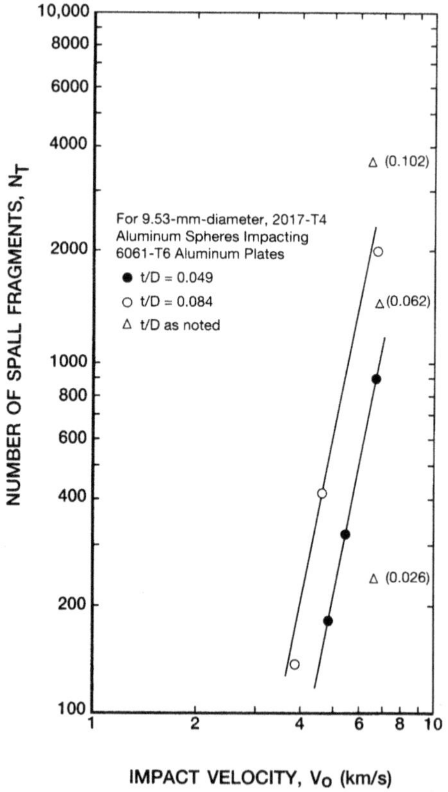

Figure 6.12. Number of spall fragments versus impact velocity for two t/D ratios.

velocity and spall-fragment, median Martin's diameter in Figs. 6.12 and 6.13, respectively. The same comments that were made regarding data presented in Fig. 6.11 apply to the data shown in these plots (i.e., the dependent variable exhibits a power-law dependence on the independent variable). Changes in t/D ratio merely change the number of fragments when impact velocity and/or the median Martin's diameter are held constant. Equations for the lines shown in Fig. 6.12 are:

$$N_T = 0.082\ V_0^{4.9} \text{ for } t/D = 0.049, \text{ and}$$

$$N_T = 0.225\ V_0^{4.9} \text{ for } t/D = 0.084.$$

Equations for the lines shown in Fig. 6.13 are:

$$N_T = 44.6\ (\tilde{d}_m)^{-6.6} \text{ for } t/D = 0.049 \text{ and}$$

$$N_T = 23.2\ (\tilde{d}_m)^{-6.6} \text{ for } t/D = 0.084$$

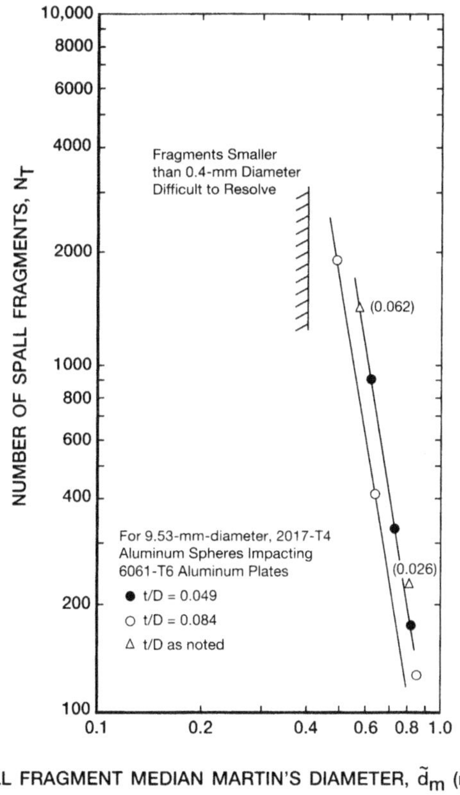

Figure 6.13. Number of spall fragments versus median Martin's diameter of spall fragments.

Table 6.2. Computed mass of spall fragments
Assumes median Martin's diameter is the diameter of a spherical fragment and the density of the fragment is the same as the unshocked projectile. $\rho_{2017-T4} = 2.791$ g/cm^3 $\rho_{1100-O} = 2.712$ g/cm^3

Shot Number	Impact Velocity, V_0 (km/s)	t/D	Median Martin's Diameter, $\tilde{d}_m$ (mm)	Mass of Fragment (mg)	Number of Fragments, N_T	Total Mass in Shell, M_T (g)
		t/D Effects, V_0 = Constant (6.70 km/s), $D = 9.53$ mm				
4-1395	6.70	0.026	0.814	0.788	240	0.189
4-1360	6.62	0.049	0.636	0.376	897	0.337
4-1359	6.78	0.062	0.600	0.316	1430	0.451
4-1289[a]	6.68	0.084	0.506	0.189	1995	0.378
4-1283	6.72	0.102	—	—	3600[b]	—
		V_0 Effects, $t/D = 0.049$, $D = 9.53$ mm				
4-1428	3.77	0.049	9.53[c]	1275[c]	1	1.275[c]
4-1433	4.71	0.049	0.807	0.768	178	0.137
4-1394	5.45	0.049	0.736	0.583	326	0.190
4-1360	6.62	0.049	0.636	0.376	897	0.337
		V_0 Effects, $t/D = 0.084$, $D = 9.53$ mm				
4-1632	3.47	0.084	—	—	10–22[d]	—
4-1631	3.64	0.084	—	—	40–60[d]	—
4-1622	3.84	0.084	0.852	0.904	136	0.123
4-1621	4.62	0.084	0.646	0.394	414	0.163
4-1289[a]	6.68	0.084	0.506	0.189	1995	0.378
		Projectile Material Effects, $D = 12.70$ mm				
4-1358	6.26	0.047	0.729	0.566	764	0.432
4-1601[e]	6.37	0.047	0.567	0.259	1376	0.356

[a] Shadows of fragments are faint on film.
[b] Fragments not distinct enough for measurement or accurate count. Value is best estimate.
[c] Projectile did not break up.
[d] Difficulty distinguishing between center and rear elements of debris cloud.
[e] Projectile was 1100-O aluminum.

The behavior of the data presented in Fig. 6.13 indicates that the mass of material removed from the sphere to form the spall shell may change as the t/D ratio and impact velocity are varied. The mass of material in the shell was estimated, for each test, by determining the mass of the median fragment in the spall shell and multiplying that mass by the number of fragments in the shell. The mass of the median fragment was determined by assuming the median Martin's diameter was the diameter of a spherical fragment and that the density of the fragment was the density of the unshocked projectile. Results of these computations are presented in Table 6.2.

The total mass of material in the spall shell, M_T, for tests at 6.7 km/s was normalized by dividing M_T by the total mass of material in the spall shell produced by the test with the thinnest bumper, $M_{T\,\min}$. This normalized mass was plotted in Fig. 6.14 as a function of normalized bumper thickness, $t/t_{\min}$,

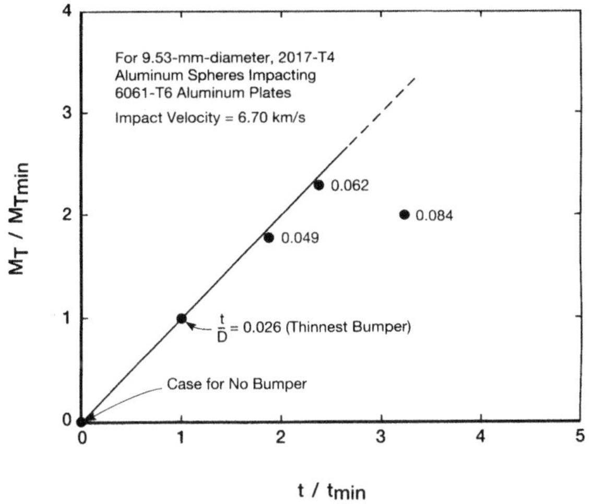

Figure 6.14. Mass of material in the spall shell versus thickness of the bumper sheet. The values are normalized by dividing by the mass of material in shell for test with thinnest bumper and the thickness of thinnest bumper, respectively.

where t_{min} is the thickness of the thinnest bumper used in the test series at 6.7 km/s. Fig. 6.14 indicates that the mass of material in the spall shell is proportional to bumper thickness, at least for t/D ratios up to 0.062. Difficulties with the resolution of small particles in the radiographs of the tests with the larger t/D ratios and impact velocities may affect the number of fragments counted and the frequency distribution of fragment sizes. Fragments smaller than those which can be resolved are likely to be present for these tests but are not counted or measured, simply because they cannot be seen. This resolution difficulty may account for the observed behavior of the results from the $t/D = 0.084$ test. A second explanation for this behavior is that some other factor or process may limit the mass of material that is contained in the spall shell.

Material properties of the projectile significantly affect the fragmentation process as evidenced in the radiographs presented in Fig. 6.9. Notable differences in the internal structure of the debris cloud were observed when the projectile material was changed from 2017-T4 aluminum to 1100-O aluminum. The diameter of the largest fragment produced by the 1100-O aluminum test was only about half the diameter of the largest fragment produced by the 2017-T4 aluminum test. Differences in the radial velocities of the center and rear elements of the debris cloud were noted when the figure was presented. The spall-fragment, median Martin's diameters for the two tests were significantly different, with the 2017-T4 aluminum producing the larger fragments—0.729-mm diameter versus 0.567-mm diameter. However, the number of spall shell fragments was considerably smaller (764 versus 1376) for the 2017-T4 aluminum test. The computed total mass of material in the spall shell was 0.432 g for the 2017-T4 aluminum test and 0.356 g for the 1100-O aluminum tests. Surmounting the difficulties with the resolution of smaller fragments in the 1100-O aluminum tests could reduce the reported median fragment diameter and increase the number of fragments counted, thereby changing the computed mass of material in the spall shell.

6.6. Summary

Flash radiography is shown to be a useful tool in the collection of debris-cloud fragmentation data. Dimensions of the largest fragment in the debris cloud and the size and number of fragments in a shell of spalled projectile fragments were obtained. Fragments as small as 0.4-mm diameter were resolved. Failure of the projectile begins with the development of a spall layer at the rear surface of the sphere. As impact velocity increases, the spall layer petals open and fragmentation of the spall layer begins. At the end of a short transition phase, during which development of the shell of spall fragments occurs, a power-law-dependent decrease in the diameter of the largest fragment and the diameter of the spall fragments is observed as impact velocity is increased. Changing the t/D ratio does not affect the power-law dependence but does alter the size of the fragments. The number of fragments that are produced increases as impact velocity and/or t/D ratio are increased. Finally, use of the fragment data to estimate the mass of material in the spall shell indicates that increasing impact velocity and/or t/D ratio increases the mass of the shell.

6.7. Acknowledgments

Three tests used in this analysis were performed under Prime Contract NAS 8-38856 on Subcontract A71447 with Martin Marietta Manned Space Systems (MMMSS). The author gratefully acknowledges Dr. Joel Williamsen of NASA Marshall Space Flight Center and Dr. Norman Elfer (MMMSS) for the use of the radiographs from these tests. The author also extends his appreciation to Dr. Robert Schmidt of Boeing Defense & Space Group for providing the thinner 6061-T6 aluminum bumper-sheet materials used in the tests.

The author is also grateful to the following associates at the University of Dayton Research Institute: The Office of the Director, for support provided for the tests and preparation of the manuscript; to Dr. Alan Berens, for performing the statistical analysis of the fragment-size frequency distributions; and to Mr. Donald Jurick, for performing the electronic image analysis.

6.8. References

Ang, J.A. Holographic Studies of Impact Fragmentation. *This volume.*

Grady, D.E. Local Inertial Effects in Dynamic Fragmentation, *J. Appl. Phys.*, **53**, pp. 322–325 (1982).

Grady, D.E. The Spall Strength of Condensed Matter. *J. Mech. Phys. Solids.*, **36(3)**, pp. 353–384 (1988).

Grady, D.E. Personal Communication (1992).

Herdan, G. *Small Particle Statistics.* Houston, TX., Elsevier Publishing Co, 1953.

Piekutowski, A.J. Properties of Largest Fragment Produced by Hypervelocity Impact of Aluminum Sphere with Thin Aluminum Sheets. (Paper No 92-1588) *Proc. AIAA Space Programs and Technologies Conference.* Huntsville, AL, 1992.

Piekutowski, A.J. Characteristics of Debris Clouds Produced by Hypervelocity Impact of Aluminum Spheres with Thin Aluminum Plates. *Int. J. Impact Engng.* **14**, 573–386 (1993).

CHAPTER 7

Pulsed Holography Diagnostics of Impact Fragmentation

J.A. Ang and B.D. Hansche

7.1. Introduction

Pulsed holography is a significant advance in state-of-the-art hypervelocity impact diagnostics. The topic of this work is the characterization of dynamic fragmentation in hypervelocity impact events. Impact events can produce material stress and strain rates sufficient to produce dynamic fragmentation in a relatively clean test environment. While other techniques for studying dynamic fragmentation such as explosives tests may also be suitable for holographic diagnostics, extensive development would be required to protect the holographic plates from fogging and physical damage in such test environments.

Unlike conventional holography, which uses exposure times on the order of seconds and thus requires vibration-isolation equipment, pulsed holography utilizes high-power pulsed lasers to provide exposure times on the order of nanoseconds. These short pulse durations are required to "freeze" the motion of the (typically) high velocity fragments created in dynamic fragmentation events. Aside from these very short exposure times, pulsed holography is similar to conventional holography. This section describes the background for the development of pulsed holography, and following sections discuss how pulsed holographic images of hypervelocity impact events are formed and captured. The critical issue for the practical application of pulsed holography is the development of holographic image analysis techniques. This issue is addressed in the last sections on image reconstruction systems and image resolution limits.

7.1.1. Benefits of Pulsed Holography

The key technical advances of pulsed holography include the ability to collect three-dimensional information about the distribution, shape, and orientation of hypervelocity impact-generated fragment clouds, and the ability to quantify the characteristics of very small fragments without the depth-of-focus restriction of conventional microscopy. The ability to capture the three-dimensional features of a hypervelocity impact event makes pulsed holography a powerful tool, especially for the validation of three-dimensional hydrocode simulations and fragmentation models. The development of these computational and analytical models is closely linked to the ability to compare model predictions to test results. With the advent of sophisticated three-dimensional hydrocodes, these computational tools are progressing beyond the ability of conventional imaging techniques to provide the necessary information for code validation. Pulsed holography has the potential to establish a diagnostic, and by extension, validation capability commensurate with the power of these three-dimensional hydrocodes.

7.1.2. Hypervelocity Impact Events of Interest

The two types of hypervelocity impact events that can be studied with pulsed holography are perforating and nonperforating impacts. Perforating impacts lead to debris cloud formation, where the debris cloud can consist of a mixture of solid fragments, liquid droplets, and/or vapor products. Nonperforating impact events produce crater ejecta and jet breakup. The study of these impact events has several applications, including understanding the formation and dispersal of behind-armor debris for survivability or lethality performance, studying the perforation of thin-walled structures for kinetic energy weapon lethality, and developing advanced debris shield designs to improve the survivability of aerospace systems. The solution to all of these problems requires the characterization of such factors as the extent, orientation, and distribution of mass in a cloud of impact-generated fragments; the solid, liquid or vapor state of particles in an impact cloud; the shape, size, velocity, and number of individual fragments; or the breakup of hypervelocity impact-formed jets into discrete fragments.

7.1.3. Present Imaging Techniques

Hypervelocity impact tests typically use conventional imaging tools such as flash X-rays, pulsed laser photography, and high speed photography with either rotating prism cameras or image-converter cameras. The use of these imaging diagnostics and their application to hypervelocity impact testing has been described in the literature [1, 2]. Unfortunately, with the exception of stereo photography, conventional imaging techniques are, by their nature, two dimensional. Therefore, they are unable to capture a significant portion of the information that is predicted by three-dimensional hydrocodes. Even stereo photography is subject to particle shadowing and masking that can limit impact event characterization. Conventional imaging techniques have practical resolution limits of about 200 μm either through a depth of focus constraint which restricts the practical image magnification, or the resolution capability of flash X-ray cassettes. As noted above, one of the primary motivations for the development of pulsed holography is to move beyond these limits and provide high fidelity information about the three-dimensional features of impact events.

Conventional holograms typically image static objects and require vibration isolation tables to limit motion of the object and optical components to less than a fraction of the wavelength of the illuminating laser light source. This motion constraint is necessary for the formation of the constructive and destructive wave interaction that produces interference fringes on the holographic plate. If the object moves more than a fraction of a wavelength during the duration of the laser exposure these interference fringes are lost and with it the holographic image of the object. However there is an exception, whereby an object that moves many multiples of the laser wavelength during the pulse duration can form "shadowgraphic" hologram images. The topic of this chapter is this exception and how it provides a key advance over conventional hypervelocity impact imaging techniques.

7.2. Measurement Objectives

There are several measurements that can characterize an impact event. For these different measurements a spectrum of resolution requirements may be defined. For example, an impact event can be characterized with macroscopic measurements to describe the overall spatial distribution of material resulting from an impact event. In the same event, microscopic measurements can focus on the impact-generated particles, their shape and volume; their state—solid, liquid or vapor; and for solid particles (fragments), their "roughness" or fractal dimension [3]. There are a number of other measurements that can fall between these macroscopic and microscopic regimens, such as the total number of particles generated, their size distribution, and their velocities.

These measurements place different accuracy requirements on the hologram analysis and tolerances for the holographic layout. Accuracy requirements for a given property measurement may be obtained by examining the resolution limits for computational and analytical predictions of that property. For purposes of illustration, assume a fragment cloud that extends over a volume of 100 mm × 100 mm × 100 mm (1,000,000 mm^3), the accuracy requirement for macroscopic measurement of fragment location may be to within 1 mm. In the intermediate regime of holographic measurements, fragment sizes and velocities may require resolution to within 100 μm. Finally in the microscopic measurement regimen, to determine the particle shape, state, or fractal dimensions may require resolutions of better than 10 μm. This spectrum of resolution requirements is illustrated in Table 7.1. This example is given to illustrate how different regimes of accuracy may coexist within a given fragmentation event. The actual requirement for microscopic resolution could be significantly higher for particular applications. If the scale of the an event of interest were larger or smaller, the

Table 7.1. Measurement objectives and estimated resolution requirements to characterize a $(100\ mm)^3$ hypervelocity impact event

Macroscopic ($\sim$1000 μm resolution)	Intermediate regime ($\sim$100 μm resolution)	Microscopic ($\sim$10 μm resolution)
Debris bubble envelope and structure	Number of particles generated	Particle shape and volume
Crater ejecta envelope and structure	Particle size distribution	Particle state: solid, liquid, or vapor
Particle position	Particle velocity	Particle roughness: fractal dimension

required resolution would also scale up or down. However, an event "scale" of 100 mm is convenient because it can be recorded on a common size for holographic plates, 102 mm × 127 mm (4 × 5 inch).

7.3. Pulsed Laser Holograms

The promise of pulsed holography for ballistic diagnostics was revealed by Hough et al. [4] in a series of tests with a small powder gun. The key breakthrough [4] was the demonstration of the ability to holographically capture information on fragments that move a distance that is many times the wavelength of the laser light during a pulse duration. For example, 1 km/s particles move 30 μm along the velocity vector during a 30 ns laser pulse. While this distance is small with respect to macroscopic dimensions, it is more than one hundred times the wavelength of visible laser light. This motion prevents the formation of a direct hologram of the impact-generated particles, but the particles can still be characterized. The key is the use of a stationary backsheet, above which the outline of the particles appear as floating shadows.

The system developed by Hough et al. [4] used a pulsed ruby laser to produce a 30 ns, 300 mJ pulse at 694 nm. A ruby laser offers several operational advantages including the ability to easily single-pulse the laser with an external trigger, the stability to provide reliable spatial and temporal beam coherence, and relatively low cost. However, there are also a number of disadvantages to ruby lasers. The primary disadvantages are that it is not possible to operate a ruby laser in a continuous wave mode for the reconstruction system and the pulse duration is relatively long for freezing the motion of particles produced in a hypervelocity impact event. If a different wavelength laser such as a helium-neon is used for continuous wave reconstruction, the resulting image is useful for qualitative viewing; however it introduces a nonlinear scale shift that complicates quantitative image analysis.

Another pulsed laser that can be used for holography is a frequency-doubled Nd-YAG laser. There are a number of pulsed, seeded, doubled

Nd-YAG lasers available that produce 5 ns, 750 mJ, coherent pulses at 532 nm. These lasers are more complex and expensive than a pulsed ruby laser, but diode-pumped, doubled Nd-YAG lasers are available that can provide a continuous wave monochromatic light source for use in a quantitative reconstruction system. Because these continuous wave lasers produce the same 532 nm wavelength as the pulsed lasers, no scale shifts are introduced that can complicate the quantitative image analysis. For pulsed, high coherence Nd-YAG systems, the seed laser is a small diode-pumped Nd-YAG laser that provides a reference oscillator cavity to increase coherence length of the main laser cavity. While ruby lasers are available with pulse energies of 2 to 3 J, holographic films are inherently more sensitive to the "green" light of a doubled Nd-YAG laser at 532 nm than to the "red" light of a ruby laser at 694 nm. A major disadvantage to the use of a doubled Nd-YAG laser is the complex triggering system that is required to obtain a reliably coherent, single pulse at the right time. This disadvantage is offset by the ability to use the same wavelength for a continuous wave quantitative image reconstruction system. Based on these considerations, the Sandia National Laboratories pulsed holography system utilizes a Q-switched, seeded, doubled Nd-YAG pulsed laser to capture "shadowgraphic" holograms, and a diode-pumped, doubled Nd-YAG laser for quantitative reconstruction.

7.4. Test Configuration

This section summarizes the experimental setup that has been used for the development of a pulsed holography hypervelocity impact diagnostics system at Sandia National Laboratories. The integration of the laser triggering subsystem into the gun control circuit for a small (12.7 mm diameter launch tube), two-stage light-gas gun was a key engineering accomplishment. Two different optical layouts have been used to capture images on holographic plates. The design of the optical layout balances the requirement for high image resolution with protection of the glass holographic plates from both impact-generated debris and potential fogging from impact flash. On all tests a pair of open shutter cameras equipped with line filters provide supplementary diagnostics.

7.4.1. Gun/Laser Synchronization and Triggering Subsystem

The main component of our pulsed holography system is a Continuum model NY-82 frequency-doubled Nd-YAG laser with a model SI-500 injection seed laser. As noted above, this type of laser was selected after considering a number of trade-offs with ruby lasers. For thermal stability, and to maintain closed-loop, frequency-locked coupling of the seed laser to the main Nd-YAG oscillator cavity, the laser requires continuous pulsing at a 10 Hz rate up to the desired shot time. Unfortunately the powder burn-time jitter of a two-stage light-gas gun prevents using the laser to trigger the gun.

The following is a description of the subsystem for triggering the pulsed Nd-YAG laser after an impact event to within a microsecond of accuracy. An external circuit triggers the laser at a 10 Hz rate in order to maintain laser coherence length. This circuit provides a trigger to the flash lamps that optically pump the Nd-YAG laser cavity. Approximately 200 μs later a second trigger is sent to the Q-switch and seed laser to generate the laser pulse. These triggers are repeated at a 10 Hz rate until an interrupt signal is received just before the gun is fired. To prevent fogging and over-exposure of the holographic plate, an aluminum shutter/beam stop blocks the stream of 750 mJ laser pulses outside the impact chamber.

When the gun fire button is pressed a number of actions occur prior to the actual firing pin release. First, an interrupt trigger is sent to the external 10 Hz pulsing circuit. Second, the shutter/beamstop is moved out of the optical path into the impact chamber, and third, the laser flash lamp capacitors are charged and held. Then the gun firing pin is released, the projectile is launched, the sabots are stripped in the flight range, and the impactor to flies through a Magnetic Velocity Induction System (MAVIS) [5] coil station placed a known distance from the target. The MAVIS provides a two-point measurement of the impactor velocity and is used with a proportional delay to trigger the Q-switch and pulse the laser at the desired time after impact. In addition, the MAVIS provides a timing point for triggering the flash lamps from 200 to 500 μs prior to the Q-switch trigger. A simple schematic diagram of this system is shown in Figure 7.1.

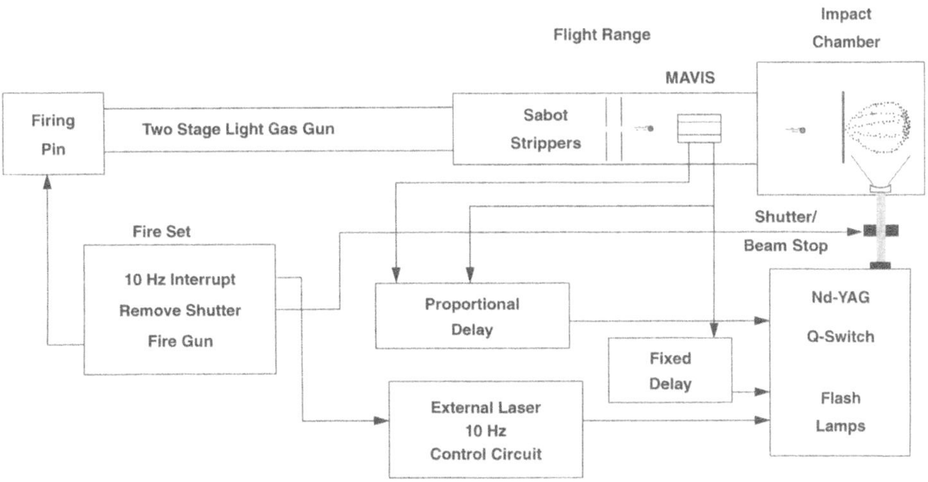

Figure 7.1. Schematic diagram of the gun-laser synchronization and triggering subsystem. The principal components include the MAVIS projectile sensing coils, the proportional delay, and the shutter/beam stop.

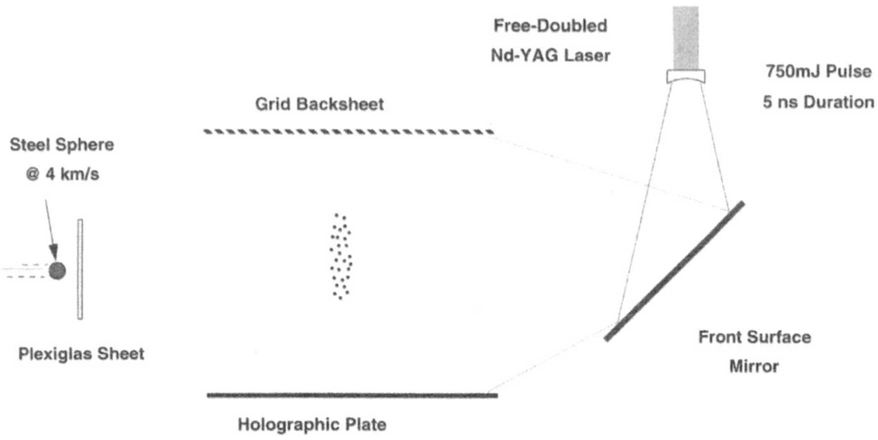

Figure 7.2. Schematic diagram of flat plate analog to LTV holography layout.

7.4.2. Holographic Layouts

Two different optical layouts that use holographic plates have been used to capture holographic images of hypervelocity impact events. The first layout is a flat plate geometry that is analogous to the cylindrical geometry that was originally developed by Hough et al. [4]. Figure 7.2 illustrates our layout of this geometry. While the cylindrical geometry used by Hough et al. produced dramatic images with the ability to examine the debris cloud from a full 180 degrees of view, this geometry is obtained with film-based holographic media. Because our effort has a primary objective of obtaining quantitative information from an analysis of the holographic images, better dimensional stability is required than can be provided with film. In order to achieve measurement accuracies that are better than 100 μm, flat, glass plate holograms are required to accurately reproduce initial reference beam geometries.

The use of flat plate holographic media led to the development of an improved layout shown in Figure 7.3. This figure shows the layout of optical components in and around the impact chamber. The view in Figure 7.3 is looking towards the gun; the projectile is coming directly out of the page. The open shutter cameras are above the impact chamber looking down. In contrast to the geometry in Figure 7.2, this layout has the benefits of (1) using the central and most intense portion of the beam, (2) eliminating the front surface mirror from the impact chamber, and (3) providing the ability to capture holographic images of nonperforating, cratering impact events. Because the front surface mirror in the layout shown in Figure 7.2 is centered along the projectile shotline, it provides a source of secondary debris that can subsequently damage the holographic plate. For non-perforating impact events, the thick target is positioned by the downstream edge of the translucent backsheet and holography plate in order to image the cratering ejecta

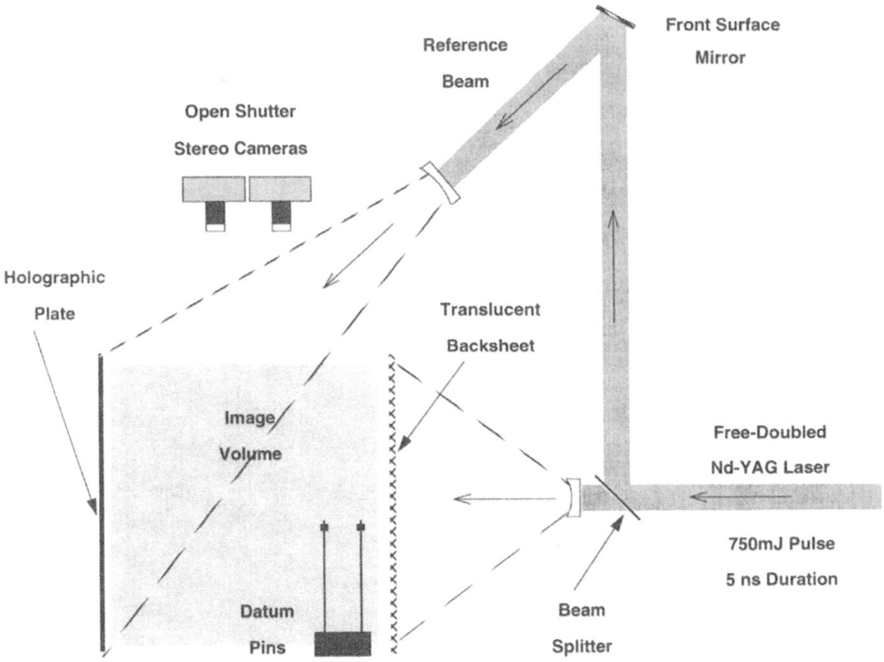

Figure 7.3. Schematic diagram of improved flat plate holography layout. Perspective is from the impact chamber looking towards the gun.

that is thrown back from the target surface. Important supplementary diagnostics used for these tests are open-shutter stereo cameras equipped with 532 nm line filters. The layouts in both Figure 7.2 and 7.3 provide a clear view from above for open shutter cameras. Figure 7.3 also illustrates a set of 3 datum pins (from this view one is hidden), that are used as fiducials to establish the shot line direction and the horizontal plane perpendicular to the film and grid surfaces.

The holographic media used in these tests are Agfa 8E56-NAH or Ilford SP 695T (102 × 127 mm) glass plates. For dimensional precision, glass plates are preferable to plastic film emulsion media; however, unless steps are taken to protect the glass, they are subject to the risk of breakage from secondary impacts. A modest amount of protection is provided by placing the holographic plate in a liquid gate. A liquid gate is a flat-walled tank with high-quality optical windows that is used to reduce the sensitivity of the holographic imaging process to nonuniformities in the plate and thickness [6]. This tank is filled with an appropriate index-of-refraction matching fluid to reduce internal reflections. But this apparatus has the additional benefit of providing improved protection for the holographic plate and reducing potential fogging from impact flash.

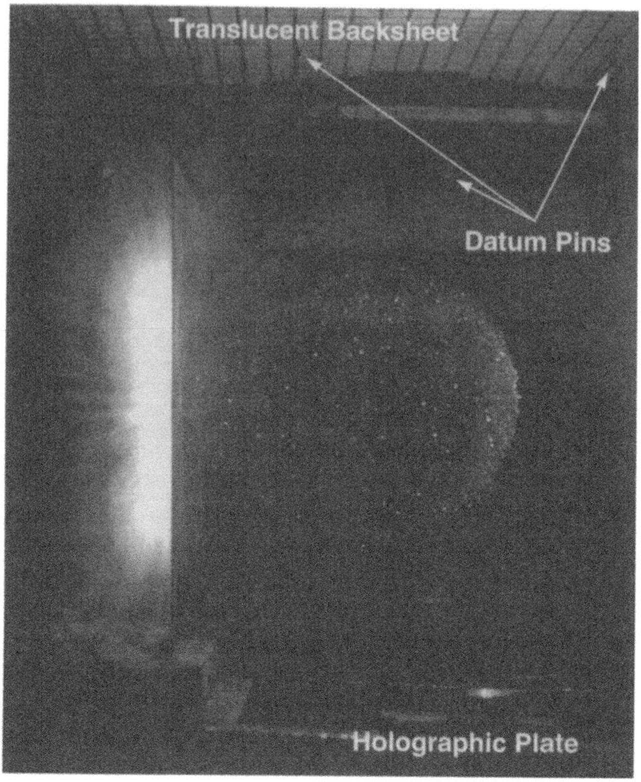

Figure 7.4. Open shutter photograph of the 4.22 km/s impact of a 6.35 mm steel ball through a 0.63 mm steel sheet 22 ms after impact with a time-integrated image of an impact flash. The holographic plate, translucent backsheet and datum pins are also shown in this photograph.

Figure 7.4 shows an open-shutter camera photograph from the perforation and breakup of a 6.35 mm steel sphere through a 0.63 mm steel sheet at 4.22 km/s. In this test the laser was pulsed approximately 22 μs after impact of the sphere on the sheet to freeze the motion of hundreds of submillimeter-sized steel fragments. As indicated in Figure 7.3, this open shutter camera has a view into the impact chamber from above. Shown at the top of the photograph is the translucent grid sheet and just below this sheet are the three datum pins. At the bottom of this image is the holographic plate. Also evident in this photo is an impact flash event. This test was not successful in capturing a holographic image because the light from the impact flash over-exposed the holographic plate. Simple shielding techniques were used to minimize the effects of impact flash in a repeat of this impact test.

Several hypervelocity impact events have been holographically imaged. Table 7.2 summarizes these impact events. The impacts of steel and copper

Table 7.2. Impact events captured with holographic images

Impactor	Velocity (km/s)	Target
6.35 mm Steel ball	4.32	1.3 mm Plexiglas sheet
6.35 mm Steel ball	4.22	3.0 mm Plexiglas sheet
6.35 mm Cu ball	4.32	1.3 mm Plexiglas sheet
6.35 mm Steel ball	4.22	0.63 mm Steel sheet
6.35 mm Cu ball	1.99	25.4 mm Cu block at 0°

Figure 7.5. Open shutter photograph of the 4.32 km/s impact of a 6.35 mm copper ball through a 1.3 mm Plexiglas sheet 144 ms after impact.

spheres through plexiglas sheets provide a clean breakup of the sphere. Similar impacts with flash X-ray diagnostics indicate that most of the metallic sphere remains in a tightly grouped cloud of fragments around the initial trajectory of the sphere. The impact of a steel sphere into a steel sheet results in a debris bubble of fine steel fragments as shown in Figure 7.4. Figure 7.5 and 7.6 are open shutter photographs from the copper-into-plexiglas shot

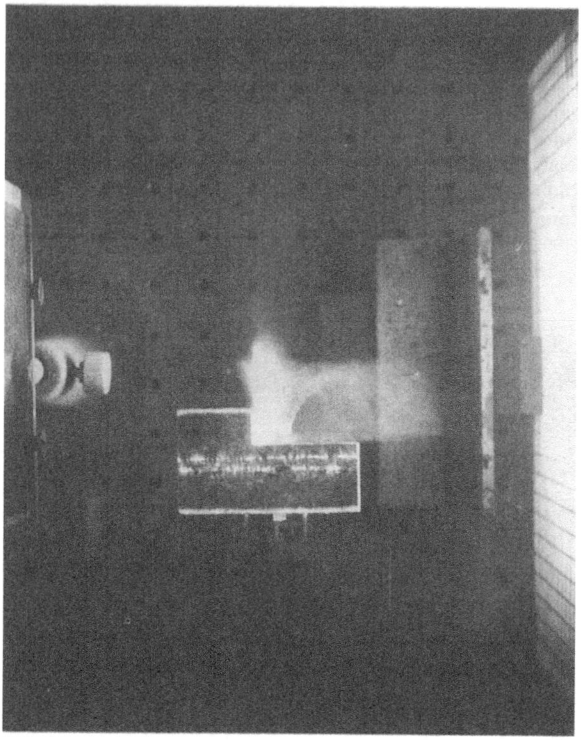

Figure 7.6. Open shutter photograph of the 1.99 km/s impact of a 6.35 mm copper ball into a 25.4 mm copper block 12 ms after impact.

and the copper-into-copper block shot, respectively. The cratering impact shown in Figure 7.6 clearly shows a step machined into the impact face of the target block. This step protected the holographic plate by deflecting the crater ejecta thrown toward the plate.

7.5. Holographic Image Reconstruction Systems

The pulsed holograms produced in this study are transmission holograms that require a monochromatic light source to reconstruct the holographic image. For qualitative viewing, this source can be a laser with a different wavelength from the pulsed laser used for generating the hologram. It is also possible to use different optics and configurations for the reconstruction beam. Figure 7.7 is a photograph of our portable system for qualitative reconstruction of holographic images. This system uses a small continuous wave 25 mW diode-pumped Nd-YAG laser operating at 532 nm, but it departs from the original reference beam geometry in interests of portability.

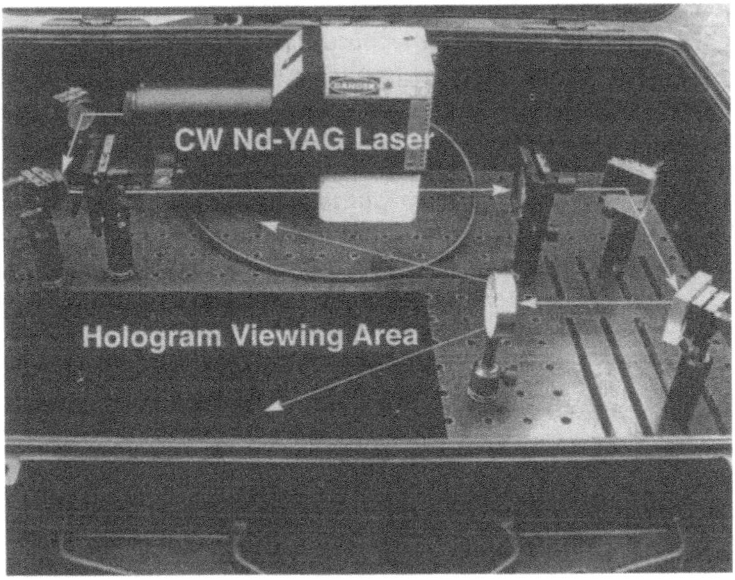

Figure 7.7. Portable, qualitative holographic image reconstruction system. Arrows indicate the beam path through the collimating and beam expanding optics. The reconstructed image is viewed above the black rectangular area.

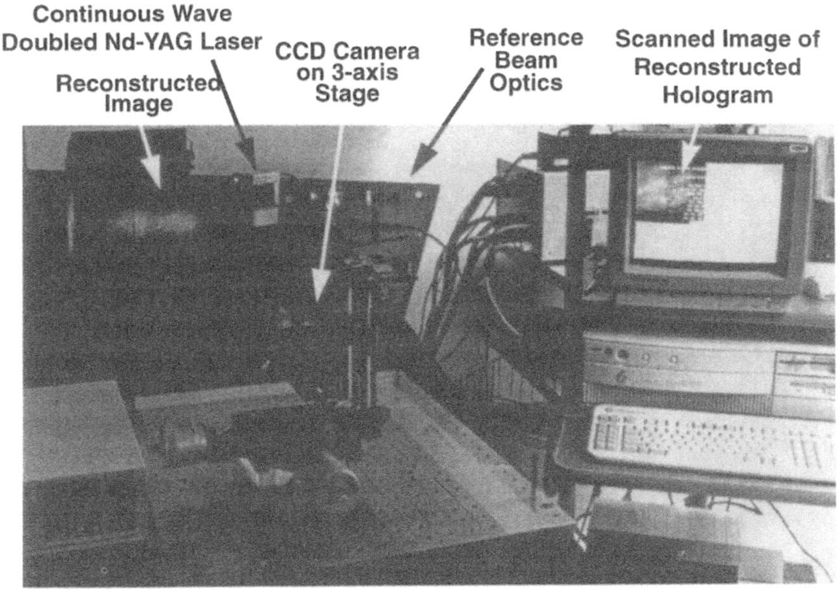

Figure 7.8. Photograph of the prototype quantitative holographic reconstruction and image processing system.

As noted previously, if the same wavelength light is used for image reconstruction, it is possible to form a reconstructed image with no length scale shifts. To achieve the high levels of resolution given in Table 7.1, the original reference beam must be accurately reproduced. A prototype image analysis subsystem with a quantitative image reconstruction subsystem has been assembled. The same 532 nm laser from the portable qualitative reconstruction subsystem has been used; however, an identical reconstruction beam was used, including the impact chamber port. The reconstructed images were imaged with a CCD camera having 768 × 493 active pixels and either a zoom lens with a focal length range of 18 to 108 mm or a matched pair of achromat 1 : 1 image transfer lenses. This CCD camera has been mounted on a manual three axis stage. The RS-170 analog video signals from the CCD camera have been digitized with a commercial image processor board for a

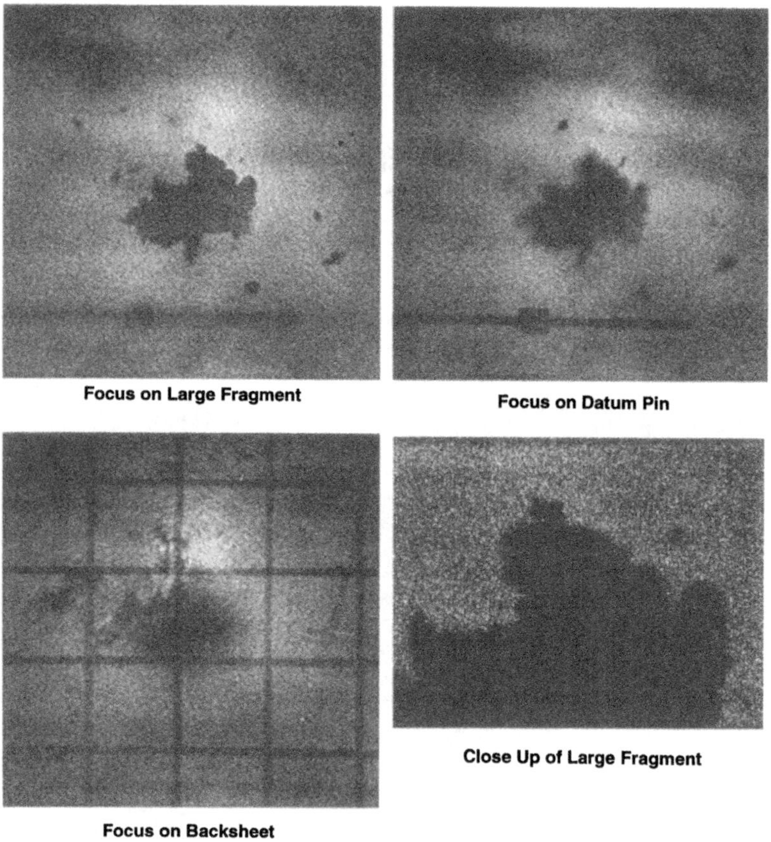

Figure 7.9. Digitized CCD camera images of the reconstructed hologram from the 4.32 km/s impact of a 6.35 mm diameter copper ball through a 1.3 mm Plexiglas sheet 144 ms after impact. This is the same impact event shown in Figure 5.

PC-compatible computer at a resolution of 640 × 480 pixels, with 8 bits intensity resolution. A photograph of this prototype image analysis subsystem is shown in Figure 7.8.

Figure 7.9 is an example of the digitized images that can be obtained from the reconstructed holographic images with the image processing system. This set of images is from the same impact event captured with an open shutter camera photograph shown in Figure 7.5. These images illustrate the ability to position the plane of focus of the CCD camera at fragments of interest, at a datum pin, or at the backsheet. The image in the lower right quadrant shows the high degree of detail that is revealed when the optics are focused on the upper edge of the large central copper fragment.

The images shown in Figure 7.10 illustrate two capabilities that the analy-

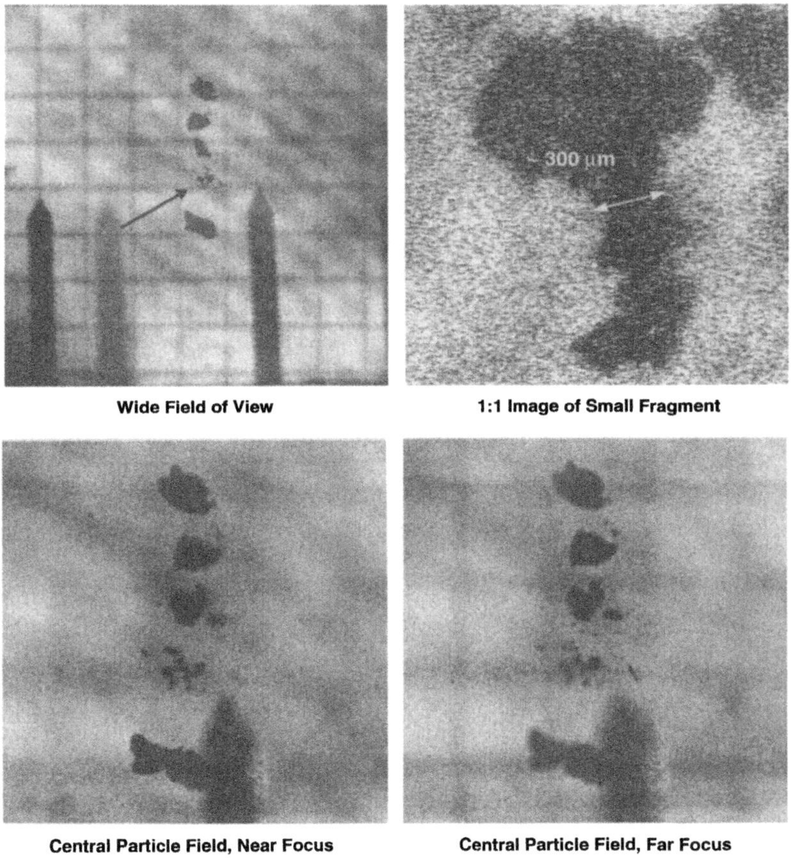

Figure 7.10. Digitized CCD camera images of the reconstructed hologram from the 4.22 km/s impact of a 6.35 mm diameter 1018 steel ball through a 3.0 mm Plexiglas sheet 134 ms after impact.

sis of pulsed holograms provide. The image in the upper left quadrant is a wide field of view that is focused on the impact-fragmented steel particles. Also shown are the out of focus datum pins and backsheet. The arrow points to a small particle below the center of the fragment field. The image in the upper right quadrant is the digitized image of this fragment using 1 : 1 image transfer optics. The waist of this fragment is approximately 350 μm in thickness. The two images in the lower quadrants demonstrate the ability to use depth of focus to throw particles into and out of focus. When coupled to a high precision three-axis stage, a narrow depth of focus provides the ability to accurately determine the depth position for a particle. Particles in the plane of focus of the camera can then be located in the horizontal and vertical directions by reading positions from the corresponding axis of the stages.

7.6. Holographic Image Resolution Limits

There are several factors that contribute to the resolution limit of these pulsed holographic images. These factors include the velocity of the particles, laser speckle and viewing optics resolution, the thickness of the particles, and the resolution of the CCD camera. Because the pulse duration is nominally 5 ns, there is also a velocity blurring effect. The velocity blurring is roughly equal to the product of the pulse width of the laser and the particle velocity. Actually, it should be possible to estimate the particle size to about half of this number; so the axial blurring resolution limit is approximately 10 to 15 μm for 5 km/s particle velocities. There is an uncertainty in determination of the edge position of a particle. This is due to a combination of lens resolution limits, laser speckle effects, and CCD camera pixel size. For an incoherently illuminated object, the approximate lens blur due to diffraction is $1.2\lambda\ s/d$, where λ is the wavelength of the light, s is the lens-to-image distance, and d is the effective lens diameter. For 1 : 1 imaging at $\lambda =$ 0.532 nm, this blur is about $1.3\ f/d$ or $1.3\ f/no.$ (in μm), where f is the lens focal length, and "*f/no.*" is the lens "*f*-number," defined as f/d. Lens aberrations will degrade this resolution somewhat, depending on the design and quality of the imaging lens. Coherent illumination of a diffuse object causes speckle, which is a random high-contrast noise field with characteristic size about the same as the blur distance. This typically degrades resolution by about a factor of 2. The lens we are currently using is an $f/5$ lens with $f = 250$ mm, so the best resolution we can expect is 13 μm. The hologram is imaged onto a CCD camera which has pixels (image samples) of 11 μm $\times$ 13 μm, which also limits resolution. Of course, this limitation can be reduced by using an image transfer system with higher magnification, at the expense of field-of-view size.

The lateral image resolution of our pulsed holographic system was measured with a standard three-bar Air Force resolution test target capable of measuring resolution in the range from 2000 μm to 2 μm. Using white-light (incoherent) illumination to project the resolution target directly through the 1 : 1 imaging system onto the CCD, a resolution of 12 μm was measured. To test the resolution of the entire holographic imaging system, a hologram was made of the stationary resolution test target, inside the test chamber. The hologram was transferred to the image analysis subsystem, and resolutions of 35 μm for the Agfa plates, and 20 μm for the Ilford plates were measured. By using a 10 × microscope objective in the original imaging process (between the resolution test target and the holographic plate), resolutions on the order of 3.5 μm have been measured for the holographic imaging process. Use of a microscope objective reduces the available test volume to the field of view of the objective—typically around 10–20 mm for a 10 × objective.

The resolution discussed above is the lateral resolution, that is, perpendicular to the viewing direction. Concerning the axial or depth dimension, the holographic imaging system has two major advantages over conventional photographic imaging systems: the capability to obtain high lateral resolutions at *any* depth within the experiment volume, and to simultaneously obtain high *depth resolution*, or the ability to measure the position of a particle in the viewing direction. Depth resolution is quantified by the "depth of field," which for a conventional lens in incoherent light is approximated by $4\lambda\ (s/d)^2$. For our current system with $s/d = 5$, we should expect a depth resolution of about 50 μm. The depth resolution can be improved by viewing a particular particle from several angles. In principle, the depth resolution should be limited by the s/d of the *hologram*, as opposed to the imaging lens, implying a possible depth resolution of a few μm.

The depth resolution of a larger particle might also affect the lateral resolution for that particle. If the particle is irregular in shape, its edges will focus at different depths. Taking the "best mean focus" approach will blur the particle image slightly, which will tend to make it appear larger (depending on the exact edge criterion used). We are currently working on techniques to optimize three-dimensional resolution.

7.7. Future Efforts

This pulsed holography development effort has achieved a number of key results. Holographic images have been captured of a number of hypervelocity impact events, including perforating and nonperforating impacts. These holographic images have been digitized with a prototype image analysis system. While the open shutter cameras are a supplementary diagnostic, the analysis of these photographs with existing image analysis tools will provide a validation of the macroscopic holographic measurements.

There are several steps that can still be taken to improve image resolution. A spatial filter can be added to the reference beam to improve the beam quality and uniformity. The pulsed laser can be replaced with a similar frequency doubled Nd-YAG laser with 50 to 150 picosecond pulse duration to reduce the velocity blurring effect. Finally, higher quality imaging optics can be used to increase the resolving power of the image analysis subsystem. Together these improvements could yield a system capable of resolving 10 μm size particles.

The quantitative analysis of the information captured in these holograms can be achieved through different approaches. The next step in analyzing the digitized images will be to use existing image processing software to define particle edges, dimensions, and areas. Sophisticated software packages are also capable of determining such factors as the image centroids. To accurately determine the location of impact-generated particles the CCD camera and image transfer optics will scan the image volume with a computer-controlled three-axis translation stage with at least one axis of rotation.

An interesting consideration is the information content potential of these holograms. Assuming the holographic image volume is 100 mm $\times$ 100 mm $\times$ 100 mm, the total potential image volume is 1.0×10^6 mm^3. If the depth of focus for the image transfer optics is 50 μm, then each 10 mm $\times$ 10 mm digitized image would cover a volume of 5 mm^3, therefore to completely record the entire image volume, 200,000 images would be required. Of course most of these images probably would not contain any useful information so sophisticated data handling and manipulating algorithms would be useful. This example indicates the role of high performance computing to handle the large volumes of data in the analysis of holographic images.

Another option for analyzing the data in these holograms will be to take stereo photographs of the reconstructed holographic image. These stereo photograph pairs, like the initial open-shutter stereo camera photographs, can be analyzed with existing stereo image analysis tools [7]. This analysis will provide cross-correlation for the macroscopic measurements of an impact event.

7.8. Acknowledgments

The key members of our development team include Carl H. Konrad and William C. Sweatt of Sandia National Laboratories, and Scott M. Gosling and Randy Hickman of Ktech Corporation. We also thank Scott Harland for his initiative in commissioning our pulsed, Nd-YAG laser and obtaining our first static holograms during his summer at our impact facility as a participant in the Sandia Outstanding Summer Student Program.

7.9. References

1. Swift, H.F., Zukas, J.A., Nicholas, T., Greszczuk, L.B., and Curran, D.R. Image forming instruments. In: *Impact Dynamics*. John Wiley and Sons, Inc., New York, 1982.

2. Isbell, W.M. Historical overview of hypervelocity impact diagnostic technology. *Int. J. Impact Engng.*, 5, 389–410 (1987).
3. Mandelbrot, B.B. *Fractal Geometry of Nature*. W.H. Freeman and Co., New York, 1983.
4. Hough, G.R., Gustafson, D.M., and Thursby, W.R. Enhanced holographic recording capabilities for dynamic applications. In: Proc. SPIE 34th Annual International Symposium, 1990.
5. Moody, R.L. and Konrad, C.H. Magnetic induction system for two-stage gun projectile velocity measurements. Sandia Report, SAND84-0638, UC-13, 1984.
6. Goodman, J.W. *Introduction to Fourier Optics*. McGraw-Hill Book Co., New York, 1968.
7. Franke, E.A., Wenzel, D.J., and Davidson, D.L. Measurement of microdisplacements by machine vision photogrammetry (DISMAP). *Rev. Sci. Instrum.*, 62, 1270–1279 (1991).

CHAPTER 8

Initiation and Propagation of Damage Caused by Impact on Brittle Materials

S.J. Bless and A.M. Rajendran

8.1. Introduction

The response of ceramics and other brittle materials to dynamic loading has been studied for several years by shock wave scientists, chiefly for defense-related applications, such as advanced ceramic armor or nuclear devices. For this reason, most studies have concerned impact scenarios in which impact stresses greatly exceed the compressive strengths of the targets. However, it is important to realize, especially in these days of "dual use" science, that impact behavior forms a continuum of which these studies are at one extreme. For example, the mechanistics-based constitutive and damage models that have been developed to help us understand the projectile penetration process into advanced ceramic armors, can also be utilized to predict failure of ceramic engine components. Understanding ceramic behavior under impact loading conditions is essential to the design of impact resistant materials for structural applications. Foreign object damage studies on ceramic turbine blades, stators, and other components, in particular, necessitate a critical evaluation of impact properties of ceramics.

To experimentally determine the failure of engine components, hundreds of experiments must be performed due to the statistical nature of brittle failure. Unfortunately, destructive tests of ceramic components are very expensive, and they do not lend themselves to scientific understanding of the complex failure process. While we may be able to construct some empirical curves from the experimental data, applications of these curves are limited to the configuration and material for which the data were collected. Therefore, when new ceramics are employed in engine components, the data base has to be extended with additional experiments. To minimize the number of experiments, we can take advantage of advanced structural dynamic finite element code analysis capabilities. The capabilities include both mechanistic and phenomenological constitutive and damage models which are essential to make predictions.

At the low-velocity end of the impact spectrum are particle impacts causing cosmetic or structural degradation; at somewhat higher velocity, we find ballistic impacts in which ceramic targets struck by small caliber bullets are shattered and penetrated. For these applications, impact stresses are less than, or commensurate with, the ceramic compressive strength. In the authors' opinion, understanding of low speed damage may benefit considerably from applications of analytical techniques developed by the shock physics community. At higher velocities, characteristic of high velocity guns, explosive-launched projectiles, or even space debris, local damage is due to flow of crushed granulated material, and stress waves cause additional damage far from the impact site. Thus, we have two aims in preparing this review article: (1) to inform our colleagues of some recent developments, in which we have participated, that help us to understand important aspects of dynamic damage; and (2) to alert our colleagues to the wider spectrum of impact phenomenology, to which they might contribute understanding.

8.2. Phenomenology

Reviews of some aspects of low-speed particle impact can be found in Evans (1979), Zukas et al. (1982), and Thompson et al. (1993). The potential use of ceramics in heat engines, and their susceptibility to catastrophic failure, has provided an incentive for much recent work. Figure 8.1, for example, illustrates an impact of a graphite ball at 278 m/s onto a silicon nitride rotor blade. Both the ball and blade tip have been destroyed. Damage like this is the opposite of "graceful failure," which is the preferred mode of engine degradation.

Low speed impacts can be very damaging because tensile cracks are relatively easy to initiate in brittle materials, and the structural strength of brittle components is extremely flaw dependent. Thus, post-impact mechanical properties can be greatly degraded even when macro-damage is not evident. In fact, recent work has emphasized the enhanced sensitivity of many ceramics to the smallest flaws (e.g., Mai and Lawn, 1987; Cook et al., 1987), so that understanding threshold conditions for initiation of impact-induced damage has significant practical implications. This has been particularly true for development of ceramic engine blades (e.g., Smyth and Morey, 1991; Akimune et al., 1989), and damage to automobile finishes (Ramamurthy et al., 1993).

At velocities above the threshold level, microcracks form in the target. The usual progression that is observed with increasing impact velocity is: ring cracks (e.g., Shockey et al., 1990; Makakawa et al., 1991, Rajendran et al., 1992); cone cracks, radial/median cracks (e.g., Evans et al., 1978; Evans and Wilshaw, 1977); and lateral cracks (e.g., ibid.; Marshall et al., 1982). These crack systems lead to strength reduction or immediate failure. Figure 8.2 shows the sequence as portrayed by Cherepanov (1979). Figure 8.3 illustrates a well developed crack system observed by Shockey et al. (1990).

In brittle materials, such as ceramics, a microcrack starts to grow when the

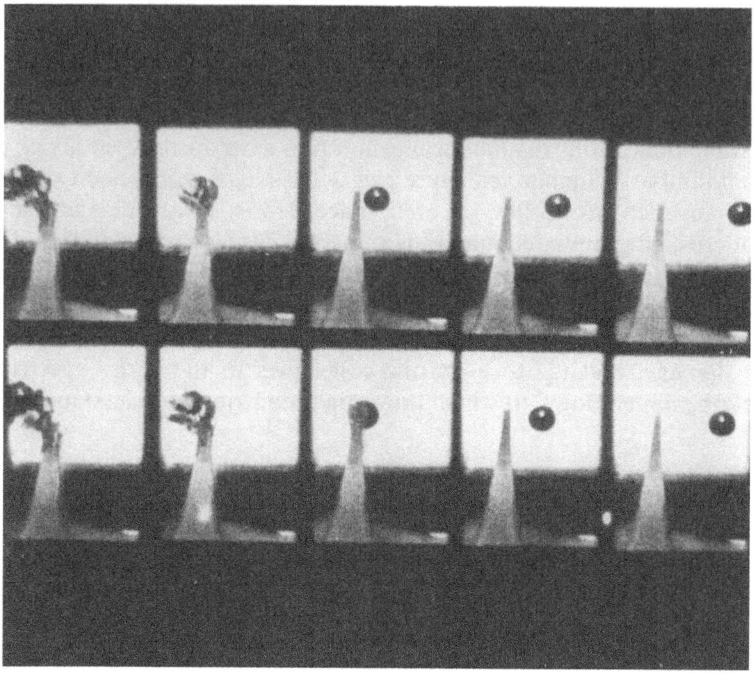

Figure 8.1. Impact of 0.5-mm diameter graphite ball on a silicon nitride rotor blade at 912 ft/s. Frames taken at 10^5/s. From Bertke (1993).

stress field at the crack tip satisfies Griffith's criterion [see Griffith (1921)]. For modeling the brittle failure of metals, Curran et al. (1987) suggested an exponential crack size distribution. Microcracks of varying sizes begin to grow at different times as the stress levels increase. At low enough strain rates, the largest microcracks begin to grow first, causing stress relaxation in local zones, thus preventing other microcracks from growing. At high strain rates, the crack tip speed is limited to a maximum value and, therefore, significant stress relaxation does not occur until nearly all the microcracks have begun to propagate.

Ring and cone cracks are often described by variations of the classic Hertzian treatment (Hertz, 1882). Initiation of the cone crack may depend on preexisting flaws (e.g., Evans et al., 1978; Bowden and Field, 1964). The orientation of the cone crack is strongly affected by sliding friction in the case of oblique impacts (e.g., Lawn et al., 1984). Analyses of ring crack formations are usually based on elastic contact theory (Timoshenko, 1928). However, in the case when the projectile fails in a plastic or brittle way, general solutions do not yet exist (see Section 8.3 below for an example).

Higher impact velocities result in penetration. This is the problem with

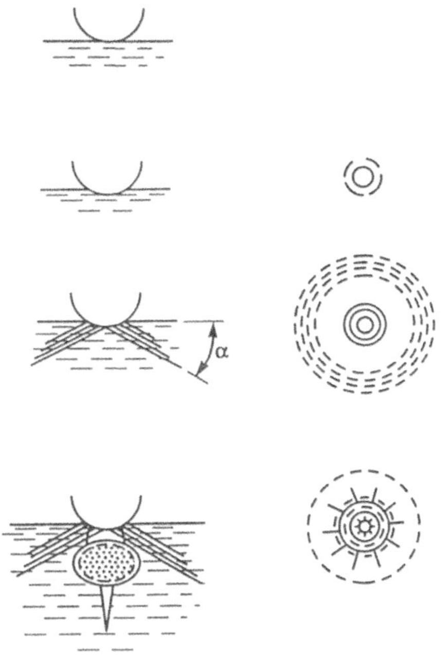

Figure 8.2. Stages of particle impact damage. From Cherepanov (1979).

ceramic armor. Generally ceramic armor mechanisms should be distinguished by "thin tile" and "thick tile" effects. Thin tile ceramic armor is the type conventionally used against small-caliber armor piercing bullets. The tile thickness is similar to the projectile diameter, and the tile width is usually at least 10 times its thickness. The armor works best against sharp bullets in which shock effects are relatively minimized (Rosenberg et al., 1990; Wilkins et al., 1971). Under these conditions, if the compressive strength of the ceramic exceeds the contact stress exerted by the bullet, given approximately by:

$$P = \tfrac{1}{2}\rho_p V_0^2 + Y, \tag{8.1}$$

(where ρ_p, V_0, and Y are projectile density, impact velocity, and strength), the bullet cannot immediately penetrate. Instead, it must wait for the ceramic to fragment. The initial stages of penetration resemble particle impact; ring or cone cracks are generated (e.g., Cheng et al., 1991; Sternberg, 1990). However, ceramic which has failed in shear but remains under highly compressive loads does not decrease in strength (Rosenberg et al., 1987).

The tile eventually fails in tension opposite the impact face. A fracture cone is developed, within which load is transferred to a backing plate, causing it to structurally fail. Recent treatments of thin tile armor penetration mechanics can be found in Woodward (1972), Rosenberg et al. (1990), and den Reijer (1991).

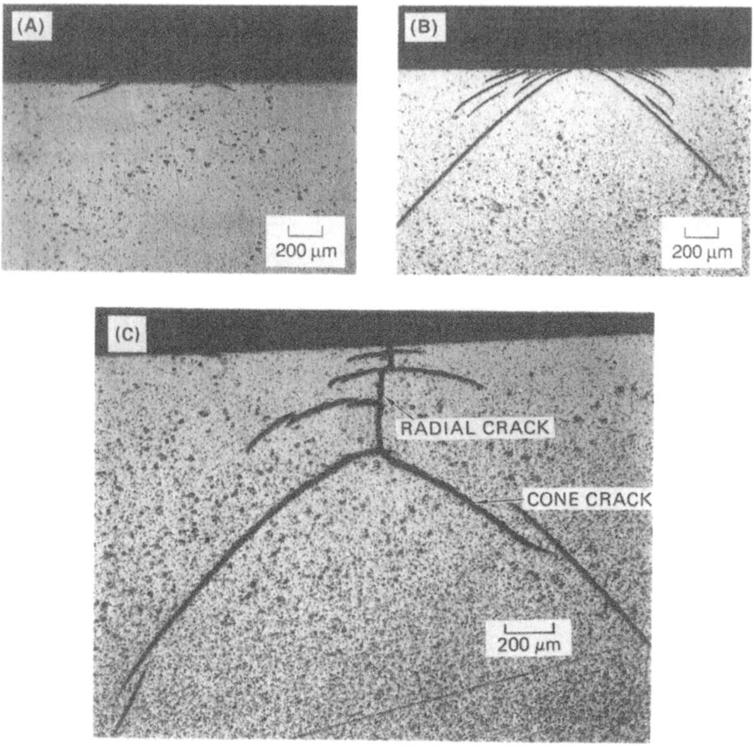

Figure 8.3. Particle impact damage in silicon nitride. From Shockey et al. (1990).

Several investigators have reported on the "dwell time" between impact and the onset of penetration (Wilkins et al., 1971; Zil'berbrand et al., 1991; Bless et al., 1992; den Reijer, 1991). This time seems to be of the order $10T/C_L$, when T is the tile thickness and C_L is the longitudinal sound speed. The dwell time is critical for defeat of the projectile, because during the dwell time the target is immobile as the projectile (especially if brittle) is destroyed. Wilkins et al. (1971) and others have reproduced the dwell phenomena in computer simulations, but only by introducing non-physical failure criteria.

Soviet investigators (Zil'berbrand et al., 1989) have proposed a qualitative theory that dwell is due to time required for a failure front to propagate from the rear surface to the impact side of the tile. In general however, a satisfactory model for the dwell time has not yet been developed. This situation has motivated the authors and others to conduct research on damage development and propagation in ceramics which is described in Section 8.4.

The penetration mechanics of thick-tile ceramic armor is also incompletely understood. This type of armor is used against long rod projectiles which erode while penetrating (see, for example, Shockey et al., 1990; Anderson

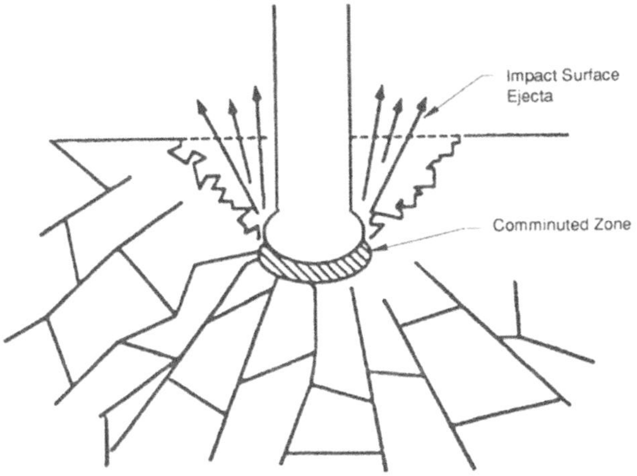

Figure 8.4. Damage in thick ceramic tile that has been penetrated by a high speed projectile. From Shockey et al. (1990).

and Morris, 1992). The impact results in failure and "flow" of the ceramic in compressive shear. Figure 8.4 illustrates the type of damage experienced by such a target. There has been only limited success conducting numerical simulation of thick tile armors, and the inability to make accurate predictions has led to the suspicion that some essential features of target failure are still not treated by the numerical approaches (Bless et al., 1993). Of particular importance is the observation that the resistance to penetration decreases for thick tiles (Hauver, 1992). The decreased resistance of thick tiles, as well as increased resistance at very high penetration velocities (Kozhushko et al., 1991), have been attributed to time-dependent failure phenomena in ceramics. The authors believe that this is one of the most important aspects of ceramic behavior that has yet to be brought into the conceptual framework of wave propagation codes. We hope that the review provided in Section 8.4 will aid future research in this field.

8.3. Damage Models

With the advent of increased capabilities and availability of advanced computers, it is now possible to perform three-dimensional finite element analyses to evaluate the damage process in ceramic materials due to both high velocity projectile impacts and low velocity particle impacts. Realistic impact-damage analysis requires accurate descriptions of material strength and failure criteria in the codes. The mathematical description of the strength variation with respect to strain, strain rate, temperature, and loading history usually is referred to as constitutive modeling. To use shock-wave-propagation based finite element/difference codes as dependable design tools, we need physically based constitutive models. In this section, we identify a few models recently developed for this purpose.

8.3.1. Background

The impact behavior of ceramics is dominated by stiffness loss. The inelastic deformations are due to microcracking and dislocation-generated microplastic flows. The approach to modeling microcracks is based on the (strain) response of a single crack to an externally applied stress field which is calculated using the appropriate stress-free boundary conditions on the crack surfaces. The derived strain components include the effects of microcracks in a brittle material through a crack density parameter. However, the bulk material itself is treated as a continuum so that all the stress–strain equations can be derived under the continuum-mechanics based theoretical framework. The microplasticity in a brittle solid is often attributed to the disloca-

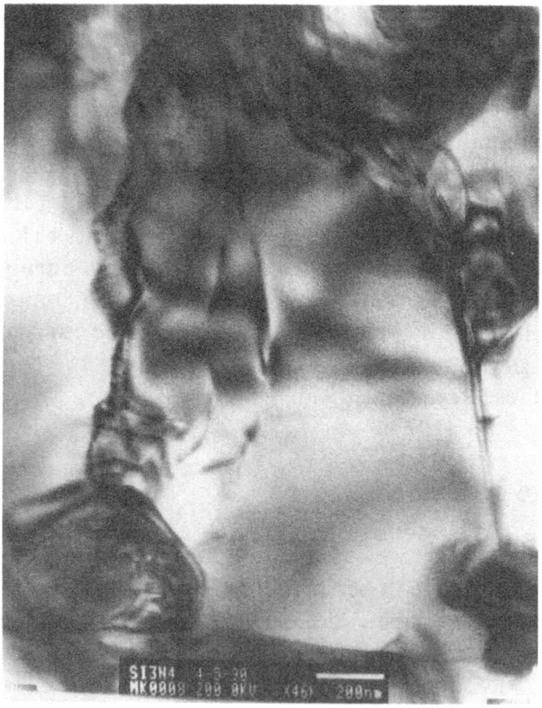

Figure 8.5. TEM picture of the impacted region of a silicon nitride bar revealing dislocation pileups.

tion motions in the vicinity of microflaw tip regions. In brittle solids, large scale grain distortions are usually absent.

A microscopic evaluation of the impacted area of a silicon nitride bar revealed dislocation pileups as shown in Figure 8.5. Several static indentation studies on silicon nitride also supported the presence of dislocation pileups. Figure 8.6 shows the dislocation pileups in 99.5 percent pure aluminum oxide (AD995) plates due to planar impact loading as found by Raiser and Clifton (1991). The presence of twinning in several ceramics was reported by Nemat-Nasser et al. (1993). These recently observed microstructural-scale plastic strains support the evidence of limited plasticity in most ceramics.

The recovered low velocity plate impact samples (AD995) of Raiser and Clifton (1992) also contained microcracks which were generated under global compressive pressure loading conditions. Microcracks in the compression-dominated regions of the AD995 plate are shown in Figure 8.7(a). The microcracks were almost closed indicating crack-growth under mode II and mode III (sliding modes); whereas the microcracks in the tension-dominated region in Figure 8.7(b) were clearly opened indicating crack-growth under mode I (opening mode). Cook (1993) also presented similar results in his studies on AD85 alumina.

Nemat-Nasser et al. (1993) recovered split Hopkinson bar ceramic samples and examined them under both transmission electron microscope (TEM) and scanning electron microscope (SEM). These samples showed

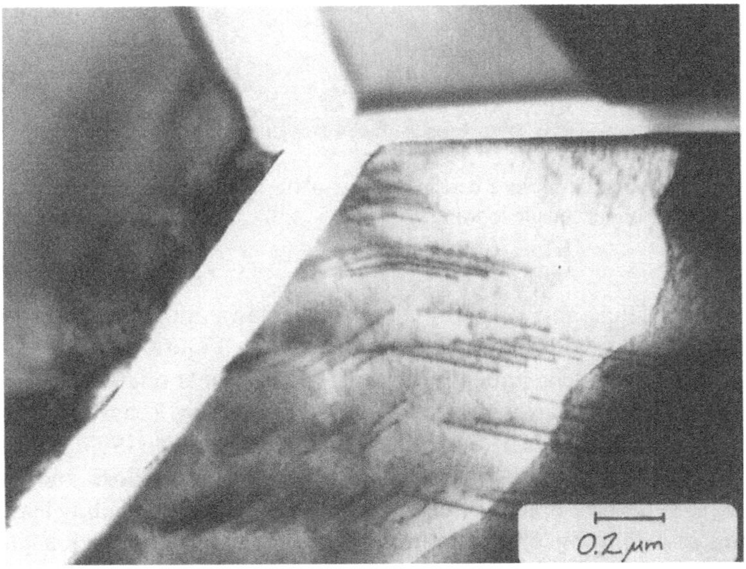

Figure 8.6. Dislocation pileups near triple point grain boundaries in AD995 alumina target plate due to flyer plate impact.

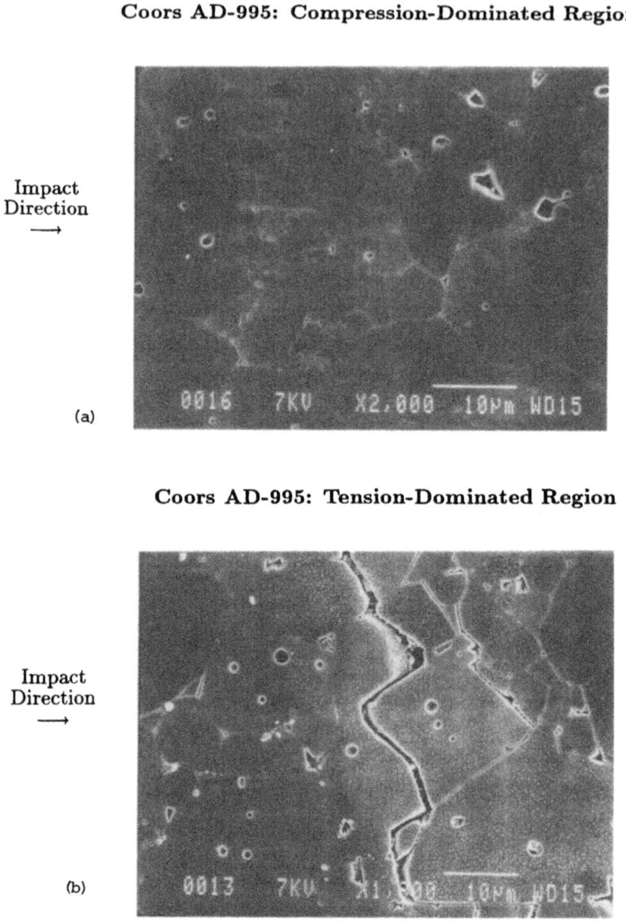

Figure 8.7. Microcracking in a ceramic target plate; a) sliding under shear loading, and b) opening under tensile loading.

dislocation pileups and twinning along microflaws and also the formation of "wing" cracks under compressive loading. In Figure 8.8, the so called "wing" cracks are schematically shown. The concepts relating to the wing (kinked or curved) cracks were originally proposed by Zener (1948), Stroh (1957), and several others (e.g., Brace and Bombolakis (1963), Hoek and Bieniawski (1965), Nemat-Nasser and Horii (1982), and Horii and Nemat-Nasser (1985). These cracks often grow under local tensile loading leading to splitting and faulting type fractures of brittle materials under a globally compressive loading.

In the absence of post-impact measurements on most ceramic materials, behavior under high velocity impact is usually inferred from analysis of

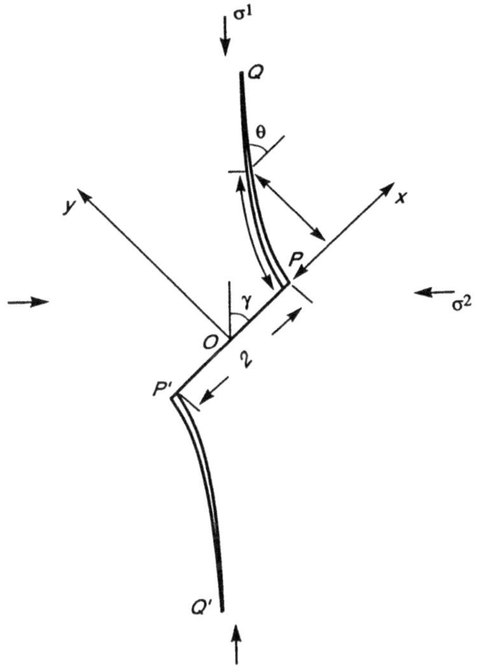

Figure 8.8. Pre-exisiting flaw PP' and curved (wing) cracks PQ and $P'Q$ and PQ'. $\sigma 1$ and $\sigma 2$ are applied axial and lateral stresses [Horii and Nemat-Nasser (1985)].

stress and velocity data from plate impact experiments. These measurements directly provide many dynamic properties, such as the Hugoniot stress and spall strength.

Rajendran and Cook (1988) conducted an in-depth literature review of impact behavior of ceramics and reported the need for ceramic constitutive and damage models in advanced hydrocodes. Subsequently, during 1988–93, several new models have been developed [Johnson and Holmquist (1993), Steinberg (1992), Rajendran and Kroupa (1989), and Rajendran (1992, 1993), Addessio and Johnson (1989)]. The recently developed constitutive/damage models for hydrocode application studies do not model the wing crack type micromechanisms. However, under large confinement pressures as in one dimensional strain (planar impact) loading conditions, these mechanisms are not dominant. Crack opening leading to dilation (volume increase) under global compressive loading is often termed as "bulking." Bulking type failures have been observed in rocks and concrete like brittle materials. Johnson and Holmqist (1992) introduced the bulking effects into their model through the pressure-volume relationship (equation-of-state).

Recently, Rajendran (1992, 1993) reported a continuum and fracture mechanics based three dimensional constitutive model to describe the impact behavior of ceramic materials. This three-dimensional, continuum damage model [see Rajendran (1992)], has been incorporated into the EPIC-2 finite element code [Johnson and Stryk (1986)]. This model was developed mainly for predicting projectile penetration into ceramic targets. However, the same model was successfully applied to predict the nucleation and propagation of ring and cone cracks in ceramic components due to particle impacts. This project demonstrated the potential "dual use" of such advanced models. This elasticplastic-microcrack strains based model is briefly described in the following section.

8.3.2. Constitutive Equations

The three stages of microcrack damage (nucleation, growth, and coalescence) occur at different times within a given elemental volume. Since current experimental diagnostics are incapable of resolving these stages of microcrack damage, incorporating this process into a constitutive theory becomes a challenging task. In the modeling of the impact behavior of ceramics, the nucleation process is not considered. The model assumes the presence of microflaws prior to any loading.

The total strain is decomposed into elastic and plastic strains as,

$$\varepsilon_{ij} = \varepsilon_{ij}^{e} + \varepsilon_{ij}^{p}. \tag{8.2}$$

The elastic strains (including crack displacements) in the microcracked material are related to the stress tensor as follows:

$$\varepsilon_{ij}^{e} = C_{ijkl}\sigma_{kl}, \tag{8.3}$$

where C_{ijkl} are the effective degraded compliance tensor components (Margolin, 1983) of the microcracked material. If C_{ijkl} is analytically inverted to the stiffness tensor components M_{ijkl} (Rajendran, 1992), the resulting stress state is:

$$\sigma_{ij} = M_{ijkl}(\gamma)\varepsilon_{kl}^{e}. \tag{8.4}$$

The elements of this matrix vary with the crack density, which is defined as:

$$\gamma = N_0^* a^3. \tag{8.5}$$

N_0^* is the average number of microflaws per unit volume in the ceramic which is a model constant, and a is the maximum microcrack size which is treated as an internal state variable. These flaws extend according to the generalized Griffith criteria [Margolin (1983) and Dienes (1986)] so crack advance is determined by the friction, μ, and fracture toughness K_I. As the microcracks extend, γ increases and the stresses relax. The increase in γ is entirely due to the increase in the crack size. The initial crack density γ_0 is determined from the relationship (8.5) using the values for the two material constants: N_0^* and the (assumed) initial value of a. The damage evolution equation is derived from a fracture-mechanics based relationship for a single crack propagation under dynamic loading conditions. An evolution law for the state variable γ is described through a strain energy release based macrocrack growth rate law:

$$\dot{\gamma} = 3n\frac{\gamma}{a}C_R\left[1 - \frac{G_c}{G_a}\right]. \tag{8.6}$$

Where C_R is the Rayleigh wave speed, G_c is the critical strain energy release for microcrack growth, and G_a is the applied strain energy release. Note that $\dot{\gamma} = 0$, when $G_a < G_c$, and n limits the crack growth rate ($n \leq 1$).

When the crack density reaches a critical value of 0.75 (Rajendran, 1992), the model assumes that the microcracks have coalesced, leading to pulverization of the ceramic.

Damage also modifies the pressure volume relationship. The equation of state is described by the following modified Mie–Grüneisen relationship:

$$P = R_K[\bar{P}_H(1 - 0.5\Gamma\mu) + \Gamma\rho_0(I - I_0)], \tag{8.7}$$

where

$$\bar{P}_H = (\bar{K}/K)(\beta_1\varepsilon_v + \beta_2\varepsilon_v^2 + \beta_3\varepsilon_v^3). \tag{8.8}$$

$\varepsilon_v (= 1 - V/V_0)$ is the elastic volume compressibility (strain) and β_1, β_2, and β_3 are the empirical parameters for the intact material. Γ is the Mie–Gruneisen parameter, ρ_0 is the initial density, I_0 is the initial value of internal energy, and I is the current internal energy. R_k is the correction factor for porosity in the ceramics. For low velocity impact (low pressure applications), β_2, β_3, and Γ are assumed to be zero. $\bar{K}$ is the degraded bulk modulus due to microcracks. K is the bulk modulus of the intact ceramic.

8.3.3. Particle Impact Configuration

Rajendran et al. (1992) considered a combined experimental and computational approach to analyze the particle-impact induced damage in silicon nitride bars. In the experiment, a silicon nitride bar (3.18-mm thick, 6.34-mm wide, 50-mm long) was impacted by silicon nitride spheres of two different diameters (0.8 and 1.6-mm). Bertke (1988) established the threshold velocities for ring crack formation in the bar specimens. The details of the experiments are provided by Bertke (1988). Two threshold velocities for ring crack formation, 210 m/s and 127 m/s, respectively, for the two sizes 0.8 and 1.6 mm of the spherical projectiles were determined from the experiments.

Using the axisymmetric option in the 1986 version of the EPIC-2 code and the parameters in Table 8.1, a 1.6-mm diameter silicon nitride sphere impacting a 50-mm diameter, 3.18-mm thick silicon nitride plate at a velocity of 127 m/s was modeled. The corresponding finite element mesh is shown in Figure 8.9(a). Since the measured average diameter of the ring crack in the experiments is about 0.5-mm, the modeling of the ring crack formation demanded a very fine mesh near the impact site. In Figure 8.9(b), the tiny area near the impact site is enlarged to show mesh details. The smallest nodal spacing is about 1/10th of the ring crack diameter. This fine resolution is necessary for an accurate calculation of the stress distribution in the target

Table 8.1. Ceramic model constants for silicon nitride

Symbol (units)	Value	Description
K_{IC} MPa$\sqrt{m}$	6	Static fracture toughness
μ	0.30	Friction coefficient
N_0^* m^{-3}	1.2 x 10^9	Total number of microcracks per unit volume
a_0 mm	0.00254	Initial maximum microcrack size
n	0.01	Controls limiting microcrack growth rate for Modes II and III

(a)

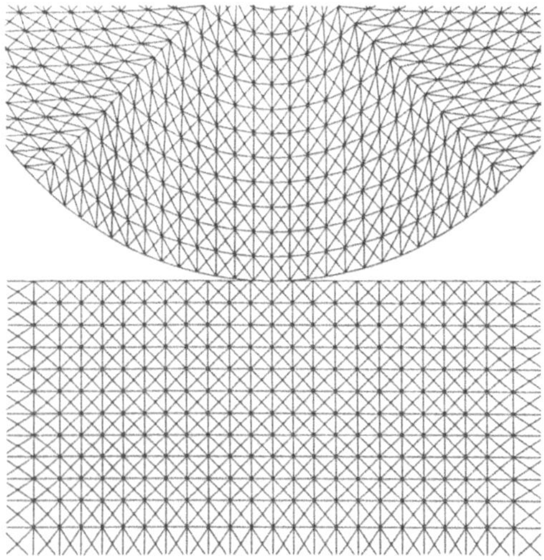

(b)

Figure 8.9. EPIC-2 mesh for a 1.6-mm diameter sphere impacting a 50-mm diameter, 3.18-mm thick disk: (a) entire mesh, and (b) mesh details near the impact size.

near the ring crack location. Modeling of the entire standard bar requires a three-dimensional code analysis. However, for model constant determination, the impact configuration was idealized as a two-dimensional axisymmetric geometry. In other words, a sphere impacting a solid circular disk was modeled.

The effects of the lateral surface on the ring crack predictions was critically examined with two simulations: (1) a 500-mm diameter disk, and (2) a 6.4-mm diameter disk, each impacted by a 1.6-mm diameter silicon nitride sphere. The results were very similar, and the damage contours matched well. The

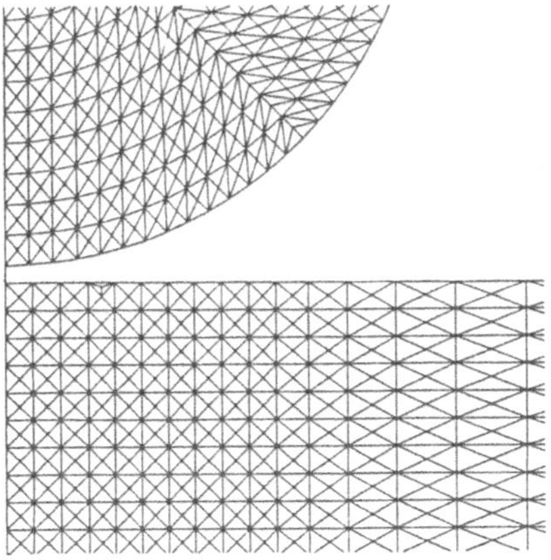

(a)

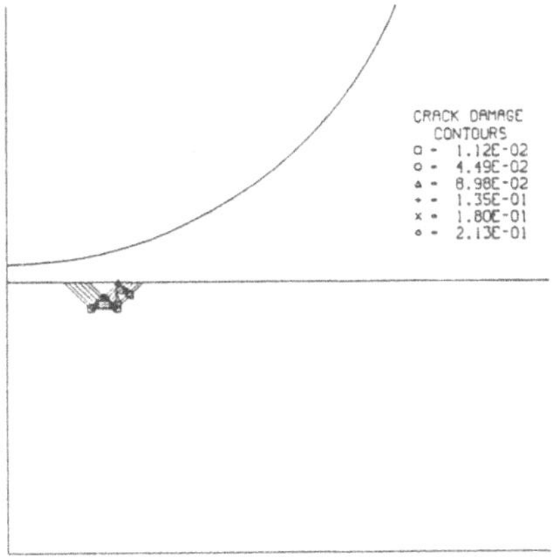

(b)

Figure 8.10. Simulated damage in the silicon nitride target resulting from a 1.6 mm diameter particle impacting at 127 m/s: (a) damaged elements, indicated by the symbol "Y," and (b) damage contours.

stress histories deviated slightly between the two cases. However, this deviation did not influence the ring crack evolution. Based on the results from these two cases, the use of a smaller radius disk to calibrate the model constants was sufficient.

The experiment in which a 1.6-mm diameter silicon nitride sphere impacting a 6.35-mm thick disk at 127 m/s (threshold velocity for ring crack formation in the experiments) was simulated. The result from this run is shown in Figure 8.10. The damage plots shown in Figures 8.10(a) and 8.10(b), clearly demonstrate the formation of the ring crack. The plots also show the development of compressive failure zones directly beneath the impact site, which qualitatively agreed with the experimental observations. The damage parameter, γ, was one for the surface element and decayed rapidly in the interior, as shown by the damage contours in Figure 8.10(b).

The detailed metallographic studies on ring crack specimens, conducted by Osborn (1991) at the University of Dayton Research Institute, revealed microdamage around the ring crack exactly as predicted. There was also a permanent indentation (or imprint) beneath the impact site where the grains were aligned perpendicular to the shock direction.

To further validate the model constants, the impact of a smaller sphere (0.8-mm) on the standard bar at a velocity of 210 m/s was modeled. The

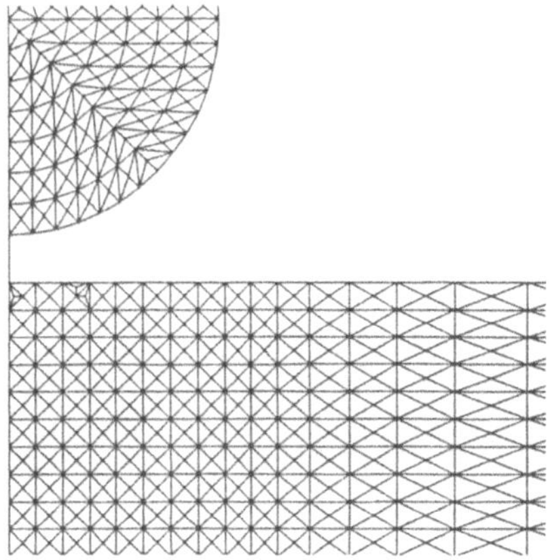

(a)

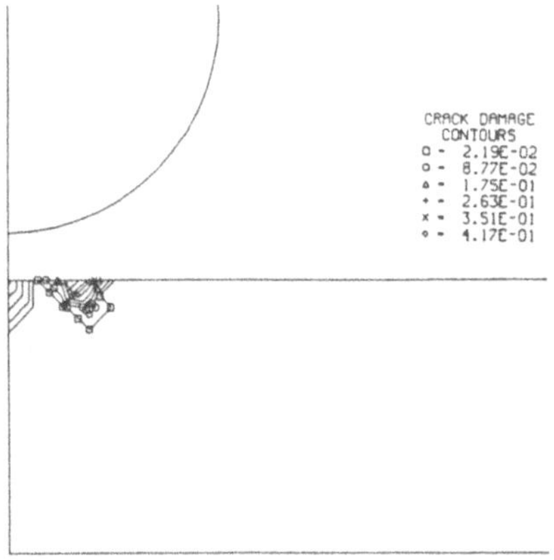

(b)

Figure 8.11. Simulated damage in the silicon nitride target resulting from a 0.8 mm diameter particle impacting at 210 m/s: (a) damaged elements, indicated by the symbol "Y," and (b) damage contours, revealing crushed zone under the impacted area.

model predicted the ring crack evolution as shown in Figures 8.11(a) and 8.11(b). Clearly, one can see the evolution of microdamage leading to the ring crack formation. A compressive damage zone beneath the impact point also can be seen from the cracked element plots in Figure 8.11(a).

To verify the 2-dimensional assumption of a 3-dimensional impact, the impact of a 1.6-mm silicon nitride sphere on the bar was modeled using the 3-dimensional version of the EPIC code. In the EPIC-3 analysis of the standard bar, the bar and the sphere were discretized with 26676 elements and 5460 nodes. At least this amount of fineness is required so that the ring crack analysis would yield dependable results. The ring crack region was modeled with extremely small 3-dimensional elements.

In Figure 8.12, a portion of the mesh configuration of a 1.6-mm silicon nitride sphere impacting the standard bar is shown. The material model constants determined from the 2-dimensional EPIC simulations were employed in these computer runs. The elements that failed, due to microcracking, are indicated by the "x" symbol inside the elements. In Figure 8.12,

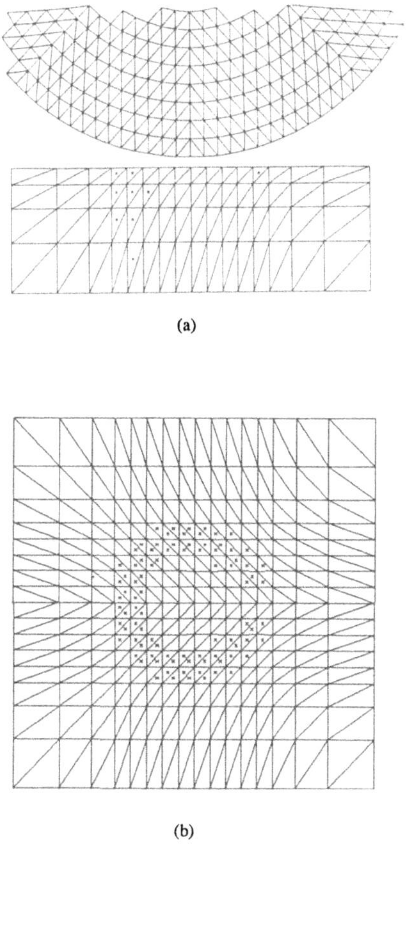

(a)

(b)

Figure 8.12. Damaged elements (indicated by the symbol "x") in the silicon nitride standard bar target for a 1.6 mm diameter particle impacting at 127 m/s: (a) view of thickness direction, and (b) top surface view of target.

several of the elements at and around the ring crack location were damaged, with damage reaching one. This observation is in agreement with the EPIC-2 and experimental results. The damage propagated as a cone crack into the interior. However, it failed to match the experimentally observed cone angle. This could be due to the poor resolution of the mesh in the thickness direction.

In general, with the advent of increasing super computing capabilities, in the future it will be possible to perform accurate 3-dimensional calculations using advanced damage models in shock-wave-propagation-based codes. These results demonstrate the possibilities of performing 3-dimensional impact analysis of ceramic engine components or for that matter any other components, such as automobile coatings or transparancies.

8.4. Failure Propagation Speed

The models described in the previous section enable simulations of impact-induced damage within the framework of continuum mechanics. As we have shown, they are very valuable for predicting the onset of damage. They can also predict the properties of grossly damaged brittle materials that are produced when impact stresses are sustained and well above strength levels. Still outside the scope of such models is the ability to predict the propagation of individual cracks and crack networks that may be initiated by impact, especially at early times when the damage may have only partially progressed through the region that will ultimately fail.

8.4.1. Time Scales

The time duration, t, of impact loads is generally rather short. For elastic impacts of spheres, the duration is determined by the time required for the contact force to reverse the momentum of the projectile; the classical solution (see, for example, Goldsmith, 1960) can be expressed as:

$$t \cong \frac{9R}{C_E^{4/5} V_0^{1/5}}, \tag{8.9}$$

where R is particle radius, V_0 is impact velocity, and C_E is an average sound velocity, defined as $\sqrt{\langle E \rangle / \rho_p}$ where $\langle E \rangle$ is the harmonic mean of the Young's moduli of the impactor and target, and ρ_p is impactor density. However, the time may be much less for a blunt elastic projectile; for the case of flat discs or rod,

$$t \cong \frac{2R}{C_L}, \tag{8.10}$$

where here R is the impactor thickness in the line of flight. If the projectile squashes up on the target (such as a bird, or a gelatin cylinder, or a ductile long rod), the impact duration is approximately:

$$t \cong \frac{2L}{V_0}, \tag{8.11}$$

where $2L$ is the projectile length in the direction of flight.

How far impact-induced damage can propagate during these short time intervals is crucial for determining the extent of target damage. In lower speed impacts, damage is in the form of independent cracks. In more severe impacts, damage is in the form of a failure front. In both cases, damage may start at or near the projectile contact, or it may start at a remote location where the impact-induced compression stress has been reflected as a tensile stress.

8.4.2. Crack Propagation

The simplest case consists of individual cracks that grow during the time that impact stress is applied. The maximum propagation speed of a single crack is usually assumed to be the Rayleigh speed, C_R, which is 0.58 C_L for Poisson's ratio equal 0.25 (see, for example, Freund, 1990). However, the velocities observed in real materials are much less than this, typically only 0.5 C_R for Mode I (tensile) cracks (see, for example, Williams and Knauss, 1985). Lawn and Wilshaw (1975) report 0.38 C_L as an upper estimate. In acrylics, cracks will not propagate faster than 0.34 C_R regardless of specimen geometry or atmosphere (Fineberg et al., 1992). Nevertheless, much higher speeds are sometimes observed under impact loading. Senf et al. (1988) report that although the normal maximum crack velocity in glass is 1460 m/s, (0.25 C_L) when cracks are driven by a diverging stress field, a maximum velocity of 4.6-mm/μs is measured (0.8 C_L). Winkler et al. (1989) found no maximum damage speed for alumina, and values exceeding 8 km/s were observed ($\sim$0.8 C_L). In traditional fracture mechanics experiments crack speeds in alumina do not exceed 0.25-mm/μs (Deobald and Kobayashi, 1992).

Thus, although isolated cracks are apparently limited to speeds well below C_R, under dynamic loading it is sometimes possible to produce cracks that propagate at considerably higher speeds. The conditions that result in higher crack speeds are not yet well defined, nor is there concensus about why conventional cracks are limited to about 0.3 C_R. Yoffe, (1951), Cotterell (1965), and others have developed explanations based on rotation of the principal stress in front of rapidly advancing cracks, causing crack branching. Fineberg et al. (1992), invoke instability concepts from nonlinear dynamics, pointing out the increasing roughness of the failure surface when limiting velocities are approached.

8.4.3. Fracture Fronts

It is possible for pre-stressed brittle materials to fail by propagation of many interacting individual cracks. This is called a fracture front. In front of a fracture front the material is intact, and behind the fracture front the material is comminuted. A large reduction in strength of brittle materials occurs at the fracture front, and the elastic energy of the intact material is transformed into kinetic energy, surface energy, and heat in the comminuted material. Thus, development of fracture fronts is often associated with explosive fracture of brittle materials. A common example is the explosive disintegration of tempered glass when the surface is broken.

Theoretical work on moving fracture fronts is relatively sparse. The first and perhaps only systematic treatment is in Cherepanov's book (Cherepanov, 1979). He derives existence criteria for self propagating failure waves: the elastic energy density of the initial material should greatly exceed that of the comminuted material, and also exceed a certain function of the fracture dissipation energy (which is assumed to be a material constant).

The theoretical propagation speed of a fracture front, C_F, is a major concern. Cherepanov finds that self propagating stable waves move at the bulk sound speed, but if the dissipation energy criterion is not fulfilled, unstable failure waves having lesser speeds may still exist. Another frequently cited Russian reference (Finkel, 1970), says $0.5\ C_L < C_F < 0.9C_T$. The fracture fronts observed by Senf et al. (1988) in glass propagated at 4.6 mm/μs, which is $0.8C_L$ or $2.3C_T$. Nikolaevski (1981) in an extended, although qualitative, discussion of failure waves assumed $C_F = C_R$. While not specifying the limit, Janach (1976) also emphasized that brittle materials can only fracture at a limiting velocity.

Rapidly advancing fracture fronts are sometimes called failure waves, and in cylindrical specimens were observed by Glenn and Janach (1977) in rock and in glass by Bless et al. (1990). They were detected in 1D strain geometry in rocks by Young and Dubugnon (1977) and in glass by Kanel et al. (1992) and Bless et al. (1992). Moving fracture fronts in glass caused by ballistic impact have been observed by Senf et al. (1988) and Hauver et al. (1991).

Bar impacts provide a relatively easy means to generate and observe this phenomenon. Bless et al. (1990, 1994) reported on projectiles striking the end of a circular cross section (12.7-mm diameter) pyrex or soda lime glass rods. Projectiles were metal plates or glass rods. Care was taken to ensure planar impact. Stress was measured with embedded manganin gauges. High speed photography of these bars revealed the presence of apparent failure waves, as shown in Figure 8.13. Two moving boundaries are visible in this image-converter camera record: there is a failure front behind which the glass becomes opaque. It is followed by a slower wave behind which the glass expands radially. This phenomenology differs slightly from the failure seen in rock bars (Glenn and Janack, op cit.), in which violent radial expansion immediately followed the failure front. In pyrex glass, the failure wave speed

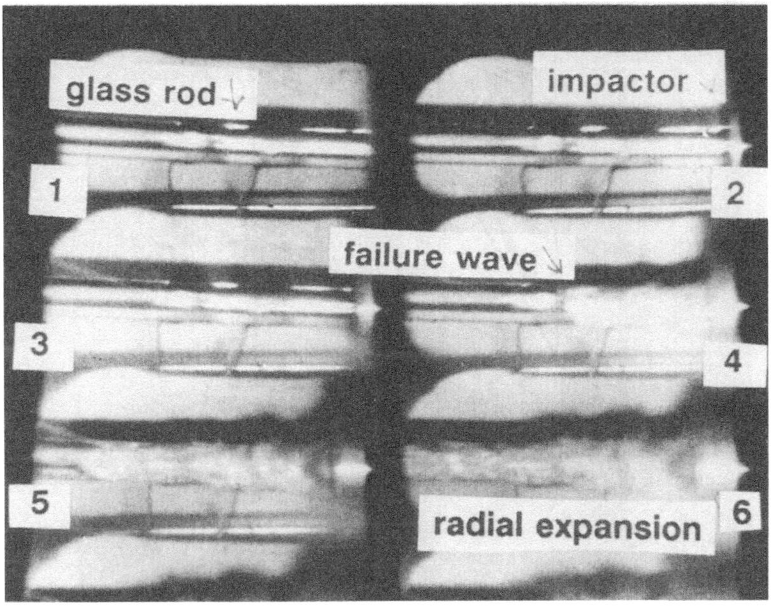

Figure 8.13. Failure waves produced in 12.7-mm diameter, one decimeter long glass bars by projectile impact. From Bless and Brar (1992).

was 2.3-mm/μs in low speed impacts, and it increased to 5.2-mm/μs at 330 m/s impact velocity. The theoretical peak impact stress in these experiments ranged from 0.8 GPa to 2 GPa. The generation of the failure front did not appear related to the ratio of impact stress to unconfined compressive strength of the bars, which, based on the manganin gauge records, was about 1.3 GPa. This suggests that the fracture front may arise from surface flaws. In experiments in soda lime glass, which used a crusher washer on the impact face, and in which impact stresses were above the compressive strength, it was observed that failure waves were initiated not only at the impact face, but also at the gauge plane. In these experiments, the two failure fronts appeared to "quench" one another, leaving a central region several diameters long that never fractured. Experiments in pyrex and soda lime glass bars gave qualitatively similar results. The detailed nature of the failure process and role of flaws on end surfaces and lateral surfaces have not yet been elucidated.

Failure waves in glass plates were first reported by Kanel et al. (1992). The phenomenology that has been deduced from his experiments and those conducted at the University of Dayton (Bless et al., 1992; Brar et al., 1991a,b) is shown in Figure 8.14. At a critical impact velocity, which is below the Hugoniot elastic limit (HEL), failure was generated at the flyer plate/target interface and propagated into the interior. In pyrex, these waves travelled at

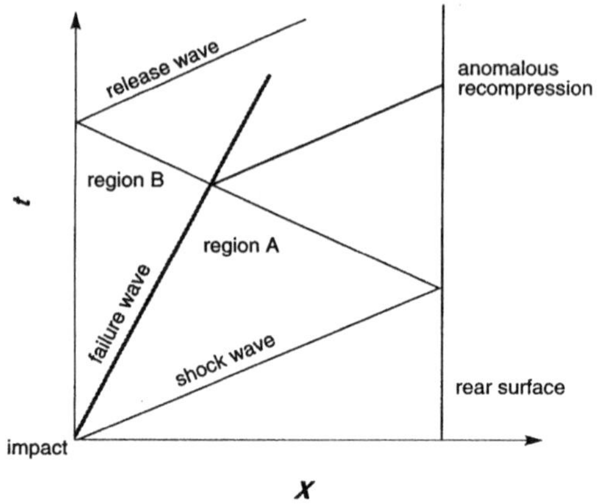

Figure 8.14. Sequence of shock and failure waves produced in flyer plate experiments. From Kanel et al. (1992) and Bless et al. (1992).

about 2 km/s. It is not clear whether this speed is constant with distance. Measurements of spall stress and transverse stress (from which the stress deviator can be computed) were made in these types of experiments. It was found that comminution of the glass occurs at the failure front, and not immediately behind the wave associated with the HEL. Thus, the HEL apparently represents densification of glass (as suggested also by Cagnoux, 1985, and Gibbons and Ahrens, 1971), and not shear failure. Only at the failure front do the spall strength and stress deviators decrease, indicating comminution. Since the stress deviator decreases but the maximum principal stress is unchanged, the failure process produces a self-pressurization of the glass.

Moving failure fronts that lagged the plastic wave in an aluminosilicate glass plates were observed by Raiser and Clifton (1994). They found impact surface roughness did not affect this phenomenon, and they present the hypothesis that (in plate impacts) the failure front is caused by a slowly propagating densification phase change.

The interpretation of moving fracture fronts is still unclear. Basically there are two approaches. Cherepanov (1979), Kanel, et al. (1992), and Bless, et al., (1992) hypothesize a failure that starts at an interface in a highly stressed unfailed material, and then propagates at a characteristic rate. Nikolaevski (1981), and Grady (1993) hypothesize a fracture process that is initiated at the compression wave front, but involves some sort of incubation process that causes the fracture to lag the stress wave front.

Implications of these measurements have yet to be incorporated into wave propagation codes. The moving failure front phenomenon is outside of the framework of conventional constitutive modeling. However, the authors believe that many problems associated with impact damage and penetration of brittle materials will not be solved until crack and failure speeds are brought into impact analysis and numerical simulations.

8.5. References

Addessio, F.L. and Johnson, J.N. Constitutive Model for the Dynamic Response of Brittle Materials, LA-UR-89-2651, Los Alamos National Laboratory, Los Alamos, NM 1989.

Akimune, Y., Katano Y., and Matoba K. Spherical-Impact Damage and Strength Degradation in Silicon Nitrides for Automobile Turbocharger Roters, *Am. Ceram. Soc.* **72**, 1422–1428 (1989).

Anderson, C.E., Jr. and Morris, B.L. The Ballistic Performance of Confined Al_2O_3 Ceramic Tiles, *Int. J. Impact Engng.* **12**, 167–188, 1992.

Bertke, R.S. UDR-TR-88-42, University of Dayton, Dayton, OH 45469, April 1988.

Bertke, R.S. Private communication, 1993.

Bless, S.J., Benyami, M., Apgar, L.S., and Eylon, D. Impenetrable Ceramic Targets Struck by High Velocity Tungsten Long Rods, Proc. Conf. Structures Under Shock and Impact, SUSI 92, ed. P.S. Bulson, Computational Mechanics Publications, 1992.

Bless, S.J. and Brar, N.S. Impact Induced Fracture of Glass Bars, 1813–1816, *High-Pressure Science and Technology—1993*, American Institute of Physics, 1994

Bless, S.J., Brar, N.S., Kanel, G., and Rosenberg Z. "Failure Waves in Glass," *J. Am. Ceram. Soc.*, **75**, 1002–1004, 1992.

Bless, S.J., Brar, N.S., and Rosenberg, Z. Failure of Ceramic and Glass Rods Under Dynamic Compression, Shock Compression of Condensed Matter, 1989, eds. S.C. Schmidt et al., pp. 939–942, Elsevier Science Publishers, 1990.

Bowden, F.P. and Field, J.E. The Brittle Fracture of Solids by Liquid Impact, by Solid Impact or by Shock, Proc. R. Soc., London, A282, 331, 1964.

Brace, W.F. and Bombolakis E.G. A Note on Brittle Crack Growth in Compression, *J. Geophys. Res.*, vol. 68, pp. 3709–3713, 1963.

Brar, N.S., Rosenberg, Z., and Bless, S.J. Spall Strength and Failure Waves in Glass, *J. de Physique IV*, pp. C3-639–C3-644, 1991a.

Brar, N.S., Bless, S.J., and Rosenberg, Z. Impact-Induced Failure Waves in Glass Bars and Plates, *Appl. Phys. Lett.*, vol. 59, pp. 3396–3398, 1991b.

Cagnoux, J. Deformation et Ruine d'un Verre Pyrex Soumis a Un Choc Intens, Thesis: University de Poitiers, 1985.

Cheng, W.L., Jen, K.C., and Langlie, S. Failure of Ceramics due to Impact Loading, Dynamic Response of Structures to High-Energy Excitations, pp. 85–91, ASME, 1991.

Cherepanov, G.P., *Mechanics of Brittle Fracture*, McGraw-Hill, 1979.

Cook, R.F., Fairbanks, C.J., Lawn, B.R., and Mai, Y.W. Crack Resistance by Interfacial Bridging: Its Role in Determining Strength Characteristics, *J. Materl. Res.*, vol. 2, pp. 345–356, 1987.

Cook, W.H. A Ceramic Fracture Model for High Velocity Impact, WL-TR-93-7023, Wright Laboratory Armament Directorate, Eglin AFB, FL, May 1993.

Cotterell, B. Velocity Effects in Fracture Propagation, *Appl. Mat'ls Res.*, pp. 227–232, Oct 1965.

Curran, D.R., Seaman, L., and Shockey, D.A. *Phys. Rept.*, vol. 147, 1987.

Dienes, J.K. Comments on 'A Generalized Griffith Criterion for Crack Propagation,' by L.G. Margolin—a Technical Note, Eng. Fracture Mechanics, 3, 23, pp. 615–617, (1986).

den Reijer, P.C. Impact on Ceramic Faced Armour, Thesis: Technical University of Delft, 1991.

Deobald, L.R. and Kobayashi, A.S. A Bar Impact Tester for Dynamic Fracture Testing of Ceramics and Ceramic Composites, *Exp. Mech.*, pp. 109–116, June 1992.

Evans, A.G. Impact Damage Mechanics: Solid Projectiles, *Erosion*, ed. C.M. Preece, Academic Press, 1979.

Evans, A.G., Gulden, M.E., and Rosenblatt, M. Impact Damage in Brittle Materials in the Elastic-Plastic Response Regime, Proc. R. Soc., London, A361, pp. 343–365, 1978.

Evans, A.G. and Wilshaw, T.R. Dynamic Solid Particle Damage in Brittle Materials: An Appraisal, *J. Materl. Sci.*, vol. 12, pp. 97–116, 1977.

Fineberg, J., Gross, S.P., Marder, M., and Swinney, H.L. Instability in the Propagation of Fast Cracks, *Phys. Rev. B*, vol. 45, pp. 5146–5154, 1992.

Finkel, V.M. *Physics of Failure*, Moscow: Metallurgiya, 109, 1970.

Freund, L.B., *Dynamic Fracture Mechanics*, Cambridge University Press, 1990.

Gibbons, R.V., Ahrens, T.J. Shock Metamorphism of Silicate Glasses, *J. Geophys. Res.*, vol. 76, pp. 5489–5498, 1971.

Glenn, L. and Janach, W. Failure of Granite Cylinders Under Impact Loading, *Int. J. Fracture*, vol. 13, pp. 301–317, 1977.

Goldsmith, W. *Impact*, 1960.

Griffith, A.A. *Trans. Roy. Soc. Lond.* Ser. A, vol. 221, p. 163, 1921.

Hauver, G., Netherwood, P.H., Benck, R.F., and Milani, A. Penetration of Shaped-Charge Jets into Glass and Crystalline Quartz, U.S. Army Report BRL-TR-3273, September 1991.

Hauver, G. Variation of Target Resistance during Long Rod Penetration into Ceramics, 13th Int. Symp. Ballistics, Stockholm, Sweden, 1–3 June 1992.

Hertz, H. *Collected Works*, vol. 1, 1882.

Hoek, E. and Bieniawski, Z.T. Brittle Fracture Propagation in Rock under Compression, *Int. J. Fract. Mech.*, vol. 1, pp. 137–155, 1965.

Janach, W. The Role of Bulking in Brittle Failure of Rocks under Rapid Compression, *Int. J. Rock Mech. Min. Sci. & Geomech. Abstr.*, vol. 13, pp. 177–186, 1976.

Johnson, G.R. and Holmquist, T.J. A Computational Constitutive Model for Brittle Materials Subjected to Large Strains, High Strain Rates, and High Pressures, 1075–1081, *Shock-Wave and High-Strain-Rate Phenomena in Materials*, ed. M.A. Meyers, L.E. Murr, K.P. Staudhammer, Marcel Dekker, 1992

Johnson, G.R. and Stryk, R.A. User Instructions for the EPIC-2 Code, AFATL-TR-86-51, Eglin Air Force Base, FL, (1986).

Kanel, G.I., Rasorenov, S.V., and Fortov, V.E. The Failure Waves and Spallations in Homogenous Brittle Materials, 451–454, *Shock Compression of Condensed Matter 1991*, Elsevier, 1992

Kozhuskhko, A.A., Rykova, I.I., and Sinani, A.B. Resistance of Ceramics to Penetration at Impact Velocities Above 5 km/s, *J. de Physique* IV, C3-117–122, 1991.

Lawn, B.R., Wiederhorn, S.M., and Roberts, D.E. Effect of Sliding Friction Forces on the Strength of Brittle Materials, *J. Materl. Sci.*, vol. 19, pp. 2561–2569, 1984.

Lawn, B.R. and Wilshaw, T.R. *Fracture of Brittle Solids*, Cambridge University Press, 1975.

Mai, Y.-W. and Lawn, B.R. *J. Am. Ceram. Soc.*, vol. 70, p. 289, 1987.

Makakawa, I. and Shin, H.-S. Damage Induced in SiC by a Particle Impact, *Engng. Fracture Mech.*, vol. 40, pp. 879–886, 1991.

Margolin, L.G. Elastic Moduli of a Cracked Body, *International Journal of Fracture*, vol. 22, pp. 65–79, 1983.

Margolin, L.G. A Generalized Griffith Criterion for Crack Propagation, *Engineering Fracture Mechanics*, vol. 19 (30), pp. 539–543, 1984.

Marshall, D.B., Lawn, B.R., and Evans, A.G., Elastic/Plastic Indentation Damage in Ceramics: The Lateral Crack System, *J. Am. Ceram. Soc.*, vol. 65, pp. 561–566, 1982.

Nemat-Nasser, S. and Horii, H. Compression Induced Nonplanar Crack Extension with Application to Splitting, Exfoliation, and Rockburst, *J. Geophys. Res.*, vol. 87, pp. 6805–6821, 1982.

Horii, H. and Nemat-Nasser, S. Compression-Induced Microcrack Growth in Brittle Solids: Axial Splitting and Shear Failure, *J. Geophys. Res.*, vol. 90, No. B4, pp. 3105–3125, March 1985.

Nemat-Nasser et al., Work at the Center of Exceleence for Advanced Materials, University of California, San Diego, La Jolla, CA 92093, 1993.

Nikolaevski, V.N. Limit Velocity of Fracture Front and Dynamic Strength of Brittle Solids, *Int. J. Engr. Sci.*, vol. 19, pp. 41–56, 1981.

Osborn, N., unpublished work, University of Dayton, Dayton, OH 45469, 1991.

Raiser, G. and Clifton, R. Performance of the Star Shaped Flyer in the Study of Brittle Materials, *J, Appl. Phys.*, vol. 72, pp. 3451–3457, 1992.

Raiser, G. and Clifton, R.J. Failure Waves in Uniaxial Compression of an Aluminosilicate Glass, 1039–1042, *High-Pressure Science and Technology—1993*, American Institute of Physics, 1994

Rajendran, A.M. High Strain Rate Behavior of Metals, Ceramics, and Concrete, WL-TR-92-4006, Air Force Report, Wright-Patterson Air Force Base, Ohio, April 1992.

Rajendran, A.M. and Kroupa, J.L. Impact Damage Model for Ceramic Materials, *J. of Appl. Physics*, 8, 66, pp. 7–20, 3560–3565, 1989.

Rajendran, A.M., Grove, D.J., and Dietenberger, M.A. A New Microphysical Model to Predict Particle Impact Induced Damage in Silicon Nitride Targets, UDR-TR-91-150, University of Dayton, Dayton, Ohio, 1992

Rajendran, A.M. Modeling the Impact Behavior of AD85 Ceramic under Multiaxial Loading, *Int'l J. Impact Engr.*, vol. 15, No. 6, pp. 749–768, 1994

Rajendran, A.M. and Cook, W.H. A Comprehensive Review of Modeling of Impact Damage in Ceramics, AFATL-TR-88-143, Air Force Armament Laboratory, Eglin Air Force Base, FL, December, 1988.

Ramamurthy, A.C., Bless, S.J., and Brar, N.S. Stone Impact Damage to Automobile Finishes, Joint AIRAPT/APS Conference, Colorado Springs, June 28–July 2, 1993.

Ramamurthy, A.C., Bless, S.J., and Brar, N.S. Stresses in Painted Steel Coupons from Stone Impact, 1091–1094, *High-Pressure Science and Technology—1993*, American Institute of Physics, 1994

Rosenberg, Z. Personal communication, 1989.

Rosenberg, Z., Bless, S.J., and Brar, N.S. On the Influence of the Loss of Shear Strength on the Ballistic Performance of Brittle Solids, *Int. J. Engng.*, vol. 9, pp. 45–50, 1990.

Rosenberg, Z., Yaziv, D., Yeshurun, Y., and Bless, S. Shear Strength of Shock-

Loaded Alumina as Determined with Longitudinal and Transverse Manganin Gauges, *J. Appl. Phys.*, vol. 62, pp. 1120–1122, 1987.

Senf, H., Rothenhausler, H., Pavel, W., Raatschen, H.-J., and Schwartz, R. Analysis of Experimental and Numerical Investigations with Rigid Projectiles Impacting Glass Targets of Differing Physical Quality, *J. de Physique*, C3, 1988.

Shockey, D.A., Marchand, A.H., Skaggs, S.R., Cort, G.E., Burkett, M.W., and Parker, R. Failure Phenomenology of Confined Ceramic Targets and Impacting Rods, *Int. J. Impact Engng.*, vol. 9, pp. 263–275, 1990.

Shockey, D.A., Rowcliffe, D.J., Dao, K.C., and Seaman, L. Particle Impact Damage in Silicon Nitride, *J. Am. Ceram. Soc.*, vol. 73, pp. 1613–1619, 1990.

Smyth, J.R. and Morey, R.E. Advanced Turbine Technology Applications Project Progress in Year Four, Proc. S.A.E. P-256, pp. 47–55, 1991.

Steinberg, D.J. Computer Studies of the Dynamic Strength of Ceramics, in Shock Compression of Condensed Matter, Elsevier Science Publishers, pp. 447–450, 1992.

Sternberg, J. Material Properties Determining the Resistance of Ceramics to High Velocity Penetration, *J. Appl. Phys.*, vol. 65, pp. 3417–3424, 1989.

Stroh, A.N. A Theory of the Fracture of Metals, *Adv. Phys.*, vol. 6, pp. 418–465, 1957.

Thompson, A.W. and Knott, J.F. Micromechanisms of Brittle Fracture, *Metall. Trans. A*, vol. 24A, pp. 523–534, March 1993.

Timoshenko, S. *Applied Elasticity*, London, 1928.

Wilkins, M.L., Landingham, R.L., and Honodel, C.A. Fifth Progress Report Light Armor Program, Lawrence Livermore Report UCRL-50980, 1971.

Williams, M.L., Knauss, W.G. *Dynamic Fracture*, particularly chapters by Dally, Dourney, and Irwin, and Ramulu and Kobayashi, 1985.

Winkler, S., Senf, H., and Rothenhausler, H. Wave and Fracture Phenomena in Impacted Ceramics, Fraunhofer Institut für Werkstoffmechanik, Report V23/89, prepared for U.S. Army Research Office, May 1989.

Woodward, R.L. A Simple One-Dimensional Approach to Modeling Ceramic Composite Armor Defeat, *Int. J. Impact Engng*, 9(4), pp. 455–474, 1990.

Yoffe, E. The Moving Griffith Crack, *Phil Mag.*, vol. 42, pp. 740–750, 1951.

Young, C. and Dubugnon, O. A Reflected Shear-Wave Technique for Determining Dynamic Rock Strength, *Int. J. Rock. Mech., Min. Sci.*, 1977.

Zil'berbrand, Y.L., Kozhushko, A.A., Polozenko, V.I., Pugachev, G.S., and Sinani, A.B. Interaction of a Ductile Projectile with a Brittle Solid Medium, *Problemy Prochnosti*, No. 9, pp. 60–62, September 1991.

Zil'berbrand, Y.L., Zlatin, N.A., Kozhushko, A.A., Polozenko, V.I., Pugachev, G.S., and Sinani, A.B. Mechanism of Interaction of a Plastic Striker with a Brittle Solid, *Sov. Phys.-Tech. Phys.*, vol. 34, pp. 1123–1124, 1989.

Zener, C. Fracturing of Metals, ASM, Cleveland, OH, pp. 3–31, 1948.

Zukas, J., Nicholas, T., Swift, H., Greszczuk, L., and Curran, D. *Impact Dynamics*, John Wiley & Sons, 1982.

CHAPTER 9

Spall and Fragmentation in High-Temperature Metals

D.E. Grady

9.1. Introduction

The high-velocity impact of a projectile with a stationary structure will lead to the catastrophic fragmentation of the projectile and some portion of the structure during the interaction. The characteristic size and velocity of fragments and the thermodynamic state of fragment debris are directly linked to the intensity of the impact-induced shock waves. Compressional or elastic potential energy introduced by the shock wave is converted to divergent kinetic energy as waves rebound from free surfaces. The magnitude of this kinetic energy determines fragment sizes and velocities. Entropy production during the shock-compression process, on the other hand, leads to elevated temperatures upon shock release, resulting in high-temperature solid, liquid, and liquid-vapor fragment debris.

The intense deposition of radiant energy on, and immediately within, the surface of a solid can result in rapid and extreme heating of the material as the photon energy is converted to thermal kinetic energy of the lattice and electronic structure. The near-isochoric nature of the deposition heating process results in an initial thermal pressure profile brought about by the characteristic Gruneisen properties of the material. The subsequent propagation of stress waves and decompression of the heated material can again result in the fragmentation of high-temperature solid, liquid, and liquid-vapor matter.

An end-state theory of dynamic fragmentation based on a balance of interfacial and dissipative energies resisting destabilization and breakup, and the energies of dynamic expansion driving fragmentation, has been proposed in earlier work [Grady, 1982; Grady, 1988]. The theory has been applied with some success to simple liquids, and to solids based on idealized models of toughness-controlled fracture (brittle spall), or plasticity-dominated hole growth and coalescence (ductile spall). In the present study we address in further detail the influence of temperature on spall and fragmentation of solid and liquid metals, and pose an extension of the theory which will focus on the destabilization and fragmentation within the region of coexistence of liquid and saturated vapor.

At temperatures above about one-half of the melt temperature, the mobility of dislocations in most metals increases significantly, so that large levels of plasticity can be accommodated with considerable ease. The precursor to fragmentation is the growth and coalescence of plastic cavities, and the local flow stress is the primary force resisting the cavity growth and breakup process. Within the energy-based theory of fragmentation the temperature and strain rate sensitivities of the flow stress at elevated temperature become the critical material parameter descriptions governing the size of fragment debris.

For liquid metals, the surface energy is the principal material property governing the characteristic length scale of fragmentation. Surface tension (the property usually measured) and surface energy are thermodynamically related. Both are functions of temperature and approach zero as the temperature reaches the critical temperature. A broad base of data exist for liquid metals, and corresponding states relations for surface tension and surface energy allow for ready prediction of fragmentation in liquid metals at any temperature.

The isentropic decompression of metals from intense shock states can lead to destabilization and fragmentation within the domain of mixed liquid-vapor coexistence. An energy-based approach to fragmentation is extended to account for this more general fragmentation problem.

9.2. High-Temperature Solids

In solid metals at elevated temperatures, transient tensile stresses can lead to the nucleation of internal incipient microscopic cavities which undergo hole growth to macroscopic sizes and subsequent coalescence through plastic flow processes due to the ease of dislocation glide. Consequently, resistance to spall fracture, and the corresponding energy dissipated in the fragmentation process, can be related to the local flow stress as cavities plastically open, link up, and fragment the body. A theory of the spall and fragmentation process based on energy balance at the end states, and an elementary model of the plastic hole growth process leads to relations for the spall stress [Grady, 1988],

$$p_s = (2\rho c^2 Y \varepsilon_c)^{1/2}, \tag{9.1}$$

and the average fragment size,

$$s = \left(\frac{8 Y \varepsilon_c}{\rho \dot{\varepsilon}^2}\right)^{1/2}. \tag{9.2}$$

In Equations (9.1) and (9.2) Y is the flow stress in simple tension and ε_c is the void volume fraction at which the maximum tensile pressure is achieved before catastrophic coalescence, tensile stress relaxation, and fragmentation occurs. Other properties include the density ρ, the sound speed c, and the global strain rate $\dot{\varepsilon}$ at the time of spall fracture. A value of $\varepsilon_c = 0.15$, independent of material and material state, has received some justification and is assumed in the present model. The reliability of this assumption has not been fully investigated.

The local flow stress Y is the principal material property governing the spall strength and fragment size in Equations (9.1) and (9.2), and can be expected to depend on temperature and strain rate in the high-temperature solid. Ashby (1972) has discussed ductile deformation mechanisms at high temperature, and noted that for the fcc and hcp metals, in which dislocation glide proceeds with reasonable ease, the flow stress is adequately described by,

$$Y = A\left(1 - \frac{\theta}{\theta_m}\right)\ln\frac{\dot{\varepsilon}}{\dot{\varepsilon}_0}. \tag{9.3}$$

In Equation (9.3) A and $\dot{\varepsilon}_0$ are constants related to dislocation obstacle characteristics and mobility frequencies, while θ and θ_m are the temperature and the melt temperature, respectively. Ashby has observed that dislocation motion in bcc and other crystal structures is more complex, and correspondingly more complicated relations for the flow stress may be required. Empirical extensions of Equation (9.3) to power law relations in the temperature have proven appropriate in some cases [Johnson and Cook, 1983]. In applying Equation (9.3) to spall in high-temperature metals we are ignoring the very high strain rate regime in which drag-controlled dislocation glide may lead to a more pronounced dependence on strain rate [Follensbee et al., 1984].

Spall strength data of Lalle [see Grady, 1988] for tin are compared with temperature dependent relations based on Equations (9.1)–(9.3) in Figure

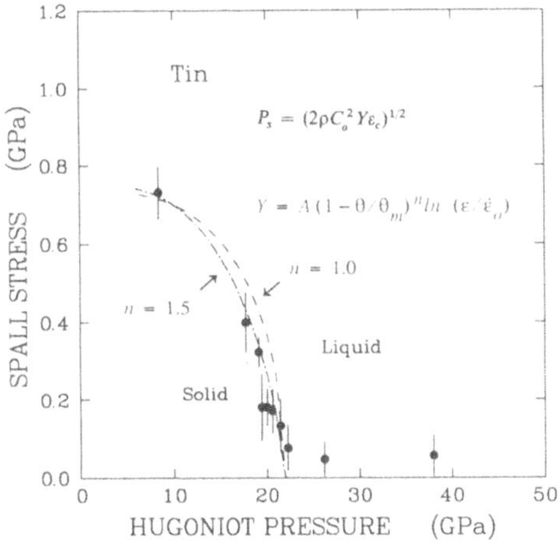

Figure 9.1. Spall data for high-temperature solid and liquid tin from Lalle [see Grady, 1988] compared with predicted spall strength relations.

9.1. Although a linear temperature dependence captures the essential behavior of the spall data, a power of $n = 1.5$ provides a more satisfactory fit. The tetragonal structure of the tin coupled with shock-induced solid phase transformation may complicate the dynamic dislocation plasticity during spall in this metal.

9.3. Fragmentation of Liquid Metals

Under more intense loading conditions the higher entropy production in the shock process can be sufficient to cause melting of metals upon dynamic pressure release. Consequently, the resulting fragment debris will be molten droplets. For example, the symmetric impact of a tin projectile on a stationary tin target at velocities of about 2 km/s and higher is sufficient to cause melt upon shock release. Similarly, velocities of 4.5 km/s and higher are needed to achieve melt in copper.

An energy-based theory of spall and fragmentation in liquids has been explored in an earlier study [Grady, 1988], and appears to be reasonably capable of correlating the relatively small base of dynamic fragmentation data available. For simple liquids, new surface energy and viscous dissipation during the spall void growth and coalescence process appear to be the principal mechanisms of energy dissipation resisting fragmentation. The energy analysis predicts that surface energy dominates at lower fragmentation strain rates with a spall strength and average fragment size given by

$$p_s = (6\rho^2 c^3 \gamma \dot{\varepsilon})^{1/3}, \tag{9.4}$$

and,

$$s = \left(\frac{48\gamma}{\rho \dot{\varepsilon}^2}\right)^{1/3}, \tag{9.5}$$

where γ is the surface energy per unit area of new fragment surface. At higher strain rates a transition to viscous-dissipation dominated fragmentation is predicted according to,

$$p_s = (2\rho c^2 \eta \dot{\varepsilon})^{1/2}, \tag{9.6}$$

and,

$$s = \left(\frac{8\eta}{\rho \dot{\varepsilon}}\right)^{1/2}, \tag{9.7}$$

where η is the Newtonian viscosity. For liquid metals with η of the order 10^{-3} Pa-s, transition to viscosity-dominated fragmentation is predicted at a strain rate of about 10^{12}/s—significantly higher than rates of interest in the preponderance of applications. Consequently, for many engineering applications, new surface energy production in liquid-metal fragmentation would appear to be the physical mechanism of primary interest.

The surface tension and the surface energy of a liquid both depend on the temperature. Consequently, the spall strength and corresponding fragment size would be expected to decrease as the temperature of the liquid increases.

For liquid metals the van der Waals relation [Allen, 1972],

$$\sigma = \sigma_o(1 - \theta/\theta_c)^n, \tag{9.8}$$

has been used successfully to describe the temperature dependence of the surface tension σ up to the critical temperature θ_c. In Equation (9.8) σ_o and n are constants. The fact that the surface tension does not achieve zero until the critical temperature is reached, which is approximately 2.5 to 4 times the boiling temperature for liquid metals [Grosse, 1961], is important to the fragmentation process in the high temperature regime.

The surface energy γ is thermodynamically related to the surface tension through the relation,

$$\gamma = \sigma - \theta\frac{d\sigma}{d\theta}. \tag{9.9}$$

In terms of the van der Waals constants, the expression for γ becomes,

$$\gamma = \sigma_o\left(1 - \frac{\theta}{\theta_c}\right)^n\left(1 + \frac{n\theta}{\theta_c - \theta}\right). \tag{9.10}$$

The van der Waals constants σ_o and n can, in principle, be determined from measurements of surface tension and its first temperature derivative at the melt temperature. Working with Equation (9.8) and its first derivative, the relations,

$$n = -(\theta_c - \theta_m)\frac{\sigma'_m}{\sigma_m}, \tag{9.11}$$

and,

$$\sigma_o = \frac{\sigma_m}{(1 - \theta_m/\theta_c)^n}, \tag{9.12}$$

are obtained. The subscript m refers to properties evaluated at the melt temperature.

Substantial surface tension data for liquid metals exist [e.g. Allen, 1972]. Unfortunately data for the temperature derivative of the surface tension σ'_m are not sufficiently accurate to satisfactorily constrain n. A more successful approach is based on a general observation due to Eötvös (1886) relating the molecular binding energy (surface tension) to the temperature decrement below the critical temperature,

$$\sigma\left(\frac{M}{\rho}\right)^{2/3} = k(\theta_c - \theta), \tag{9.13}$$

where M is the molecular weight and k is a constant of proportionality.

Table 9.1

Metal	ρ_m[a,b] Mg/m^3	α_m[a] 10^{-6} K^{-1}	θ_m[a] K	θ_c[c] K	σ_m[a] N/m	σ_o[d] N/m	n[d]
Lithium	0.513	164	453	3040/—	0.398	0.489	1.28
Beryllium	1.69	69	1550	—/—	1.390	—	—
Sodium	0.926	237	371	2390/—	0.191	0.239	1.32
Magnesium	1.59	167	923	2870/—	0.559	0.897	1.22
Aluminum	2.38	114	933	5220/7151	0914	1.188	1.33
Silicon	2.52	145	1683	—/—	0.865	—	—
Potassium	0.823	268	337	2160/—	0.115	0.144	1.33
Calcium	1.37	162	1111	3290/4590	0.361	0.602	1.24
Titanium	4.11	170	1941	7030/—	1.650	2.749	1.58
Vanadium	5.75	—	2173	7590/11325	1.950	3.085	1.36
Chromium	(6.46)	—	2148	6100/—	1.700	2.937	1.26
Manganese	5.73	127	1518	4640/3408	1.090	2.160	1.16
Iron	7.01	119	1810	6330/9340	1.872	2.940	1.34
Colbalt	7.67	160	1768	6510/—	1.873	3.022	1.51
Nickel	7.77	142	1726	6520/9576	1.778	2.777	1.45
Copper	7.99	100	1356	5330/7625	1.360	1.975	1.27
Zinc	6.57	177	693	2680/3170	0.782	1.130	1.23
Gallium	6.09	100	303	4500/7043	0.718	0.785	1.28
Germanium	5.49	94	1210	—/9803	0.621	0.761	1.54
Selenium	4.01	274	490	—/—	0.106	—	—
Rubidium	1.47	272	312	2060/—	0.085	0.106	1.32
Strontium	2.38	110	1041	—/—	0.303	—	—
Zirconium	6.06	—	2125	8840/—	1.480	2.205	1.45
Columbium	(7.83)	—	2741	—/—	1.900	—	—
Molybdenum	(9.35)	—	2883	9470/14588	2.250	3.795	1.44
Ruthenium	(10.9)	—	2523	9390/—	2.250	3.553	1.46
Rhodium	(11.1)	—	2239	8120/—	2.000	3.132	1.39
Palladium	10.52	121	1825	5850/8301	1.500	2.457	1.32
Silver	9.33	97	1234	4660/6410	0.903	1.314	1.22
Cadmium	8.02	134	594	2440/2619	0.570	0.790	1.17
Indium	7.03	115	429	4140/5823	0.556	0.640	1.28
Tin	7.00	88	505	4890/7297	0.544	0.624	1.26
Antimony	6.48	85	904	—/6370	0.367	0.448	1.31
Tellurium	5.80	96	723	9580/—	0.180	0.204	1.57
Cesium	1.838	283	328	2000/—	0.070	0.089	1.32
Barium	3.32	83	329	—/4154	0.277	0.306	1.21
Lanthanum	5.95	40	1193	6570/—	0.720	0.905	1.14
Cerium	6.69	34	1077	6460/9089	0.740	0.908	1.12
Neodymium	6.68	79	1292	5240/—	0.689	0.970	1.21
Gadolinium	(714)	—	1585	6150/—	0.810	1.193	1.30
Hafnium	(12.0)	—	2495	8990/—	1.630	2.595	1.43
Tantalum	(15.0)	—	3269	11080/17329	2.150	3.658	1.52
Tungsten	(17.5)	—	3683	11950/18538	2.500	4.426	1.55
Rhenium	(18.7)	—	3453	10900/17293	2.700	4.781	1.50
Osmium	(20.1)	—	3283	10630/—	2.500	4.335	1.49
Iridium	(20.0)	—	2727	9630/—	2.250	3.658	1.46
Platinum	18.91	153	2042	8280/12526	1.800	2.888	1.67
Gold	17.29	170	1336	5750/8267	1.140	1.695	1.50
Mercury	13.70	174	235	1779/—	0.498	0.589	1.18

Table 9.1 (cont.)

Metal	ρ_m[a,b] Mg/m³	α_m[a] 10^{-6} K^{-1}	θ_m[a] K	θ_c[c] K	σ_m[a] N/m	σ_o[d] N/m	n[d]
Thallium	11.29	115	576	3340/4353	0.464	0.583	1.21
Lead	10.68	123	600	3530/4668	0.468	0.590	1.24
Bismuth	10.07	125	544	2440/5395	0.378	0.506	1.16
Thorium	(10.5)	—	2023	8430/14257	0.978	1.448	1.43
Uranium	17.90	58	1405	7750/13043	1.550	1.990	1.25
Plutonium	16.65	91	913	5550/—	0.550	0.692	1.28

[a] Values in parentheses are estimated from thermal expansion data (see Allen (1972)).
[b] Data are from Allen (1972).
[c] Data on the left are from Martynyuk (1977)—that on the right are from Young and Alder (1971).
[d] To calculate values for σ_o and n (Equation (12) and (14)) the lowest value of the critical temperature was used. Where thermal expansion data was not available a value of $\alpha_m = 100$ was arbitrarily used.

Using Equation (9.13), one obtains

$$n = \frac{2}{3}\alpha_m(\theta_c - \theta_m) + 1, \tag{9.14}$$

where α_m is the coefficient of thermal expansion at the melt temperature. Values of n ranging from about 1.1 to 1.7 are calculated for liquid metals and provide a reasonable description of the temperature dependence of surface tension and surface energy in these materials.

Material data for liquid metals are provided in Table 9.1. Liquid density, thermal expansion, melt temperature, and surface tension data are from Allen (1972). The first critical temperature data (left number) in Table 9.1 is from Martynyuk (1977) measured by an exploding foil technique. The second set of critical temperature values reported in Table 9.1 are from Young and Alder (1971), calculated from a hard sphere model. The latter values are in reasonable agreement with work of Grosse (1961). Improved calculations of critical constants for selected metals are also reported by Young (1977) using a soft sphere model. The range of critical temperature values generally indicates the level of uncertainty in this property. Surface tension constants σ_o and n for Equations (9.8) and (9.10) have been calculated arbitrarily based on the lower critical temperature value.

A plot of the surface tension and surface energy for liquid tin is shown in Figure 9.2. Although surface tension decreases markedly with temperature, surface energy up to and beyond the boiling temperature remains relatively insensitive to temperature. Both properties reach zero at the critical temperature.

There is very little quantitative experimental data on impact-induced fragmentation of liquid metals. Order-of-magnitude estimates of particle sizes

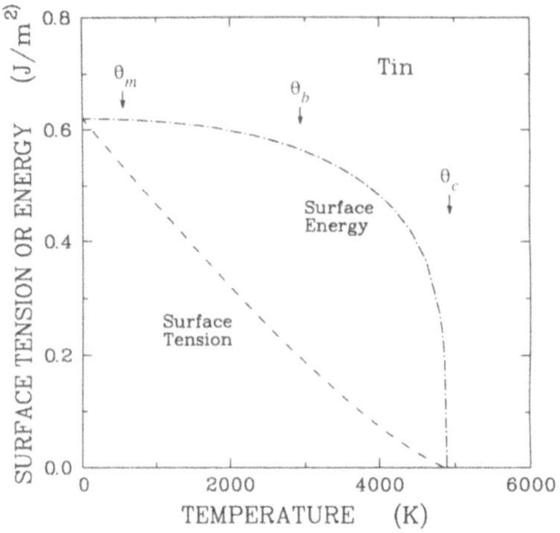

Figure 9.2. Surface energy and surface tension for liquid tin based on the equation of van der Waals.

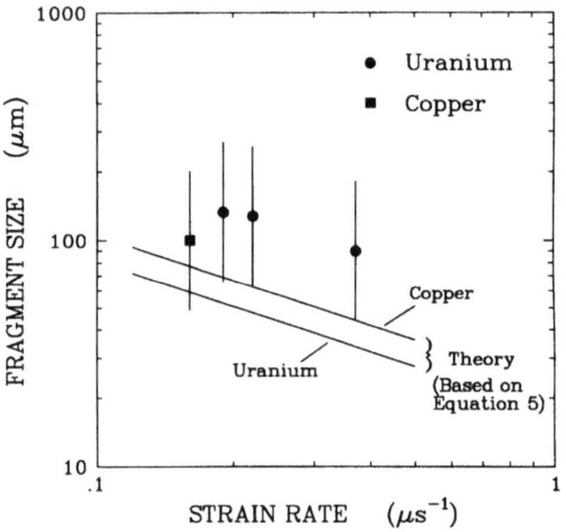

Figure 9.3. Fragment size data for liquid copper and uranium, and comparison with theoretical prediction based on Eq. (9.5). (See text for references.)

are readily calculated from the present development, however. Typical impact experiments involve projectiles and targets with characteristic dimensions of several millimeters. Impact velocities sufficient to cause shock-induced melting on release result in debris expansion velocities of several kilometers per second. Consequently, fragmentation strain rates will be of the order of $\dot{\varepsilon} \sim u/r \sim 10^6$/s. With a typical metal density of 10^4 kg/m^3 and a liquid surface energy of 1 J/m^2, particle sizes predicted from Equation (9.5) will be on the order of tens of micrometers.

Some fragment size data for several liquid metals are shown in Figure 9.3. The data points on uranium were obtained from implosion experiments on cylindrical uranium shells [Grady and Kipp, 1985]. Particle size data was inferred from impact craters on lead witness plates. Liquefaction of uranium occurred through entropy production in the shock process and was confirmed through computational simulations of the implosion events. The single fragment size data point for copper was obtained in a high velocity (4.5 km/s) impact of a copper projectile on a stationary target plate [Grady and Kipp, 1994]. Again melt was achieved through shock entropy production. The nominal fragment size was inferred from penetration holes in thin aluminum witness foils. Large uncertainties in fragment size are ascribed to difficulties in relating crater or hole size to particle size and to the treatment of the size statistics. There is also substantial uncertainty in estimating the strain rate at fragmentation. Predictions of fragment size in Figure 9.3 are based on Equation (9.5) using properties for γ and ρ at the melt temperature. Predictions appear to be in reasonable accord with the scant fragment size data on liquid metals.

9.4. Liquid–Vapor State Fragmentation

As one continues to examine the consequences of increasingly intense loading conditions on a solid target—going beyond events considered in the last sections which led to high-temperature solid or liquid states upon decompression—the shock compression process becomes increasingly non-isentropic. Consider for example current experimental capabilities in which it is possible to accelerate and launch projectiles weighing approximately 1 g to velocities approaching 10 km/s. When such projectiles are caused to impact stationary targets, intense shock waves are produced which accommodate transient pressures near several hundred GPa and compressions of about a factor of two. Such transient states are typically maintained for a few tenths of a microsecond.

During shock compression to the states mentioned above, temperatures approaching tens of thousands of degrees can be achieved at the shock state. Upon release of the shock pressure, the material will remain heated to an extent which is determined by the level of dissipation produced within the shock process. For sufficiently intense shock waves, vaporization of the material can be achieved during pressure release [Zel'dovich and Raizer, 1967].

As an alternative example, consider the radiation of a high-energy laser beam focused on the surface of a solid target. At intensities below about 10^8 Watts/cm^2, energy levels are not sufficient to produce an immediate plasma cloud at the surface, shielding the solid target from further radiation. Instead, energy is deposited at depth within the material, leading to constant volume heating and accompanying pressure rise. Rapid material decompression occurs during the subsequent propagation of pressure release waves inward from the illuminated free surface. At appropriate energy levels material vaporization can occur during the dynamic decompression process.

The mechanisms of shock-induced vaporization are not well understood. Vaporization initiates when a strong shock wave emerges at a sample free

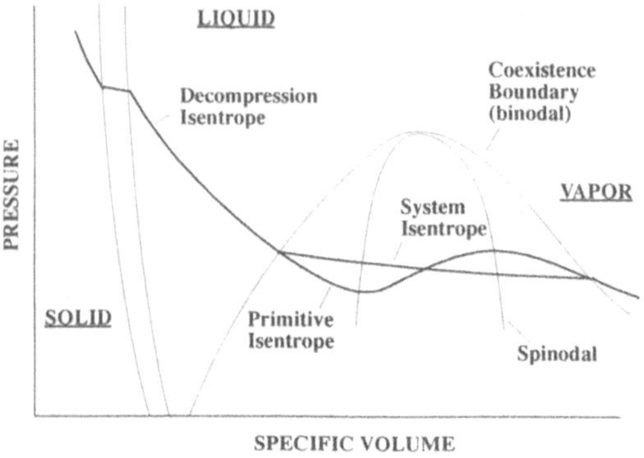

Figure 9.4. Phase diagram in the pressure versus specific volume plane for a typical metal.

surface. The reflected release wave leads to rapid isentropic decompression as this wave propagates into the material. It should be noted that for sufficiently strong shock compression, isentropic decompression can carry the material into the liquid–vapor coexistence region on the condensation side of the critical point. The present study, however, will be restricted to the relatively cooler release isentropes which enter the vapor dome on the boiling side of the critical point such as that illustrated in Figure 9.4. Due to the extreme rates of decompression, it has been suggested that the vaporization process may be highly nonequilibrium because of limiting kinetics associated with the second phase nucleation and growth processes [Zel'dovich and Raizer, 1967].

It is common, however, to assume equilibrium thermodynamics in the computational treatment of shock vaporization [Bjork, 1987; Holian and Burkett, 1987; Trucano and Asay, 1987], and expansion isentropes from shock-compressed states for lead have been determined by Al'tshuler, et al. (1977). The latter method is not very sensitive to details of the vaporization process, however. A newer method [Trucano and Asay, 1987], in which the stagnation pressure of the expanding vapor debris is measured, offers possibilities for extracting further details of shock-induced vaporization.

In the present investigation attention is focused on aspects of the shock vaporization process which might lead to deviations from the isentropic decompression path predicted through equilibrium thermodynamics. Those aspects of the problem which we wish to pursue are best introduced through a more careful consideration of the shock decompression process. Decompression initiates when a compressive shock wave emerges at a free surface and is completed through the penetration of a centered rarefaction wave into the shock-compressed material. Although the rarefaction wave is initially infinitely steep, it disperses with propagation distance. For a typical metal, as isentropic decompression approaches the vapor dome as illustrated in Figure 9.4, expansion strain rates are of order 10^8/s by the time the rarefaction wave has penetrated about 10 μm and are down to order 10^6/s at a depth of one mm. For most applications, therefore, we will be interested in decompression strain rates within the range of about 10^5/s to 10^8/s.

Thus, consider a material undergoing rapid isentropic decompression on a pressure–volume path which penetrates the liquid–vapor coexistence curve as shown in Figure 9.4 at a specific volume somewhat less than the critical specific volume. A van der Waals equation of state within the coexistence region has been qualitatively successful in describing the continuous material response, from the liquid regime in which intermolecular forces predominantly determine the pressure, to the gaseous regime in which pressure is principally determined by thermal motion of the molecules. Such a description entails multi-valued pressure-volume isotherms (van der Waals loops) and consequently homogeneous states are either metastable, if they reside within the binodal (liquid–vapor coexistence curve), or absolutely unstable if they reside within the spinodal (locus of isothermal extrema), with respect to decomposition into a mixture of liquid and gaseous species. Thus, at any specific volume within the vapor dome, two material states consistent with isentropic decompression are allowed—one that is homogeneous and is either conditionally stable or unstable (called here the primitive isentrope) and one which is a thermodynamically stable heterogeneous mixture of liquid and vapor (the system isentrope).

Consequently, on the initial unloading isentrope, contact at the coexistence curve identifies a point of bifurcation. Continued decompression may continue along the equilibrium system isentrope, if conditions needed to achieve equilibrium are met, or decompression along the metastable primitive isentrope, if kinetic barriers prohibit equilibration under the decompression rates of interest. The kinetic processes required to achieve equilibrium involve the nucleation and growth of vapor bubbles within the dominant liquid phase. Nucleation involves the spontaneous activation of a population of bubble nuclei of sufficient radius to overcome the excess interfacial energy created in the activation process. Growth kinetics involves the passage of molecules from the liquid to the gaseous state through the intervening interface.

In the present investigation, attention is directed to the interfacial energy created in the transition from a homogeneous body to a liquid–vapor mixture. Calculations of equilibrium thermodynamic mixture states within the coexistence region typically ignore the interfacial energy in comparison with the appropriate free energy densities. Under the present extremely rapid decompression rates, however, it is intuitively clear that the characteristic size of particulate during transition to a liquid–vapor mixture must be small and consequently the possibility of creating a large specific interfacial area (and energy) should be considered. As an example suppose that the liquid–vapor mixture created on decompression was composed of particles of size of order 10 μm. Liquid–vapor interfacial energies for metals are of order 1 J/m^2. Consequently a compressional energy comparable with this magnitude of specific interfacial energy would imply $p^2/2K = \gamma/s$. Using typical values for metals provides a pressure of 100 MPa. This pressure suggests the magnitude of pressure excursion from equilibrium which would be required to account for this amount of interfacial energy.

To better account for the fragmentation process occurring during rapid decompression, we draw on a theory of fragmentation which has previously been proposed [Grady, 1988], and extend its application to liquid–vapor breakup. This theory is based on a balance of the interfacial energy which resists destabilization and breakup, and the energies associated with the expansion process which will drive fragmentation. The theory predicts the size of particulate necessitated by the dynamics of the event and constrains the extent of interfacial energy which will be generated in the heterogeneous mixture. Through this approach a lower bound for the nonequilibrium behavior will be determined which is consistent with expansion rates experienced in the shock vaporization process. Kinetics associated with mass transport between liquid and vapor states, which would clearly contribute to the nonequilibrium excursion, are not treated in this work.

Features of the fragmentation process and vaporization kinetics have also been observed in molecular dynamic simulations of rapid isentropic decompression of a liquid [Holian and Grady, 1988]. In that work selected dynamic expansion rates were impressed on a Lennard-Jones liquid. The ensemble of particles were observed to undergo immediate fragmentation into an exponential distribution of particle clusters. Size and distribution of cluster size were consistent with predictions based on energy-balance fragmentation. Subsequent evolution of the fragment (cluster) distribution was believed to reflect vaporization kinetics as elemental particles were ejected from surfaces of larger cluster fragments.

Thermodynamic Path of a System with Interfacial Energy

As an isentrope crosses the binodal curve on decompression, a trajectory continuing down the primitive isentrope represents an unstable process with respect to a two-phase thermodynamic description of the liquid–vapor system. Such a thermodynamic description, however, ignores the interfacial energy associated with the interface between liquid and vapor phases. This interfacial energy is accounted for in the present development, and provides the means for characterizing the fragmentation process as the material enters the coexistence region during rapid decompression. Nonequilibrium kinetic effects are ignored. The fragmentation theory is based on concepts of equilibrium thermodynamics.

The method is illustrated in Figure 9.5. As the isentropic trajectory enters the region of coexistence, decompression will continue along the primitive isentrope indicated by path AB in Figure 9.5. Points on the primitive isentrope continue to represent stable states because corresponding points on the system isentrope (same specific volume), accounting for interfacial energy are higher energy states. Point B on the primitive isentrope corresponds to the state at which the excess compressional energy is just sufficient to balance the interfacial energy required for transition to the system isentrope. Consis-

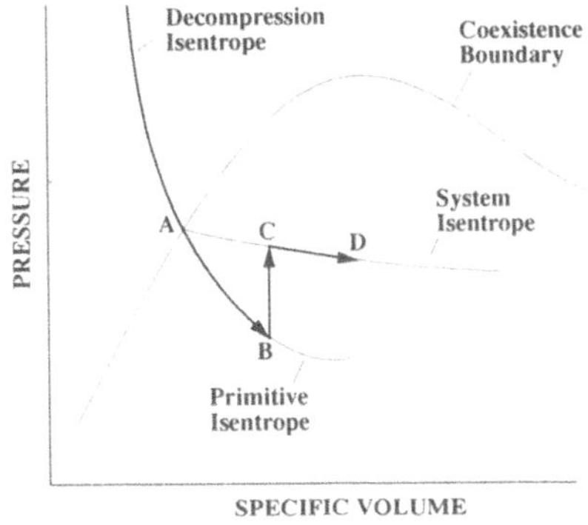

Figure 9.5. Path of metastable entry and fragmentation in the liquid-vapor mixed phase region.

tent with equilibrium thermodynamics, the transition (fragmentation) from the primitive isentrope to the system isentrope (path BC) is assumed to occur instantly when this condition is met. Continued decompression is then assumed to proceed along the system isentrope (path CD).

The criterion for phase fragmentation is consistent with thermodynamic principles. Upon entering the coexistence region the primitive and system isentropes represent a bifurcation of the primary decompression isentrope. At a specified specific volume within the coexistence region the stable state corresponds to that of lowest internal energy. Accounting for interfacial energy, the system isentrope is initially a higher energy path and will remain so until the difference in compressional energy between the two isentropes is sufficient to account for the interfacial energy.

If we identify W with this excess compressional energy, then W will be a monotonic increasing function of the fractional specific volume change, $\eta = 1 - v_c/v$ where v_c is the specific volume at the point at which the decompression isentrope enters the coexistence region. At this point in the development a necessary criterion for fragmentation (path BC in Figure 9.5) would suggest that $W(\eta) \geq \Gamma$ where Γ is a measure of the interfacial energy per unit volume.

Due to the high rate of decompression, there is an additional energy available to assist in the fragmentation process. The source of this additional energy is a consequence of a redistribution and change in the local kinetic energy associated with the initially uniform expansion due to clumping during the fragmentation process [Grady, 1982, 1988]. This kinetic energy, denoted by T, is a function of the rate of expansion $\dot{\eta}$. Including the kinetic energy, the necessary criterion for phase fragmentation would require that,

$$W + T \geq \Gamma. \tag{9.15}$$

The assumption of thermodynamic equilibrium (fragmentation transition when the energy criteria is just satisfied) would require an equality in Equation (9.15).

The rate of isentropic decompression is an essential parameter in the theory of phase fragmentation. The material point of interest is an interior point of the body well removed from free surfaces. Prior to fragmentation the gradients in material density are assumed to be small. Consequently, the single parameter $\dot{\eta}$, characterizing the rate of dilatation, uniquely describes the kinematic state of the material.

Excess Potential Energy of the Super-Cooled Liquid

The excess potential energy of the rapidly decompressing liquid as it enters the coexistence region and continues down the metastable primitive isentrope is readily calculated by integrating the difference between the primitive and system isentrope from the specific volume v_c at the entry bifurcation point to the present specific volume,

$$W = \rho \int_{v_c}^{v} (p_{sys} - p_{prm})\, dv. \tag{9.16}$$

In the present development, excursions into the coexistence region are assumed to be sufficiently small before fragmentation occurs that a linear representation of Equation (9.16) is adequate to describe the behavior. Consequently, an expansion of Equation (9.16) results in,

$$W = \rho \int_{v_c}^{v} \left(\left.\frac{dp_{sys}}{dv}\right|_s - \left.\frac{dp_{prm}}{dv}\right|_s \right)(v - v_c)\, dv, \tag{9.17}$$

where the subscript s refers to the isentropic derivative on the system and primitive isentropes at the entry point, respectively. Equation (9.17) is then written in the form,

$$W = \int_0^{\eta} \Lambda_s \eta \, d\eta, \tag{9.18}$$

where η is the fraction volume expansion defined earlier, and Λ_s is the corresponding isentropic modulus associated with the pressure difference or underpressure,

$$\delta p = p_{sys} - p_{prm} = \Lambda_s \eta. \tag{9.19}$$

Both Λ_s and the isentropic derivatives are evaluated on the coexistence boundary at the point of entry ($\eta = 0$).

The derivative of the system isentrope is required and this term has been obtained by others [Duvall and Horie, 1965]. A particularly transparent derivation follows from consideration of the specific volume and specific entropy in the forms,

$$v(p, \theta, \lambda) = v_v(p, \theta)\lambda + v_l(p, \theta)(1 - \lambda), \tag{9.20}$$

$$s(p, \theta, \lambda) = s_v(p, \theta)\lambda + s_l(p, \theta)(1 - \lambda), \tag{9.21}$$

where λ is the mass fraction of the saturated vapor and the subscripts v and l refer to the vapor and liquid phases, respectively. The differential forms for specific volume and entropy then become,

$$dv = \frac{\partial v}{\partial p}\,dp + \frac{\partial v}{\partial \theta}\,d\theta + \frac{\partial v}{\partial \lambda}\,d\lambda, \tag{9.22}$$

$$ds = \frac{\partial s}{\partial p}\,dp + \frac{\partial s}{\partial \theta}\,d\theta + \frac{\partial s}{\partial \lambda}\,d\lambda. \tag{9.23}$$

A third differential relation is obtained by noting that the path is an isentrope,

$$ds = 0, \tag{9.24}$$

and a fourth from phase coexistence,

$$\frac{dp}{d\theta} = r(\theta). \tag{9.25}$$

Equation (9.25) is a differential form for the Clausius–Clapeyron condition in the pressure-temperature plane.

Equations (9.22) through (9.25) constitute four linear equations in the five differential variables (dv, ds, dp, $d\theta$, $d\lambda$) which can be solved for the system isentrope. Observing from Equations (9.20) and (9.21) that $(\partial s/\partial\lambda)/(\partial v/\partial\lambda) = dp/d\theta$, and making use of the Maxwell relation $\partial s/\partial p|_\theta = -\partial v/\partial\theta|_p$, the variables ds, $d\theta$, and $d\lambda$ can be eliminated to obtain an expression for the system isentrope,

$$\left.\frac{dp_{sys}}{dv}\right|_s = \frac{r^2}{r^2\dfrac{\partial v}{\partial p} + 2r\dfrac{\partial v}{\partial \theta} - \dfrac{\partial s}{\partial \theta}}. \tag{9.26}$$

The isentropic modulus,

$$\Lambda_s = v\left(\left.\frac{dp_{sys}}{dv}\right|_s - \left.\frac{dp_{prm}}{dv}\right|_s\right), \tag{9.27}$$

can be solved for by making use of the thermodynamic expressions,

$$\frac{\partial s}{\partial \theta} = \frac{c_p}{\theta}, \tag{9.28}$$

$$\frac{\partial v}{\partial \theta} = \alpha v, \tag{9.29}$$

$$\frac{\partial v}{\partial p} = -\frac{v}{\beta_s} - \frac{\theta}{c_p}(\alpha v)^2, \tag{9.30}$$

where c_p, α, and β_s are the specific heat, coefficient of thermal expansion, and isentropic bulk modulus of the liquid state at the entry point on the coexistence boundary. This yields,

$$\Lambda_s = \beta_s \left[\frac{\vartheta(\xi - 1)^2}{1 + \vartheta(\xi - 1)^2} \right], \tag{9.31}$$

where ϑ and ξ are the dimensionless thermodynamic functions,

$$\vartheta = \frac{\beta_s c_p}{\theta v r^2}, \tag{9.32}$$

$$\xi = \frac{\alpha \theta v r}{c_p} = \frac{\gamma_g \theta r}{\beta_s}. \tag{9.33}$$

In Equation (9.33) the Gruneisen parameter, $\gamma_g = \alpha\beta_s v/c_p$, has been introduced.

Fragmentation in the Liquid-Vapor Region

The necessary relations have now been developed to predict particle size assuming that the equilibrium energy conditions outlined in the paragraphs leading up to Equation (9.15), namely $W + T = \Gamma$, are satisfied. This expression becomes,

$$\frac{\delta p^2}{2\Lambda_s} + \frac{1}{120}\rho\dot{\eta}^2 s^2 = \frac{6\gamma}{s}. \tag{9.34}$$

The first term in Equation (9.34) is the excess potential energy acquired as the metastable liquid enters the coexistence region. The second term is an expression for the local expansional kinetic energy where $\dot{\eta}$ is the rate of volumetric expansion and s is the particle size.

The final term is the interfacial energy density $\Gamma = 6\gamma/s$ associated with a nominal particle size s and interface energy γ. Equation (9.34) is formally the same as that developed in Grady (1988) for surface-energy dominated fragmentation in liquids. The principal difference is the potential energy in the leading term which accounts for the nature of the metastable isentropic decompression in the mixed liquid–vapor phase.

Following the development of the theory of equilibrium energy-based fragmentation Grady (1988), the local sound speed establishes a horizon (characteristic length) which bounds the fragment size,

$$s = 2ct_c. \tag{9.35}$$

In Equation (35) $c = \sqrt{v\beta_s}$ is the isentropic sound speed on the coexistence boundary and t_c is the time from entry into the coexistence boundary to fragmentation (transit from A to B in Figure 9.5) established by the expansion rate $\dot{\eta}$. Lastly, an expression for the metastable underpressure at fragmentation is given by,

$$\delta p = \Lambda_s \dot{\eta} t_c. \tag{9.36}$$

Equations (9.34), (9.35) and (9.36) can be solved for the three unknowns δ_p, s and t_c. In particular, the underpressure and characteristic particle size are,

$$\delta p = \left[\frac{6\gamma\Lambda_s^2\dot{\eta}}{c\left(1 + \frac{1}{15}\frac{\rho c^2}{\Lambda_s}\right)}\right]^{1/3}, \tag{9.37}$$

$$s = \left[\frac{48\gamma c^2}{\Lambda_s\dot{\eta}^2\left(1 + \frac{1}{15}\frac{\rho c^2}{\Lambda_s}\right)}\right]^{1/3}. \tag{9.38}$$

Equations (9.37) and (9.38) predict that the fragmentation properties in the liquid–vapor regimen depend both on the rate of isentropic decompression and the thermodynamic properties at the entry point of the coexistence region.

Equation (9.37) and (9.38) and the relations leading to them have not been explored in detail although some work with an analytic equation of state for tin has been pursued. Below temperatures of about 30% of critical Λ_s is well approximated by β_s for the liquid (see Equation 9.31), and the fragment size expression for liquid breakup readily applies. At the 30% level tin enters the coexistence region at about 0.1 GPa. At higher temperatures Λ_s becomes increasingly less than β_s, approaching zero at the critical temperature. The surface energy γ also decreases within the same temperature range further complicating Equations (9.37) and (9.38).

In closing the present section it is noted that Equation (9.38) provides a measure of the characteristic length (or fragment size) associated with the liquid–vapor fragmentation process. It is based on a balance of the interfacial energy created during breakup, and the excess elastic and kinetic energy fueling the fragmentation process. The method tacitly assumes that the kinetics associated with vapor fraction nucleation and growth are sufficient to accommodate the process. Details of the vaporization kinetics is expected to affect both the process of fragmentation, as well as the persistence of condensed-matter fragments during continued expansion following the breakup event.

In addition, the assumption of small excursions into the binodal coexistence region, and linearity during this excursion, avoids considerable conceptual complications. It can be expected under certain thermodynamic conditions and sufficiently high expansion rates that excursions on the primitive isentrope into the spinodal would be achieved without satisfying the energy-based fragmentation criterion. At this state, the local sound velocity goes to zero, and the mechanism of wave propagation which provides for regions of matter to achieve coherent motion in the fragmentation process becomes meaningless. It is expected that different processes will govern the disintegration behavior within the coexistence spinodal region when sound velocities are imaginary.

9.5. References

1. Allen, B.C. The Surface Tension of Liquid Metals, in *Liquid Metal Chemistry and Physics*, S.E. Beer, ed., Marcel Dekker, pp. 151–212 (1972).
2. Al'tshuler, L.V., Bakanova, A.A., Bushman, A.V., Dudoladov, I.P., and Zubarov, V.N. Evaporation of Shock Compressed Lead in Release Waves, *Soviet Phys. JETP* **46**, pp. 980–983 (1977).
3. Ashby, M.F. A First Report on Deformation-Mechanism Maps, *Acta Met.* **20**, 887–899 (1972).
4. Bjork, R.L. The Physics of Hypervelocity Lethality, *Int. J. Impact Eng.* **5**, 129–154 (1987).
5. Duvall, G.E. and Horie, Y. Shock Induced Phase Transitions, in *Proceedings of the Fourth Symposium on Detonation*, U.S. Naval Ordinance Laboratory, White Oak, Maryland, pp. 248–259 (1965).
6. Eötvös, R. *Pogg. Ann. Phys. Chem.* **27**, 448 (1886).
7. Follensbee, P.S., Regazonni, G., and Kocks, U.F. The Transition to Drag Controlled Deformation of Copper at High Strain Rates, in *Mechanical Properties at High Strain Rates—1984*, Institute of Physics Press, London, 71–80, 1984.
8. Grady, D.E. Local Inertial Effects in Dynamic Fragmentation, *J. Appl. Phys.* **53**, 322–325 (1982).
9. Grady, D.E. The Spall Strength of Condensed Matter, *J. Mech. Phys. Solids* **3**, 353–384 (1988).
10. Grady, D.E. and Kipp, M.E. Mechanisms of Dynamic Fragmentation: Factors Governing Fragment Size, *Mech. Materials* **4**, 311–320 (1985).
11. Grady, D.E. and Kipp, M.E. Experimental and Computational Simulation of the High Velocity Impact of Copper Spheres on Steel Plates, *Int. J. Impact Eng.* **15**, 645–660 (1994)
12. Grosse, A.V. The Temperature Range of Liquid Metals and an Estimate of Their Critical Constants, *J. Inorg. Nucl. Chem.* **22**, 23–31 (1961).
13. Holian, K.S. and Burkett, M.W. Sensitivity of Hypervelocity Impact Simulations to Equation of State, *Int. J. Impact Eng.* **5**, 333–341 (1987).
14. Holian, B.L. and Grady, D.E. Fragmentation by Molecular Dynamics, *Phys. Rev. Lett.* **60**, 1355–1358 (1988).
15. Johnson, G.R. and Cook, W.H. A Constitutive Model and Data for Metals Subjected to Large Strains, High Strain Rates and High Temperatures, in *Proceedings of the Seventh International Symposium on ballistics*, The Hague, Netherlands, pp. 541–551 (1983).
16. Martynyuk, M.M. The Critical Parameters of Metals I, Critical Temperature, *Russian J. Phys. Chem.* **51**, 705–706 (1977).
17. Trucano, T.G. and Asay, J.R. Effect of Vaporization on Debris Cloud Dynamics, *Int. J. Impact Eng.* **5**, 645–653 (1987).
18. Young, D.A. and Alder, B.J. Critical Point of Metals from the Van der Waals Equation of State, *Phys. Rev.* **A3**, 364–371 (1971).
19. Young, D.A. A Soft Sphere Model for Liquid Metals, Lawrence Livermore Laboratory Technical Report UCRL-52352, November, 1977.
20. Zel'dovich, Y.B. and Raizer, Yu. P. *Physics of Shock Waves and High-Temperature Hydrodynamic Phenomena—Volume II*, Academic Press, 1967.

CHAPTER 10

Disorder, Percolation, and Wave Propagation Effects in Ductile Fracture

D.L. Tonks

10.1. Introduction

During the last few years, physicists have been performing computer simulations of damage in brittle elastic systems, resistor networks, and fuze networks [1–8]. Many of the results are similar for the three kinds of simulations. We will address common aspects of all of these calculations but will use the terminology of the brittle elastic systems for simplicity. The simulations typically involve beam models of a solid. In these models the beams have elastic deformation properties and thresholds for breaking. The systems are loaded quasistatically until a beam breaks. Then the beam is removed, the system is equilibrated, and the process is repeated. The interest here is on the spatial pattern of the damage during evolution and at fracture. The questions typically asked are whether the damage is "brittle like," i.e., spatially localized, or "ductile like", i.e., spatially diffuse.

It has been found that the damage is localized or diffuse depending on the disorder of breaking thresholds initially present. For small disorder, for example if the breaking thresholds lie in a narrow band bounded far away from zero strength, the damage at fracture is very localized in a single crack-like cluster that passes through the sample. For large disorder, for example if the breaking thresholds extend completely down to zero strength, the damage at fracture is diffuse throughout the sample. A great deal of random-like, separated damage occurs during the earlier sample straining before enough damage collects to span the sample. This diffuse damage behavior results in greater toughness than the very localized behavior, because more energy-absorbing damage occurs before fracture.

The crossover between the localized and diffuse behavior can be thought of as a competition between the disorder in damage thresholds and the disorder in local stresses that drive the damage generation. If the former dominates, new damage will tend to occur in new places not connected to damage already present. If the disorder in local stresses dominates, there will be places of intense local stress, caused by damage concentrations, and new damage will tend to occur in these areas. This will give rise to catastrophic damage growth and localized, highly organized damage clusters that quickly expand to break the sample.

The above concepts will be useful later in understanding very-high-strain-rate ductile damage where wave propagation times limit the stress or strain enhancement around large damage clusters. This effect severely limits the local "stress disorder" and promotes diffuse damage.

Even in diffusely damaging systems, there is some sort of catastrophic damage process at the end of the fracture process. During the early phases of the straining process, damage is diffuse and new damage occurs in new places, often somewhat randomly. During this stage, the local stress disorder is limited; the local stresses can be approximated as a mean field. The damage will eventually progress to the point, however, where cluster interactions become important. This final phase, that results in breaking, is catastrophic and the big cluster that finally breaks the sample is "brittle like", but appears, of course, only after a period of diffuse damage growth.

These elastic systems exhibit size scaling effects [4]. The diffuse portion of their stress strain curves can be scaled using the system size to form universal curves. The scaling exponents seem to be about the same for a great variety of different models. This behavior is close to being "universal" in the sense of phase transitions, i.e., the exponents in the scaling laws depend little on system details.

There is qualitative similarity between damage evolution in the above systems and the void-linking process in ductile fracture. The nodal points of the beam system correspond to the voids in the void system, and the breaking of beams corresponds to the linking of the voids. The variation in beam breaking thresholds is analogous to the variation in void–void separation. A greater separation between two voids makes it harder to link them, or to break the solid ligament between them. The local stress in the elastic beam system corresponds to the void linking range in the void system. In the beam system, the applied stress drives the local stresses, while in the void system, the applied strain increases void size and void linking ranges. In both systems, the clustering of damage leads to local stress enhancement or to enhancement of the local void linking range.

Because of this qualitative similarity, the concepts described above are useful in understanding the void system. There are many differences in detail between the elastic beam simulations and the void system, however. For example, the beam simulations are usually done for a perfect lattice, i.e., the beam ends form a regular spatial array. Growth of a damaged cluster is limited to steps involving one nearest neighbor beam. However, the void system constitutes a "random lattice" that has greater disorder than the regular lattice. Also, the damage growth caused by a given cluster is not limited to the closest adjacent voids. The void linking range of a cluster could encompass voids lying at a variety of separations and link to all of them in one step. Hence, the void system differs in important ways and the results and analytical theories pertinent to the elastic beam configuration cannot be automatically taken over in detail to the void system.

Also, the elastic beam research is all being done in the quasi-static limit. Here, we are concerned with dynamic effects at very high strain rates and high stresses. Wave propagation and wave relaxation effects will enter here. Also, the dynamic systems will mostly be in plastic flow and not elastic. Hence, the dynamic ductile problem considered here is more general than the elastic beam problem.

The work here is the first effort that the author is aware of to address dynamic effects within the general context described above. Simplifications have been necessary to attack the problem, but the results should be qualitatively correct. In particular, two-dimensional simulations have been performed and the results compared with analytic models. Three-dimensional (3D) analyses in progress should eventually fill out the picture.

10.2. Selected Topics in the Phenomenology of Ductile Fracture

In this section, various experimental results of ductile fracture will be presented by way of background. The treatment will not be complete, but will involve those aspects needed to support the modeling described later on. No void nucleation is included in the simulations so this topic is not mentioned here. Void nucleation effects in this context will be studied in the future.

Ductile fracture is characterized by the nucleation, growth, and coalescence of voids in a plastically flowing material [9–11]. The void evolution generates new internal free surface by plastic flow and coalescence until enough surface exists to separate the sample. Ductile fracture surfaces examined microscopically exhibit a frothy appearance from many dimples where voids grew to coalescence. This process is the basic mechanism of the fracture described here. It is quite different from brittle fracture in which new internal surface is formed by cleavage in an elastic medium. The term "brittle like," occasionally used here to describe the spatial form of void clusters, should not be construed to mean that the microscopic damage mechanism has changed to cleavage.

In order to form a continuous internal surface that separates the sample, the growing voids must come together in some way, or coalesce. We will use the term "void linking" here. In dynamic situations, this will occur when the intervoid ligament undergoes a mechanical localized instability that rapidly thins the ligament in a localized region at the expense of an unloading and reduction of deformation in the surroundings [12–14]. This process could take more than one form. A local shear band is one possibility. Another is the localized tensile necking described by Thomason [12–14]. The modeling here is fairly independent of the actual mechanism. We assume only that some sort of localized flow quickly destroys the stress carrying ability of the intervoid ligament when the instability is triggered. We make the approximation that the loss of strength happens instantaneously. In reality, some local strain must occur first, but the local strain rate is greatly amplified by the linking instability, so that the loss of strength should happen rapidly compared to the macroscopic straining. For example, for the quasistatic case, Thomason [12] envisions a whole sheet of voids whose intervoid ligaments undergo flow localization all at once. He states that, in this case, the elastic unloading extends all the way to the sample boundaries, so that most of the sample straining becomes concentrated in the very narrow linking void sheet, which then breaks almost instantly with very little additional macroscopic strain. We are using this same general idea here, but on a local scale.

To model void linking, the triggering conditions must be known. We will make the approximation that the linking process is triggered when voids are closer than a critical distance. We will give each void a linking range such that when the ranges of two voids overlap, the ligament necking process is triggered. The range of a spherical void is roughly the size of its diameter [12]. Thomason has justified this model for the quasistatic situation with slip line fields and instability theory in both two and three dimensions [12–14]. We have generalized this modeling for the dynamic case by adding wave propagation effects in an approximate way to be described below.

We have also added the effect of voids first linking into local units which then further link up. To do this, one must deal with individual voids linking to larger void groups comprising voids that are already linked. It is apparent that a large linked void group constitutes a big void with an enhanced linking range. This is apparent in uniaxial stress experiments done on sheets with drilled-in holes [15–16]. One can deduce from the analytic fits to the data [15] that the linking range of a long, narrow void cluster is enhanced at the ends of the cluster, in a direction along the axis of the cluster, by a factor roughly equal to the square root of the cluster length. (The stress applied in the thin sheet experiments was uniaxial tension and, so, linking of single voids was enhanced in the direction perpendicular to the stress, which gave rise to clusters extending in the perpendicular direction, which gave rise, via the size enhancement described above, to even longer clusters extending in the perpendicular direction, etc.) Thomason's approximate slip line solutions for rectangular voids in plane strain under uniaxial tension can also be used to show the plausibility of an approximate square root dependance of the void linking range on the void length at the at the ends of a long, slender void, or void cluster. This will be shown in a future publication.

In the simulations done here, the square root size enhancement will be used as a realistic approximation. The general results should not depend on the exact power. Uniaxial tension will be used exclusively, so, for the quasistatic case, only long slender void clusters will be produced. This situation is not as restrictive as it might seem, because there is evidence for the quasistatic case [12] that ductile void linking driven by more general stress fields still occurs in 3D in sheets, or in narrow bands in 2D, more perpendicular than not to the axis of greatest principal stress. For high strain rates, on the other hand, the void linking is much more uniformly random; its spatial pattern depends less on the triaxiality of the stress state. (The rate of individual void growth depends a great deal on the presence of negative pressure, however.)

10.3. 2D Plane Strain Simulations

In this section the 2D void-linking simulations and their results will be described. The applied stress used is uniaxial tension. No void nucleation is included. The growth of a random spatial distribution of equal-sized voids is studied. Wave propagation effects in the void-linking process are included in an approximate way. The general effects apparent in the results should apply qualitatively to 3D configurations and to other kinds of driving stress. The void cluster morphologies that occur in these simulations are rather elongated, i.e., long and slender, but this morphology is probably common in practice, perhaps generalizing to sheets of voids in 3D, as envisioned by Thomason [12].

The effects of high triaxiality, i.e., volumetric tension effects, such as those that occur in spallation, are not studied here. However, the qualitative features of the high strain rate limit enumerated here are very pertinent to spallation.

The action of stress release waves caused by a given growing cluster on nearby clusters has not yet been studied in any detail. The simulations described here do not include it. A few initial results that do include it will be briefly described. The action of stress release waves from a cluster on the cluster itself is treated, of course, by means of the wave propagation modeling of the void linkage process. This self action is responsible for the enhancement of the void-linking range.

The simulation results given include snapshots of the void arrays, the effect of strain rate on breaking strain, the effect of wave propagation on void linking ranges, statistics of void clusters, the size effect of breaking, and the size effect of total damage.

The modeling of void linking range and void growth used here is reminiscent of that found for holes in thin sheets by Magnusen et al. [15]. For example, our void linking is enhanced in the direction perpendicular to the tensile axis, as in their thin sheet experiments. For thick sheets, i.e., plane strain, Magnusen et al. [16] found the link up direction to be at about 45° to the tensile axis. Hence, the direction our long clusters take is more like plane tension, but our assumption of fully plastic flow is like plane strain, i.e., thick sheet, deformation.

As Thomason points out [12] in thin sheet deformation, with increased loading, most of the sample remains elastic when the intervoid ligaments undergo local plastic failure, i.e., before general plastic loading. In plane strain deformation, on the other hand, the void population undergoes extensive general plastic straining before coalescence and rupture [12]. For these reasons, plane strain deformation with general plastic loading (as modeled here) is closer to 3D than thin sheet deformation [12].

Our work differs from that of Magnusen et al. [15] who have done void linking simulations for thin sheets based on their experimental findings. We include, in addition, wave propagation effects, cluster analysis of the void clusters, and the application of percolation theory. This theory, suitably generalized when necessary, is a good tool for understanding the statistics of cooperative void behavior.

The simulations here are not hydrodynamic calculations in the usual sense. Stresses and strains are not calculated using the dynamic equations of motion with constitutive relations. Rather, a global uniform driving strain field is assumed, together with void growth laws and void linking laws that are drawn qualitatively from experimental work like that of Magnusen et al. [15] and slip line theory such as that of Thomason [12] and Melander and Stahlberg [17–18]. Thus, here, the uniform strain field is imposed and the voids grow and link according to the laws imposed. Hence, most of the physics enters into the calculation via the void growth and linking laws. The imposed uniform strain field is in the spirit of a "mean field theory" in which the local strain field seen by a cluster is an averaged imposed one. For this reason, the behavior will be predicted accurately only until a certain advanced level of void linkage, at which point local void interactions significantly alter the imposed uniform strain field. The range of behavior accurately predicted will extend almost to the final break-up, however, and should be adequate for the purposes here, i.e. general constitutive modeling and not a detailed study of the final half percent of strain.

Inertia is included only as it affects the wave propagation retardation of void linking. Inertia effects might not be important for the individual void growth laws, however, for which the spatial length scale is small. For void growth calculations for spallation in Cu, for example, Johnson [19] found that inertia was not important compared to the strain rate dependence of the flow stress.

The mean field approach used here should give a good estimate of the final point of breaking. Approximations of this sort in other fields have shown, for example, that mean field assumptions can qualitatively predict critical behavior in phase transitions, which are roughly analogous to breaking. The critical region is usually very close to the phase transition, or, in this case, the point of fracture.

A full hydrodynamic code calculation would, indeed, more accurately model the behavior considered here, but the work and expense involved would be enormously greater. Indeed, including 10,000 voids in such a calculation, as was done here in a few cases, may well be impossible at the present time. The assumptions made in this work still have enough reality in them to produce qualitatively realistic results. Indeed, this approach to determining constitutive behavior should produce far better models than many ad hoc, unfounded models seen today.

Rectangles are used for the void shapes, instead of circles. This approximation is made for calculational simplicity and should not affect the results. For example, many percolation results depend only on the area of the objects involved and not very much on their shapes.

The void growth laws used are the following. The width of a void in the direction perpendicular to the stress axis stays at the constant value D_0. (The perpendicular direction will henceforth be referred to as the horizontal direction, for brevity.) The height of a void, in the direction of the stress axis, grows as follows: $D = D_0\,(1 + \alpha e_1^a)$, where D_0 is the initial void size, e_1 is the longitudinal strain, and a and α are constants. The values used for a, α, and D_0 are 1.3 and 12.1 and 8.0×10^{-4} cm, respectively. Except for D_0, they are reminiscent of the thin sheet results of Magnusen et al. [15] for 1100-0 aluminum. Hence, the modeling here is typical of a soft metal with little workhardening and fairly vigorous void linking behavior. (Strong workhardening inhibits void linking.) The large void size is reminiscent of large inclusions. The law used for r, the link range of a void in the horizontal direction is $r = D\sqrt{(W/D)}$, where W is the horizontal width of the cluster to which the void belongs. Thus, r is scaled by the void size. The link range r, is given the lower bound, D. As a consequence, the horizontal range of an isolated void is D. The link range of all voids in the direction along the stress axis is taken to be simply $D/2$ [15].

For very short times, times less than about $D_0/(2C_1)$, $D_0/2 + C_1 t$ is taken as the range. This is done so that, at zero time, the void's linking outline is approximately the physical void size. After this expression for r becomes larger than that previously described, which it does at about the time $D_0/(2C_1)$, the earlier version is used.

The cluster-size enhancement of the void linking range used here involves only the horizontal linking range, and not the vertical range, being enhanced by the cluster width. This is an approximation that held for the thin sheet with holes [15]. It probably worked in that case because the stress was uniaxial. For more general stress states, the linking range should be enhanced in more than one direction. This is allowed by the present theory; but was not done in the work here, which was intended to study the uniaxial stress case. The long and stringy cluster morphologies seen here are a direct result of the restriction of the cluster-size range enhancement to the horizontal linking range. This should be kept in mind.

Stress wave propagation effects are introduced into the cluster size enhancement of the linking range by appropriately limiting the cluster width, W, used in the formula for r. The idea is that a given void in a cluster will only "know about" other member voids if it is within the propagation distance of the release stress waves that originated when the other member voids linked to their neighbors. The stress enhancement seen by a given void due to others in its cluster arises from the stress relaxation waves originating from the intervoid ligaments of the other voids when those ligaments failed or localized. It takes time for these relaxation waves to propagate. If not enough time has passed for such waves to travel along a cluster, a void in the cluster will not "see" the full-width enhancement of its linking range.

This wave propagation process is quite complicated to model. Here, we use an approximation to the time available for wave propagation. For quasi-static straining, of course, this approximation is not pertinent.

The actual algorithm used for the W in the range formula is as follows. An effective wave-speed-limited cluster length, W, is calculated and updated for each cluster. This length is nominally incremented at each time step, dt, by dt times C_1, the longitudinal sound speed; but the effective length is given the upper bound at any given time of the actual, i.e., instantaneous, cluster length. Hence, when clusters whose effective lengths equal their static lengths suddenly join, the effective length of the new combined cluster grows with the wave speed, C_1, until it reaches the new, actual length. (When the clusters join, the current effective lengths of the members are replaced by an average

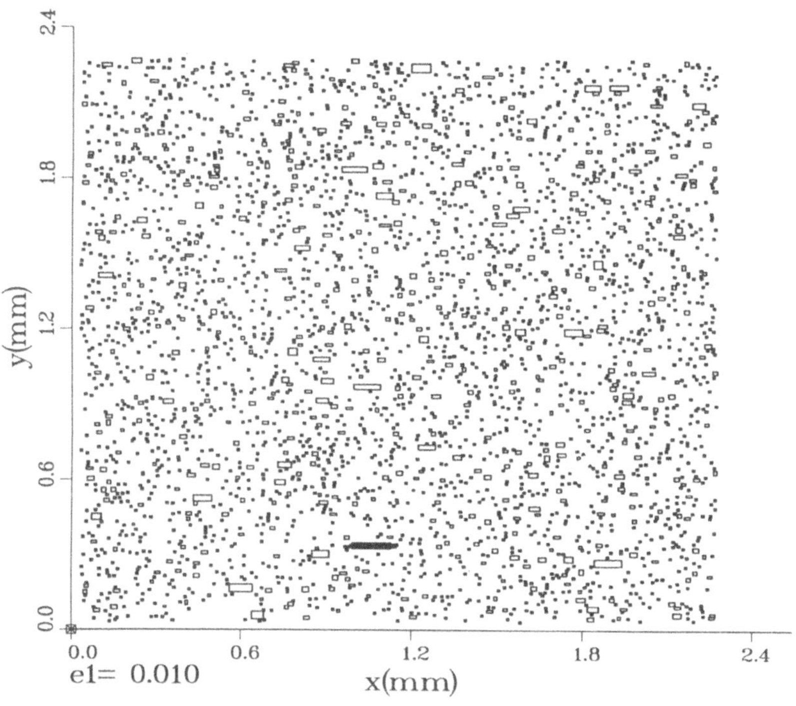

Figure 10.1. Simulation snapshot for a strain rate of 1/s, i.e., quasistatic, at a strain (e_1) of 0.01. Rectangles cover, but do not exactly outline, void clusters. The smallest covering rectangles correspond to individual voids, for which the rectangles are identical to the single voids themselves. The biggest cluster is broken down into the link-up range figures (rectangles) of individual voids.

with respect to the sizes of the old clusters. This is an approximation to simplify the bookkeeping.) Hence, the effective cluster length, W, roughly accounts for the time delay for joining clusters to "find out" about each other. A lower bound of D is applied to W, which has an effect only at small times.

The precise effect of these wave limiting approximations is difficult to visualize, except to see that they limit the effective cluster sizes in some way by $C_1 t$ when clusters form quickly. The effects can be understood qualitatively from the simulation results, however.

The value used for C_1 was 4×10^5 cm/s. The linking range along the stress axis is always taken to be $D/2$, which is the physical void boundary in this direction. Thus, the linking figure of a void is a rectangle of height D and width $2r$, centered on the void center. (The linking figure is the outline which reflects the range of a void. Two voids link when their figures overlap.)

Figures 10.1 through 10.7 show snapshots of the void linking configuration for two simulations done at strain rates of 1/s and 10^7/s, respectively. The initial porosities are 0.05, and 4,000 voids were used. The initial void configurations are the same. Fig. 10.1 shows the configuration for a strain, e_1, of 0.01.

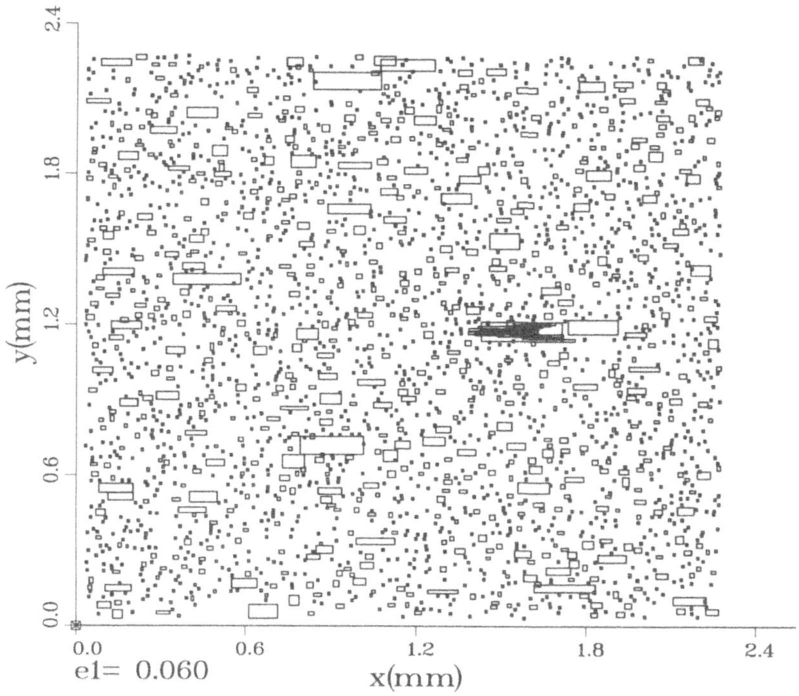

Figure 10.2. Same as Fig. 10.1, but at a strain of 0.06.

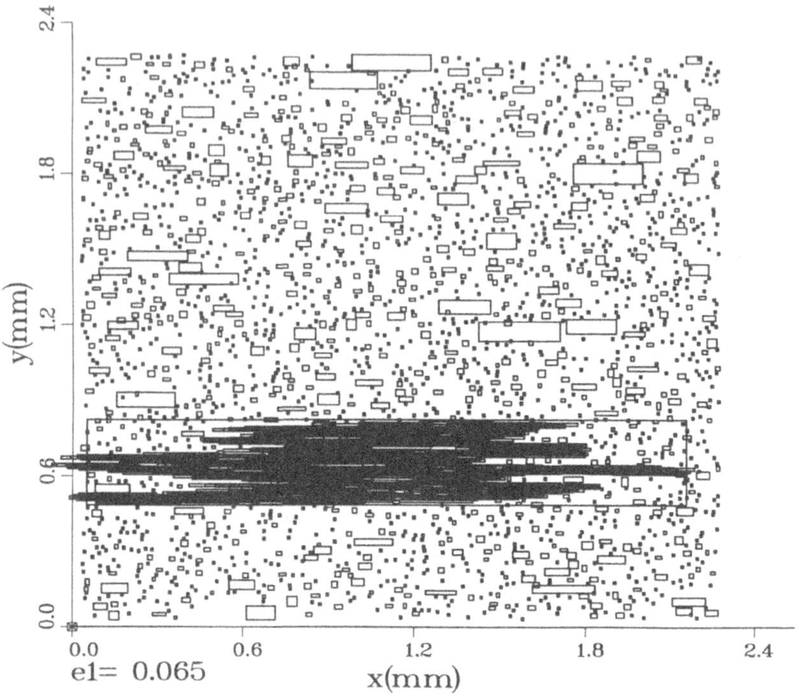

Figure 10.3. Same as Fig. 10.1, but at the breaking strain of 0.065. Note that the biggest cluster has extended across the sample to break it. The smaller clusters form a background that is much less interconnected than the largest cluster.

Except for the largest cluster, only rectangular outlines enclosing the void clusters are shown. For clusters with more than one void, the actual cluster outline falls inside this rectangle, as is seen for the largest cluster. For clusters with a single void, the outline is identical with the physical boundaries of the void. Hence, the smallest outline rectangles in the figures correspond to single voids.

Note that in some cases, a large cluster outline includes other, smaller cluster outlines inside, which reside in the extra space between the actual large cluster and its outline.

To provide more information, the largest cluster is broken down into the individual void range figures surrounding each member void. The heights of the smaller (but long) rectangles for the largest cluster correspond to the actual void heights, but the lengths correspond to the horizontal void ranges.

This graphical scheme is used to bring out the void cluster structure, which is quite important to this discussion. If only the voids themselves were shown, the points to be made would not be nearly as obvious. The connectivity of the voids would not be apparent.

The two crossed boxes along the axes in Fig. 10.4 form a ruler showing the

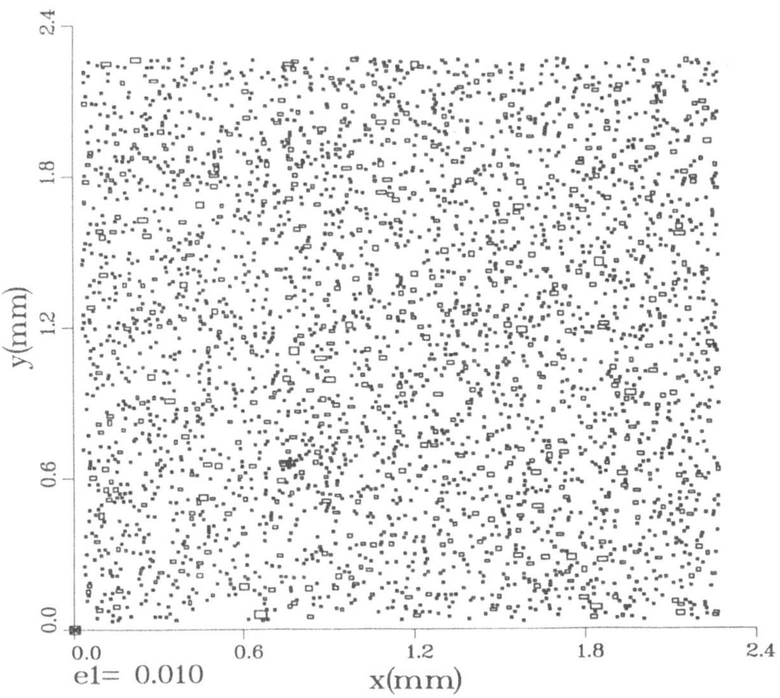

Figure 10.4. Simulation snapshot for a strain rate of 10^7/s and a strain of 0.01. The two crossed boxes along the x-axis indicate the distance traveled by a stress wave since the initial time. (Clearer in Fig. 10.5.)

wave propagation distance, $C_1 t$. The ruler shows the distances over which voids "know about" each other for the cluster size enhancement of the void linking. Roughly speaking, each void "thinks" its cluster is of length $C_1 t$ or the actual length, whichever is smaller.

Notice in the low strain rate simulation illustrated in Figures 10.1 through 10.3, that the ruler extends beyond the system. The linking there is not wave speed limited. Notice that general background damage evolves, but that a large cluster eventually outstrips its fellows and grows to break the system. The cluster size enhancement becomes very strong indeed for the larger clusters.

In the high strain rate simulation, illustrated by Figures 10.4 through 10.7, the initial situation is the same as for the quasistatic case, and the initial background damage seems qualitatively similar, but the time evolution does not give rise to a single large crack. This is apparent by comparing the second snapshot in each strain rate sequence, Figures 10.2 and 10.5, which are both at a strain of 0.06. The wave speed limitation of the cluster size range enhancement does not favor the largest initial cluster at high strain rates, but permits many clusters to grow large together, as seen in Fig. 10.6. When the

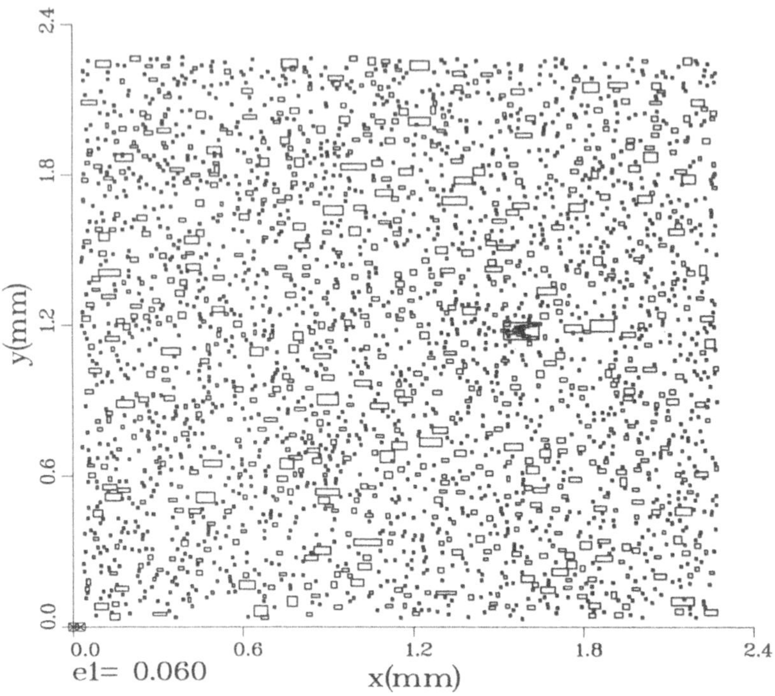

Figure 10.5. Same as Fig. 10.4, but at a strain of 0.06, close to the breaking strain of the quasi-static case.

system finally breaks, in Fig. 10.7, there are many large clusters seemingly situated at random. At the end, there is a strong preference for long cluster shapes due to the uniaxial nature of the modeling, but the overlying randomness is also apparent.

The final strain at breaking for the high strain rate case is 0.197, significantly larger than that of 0.065 for the quasistatic case. This is because cluster size enhancements of the range did not become important for the high strain rate case. The damage was not organized into long, far reaching structures and, so, more of it had to occur until enough of it was present to break the system.

These two simulations illustrate well the transition from a quasi-static process in which the damage can organize and extend itself, to the high strain rate process in which the damage occurs almost randomly and finally breaks the system by simply overwhelming it everywhere.

This latter behavior, the high strain rate limit, is an example of random continuum site percolation theory [20–23]. In this theory, objects are placed at random and objects that overlap are said to be connected and to form clusters. The properties of the resulting clusters are studied. In particular, near the condition when a cluster exists which is large enough to span the

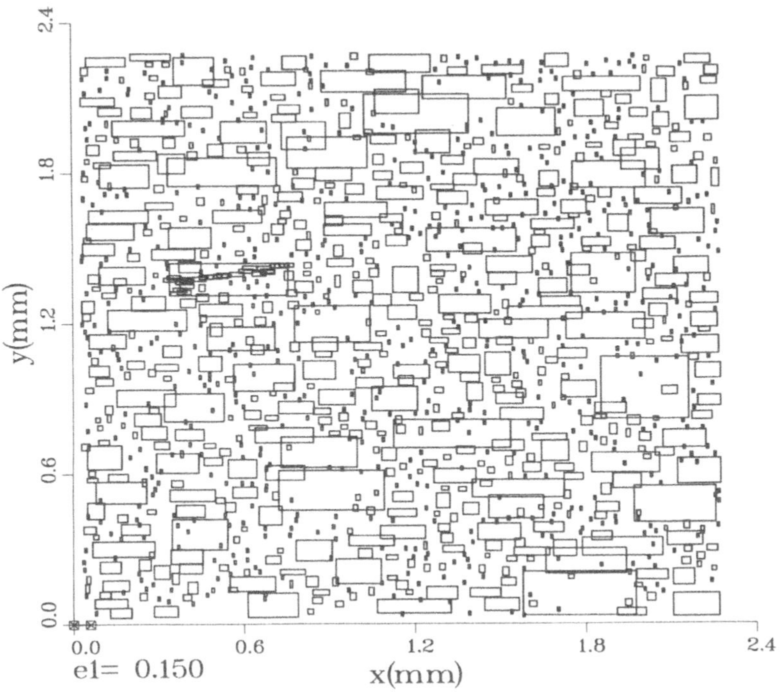

Figure 10.6. Same as Fig. 10.4, but at a strain of 0.15. Note that many clusters are growing isolated from each other. There is no enhancement of linking range due to cluster size; it is being repressed due to wave speed limiting.

system, a critical condition is said to exist. This means, for example, that clusters of arbitrarily small and large sizes exist. This also means that no average cluster size exists and that the cluster size population, the number of clusters per size (number of voids), obeys a power law. The clusters are very ramified in shape, i.e. filled with lots of space and not compact.

This condition must obtain in the high strain rate limit. In a real situation, however, the limit may not be realized closely enough for the critical behavior to be clear. For small systems, the limit may also not be apparent.

In the simulations presented here, percolation behavior is evident, as the high strain rate snapshots show. Also, Fig. 10.11, to be described shortly, shows a transition to a power law cluster distribution, which is typical of the approach to critical percolation behavior. Also, the value of the final breaking porosity given by continuum percolation theory is well realized here for the highest strain rates.

It has been found by researchers in simulations for many common shapes of the same size, including rectangles and ellipses, that the point where an infinite cluster forms depends only on the porosity and not on the kind of shape. This is even true to a good approximation for objects of variable size,

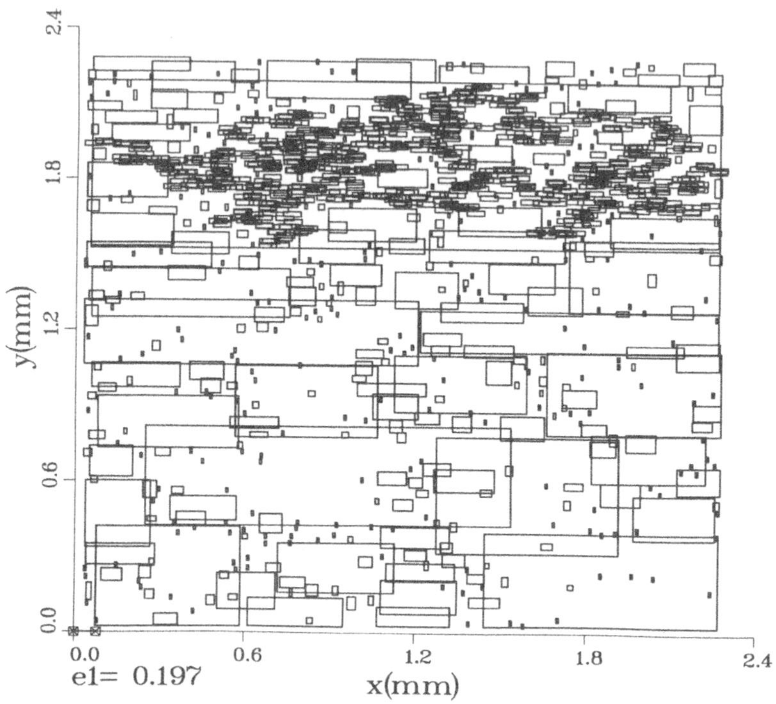

Figure 10.7. Same as Fig. 10.4, but at the breaking strain of 0.197. Note that the cluster that broke the sample is surrounded by many large clusters. This behavior is similar to random continuum percolation.

if a suitable average over sizes is used [21]. The porosity usually quoted is the "bare porosity," Φ_0, given by the total area of objects without regard to overlaps. φ_0, the porosity actually seen in the system is given in terms of Φ_0, for random placement of the objects, by $1 - \exp(-\Phi_0)$. A reasonable averaged bare porosity for the present simulations is $2\langle r \rangle D\rho_0$, where $\langle r \rangle$ is the average void range. We will call this quantity, evaluated from the simulations, Φ_p. According to the continuum percolation simulations, when Φ_0 reaches about 1.1 an infinite cluster should form [23]. In the simulations at high strain rate, the value of $2\langle r \rangle D\rho_0$ when a breaking cluster forms is usually very close to this value, as we will see. The value at breaking could depart from the theoretical one due to small system effects and the usually slight effects of averaging over a range of object sizes. Another possibility is that the percolation limit has not quite been reached.

We note that the static void array calculations of Melander [18] did not include the enhancement of linking range due to cluster size. In this sense, his results are more typical of high strain rate behavior than static behavior. This is probably why he found that a random void array required more strain to break than a regular array, contrary to experimental results [15, 16].

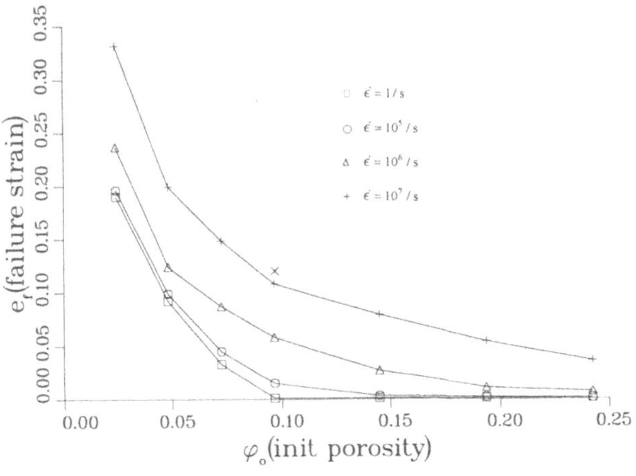

Figure 10.8. Failure strain versus actual initial porosity for a system of size 0.832 mm × 0.832 mm at various strain rates. The squares, circles, triangles, and pluses are at strain rates of 1, 10^5, 10^6, and 10^7, respectively. The lines are to guide the eye. The *X* is a prediction of an analytic theory.

Results will be given next for the breaking strain to failure for various initial porosities and strain rates. Figure 10.8 shows the strain to failure versus the initial porosity for a sample of size 0.832 mm × 0.832 mm. Various initial porosities imply differing numbers of voids. There were 3000, 2327, 1695, 1099, 813, 535, and 264 voids, ranging from the largest to the smallest initial porosity, for the initial porosities shown in Fig. 10.8. Note that the quasistatic curve involves the least strain to failure. The place where it intersects the x-axis, at about 10% porosity, is the initial porosity at which failure will occur as soon as the stress waves have time to propagate and fail the intervoid ligaments. (The initial porosity at which the system falls apart without ligament failure, i.e. simply by virtue of being full of holes, is at about 67%.) In general, the failure curves for higher strain rates fall at ever increasing strain. As mentioned earlier, this is because more strain with more random damage is necessary before the sample breaks when size enhancement of void linking is restricted by the requirements of the wave transit time. The curve portions that lie to the right of the static curve x-axis intercept, which involve higher strain rates, have non-zero strains to failure due to the necessity of wave motion before intervoid ligaments neck down. That is, there is enough porosity for the sample to break quasistatically, but at high strain rates, some overall strain occurs while the stress waves are moving to cause the failure of the intervoid ligaments.

Judging both by cluster morphology and the value of $2\langle r\rangle D\rho_0$, (which is Φ_p), the random percolation limit is approximated well for the curve with strain rate of 10^7/s and somewhat less well for the curve with strain rate of 10^6/s. The Φ_p values at breaking for the curve of strain rate 10^7/s, ranging

Table 10.1. Φ_p and Void Range versus Strain Rate

$\dot{\Psi}(s^{-1})$	Φ_p	$\frac{r_m}{\sqrt{C_l tD}}$	Cluster Morphology at Breaking
1	0.56	0.0012	One large cluster
10^5	0.57	0.30	One large cluster
10^6	0.72	0.58	Several large clusters
10^7	0.93	0.86	Many large clusters

from low to high initial porosity, were 1.070, 0.931, 0.921, 0.848, 0.970, 0.953, and 0.905, respectively. The 10^5/s strain rate curve involves one large breaking cluster except for initial porosities larger than 0.10, where percolation-like cluster morphology reappears. Here there is enough initial porosity for percolation-like fracture as soon as the stress waves have time to reach out to nearby voids. The cluster morphology for the strain rate of 1/s always involved one large breaking cluster.

The transition from the high strain rate percolation limit to quasistatic behavior is illustrated in Table 10.1, where the value of Φ_p, the average bare porosity of the individual voids surrounded by their ranges, is given at breaking for various strain rates. The simulations used for Table 10.1 are the same ones used above in Fig. 10.8 for the initial porosity of about 0.05. There were 535 voids. The maximum cluster range, r_m, seen in the simulation at the time of breaking is also given there divided by the square root of $C_1 tD$ at the time of breaking, which is the wave speed limit of the linking range. When this ratio approaches one, the ranges of the biggest clusters are being strongly limited by wave speed effects. As Table 10.1 shows, this is the case for the higher strain rates, where, also, the value of Φ_p is close to the percolation limit. When this ratio is much less than one, the stress waves at breaking have already fully swept out the clusters before anything much happens and the cluster ranges are determined by the static void configurations alone. It is seen that this behavior sets in at the strain rate of about 10^5/s. This coincides with the fall of Φ_p from the percolation limit of 1.1.

Some other simulation results will now be given that involve mostly cluster statistics. The quasistatic results can be compared with those of the elastic beam simulations mentioned earlier. Some strain rate effects will be mentioned.

First, Fig. 10.9 shows the cluster size population, i.e., the number of clusters versus the number of voids in a cluster, for the same quasistatic simulation illustrated earlier by the snapshots, i.e. initial porosity of 0.05, strain rate of 1/s, and 4,000 voids. For all but the last curve, with the diamond symbols, for which the system was near breaking, the initial portions of the curves roughly obey the law that the number of clusters is exponential in the

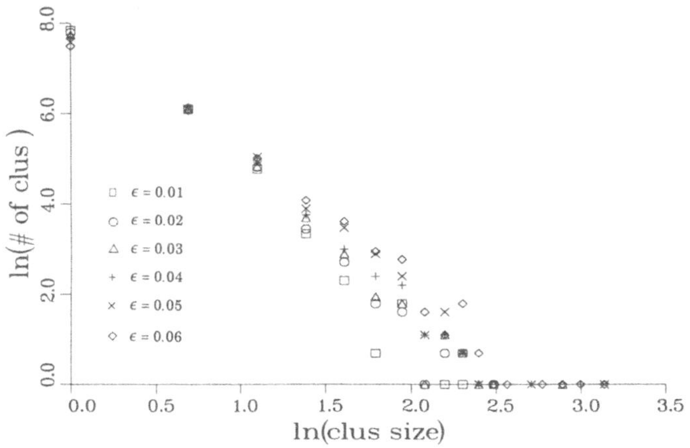

Figure 10.9. Log-log plot (natural logs) of the cluster population density versus the cluster size (number of voids) for various strains at a strain rate of 1/s. The squares, circles, triangles, pluses, crosses, and diamonds correspond to strains of 0.01, 0.02, 0.03, 0.04, 0.05, and 0.06, respectively. The initial actual porosity is 0.05; there were 4000 voids.

cluster size. This sort of behavior is given by the random positioning and overlap of voids [7]. When the curves were scaled with the average cluster size (not shown), their initial portions all fell on the same roughly exponential line, except for the curve that was near breaking. Note that most of the points on the graph come from the smaller background clusters and not from the largest clusters, which are rare. The population behavior for the larger clusters is chaotic.

In Figure 10.9, the curve with the diamond symbols at a strain of 0.06 taken just before failure has mostly straightened out on the log-log plot to a power law. Thus, at this strain, many of the large clusters are losing their "random" character, which gives rise to the exponential law mentioned above, and are showing signs of organization. This corresponds, in the language of the elastic beam simulators, to the period of diffuse growth giving way to the catastrophic regime in which the damage becomes much more organized. The change to a power law was also seen by Curtin and Scher in simulations in a different system [24–25]. It did not mean there, as it does not mean here, a change to a random percolation critical state, in which the cluster population also becomes a power law, but a change to more organization.

This behavior might seem a bit tenuous in Fig. 10.9, but it also appeared very consistently in many other simulation results.

The same initial configuration simulated at a strain rate of 10^7/s did not produce a straight line at the strain of 0.06, as the curve with the diamond symbols in Fig. 10.10 shows. The results for a strain of 0.07 are also plotted to show the smooth *curve*, i.e., not a straight line, that results. This curve is exponential-like, as expected.

Results for higher strains for the high strain rate of 10^7/s, on up to the percolation-like failure, are shown in Fig. 10.11. These results illustrate that a straight line develops at higher strains, i.e., the cross symbols, somewhat

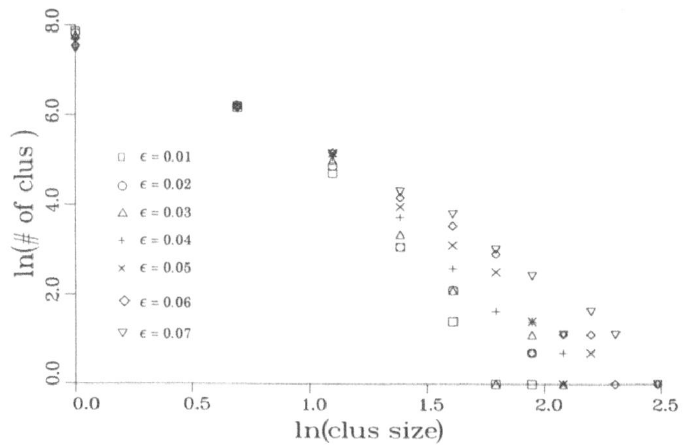

Figure 10.10. Same as Fig. 10.9 but for a strain rate of 10^7/s. One additional strain is included: the upside-down triangle corresponds to a strain of 0.07.

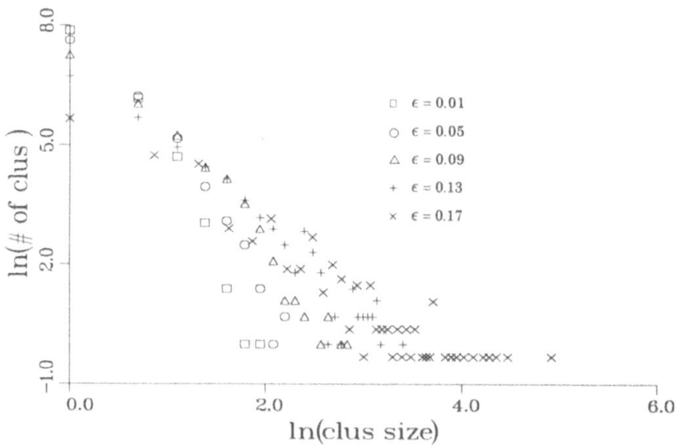

Figure 10.11. Log-log plot (natural) of the cluster population density versus the cluster size at various strains for a strain rate of 10^7/s. The squares, circles, triangles, pluses, and *x*'s correspond to strains of 0.01, 0.05, 0.09, 0.13, and 0.17, respectively. This is for the same simulation as in Fig. 10.10, but more advanced strains are shown.

before the coalescence of clusters just before fracture destroys the trend, i.e., the x symbols. The exponent of the power law portion (the initial portion) of the curve with cross symbols is about 2.0, which is close to the theoretical value of 187/91 (which equals 2.055) for τ, the corresponding exponent in percolation theory. This change to a power law is expected for percolation behavior close to the critical point. It is indicative that an average cluster size no longer exists, or at least is very large, contrary to the case for the exponential law. This change to a power law is apparently different in nature from that seen at lower strains, i.e. Figure 10.9, for the quasistatic case. The value of the exponent there is more like 8/3. This matter is just becoming better understood [24–25].

Most of the structure of the cluster size population curves is from the diffuse background. The large, organized clusters form an erratic tail. Hence, even the quasistatic systems here are not really "brittle like", in the sense that the elastic beam systems with disorder are not "brittle like"; their fracture involves much diffuse damage, even though a large "crack" takes over eventually for the quasistatic case.

Figure 10.12 shows on a log-log plot, the number of linked voids minus the number of clusters, i.e., a measure of the linking, versus $\rho_0 D^2$, which increases with strain, for the same quasistatic simulation, which involved 4,000 voids. Only for 1D clusters, roughly the situation here, would this quantity be the actual number of void–void links. Note that on the log-log plot, this number is roughly linear in the strain until fracture. This behavior, again, comes mostly from the diffuse background damage in the form of small clusters linking. In fact, this behavior can be roughly predicted from void pair linking of randomly situated voids as follows. The probability that the nearest neighbor of a void lies at a radius r within the interval dr is:

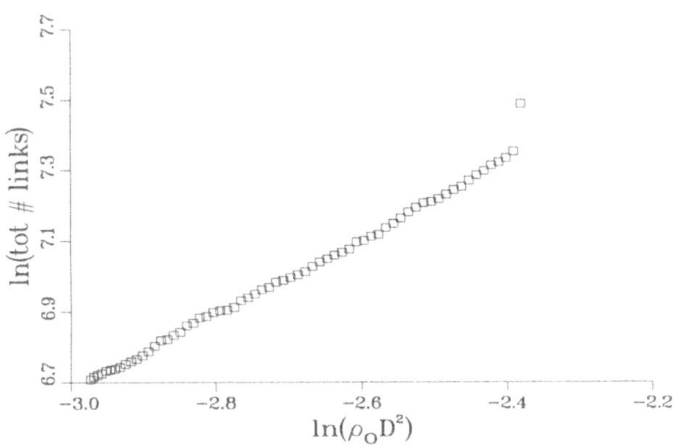

Figure 10.12. Log-log plot (natural) of the number of voids linked (minus the number of clusters), versus $\rho_0 D^2$ for the simulation with strain rate 1/s, initial actual porosity 0.05, and 4000 voids.

$2\pi\rho_0 r(\exp(-\pi r^2\rho_0))\,dr$. When this quantity is integrated from zero to $1.5D$, which approximates the single void ranges used here, and multiplied by $N/2$, one-half of the total number of voids (the 1/2 is to avoid double counting of voids), one obtains, roughly, the number of paired void–void links. If one uses the approximation: $1 - e^{-y} \approx y$ in the result, which should obtain for small porosities, one obtains, approximately, $1.12\pi N\rho_0 D^2$ for the number of links. Figure 10.12 follows this relation fairly well, both in the slope and in the y-axis intercept.

Figure 10.13 shows on a log–log plot, the same quantity, the number of linked voids (minus the number of clusters) just before failure, versus the system size, for the same quasistatic strain rate of 1/s for systems of various sizes. The slope of the line in Fig. 10.13 is about 1.9. This value is similar to (but not exactly the same as) the corresponding quantity for various elastic beam simulations. In that case, the corresponding quantity was rather independent of model details. A value near two shows that our system exhibits toughness, i.e., the damage necessary to break it is not localized but is spread throughout. Most of this damage is coming from the diffuse background and not the largest cluster. If the fracture process were "brittle-like," or more localized, this exponent would be significantly smaller. Its value would be close to zero if a single long cluster could break the sample with no other damage. This is another illustration of the diffuse nature of the background damage in this system, even though a large cluster eventually breaks the system at these low strain rates.

Figure 10.14 shows, for the strain rate of 1/s, system size effects in quasistatic fracture, i.e. it shows $\rho_0 D^2$, the "bare void porosity," at fracture versus the logarithm of the system size. The log of the system size seems to work

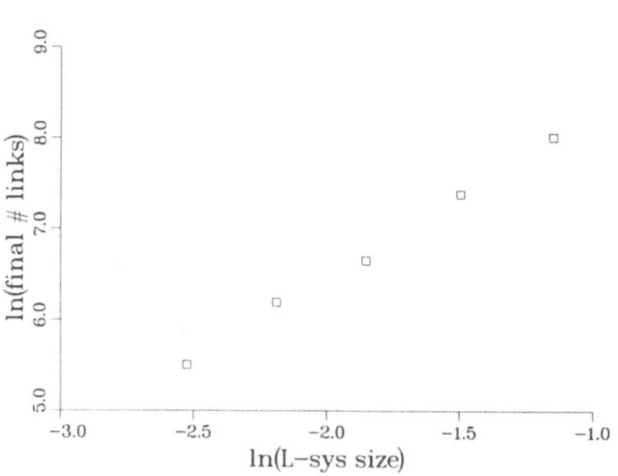

Figure 10.13. The log of the number of linked voids at the point just before failure (minus the number of clusters), versus the log of the system size (edge length) for various systems.

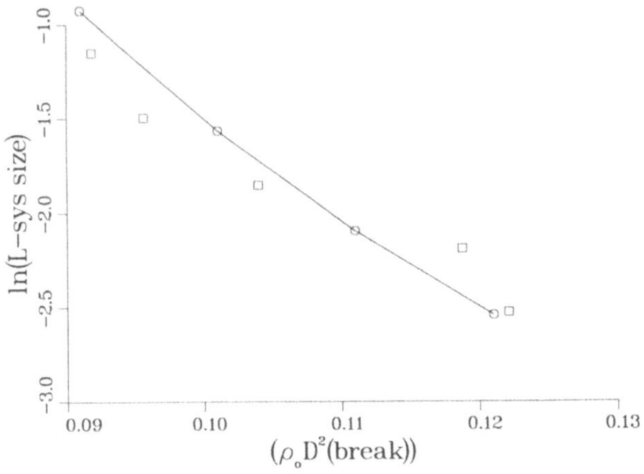

Figure 10.14. The squares are the logarithms of the system size (edge length) versus $\rho_0 D^2$ at breaking for various systems. The circles are from the analytical approximation illustrated in Fig. 10.15. The line is to guide the eye.

fairly well here to straighten out the plotted line. This is similar to the elastic beam systems [3]. The range in bare porosity is too small, however, to determine whether $\rho_0 D^2$ or its logarithm produces a straighter curve.

10.4. Analytical Approximations

This section concerns certain analytical approximations to the 2D simulations. The emphasis will be on predicting the point of fracture. The analytical theory applies only qualitatively to the simulations because of certain approximations necessary to facilitate analytical treatment. Nevertheless, with adjustment of one parameter, the analytical expressions predict the point of fracture of the simulations fairly well. The analytical expressions, hence, constitute a model for the 2D behavior.

Another benefit from the analytical expressions is a more fundamental understanding of the behavior occurring in the simulations. Simulations by themselves are often difficult to learn from. The derivation of analytic expressions that predict the simulation results indicate that a deeper understanding has been arrived at. The analytical theory has various concepts and approximations that are demonstrated to be correct by predicting the simulations.

The basic analytical tool to be used here is an integral equation for the probability, P, that a given span in the horizontal direction (perpendicular to the tensile stress axis) is covered by a void cluster. By "cover" is meant that the linking range of the cluster or the cluster itself covers the end of the interval. It is assumed that a seed void exists at the beginning of the span. Included in the probability are clusters longer than the interval, so the spanning probability, P, is really the integral from infinite cluster size down to a certain minimum cluster size, of a cluster population versus cluster length. (The clusters here are assumed to be long and to lie in the horizontal direction.) The quantity, P, is useful in determining when enough damage exists to break a 2D sample. A similar quantity is dealt with by Domb [26]. Our situation is more difficult in that it involves a variable linking range.

The word "span" includes the flexibility that the clusters involved can meander in the vertical direction, i.e., they don't have to be exactly straight. This meandering is taken care of by the factor, f, in the equation, and only the linking horizontally is treated in detail.

The integral equation for P is:

$$P(l) = \phi \int_{l_l}^{l} e^{-\phi(l-l')} P(l')\, dl', \tag{10.1}$$

where l is the horizontal distance (span) of the interval divided by D, the void size, and ϕ is the "bare porosity" $\rho_0 D^2$ times an adjustable factor, f. The factor f was adjusted to obtain fits with the simulation results. It accounts for, among other 2D effects, the 2D effect that a given void linking to the end void in a void cluster need not have the same vertical coordinate. Its vertical coordinate can be within a range of "2D" and still overlap. Hence, $f \geq 2$. In the fitting done here, it is also used to account for the process of a void cluster forming not just one thin one-dimensional like "needle" but, perhaps, many, formed by part of a needle extending vertically to start another needle, etc. The simulation snapshots show that this does, indeed, occur.

The analytical theory accounts for just one needle-like cluster and must include such effects through adjustment of f. An f-value of 5.5 was found to fit the quasistatic simulation results. Thus, the 2D effects included through f are not trivial, but, with this f-value the analytical theory is successful. This is probably so because the clusters, even though consisting of more than a single needle, are still long and narrow.

Underlying Eq. 10.1 above, is the assumption that a seed void exists at $l = 0$. The quantity l_1 in Eq. 10.1 above is the length of the shortest cluster, starting from the seed void and extending to a void at its other end, whose range still reaches the horizontal coordinate l. We will use the equation $l = l_l + \sqrt{l_l}$, modified for small l, to determine l_l. For $l = 2$, we have $l_l = 1$. For smaller l, the value $l - l$ will be used for l_l. The "last" void in the cluster (to be described below, not the seed void) will reach this far by itself. One could possibly argue for somewhat more range for these short clusters, i.e. $l < 2$ but the particular ranges used should not affect the results for very long clusters, which are the principal interest here.

The meaning of the integral equation is as follows. The probability for an interval of length l being covered is broken down into a sum over the various positions of the last void in the interval before the length l is reached. The different values of this position correspond to mutually exclusive events. The probability for a given position equals the product of three probabilities: (1), that there are no voids in the interval remaining from the void to the end of the range, (2) that there is a void within the interval dl at l, and (3) the probability that the last void position is covered. The first probability is the exponential in the integral equation, the second is the factor ϕdl, and the third is the factor $P(l)$. As mentioned previously, the factor ϕ includes 2D effects.

An approximation to $P(l)$ will illustrate its meaning. $P(l)$ is calculated at a set of discrete values l_i obtained from the relations $l_l = 1$ and $l_i = l_{i-1} + (0.5)\sqrt{l_{i-1}}$. Each new l_i extends only half of the range from the previous one, the series begins with $P(l_l) = 1$, and the values of P at subsequent l-values are given as follows: $P(l_i) = p(l_{i-1})(1 - \exp(-\phi\sqrt{l_{i-1}}))$. The idea behind this approximation is that the factor with the exponential is the probability that at least one void is in the range-interval extending a distance of $\sqrt{l_{i-1}}$ in front of a cluster with length l_{i-1}. If no more than one void is likely to appear in the interval, which is roughly the case for the smaller ϕ-values and system sizes considered here, the average extension of the cluster is one half of the range. This explains the factor 0.5 in the equation for l_i. The concept here is clusters, rather than the covering interval of the continuum case, but the fact that no "ending" probability factor is used, i.e., all subsequent processes are included, makes this P roughly equivalent to the continuum P.

The usefulness of this approximation is mostly pedagogical, since, as seen in Fig. 10.15, it does not do very well when compared to the continuum result for large l. Its accuracy improves for smaller ϕ, however.

The quantities ϕ and l, which equal $f\rho_0 D^2$ and the interval divided by D, respectively, appear naturally in the analytical expressions above. This indicates that they are natural scaled variables to use in any treatment of the problem. This insight is one example of the benefit of the analytical work.

The calculated P is used to predict the breaking point in the following way. Each void in the system is taken to be a seed void which gives rise to a cluster family whose covering properties are given by $P(l)$. The number of seeds is $L^2\rho_0$, where L is the system width. $L^2\rho_0$ times $P(L/D)$, is, thus, the probability that at least one of the seeds grows to a cluster that spans the system. Thus, when this expression equals one, the system should break. Only one direction of growth is assumed, which is appropriate for clusters almost as large as the system.

In Fig. 10.15 we have plotted $P(l)$ and $l/(l^2\rho_0 D^2)$ versus l. (Note that $l^2\rho_0 D^2$ equals $L^2\rho_0$.) The intersection of the line with a curve for $P(l)$ determines the l-value, or system size, that breaks for the given value of ϕ. Plots for several ϕ values are given to show the variation in breaking size with ϕ. The dashed curve is the approximation to $P(l)$ described above. It corresponds to the top most curve for $P(l)$. It can be seen that the approximation

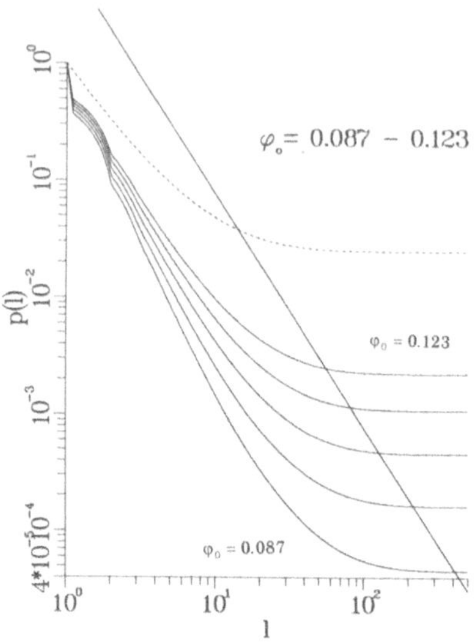

Figure 10.15. Log-log plot of $P(l)$, the void covering probability for the interval of length l, versus l, for five actual porosities ranging from 0.087 to 0.123. The dotted line is an approximation to $P(l)$ described in the text for the porosity 0.123. The straight line is a plot of $l/(l^2 \rho_0 D^2)$. The intercept of this line with the solid curves gives the breaking sizes of l for the porosities involved. Four of these are plotted in Fig. 10.14.

qualitatively follows $P(l)$, but is not accurate enough for good numerical results.

An f factor of 5.5 for ϕ in the above analytical theory produced the calculated points for breaking size versus $\rho_0 D^2$ in Fig. 10.14. The trend of the simulation points there is fairly well reproduced by the calculation. No attempt was made to optimize the fit by finer adjustment of ϕ.

Wave propagation effects can be approximately introduced into the above theory in a way similar to the simulations. The lower bound in the integral equation can be taken to be the upper bound minus the minimum of the static range and the square root of $C_l t$. (Special treatment is necessary for small cluster sizes.) This has the effect for high strain rates, in the calculation of $P(l)$, of eliminating the horizontal asymptote seen in Fig. 10.15; the new version of $P(l)$ (not shown) falls precipitously without leveling off. The static type fracture mode is severely curtailed. The calculated $P(l)$ can then be used to calculate a version of Φ_p, to see when it reaches 1.1 to predict when the percolation fracture mode takes over. This theory can roughly predict the percolation like fracture seen in the simulations. The X in Fig. 10.8 is a prediction using this approach for the same strain rate as the nearby curve. At this writing, this line of work is not complete, however.

10.5. Application to Spallation

2D spallation phenomena are expected to approach the high strain rate limit, described above. In spallation, the void linking geometry will be isotropic, instead of directional, as used here, but the percolation limit will be similar to that here. That is, if the strain rate is high enough, a continuous path of linked voids will form in the system when ρA approximately equals 1.1, where ρ is the density of object centers and A is the area or average area of their linking figures, i.e., the outer boundary of an isolated void extended by its void linking range.

In three dimensions, there arises an additional complication not present in 2D. In 2D, the porosity at which a cluster of linked voids just spans the system is also the smallest porosity at which the system can fall apart. In 3D, however, these two porosities are different. Due to added dimensional freedom in 3D, at the porosity when a spanning cluster of linked voids first appears, the solid part of the system is still connected. It extends continuously from one boundary to another. Only at a second percolation point at higher porosity does the solid part of the system cease to be connected. This is the percolation point of interest in studying fracture.

Of course, after the first percolation point, or the point with a path of linked voids, is reached, the system has suffered substantial damage and will probably "soon" break up. Thus, the use of this first percolation point to establish a fracture point in 3D, as is commonly done, is valid in a sense, but not strictly correct.

Work is currently underway to investigate the second percolation point, or point of solid disconnect. The results will be reported in a future publication.

10.6. Summary and Conclusions

In this article, we have presented some essentials of a probabilistic approach to ductile fracture, and have illustrated this approach with 2D simulation results and analytic modeling. The basis of this approach is the concept of void linking, or the loss of the stress carrying ability of the intervoid ligaments due to necking down or shear banding. This process produces clusters of linked voids which act, in a sense, like bigger voids. An important part of the theory is the enhancement of void linking range with cluster size, i.e., a void member of a linked cluster will have an enhanced range in linking to non-cluster voids due to its membership in a linked cluster of voids. Another important part is the time limiting of this void linking enhancement due to wave propagation effects.

The modeling of these effects is approximate and limited to 2D at present. However, the general results should be qualitatively correct. In the case of the brittle system simulations described earlier, for example, general behavior emerged that was fairly independent of modeling details. This should also be the case here.

For low strain rates and for a uniaxial imposed strain field, it has been found that a diffuse background of void linking occurs, producing small clusters until a large cluster finally grows "explosively" to break the system. Due to the cluster size enhancement of void linking, the growth of a large cluster eventually far outstrips that of all the others. A quasi 1D analytic theory, which can be fitted to the results fairly well, provides understanding of this process in terms of the growth of long, narrow clusters. At high strain rates, there isn't time for stress release waves to propagate to bring about the void linking enhancement. The clusters in this case are much shorter and many more of them must form to break the system by sheer numbers. This behavior is approximately describable by random percolation theory, as illustrated above.

Stress relaxation due to void linking and stress release wave propagation is only included by the time limiting of the void linking enhancement, and not in the creation of zero stress "shadows" around clusters. Some preliminary work including this effect has shown a narrowing of the clusters and increased strain to failure.

Also, the imposed strain rate here is taken to be uniform throughout the sample. Hence, the spatial variations of driving waves in real systems are not included here. This treatment, therefore, applies to small computational cells over which these driving waves are reasonably uniform.

This approach to fracture appears to be promising and future work is planned to remove the limitations of the present work.

10.7. Acknowledgment

I am pleased to acknowledge support from the Joint DoD/DOE Munitions Technology Development Program and the DOE.

10.8 References

1. Charmet, J.C., Roux, S., and Guyon, E., eds. *Disorder and Fracture*, Plenum, New York (1990).
2. Herrmann, H.J. and Roux, S. *Statistical Models for the Fracture of Disordered Media*, North-Holland, New York (1990).
3. Sahimi, M. and Arabi, S. Mechanics of Disordered Solids. III. Fracture Properties, *Phys. Rev.* **B47**, 713 (1993-II).
4. de Arcangelis, L., Hansen, A., Herrmann, H.J., and Roux, S. Scaling Laws in Fracture, *Phys. Rev.* **40**, 877 (1989).
5. Herrmann, H.J., Hansen, A., and Roux, S. Fracture of Disordered, Elastic Lattices in Two Dimensions, *Phys. Rev.* **B39**, 637 (1989).

6. Hansen, A., Roux, S., and Herrmann, H.J. Rupture of Central-Force Lattices, *J. Phys.* **50**, 733 (1989).
7. Kahng, B., Batrouni, G.G., Redner, S., de Arcangelis L., and Herrmann, H.J. Electrical Breakdown in a Fuse Network with Random, Continuously Distributed Breaking Strengths, *Phys. Rev.* **B37**, 7625 (1988).
8. Duxbury, P.M., Leath, P.L., and Beale, P.D. Breakdown Properties of Quenched Random Systems: The Random-Fuze Network, *Phys. Rev.* **B36**, 367 (1987).
9. Thompson, A.W. Modeling of Local Strains in Ductile Fracture, *Metal. Trans. A*, **18A**, 1877 (1987).
10. Bourcier, R.J., Koss, D.A., Smelser, R.E., and Richmond, O. The Influence of Porosity on the Deformation and Fracture of Alloys, *Acta Metall.* **34**, 2443 (1986).
11. Hancock, J.W. and Mackenzie, A.C. On the Mechanisms of Ductile Failure in High-Strength Steels Subjected to Multi-Axial Stress-States, *J. Mech. Phys. Solids* **24**, 147 (1976).
12. Thomason, P.F. *Ductile Fracture of Metals*, Pergamon, New York (1990).
13. Thomason, P.F. A Three-Dimensional Model for Ductile Fracture by the Growth and Coalescence of Microvoids, *Acta Metall.* **33**, 1087 (1985).
14. Thomason, P.F. A Theory for Ductile Fracture by Internal Necking of Cavities, *J. Inst. of Metals* **96**, 360 (1968).
15. Magnusen, P.E., Srolovitz, D.J., and Koss, D.A. A Simulation of Void Linking During Ductile Microvoid Fracture, *Acta Metall. Mater.* **38**, 1013 (1990).
16. Magnusen, P.E., Dubensky, E.M., and Koss, D.A. The Effect of Void Arrays on Void Linking During Ductile Fracture, *Acta Metall.* **36**, 1503 (1988).
17. Melander, A. and Stahlberg, U. The Effect of Void Size and Distribution on Ductile Fracture, *Int. J. Frac.* **16**, 431 (1980).
18. Melander A. Computer Simulation of Ductile Fracture in a Random Distribution of Voids, *Mat. Sci. Eng.* **39**, 57 (1979).
19. Johnson, J.N. Dynamic Fracture and Spallation in Ductile Solids, *J. Appl. Phys.* **52**, 2812 (1981).
20. Stauffer, D. *Introduction to Percolation Theory*, Taylor and Francis, London (1985)
21. Balberg, I., Anderson, C.H., Alexander, S., and Wagner, N. Excluded Volume and its Relation to the Onset of Percolation, *Phys. Rev.* **B30**, 3933 (1984).
22. Gawlinski, E.T. and Stanley H.E. Continuum Percolation in Two Dimensions, Monte Carlo Tests of Scaling and Universality for Non-Interacting Discs, *J. Phys.* **A14**, L291 (1981).
23. Pike, G.E. and Seager, C.H. Percolation and Conductivity: A Computer Study. I, *Phys. Rev.* **B10**, 1421 (1974).
24. Curtin, W.A. and Scher, H. Algebraic Scaling of Material Strength, *Phys. Rev.* **B45**, 2620 (1992-II).
25. Curtin, W.A. and Scher, H. Analytic Model for Scaling of Breakdown, *Phys. Rev. Lett.* **67**, 2457 (1991).
26. Domb, C. On Hammersley's Method for One-Dimensional Covering Problems, in *Disorder in Physical Systems*, edited by G.R. Grimmett and D.J.A. Welsh Clarendon, Oxford (1990).

CHAPTER 11

Maximum Entropy Principles in Fragmentation Data Analysis

R. Englman

11.1. Introduction

It is widely thought that probability theory, laws of inference, and predictive procedures are among the most debated of disciplines. They have been such since the days of the Bernoullis and have remained such up to the present in terms of fundamentals, in spite of remarkable technical advances in the use and manipulation of the rules [1–3]. Thus the traditional controversy between frequencists and Bayesians continues unabated, with each side holding to its position tenaciously.

The use of Maximum Entropy (ME) as a predictive tool shares the trait of being disputed in principle and effective in practice [2]. The formal apparatus of the method is briefly set out in Appendix 11.1. Stated in verbal language, the method predicts that the "best" probability $p(s)$ for the occurrence of the value s for a variable is obtained by maximizing the information entropy, subject to constraints (see appendices and the next section for quantitative definitions). The method has been used in a wide range of subjects, e.g., statistical mechanics, molecular biology, pattern recognition.

The present author has applied the method to distribution of sizes obtained by a variety of fragmentation processes, where the issue was: supposing you know the constraints (regularities, conservation laws) on the distribution, what $p(s)$ do you anticipate [4, 5]?

In this article we address ourselves to two issues of principle in the use of the ME method: First, given an empirical distribution $p(s)$ what can you say about the regularities (constraints, laws) that underlie the processes which have led to the distribution? Second, we treat the question of priors (probabilities or weights that are attached to each event before receiving experimental information). Both subjects will be treated from the viewpoint of fragmentation, though their implications are far wider.

11.2. Deductions of and from Maximum Entropy

In this section we shall pose and discuss the following two questions (application will be given in Section 3 and particular instances will be worked out in Appendix 11.2):

(i) If our fragmentation experiments have resulted in a distribution of size s in the form $p(s)$, is there a way to establish that $p(s)$ is a distribution with maximum entropy?
(ii) What information can be gleaned from a ME distribution?

The immediate answer to question (i) is in the negative: there is no way of learning from a single, experimentally derived distribution whether or not it is in accordance with ME. For suppose that experiments provide a set of numbers, e.g., by sorting about 10^4 fragments into $10^{1.5}$ bins representing sizes. Writing

$$\log p(s) = -g(s) - Q \tag{11.1}$$

(where Q is a normalizing constant) has the "appearance" of a ME distribution, especially if $g(s)$ has a simple form, like being linear in s, namely

$$g(s) = Gs. \tag{11.2}$$

It then appears natural to conclude that G is the Lagrange multiplier and $\langle s \rangle$ is the invariant for the distribution. However, this is illusory. To see this, we note that we can rewrite (11.2) as

$$\begin{aligned} g(s) = Gs &= G_1(s - as^2) + G_2 s^2, \qquad (G_2 = aG_1) \\ &= G_1(s - as^2) + G_2(s^2 - bs^3) + G_3 s^3, \qquad (G_3 = bG_2) \end{aligned}$$

and in many other ways that would indicate other invariants and multipliers consistent with the data. Therefore, to summarize, a single set of experiments cannot establish the ME nature of the distribution. To achieve this we have to perform the experiments repeatedly under different experimental conditions that will generate different distributions. The quantities that we wish to vary through varying the experimental conditions are primarily the averages and then the entropy S. The analysis is as follows:

Suppose that we suspect the logarithm of $p(s)$ to be composed of several linearly independent functions $g_i(s), g_r(s) \ldots g_R(s)$. ($R$ cannot be larger than the number of bins.)

$$g(s) = \sum_{r=1}^{R} G_r g_r(s) \tag{11.3}$$

We next compute the entropy and the averages

$$S = -\sum_s p(s) \log p(s) \tag{11.4}$$

$$\langle g_r \rangle = \sum_s p(s) g_r(s) \tag{11.5}$$

from the distributions that are obtained under several different conditions; in particular, under conditions that the averages $\langle g_r \rangle$ change infinitesimally so that one can obtain the derivatives of S with respect to $\langle g_r \rangle$. For precision's sake we shall write the derivative in the form customary in thermodynamics:

$$\left(\frac{\partial S}{\partial \langle g_r \rangle}\right)_{(\langle g_s \rangle, s \neq r)} \equiv G_r^* \tag{11.6}$$

with the averages in the subscript being kept constant.

We now assert and shall presently prove that $p(s)$ is a ME distribution if and only if the entropy derivatives are all equal to the corresponding coefficients in the exponent, or

$$G_r^* = G_r \tag{11.7}$$

If these quantities are linearly independent functions of the averages, then the averages are invariants for the generation of distributions by experiment. If there are some linear relations between the G_r's, as functions of the means $\langle g_r \rangle$, then the invariant functions are superpositions of the g_r's. The simplest example of this is when, say,

$$(G_2^* =)G_2 = G_4(= G_4^*)$$

then an invariant function is $h(s)$ where

$$g_2(s) + g_4(s) = h(s) \tag{11.8}$$

If the equality (11.7) is not achieved with the chosen set $g_1(s), \ldots, g_R(s)$, then one should choose other sets of g's to see whether (11.7) holds. Formally, one can expand the logarithm of $p(s)$ in terms of complete sets, whose number of components cannot exceed the number of bins $\sim 10^{1.5}$, [6], but this seems to be prohibitively laborious. Clearly, we do not claim that the observed distributions are necessarily ME distributions.

However, if they are such, then Eq. (11.7), can be proved as follows: Let the experimental conditions be described by a set of constraints on the functions $g_r(s)$ so that their averages are given and equal to $\langle g_r \rangle$. The ME distribution is

$$\begin{aligned} p(s) &= \exp[-\Sigma G_r g_r(s) - Q] \\ &= \exp[-g(s) - Q] \end{aligned} \tag{11.9}$$

having simplified by using (11.3). The normalizing constant is

$$Q = \log \sum_s \exp[-g(s)] \tag{11.10}$$

In (11.9) the coefficients G_r are so chosen as to reproduce the averages $\langle g_r \rangle$. We assume that this can be done uniquely. (If not, we restrict our considerations to regions neighbouring the ME choice for the Lagrange multipliers G_r). A non-ME probability $p(s)$ is represented by the same functions $g_r(s)$ whose averages are fixed by the values $\langle g_r \rangle$ and an additional function $\delta g(s)$. To maintain values of $\langle g_r \rangle$ as before the coefficients G_r change to, say, $\tilde{G}_r$. With adaptation of the notation in (11.3) we write

$$\tilde{g}(s) = \sum_r \tilde{G}_r g_r(s) \tag{11.11}$$

$$\tilde{p}(s) = \exp[-\tilde{g}(s) - \delta g(s) - \tilde{Q}] \tag{11.12}$$

$$\tilde{Q} = \log \sum_s \exp[-\tilde{g}(s) - \delta g(s)] \tag{11.13}$$

Furthermore

$$S = \sum_r G_r \langle g_r \rangle + Q \tag{11.14}$$

$$\tilde{S} = \sum_r \tilde{G}_r \langle g_r \rangle + \tilde{Q} + \langle \delta g \rangle \tag{11.15}$$

Differentiating first the expression (11.14) for the ME case

$$\frac{\delta S}{\partial \langle g_r \rangle} \equiv G_r^* = G_r + \sum_m \frac{\partial G_m}{\partial \langle g_r \rangle} \langle g_m \rangle + \frac{\partial Q}{\partial \langle g_r \rangle} \tag{11.16}$$

However, from (11.10)

$$\frac{\partial Q}{\partial \langle g_r \rangle} = -\sum_m \sum_s g_m(s) \frac{\partial G_m}{\partial \langle g_r \rangle} e^{-g(s)} / \sum_s e^{-g(s)} \tag{11.17}$$

$$= -\sum_m \langle g_m \rangle \frac{\partial G_m}{\partial \langle g_r \rangle} \tag{11.18}$$

so that the last two terms in (11.16) cancel and Eq. (11.17) is proven for a ME distribution.

In the entropy derivatives $\partial \tilde{S} / \partial \langle g_r \rangle$ for a non-ME distribution two terms, corresponding to the last two terms in Eq. (11.16), again cancel but one is left with a further term, as follows:

$$\frac{\partial \tilde{S}}{\partial \langle g_r \rangle} = \tilde{G}_r + \frac{\partial \langle \delta g \rangle}{\partial \langle g_r \rangle} \tag{11.19}$$

The additional term in (11.15) can be rewritten as the covariance

$$\operatorname{cov}(g_r(s) - \langle g_r \rangle, \delta g_{,r}(s) - \langle \delta g_{,r} \rangle) \tag{11.20}$$

where the comma notation represents

$$\delta g_{,r}(s) \equiv \partial \delta g(s) / \partial \langle g_r \rangle \tag{11.21}$$

and the differential operates on the coefficients $\tilde{G}_m$.

Though we have not been able to prove this rigorously, the cov are extremely unlikely to vanish for all the indexes r. Their vanishing would mean that the coefficients G_r are unaffected by the addition of the new function δg, while still maintaining the averages $\langle g_r \rangle$.

We have thus shown that the entropy derivatives are equal to the corresponding coefficients in the probability exponent for a ME distribution and fail to be such in a non-ME distribution.

A corollary to the proof answers question (ii), in that the functions $g_r(s)$ for which (11.7) holds have invariant averages or, in other words, reflect some physical, geometrical, etc. law or constraint. The introduction of entropy-derivatives is a key step in the development. The second derivatives of the entropy

$$\frac{\partial^2 S}{\partial\langle g_r\rangle \partial\langle g_m\rangle} \equiv S_{,rm}$$

are also important. It can be shown that the determinant formed out of these derivatives vanishes identically as function of the averages, when a smaller number of the functions $g_r(s)$ can be found (by some linear superposition of the original functions) which form invariants. Thus the nonvanishing of the determinant

$$|S_{,rm}| \neq 0$$

is a further necessary condition.

It is now possible to develop "Statistical Dynamical" formulae equivalent to the familiar thermodynamical ones. The conjugate variables in the latter (e.g., p and $-v$) are replaced by the twin set $\{G_r, \langle g_r\rangle\}$ and the Legendre-transformed functions are:

$$S(\{\langle g_r\rangle\}),\ Q(\{G_r\}) \tag{11.22}$$

connected by the relation (11.14) as functions of the averages [11, 7]. Alternatively, as functions of the Lagrange multipliers

$$Q(\{G_r\}) = S[\langle g_m\rangle(\{G_r\})] - \sum_m \langle g_m\rangle(\{G_r\})\cdot G_m \tag{11.23}$$

where, on the right-hand side, the averages have been expressed as functions of the multipliers G_r. The relations

$$\frac{\partial Q}{\partial G_r} = -\langle g_r\rangle \tag{11.24}$$

complement the derivative relations

$$\frac{\partial S}{\partial\langle g_r\rangle} = G_r \tag{11.25}$$

with which we have been busy.

Maxwell's relations take the forms

$$\frac{\partial G_r}{\partial\langle g_m\rangle} = \frac{\partial G_m}{\partial\langle g_r\rangle} \tag{11.26}$$

$$\frac{\partial\langle g_r\rangle}{\partial G_m} = \frac{\partial\langle g_m\rangle}{\partial G_r} \tag{11.27}$$

The equivalent of the Gibbs–Duhem relations is given by (11.17).

It has to be emphasized that the validity of these relations ("Statistical Dynamics") does not depend on an equilibrium (indeed the processes leading to the distributions can take place far from equilibrium), but only on the continuous variation of the data with the parameters of the experiment. (e.g., a continuous change of the outcome of a biased-coin tossing experiment with the height of the toss.) Occasionally continuity does not obtain: then we have the analogy of phase transitions, which, however, are not treated here. What is important to note is that we work in a space (or subspace) of the data base, without direct reference to the physical processes that generate the data.

11.3. Application in Target Ballistics

In experiments by Matsui et al. [8] projectiles of mild steel and of rocks having velocities between 17 and 270 m/s were made to impact on four kinds of rocks (tuff, basalt, granite and dunite) with various shapes and sizes. The size distributions of fragments were plotted for 20 target samples out of the 60 or so studied. (We emphasize these numbers to focus on our earlier assertion that a statistical analysis requires the availability of sets of data under different conditions, rather than a single set.) Their data are shown in Figures 11.1(a)–11.1(d) and a numerical representation is exhibited in Table 11.1. The authors of Ref. [8] fit their data to an inverse power-law depen-

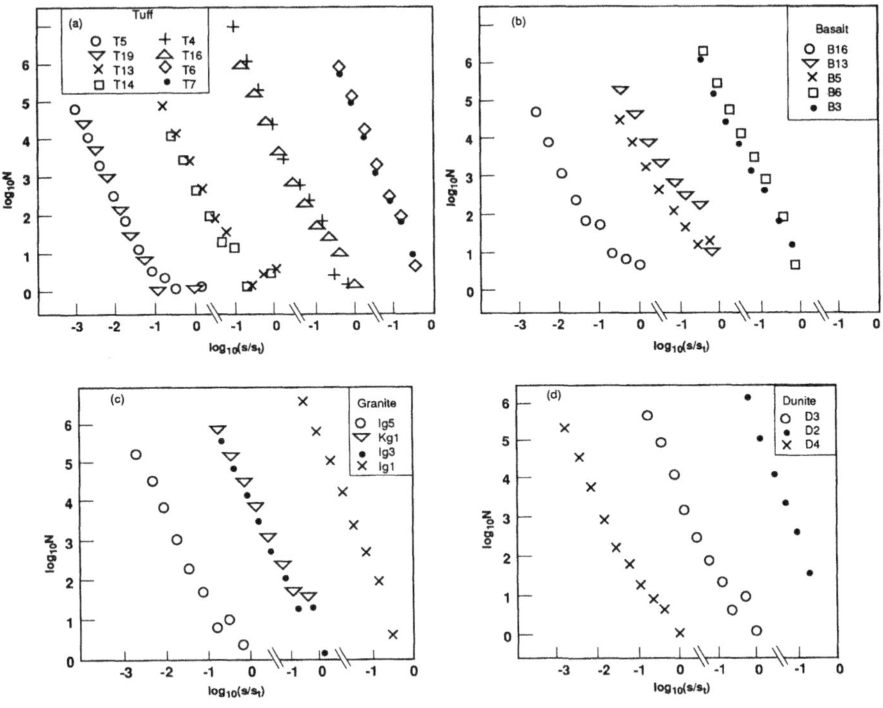

Figure 11.1. Fragment size distributions for four target types after high velocity impact obtained by Matsui et al. [8]. In the log-log plots N is the number of fragments per unit volume size s. s_t is the target volume. The different symbols relate to different samples, whose data appear in Table 11.1.

Table 11.1. Analysis of rock fragmentation data [8]

Sample		$\langle z \rangle \times 10^3$	$\langle \log z \rangle$	Cross-entropy	z_0 (minimum size) $\times 10^3$
Tuff	T5	1.642	−6.524	−0.280	1.270
	T19	2.610	−6.069	−0.246	1.953
	T13	2.533	−6.095	−0.283	1.953
	T14	4.245	−5.622	−0.226	3.008
	T4	1.760	−6.410	−0.328	1.465
	T16	2.474	−6.098	−0.278	1.914
	T6	6.518	−5.086	−0.362	5.625
	T7	6.820	−5.055	−0.307	5.625
Basalt	B16	4.961	−5.452	−0.307	3.750
	B13	4.416	−5.650	−0.1964	2.851
	B3	2.767	−5.992	−0.298	2.187
	B5	4.341	−5.583	−0.297	3.281
	B6	4.004	−5.627	−0.285	3.125
Granite	I5	4.172	−5.607	−0.222	3.047
	I3	4.172	−5.386	−0.219	3.789
	I1	3.927	−5.633	−0.255	3.047
	K1	3.293	−5.853	−0.200	2.344
Dunite	D4	2.564	−6.081	−0.263	1.953
	D3	2.302	−6.171	−0.272	1.797
	D2	2.536	−5.704	−0.324	2.969

Computed averages are over observed distribution points z_n. $z \equiv s/s_{\text{total}}$ is the mass variable normalized to total mass. A prior $\Pi_n = z_n^{-1}$ was used to calculate the cross-entropy $(=\sum_n p_n \log p_\pi/\pi_n)$. z_0 is the minimum normalized size that was sampled.

dence of distribution p on the size s relative to the total target size. In our terminology this has the form

$$p(s) = s^{-1} \exp(-G \log s - Q) \tag{11.28}$$

where Q is a normalizer, s^{-1} is the prior, and G the Lagrange multiplier. In the choice of the prior we have followed Jeffrey's argumentation [9], based on the scale invariance of the distribution. The multipliers G $(=\alpha - 1$ in the notation of [8]) are given in Table 11.1; they are not constant but depend mildly (logarithmically) on the kinetic energy of the projectile. For thoroughly fragmented targets G ranges between 1 and 2. (The values of G are close to the fractal exponents of Turcotte [10] found for a wide assortment of fragmentation processes). As shown in the previous section, the verification of (11.28) being a ME distribution is contingent to satisfying

$$\frac{\partial S}{\partial \langle \log s \rangle} = G \tag{11.29}$$

To form derivatives from numerical data we require data in which two values of $\langle \log s \rangle$ are close. Inspection of Table 11.1 shows that there are four pairs of targets in which $\langle \log s \rangle$ fall within 0.03 of each other. The following

Table 11.2. Entropy derivatives computed from data points [8] for pairs of rocks, compared with corresponding coefficients G (T = tuff, B = basalt, IG = granite); also, smallest relative fragment size z_0. ($z = s/s_{\text{total}}$, $\bar{x} = \langle \log z \rangle$.)

Rock 1/Rock 2	$dS/d\bar{x}$	G	z_0 ($\times 10^3$)
T19/T13	1.42	1.7/1.4	1.95/1.95
T6/T7	1.94	1.94/1.84	5.6/5.6
B13/B6	−4	0.87/1.28	2.85/3.1
IG5/IG3	1.25	1.4/1.5	3.05/3.05

table (Table 11.2) shows that for three pairs (two in tuff and one in granite) Eq. (11.29) is satisfied to within ten per cent, while in the remaining pair (in basalt) the discrepancy is large (even the signs of the two members differ).

One might thus conclude that only in tuff and granite is the distribution that of ME, but not in basalt. In fact, however, the discrepancy in the last is due to another cause: the minimum sizes (the lower cut-off) in B13 and B6 differ (column 4 in Table 11.2).

Now, the cut-off s_0 can be incorporated in the ME formalism by a constraint of the form

$$\int H(s_0 - s)p(s)\,ds = 0 \tag{11.30}$$

where H is the Heaviside step function. (The corresponding Lagrange multiplier is infinite). The differentiation in (11.29) implies [as in Eq. (11.6)] the constancy of all other invariants (or invariance of s_0). It appears that violation of this requirement has led to the failure of satisfying Eq. (11.29).

In summary, then, the experimental data of Mitsui et al. are representable, by Eq. (11.28), as ME distributions with the invariant being $\langle \log s \rangle$. It is not clear at this stage what it is in the physical process which brings about fragmentation that causes $\langle \log s \rangle$ to be invariant. However, only very recently has a suggestion been made concerning a physical cause for a self-similar distribution [11] and apparently much work needs to be done before relations between distributions and their governing processes will be established.

11.4. Priors and Simple Constraints

A source of difficulty in the use of ME are the priors $\Pi(s)$ (described formally in Appendix 11.1). They are weights ascribed a priori (that is to say, without regard to observation) to each event and are analogous to "measure" in set theory. In the words of Skilling [2]: "The prior is the Lebesgue measure associated with [the variable], which must be given before an integral can be defined." An alternative, probabilistic interpretation identifies priors with subjective beliefs as regard to outcome of experiments or with relative reliance attached to different events. Only in quantum mechanics is the choice of priors ("equal weight to all states") incontrovertible, and even there only for discrete Hilbert spaces; whereas for distributions in a classical or in a continuous domain prior choices are flexible. Attempts to find "best" priors were based on invariance principles [9] or the weighting of several expert opinions in view of new evidence [12]. (Bayesians tend to use "priors" in the sense of hypotheses that are already there and need not to be invented, but have to be verified by observation [13]).

A different understanding of priors and indeed of the ME method was provided by Levine and coworkers, mainly in the realm of molecular dynamics [14], and extended recently to classical processes (fragmentation) [15]. In this approach, the factorization of the distribution function $P(s)$ into two factors, as shown in Eq. (A8) of Appendix 11.1, is based on the philosophical principle that Nature is generally kind enough to make the second exponential factor short and simple. This happens, e.g., in a canonical distribution, where the prior is the multiplicity or statistical weight of the energy shell and the exponent is simply: $-(\text{Energy}/k_B \times \text{temperature})$. Molecular collision and photofragmentation experiments also lead to simple exponential factors (called also constraint or likelihood factors) [14], provided that the statistical or probabilistic information is condensed in the prior. Then the constraint factor incorporates the characteristics of the *process* that led to the distribution. This process is almost invariably very complex and it is only through Nature's so far unfathomed benevolence that the experimental data can be approximated by constraint factors that are simple.

Simple exponential factors rule many empirical distributions, e.g., gaussians and log-normal ones. They also work for distributions that arise from numerical simulations. For example, distributions of polygons in cellular networks can be well represented by the product of a probabilistic prior and a simple constraint factor [16, 17]. In the following section we employ a ME distribution function, consisting of a prior and a simple constraint factor, to describe the statistics of broken bonds in a simulated Voronoi—network that has been fragmented by the application of tensile forces.

11.5. Fragmentation in a Disordered Network

To model fracture in granular materials with geometrical disorder a two dimensional model of solids consisting of polygonal grains has been constructed by Voronoi-tessellation [18]. This means that (up to 2000) points were chosen randomly in a plane and, by joining up the perpendicular bisectors of bonds that link neighbouring points, the plane was divided into polygons of different sizes and shapes (Figure 11.2). The average number of sides of the polygons is 6. A theory of size distribution in 2D and in 3D has

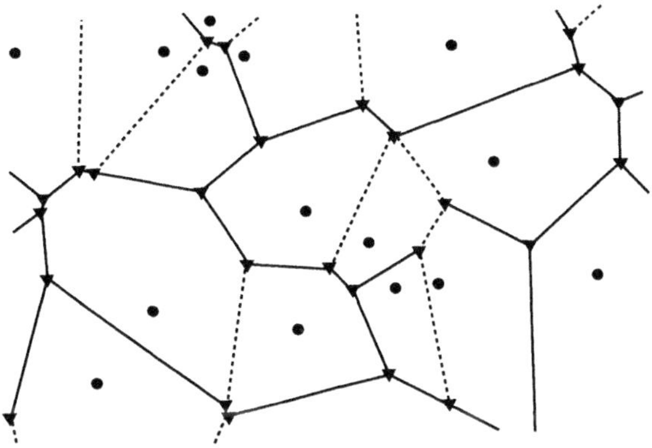

Figure 11.2. Voronoi-tessellation of a plane. (Part of the plane is shown). The lines are perpendicular bisectors of the bonds (not shown), joining neighboring points (thrown at random). The broken lines are bisectors of those bonds that have been fractured by the application of tensile strain at the boundary points.

been given in [19]. The present, Voronoi-tessellation is a reasonable approximation to disordered multiphase structures containing irregular grains. More precisely, it is deemed superior to alternative models in which disorder enters through randomness of bond strengths in a geometrically regular (e.g. square or triangular) lattice [20, 21].

In this simulation fracture is brought about by applying either a fixed stress or a fixed elongation (constant strain) at two parallel edges and calculating the elongation in each bond for a harmonic, central force model of springs.

One or more bonds break when their calculated relative extensions (defined as the ratio of the extension to the original bond length) exceed a certain value. Then the stress or strain is reapplied or is raised until newer bonds breaks (and in fact they do so with greater facility since the broken bonds no longer support the stress). The procedure is repeated until one gets a continuous domain of broken bonds cutting across the sample, which is then considered fractured.

Several parameters of the fracture process are computed, such as:

(a) the least stress or strain which causes fracture (the fracture stress or strain),
(b) the number of bonds that are broken at each stage of the fracture process,
(c) the location of the broken bonds as function of their length,
(d) the distribution of broken bonds as function of their length (Fig. 11.3) under condition that a fixed elongation ΔL_x (in the x-direction, say) was imposed on two edges.

A simple ME analysis of the data shown in Fig. 11.3 is as follows:

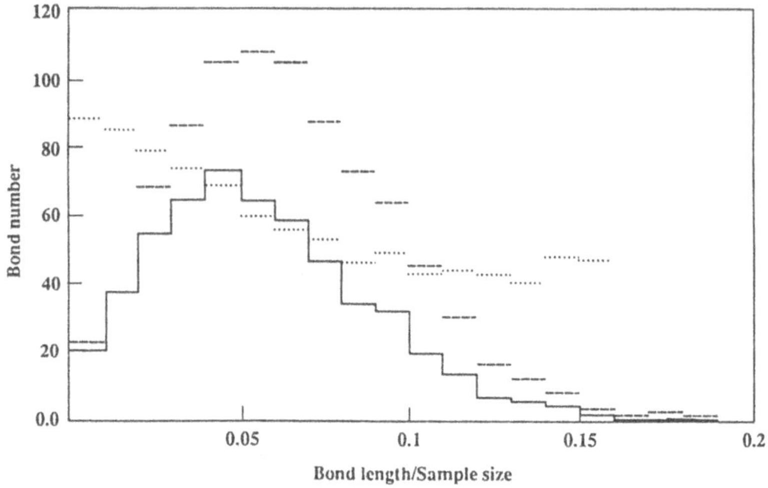

Figure 11.3. Histogram of broken bonds as function of the original bond length at fracture. The data represent averaging over several runs. The criterion for bond rupture is that the relative elongation $\delta l/l$ of bonds exceeds 0.2. The distribution $N_0(l)$ of original lengths is also shown by broken lines and the percentage of bonds broken $N_F(l)/N_0(l)$ by dotted lines. In our model short bonds have a preference for being broken.

The distribution $q(\delta l_x)$ of elongations δl_x of bonds in the x-direction ($\|\Delta L_x$) can be obtained from the conservation condition for the elongation component δl_x^i of the i'th bond:

$$\Delta L_x = M^{-1}\sum_i \delta l_x^i \tag{11.31}$$

where M is the number of instances that the elongation has been applied. The ME distribution then follows:

$$q(\delta l_x) = (N/M\Delta L_x)\exp[-\delta l_x/(M\Delta L_x/N)], \tag{11.32}$$

where N is the total number of bonds.

The fracture criterion used in the simulation was for the elongation δl^i of the i-bond of original length l_i

$$\frac{\delta l^i}{l_i} \geq \varepsilon_f \qquad (=0.2) \tag{11.33}$$

Assuming that the bond-directions between points in a Voronoi-tessellation are isotropic, we can replace the above relation by the following relation (true in the mean) for the elongations in the x-direction

$$\frac{\delta l_x^i}{l_i} \geq \frac{\varepsilon_f}{\sqrt{2}}$$

Then the number of fractured bonds of length l is

$$N_F(l) = N_0(l) \int_{l\varepsilon_f/\sqrt{2}}^{\infty} dx\, q(x)$$
$$= N_0(l) \exp[-l\varepsilon_f/(\sqrt{2}M\Delta L_x/N)] \tag{11.34}$$

The factor $N_0(l)$ is the number of bonds of length l created originally and is shown in Fig. 11.3 as the full-lined histogram. $N_0(l)/N_{\text{total}}$ serves as the prior for the distribution $N_F(l)/N_{\text{total}}$ of broken bonds. The percent of broken bonds $[N_F/N_0 \times 100]$ is shown in Figure 11.4 for three simulations in which elongations $\Delta L_x(\sim 0.04)$ were applied to the end points (numbering about 20) several times (about 5 times) until the bond structure, containing about 800 bonds ($N \approx 800$), became fragmented. The least square fits of the data to exponential curves (seen in Fig. 11.4) were reasonably accurate. In terms of our earlier discussion of the roles of the prior and of the constraint (or likelihood) factor in a ME distribution the present simulations confirm the expectation that the exponent in the constraint factor takes a simple form (linear in the bond lengths l), in spite of the complexity of the fragmentation process.

The fitted exponents in Fig. 11.4 are about 10 l. Comparing this with the exponent in Eq. (11.34), using also the appropriate values of the symbols

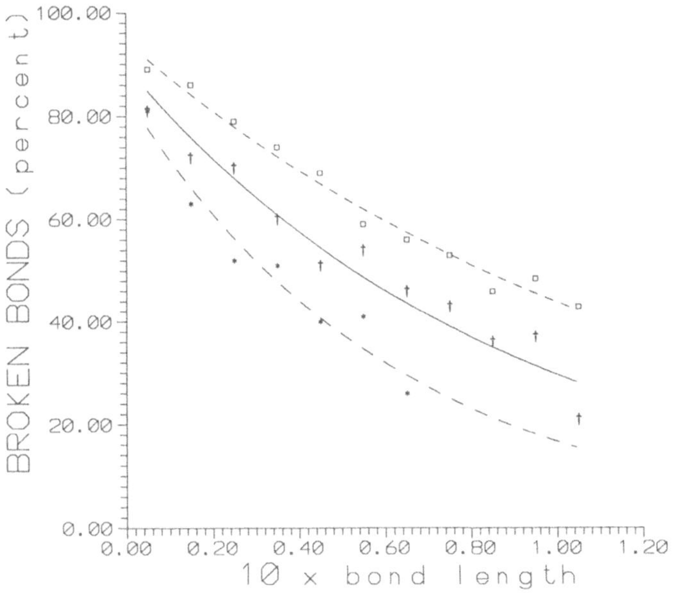

Figure 11.4. Fraction of broken bonds (percent of the original number) as function of original bond length at the moment when the network becomes fragmented. Data points from three simulations are shown together with fitted exponentials.

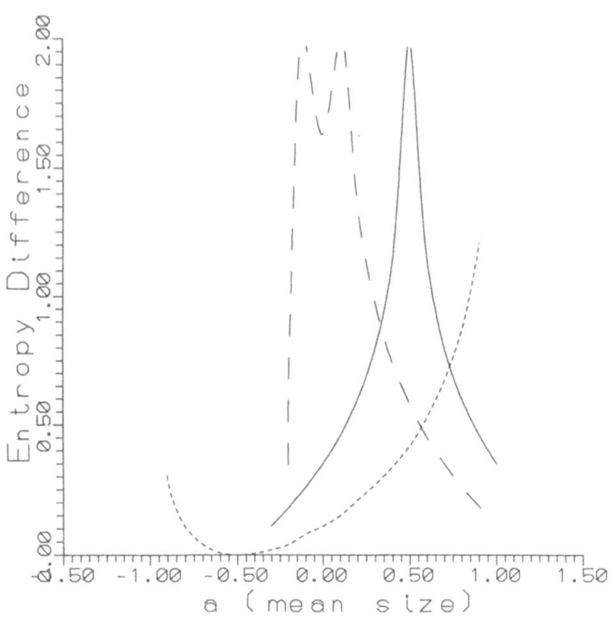

Figure 11.5. The entropy difference $\Delta S_{1,2}$ in (11.A13). It is non-negative since the distribution in (11.A12) has less constraint (more disordered) than that in (11.A10). The three curves are for $\sigma^2 = 1$ (broken line), 5 (full line), 1 (dotted line).

(M, N, ε_f, ΔL_x), the latter predicts a much faster decay (30 l) of the distribution curve.

11.6. Conclusion

Fragmentation is a complex phenomenon whose outcome, the fragment distribution, is expected to reflect this complexity. The ME approach has been used in the past to impose regularities on the distribution [4, 5, 14]. In this article we set out to *uncover* regularities from the data. The key to this effort are the derivatives of the entropy with respect to the invariants. The procedure leads to a "Thermodynamics" of the fragment—(or any other) data. Application has been made to relatively simple case of ballistic fragmentation of rock targets, where only one invariant $\langle \text{logs} \rangle$ is in evidence. For future uses of the procedure, when more than one invariant is expected, further development of the method is needed to make it practical. This appears possible.

A prior-constraint function decomposition of distributions has been carried out in this article for the broken bonds in a disordered network at the moment of its falling to pieces. The aim was to see whether the constraint function is simple, as anticipated in the ME treatment. This was found.

Additional interest is attached to possible changes in the distribution as the fragmentation threshold is approached.

Appendix 11.1

The Formalism

In deriving the distribution $p(s)$ of fragment sizes s the ME method maximizes the information entropy (or missing information) given by

$$S = -\sum_{s} p(s) \log p(s) \tag{11.A1}$$

One allows for R observational inputs or constraints in the form of averages whose values are given:

$$\langle g_r \rangle = \sum_{s} p(s) g_r(s) \qquad (r = 1, \ldots, R) \tag{11.A2}$$

Then one maximizes with respect to $p(s)$ the quantity

$$S - \sum_{r} G_r \langle g_r \rangle \tag{11.A3}$$

where G_r are Lagrange multipliers, whose role is to ensure that with the solution $p(s)$ the constraints are satisfied. This solution is

$$p(s) = \exp\left[-\sum_{r} G_r g_r(s) - Q\right] \tag{11.A4}$$

where Q normalizes p. Exp Q is also the partition function.

When there are "priors" $\Pi(s)$, discussed in Section 11.4 that are not uniform, the previous quantities change in the following way:

$$S = -\sum_{s} \pi(s) p(s) \log p(s) \tag{11.A5}$$

$$\langle g_r \rangle = \sum_{s} \pi(s) p(s) g_r(s) \tag{11.A6}$$

and the distribution is

$$p(s) = \pi(s) p(s) \tag{11.A7}$$

$$= \pi(s) \exp\left[-\sum_{r} G_r g_r(s) - Q\right] \tag{11.A8}$$

Appendix 11.2

On Gaussian Probabilities

These illustrate different cases that are treated in the text in a general manner. (For convenience averages are shown with bars over the variable).

Suppose that we find that the experiments yield a distribution p of a quantity z that can be represented as

$$p(z) = \exp[-G_1 z - G_2 z^2 - Q]. \tag{11.A9}$$

Several possibilities arise.

(1) Two constraints (in addition to normalization).

These are on the mean and the variance, in the form

$$g_1 = z, \qquad \bar{g}_1 = a$$

$$g_2 = z^2, \qquad \bar{g}_2 = \sigma^2$$

Then p in (11.A9) has the form:

$$p(z) = [2\pi(\sigma^2 - a^2)]^{-1/2} \exp[-\tfrac{1}{2}(z - a)^2/(\sigma^2 - a^2)] \tag{11.A10}$$

From this we find immediately

$$\begin{aligned} G_1 &= -a(\sigma^2 - a^2)^{-1}, \qquad G = \tfrac{1}{2}(\sigma^2 - a^2)^{-1} \\ Q &= \tfrac{1}{2}\sigma^2(\sigma^2 - a^2)^{-1} + \tfrac{1}{2}\log[2\pi(\sigma^2 - a^2)] \end{aligned} \tag{11.A11}$$

The entropy is

$$\begin{aligned} S &= G_1\bar{z} + G_2\overline{z^2} + Q \\ &= \tfrac{1}{2} + \tfrac{1}{2}\log[2\pi(\sigma^2 - a^2)] \end{aligned}$$

The partial derivatives of S with respect to the averages are:

$$\frac{\partial S}{\partial a} \equiv G_1^* = -a/(\sigma^2 - a^2)$$

$$\frac{\partial S}{\partial \sigma^2} \equiv G_2^* = \frac{1}{2}(\sigma^2 - a^2)^{-1}$$

which are respectively equal to G_1 and G_2. This verifies that (11.A10) is a ME distribution.

(2) One constraint

(2a) $$g' = z + z^2, \qquad \overline{g'} = a + \sigma^2$$

but no constraint on z and z^2 separately.

In (11.A9) we have now $G_1 = G_2$.

The ME distribution satisfying the one constraint is

$$p(z) = [2\pi(\sigma^2 + a + \tfrac{1}{4})]^{-1/2} \exp[-\tfrac{1}{2}(z + \tfrac{1}{2})^2/(\sigma^2 + a + \tfrac{1}{4})] \tag{11.A12}$$

from which the coefficient in the exponent is

$$G' = [2(\sigma^2 + a + \tfrac{1}{4})]^{-1}.$$

$$S = \tfrac{1}{2} + \tfrac{1}{2}\log[2\pi(\sigma^2 + a + \tfrac{1}{4})]$$

Differentiating this

$$\frac{\partial S}{\partial \overline{g'}} \equiv G'^{*} = [2(\sigma^2 + a + \tfrac{1}{4})]^{-1} = G'$$

so that (11.A12) is also a ME distribution.

However, (11.A10) satisfying two constraints has a lower entropy than (11.A12) satisfying one constraint (since, in general, adding constraints makes the distribution less uniform). Indeed

$$\Delta S_{1,2} \equiv S_{G'} - S_{G_1,G_2} = \tfrac{1}{2}\log[(\sigma^2 + a + \tfrac{1}{4})/(\sigma^2 - a^2)] \quad (11.A13)$$

This entropy difference is plotted in Fig. 11.5 as function of a for different choices of σ^2 and is seen to be non-negative (for all values of a that make sense: $a^2 \leq \sigma^2$). (Note, however, that the entropy difference can be negative for certain choices of the priors, in particular of non-uniform ones)

(2b) $$g' = z^2, \qquad \overline{g'} = \sigma^2$$

without constraining z

$$p(z) = (2\pi\sigma^2)^{-1/2}\exp(-\tfrac{1}{2}z^2/\sigma^2) = \exp(-G'z^2 - Q) \quad (11.A14)$$

$$G' = (2\sigma^2)^{-1}, \qquad Q = \tfrac{1}{2}\log(2\pi\sigma^2)$$

$$S = \tfrac{1}{2} + \tfrac{1}{2}\log(2\pi\sigma^2)$$

$$\frac{\partial S}{\partial \overline{g'}} \equiv G'^{*} = (2\sigma^2)^{-1} = G'$$

demonstrating the ME nature of the distribution (11.A14). The entropy difference compared to the 2-constraint distribution is again positive:

$$\Delta S = \log(\sigma/\sqrt{\sigma^2 - a^2}) \geq 0$$

(3) NME Distribution

We now consider a non-ME (NME) distribution and show inter alia that partial derivatives of the entropy with respect to averages are not, in general, equal to the corresponding coefficient in the exponent of the distribution p, i.e., $G_i^* \neq G_i$.

(3a) Assume that there is just one constraint.

$$g = z, \qquad \bar{g} = a$$

and no constraint on z^2. Suppose further that the observation can be represented (somewhat to our surprise) by the distribution

$$p(z) = \exp[-\tilde{G}z - Cz^2 - Q] \quad (11.A15)$$

To satisfy normalization and $\bar{z} = a$, one must have

$$\tilde{p}(z) = (2\pi/C)^{-1/2}\exp[-C(z - a)^2]$$

whence

$$\tilde{G} = -2Ca$$

$$\tilde{Q} = Ca^2 + \tfrac{1}{2}\log(\pi/C)$$

$$\tilde{S} = \tfrac{1}{2} + \log(\pi/C)$$

$$\tilde{G}^* = \frac{\partial \tilde{S}}{\partial \bar{z}} = -\frac{1}{\partial C}\frac{\partial C}{\partial a} \tag{11.16}$$

Since the variation of C with a or $\bar{z}$ is not prescribed $\tilde{G}^*$ will not in general equal $\tilde{G}$. Should one find experimentally such dependence of C on a that $\tilde{G}^* = \tilde{G}$ is satisfied, then a new constraint on z^2 has been discovered.

(3b) Constraint on $g = z^2$: $\bar{g} = \sigma^2$, no constraint on z.

Suppose that observations yield the non-ME distribution

$$\tilde{p}(z) = \exp[-Az - \tilde{G}z^2 - Q].$$

Satisfaction of the constraint leads to the quadratic in $X \equiv \tilde{G}^{-1}$,

$$\tfrac{1}{2}A^2X^2 + \tfrac{1}{2}X - \sigma^2 = 0$$

$$\tilde{Q} = A^2/4C + \tfrac{1}{2}\log(\pi/C)$$

$$\tilde{S} = \tfrac{1}{2} + \tfrac{1}{2}\log[2\pi(\sigma^2 - \tfrac{1}{4}AX^2)]$$

$$\tilde{G}^X \equiv \frac{d\tilde{S}}{d\sigma^2} = \frac{\partial \tilde{S}}{\partial \sigma^2} + \frac{\partial \tilde{S}}{\partial X}\frac{dS}{d\sigma^2} + \frac{\partial \tilde{S}}{\partial X}\frac{\partial X}{\partial A}\frac{dA}{d\sigma^2} + \frac{\partial \tilde{S}}{\partial A}\frac{dA}{d\sigma^2} \neq \tilde{G}$$

The last non-equality is again due to the arbitrary dependence of A on σ^2.

References

1. Landsberg, P.T. *Thermodynamics and Statistical Mechanics.* Oxford University Press, Oxford (1978).
2. Skilling, J. *Maximum Entropy and Bayesian Methods*, Kluwer, Dordrecht (1989).
3. Jaynes, E.T. *Papers on Probabilities Statistics and Statistical Methods.* R.D. Rosenkrantz (Ed.). Reidel, Dordrecht (1983).
4. Englman, R., Rivier, N., and Jaeger, Z. *Phil. Mag.* **B56**, 751 (1988).
5. Englman, R. *J. Phys. Condens. Matter* **3**, 1019 (1991).
6. Tikochinsky, Y., Tishby, N.Z., and Levine, R.D. *Phys. Rev. Lett.* **52**, 1357 (1984).
7. Balian, R. *From Microphysics to Macrophysics* Part 1, §6.3. Springer, Berlin (1991).
8. Matsui, T., Waza, T., Kani, K., and Suzuki, S. *J. Geophys. Res.* **87**, 10,968 (1982).
9. Jeffreys, H. *The Theory of Probability*, 3rd edition, Chap. III. Oxford University Press, Oxford (1961).
10. Turcotte, D.L. *J. Geophys. Res.* **91**, 1921 (1986).

11. Sommerer, J.C. and Ott, E. *Science* **259**, 335, 1993.
12. Rivier, N., Englman, R., and Levine, R.D. in *Maximum Entropy and Bayesian Methods*, P.F. Fougère (Ed.). Kluwer, Dordrecht (1990).
13. Helstrom, C.W. *Statistical Theory of Signal Detection*. Pergamon, Oxford (1960).
14. Levine, R.D. Statistical dynamics. in *Theory of Chemical Reaction Dynamics*, Vol. 4, p. 1, M. Baer (Ed.). C.R.C., Boca Raton, FL (1985).
15. LeCaër, G. and Delannay, R. *J. Phys. A: Math. Gen.* (1993).
16. Englman, R. Unpublished (1992).
17. Lemanska, M. Unpublished (1990–93).
18. Rivier, N. *Phil. Mag.* **B52**, 795, (1985).
19. Duxbury, P.M. Breakdown of diluted and hierarchical systems. in *Statistical Models for the Fracture of Disordered Media*. H.J. Herrmann and S. Roux (Eds.). Elsevier–North Holland, Amsterdam (1990).
20. de Arcangelis, L., Hansen, A., Herrmann, H.J., and Roux, S. *Phys. Rev.* **B40**, 877, 1989.

CHAPTER 12

Experimental and Numerical Studies of High-Velocity Impact Fragmentation

M.E. Kipp and D.E. Grady

12.1. Introduction

Some basic theories have emerged within the past 10 years for predicting the consequences of dynamic fragmentation brought about by high-velocity impact or explosive events. These theories have focused principally on the prediction of mean fragment size through energy and momentum balance principles (e.g. Grady, 1982; Kipp and Grady, 1985; Grady, 1988; Glenn and Chudnovsky, 1986) and on the statistical issues of fragment size distributions (e.g. Brown, 1989; Englman, et al., 1984; Grady and Kipp, 1985; Grady, 1990). This theoretical basis has provided the underlying framework for a number of computational algorithms employed to analyze complex fragmentation events (e.g. Johnson, et al., 1990; Melosh, et al., 1992; Smith, 1989).

High-velocity impact represents a mechanical event which is ideally suited to treatment with large scale computational wave-propagation codes. Through parametric computational studies involving material properties, geometry and impact velocity, much can be learned about the high-velocity impact process. Such computational studies can also prove useful in guiding the development of models which provide the components for more specific engineering solution methods of the impact process. Because of the high velocities and intense stress states achieved on impact, a significant portion of the motions are hydrodynamic in nature in spite of the high-strength materials which may be involved in the interaction. Nonetheless, critical aspects of the impact interaction are strongly dependent on strength properties of the materials involved. The total mass and size distribution of ejecta material, along with the size and structure of the impact or penetration crater, depend sensitively on the strength properties.

High-velocity impact also exercises the full thermodynamic equation-of-state of materials. Entropy production due to both the initial shock wave and the dynamic shear deformation will lead to a broad range of temperature states. Melting and vaporization of some fraction of the impact debris will commonly occur above impact velocities of about 5 km/s.

There are, nonetheless, a number of unresolved issues within the development of statistical energy-balance theories of fragmentation, and there is a pressing need for high-resolution experimental data focused on the validation of these theories. Furthermore, implementation of these theories into Eulerian wave propagation codes is important for the purpose of investigating dynamic fragmentation in large scale catastrophic events.

The present studies focus on the development of both numerical and experimental capabilities for characterizing the debris spray produced in penetration events. A systematic fragment-debris database is essential for the continued development of a theoretical understanding of fragmentation and the associated computational model development and verification. This investigation of impact-induced fragmentation was undertaken to provide such an experimental base of high-resolution impact fragmentation data for evaluating models and determining the accuracy of current computational fragmentation analysis techniques.

The primary experimental objective in this study was to investigate the dynamic fragmentation characteristics of a high-strength steel through controlled impact experiments, using flash-radiography diagnostics. Experiments of this type usually involve the high-velocity interaction of a metal projectile with a stationary target (plate) of similar or dissimilar metals. We report high-velocity impact experiments in which copper spheres strike 4340 steel target plates at velocities between 3 and 5 km/s, and at both normal and oblique impact angles. In addition to flash radiography, witness plate and fragment capture methods were used to evaluate size and trajectory statistics of ejecta fragment debris.

Radiographic diagnostics of the fragmentation event cannot readily discern projectile fragments from target fragments, however. In a second suite of experiments, a high-velocity steel sphere, accelerated to a velocity in the range of 3 to 5 km/s, undergoes normal impact on a thin stationary plastic plate. The plate imparts to the steel sphere a controlled impulse of magnitude and duration determined by impact velocity and plate thickness. The radiographic diagnostic exclusively images the fragmented steel sphere, since the target plastic is transparent to the x-ray beam. This technique offers valuable analysis and interpretation features of the fragmentation event not available in multi-metal impact experiments, foremost of which is the association of the known mass of the sphere with the debris in the radiograph. Several secondary issues were also investigated in the experimental aspects of this study. An aluminum target plate was placed at a fixed position behind the primary plastic plate and intercepted the flux of steel fragments. The objective was to assess secondary target damage by a noncontinuum fragment debris pattern. The experimental consequences of the secondary impact event were evaluated. Finally, interesting damage modes were noted in the primary plastic plate and these results have also been documented.

The numerical tool chosen for development of fragmentation prediction capability is the three-dimensional Eulerian wave propagation code, CTH (McGlaun, et al., 1990). A post-processor was developed to determine local average fragment sizes from strain-rate and temperature information, using the dynamic fragmentation theories mentioned previously.

12.2. Steel Sphere Impact on PMMA Targets

The first major series of experiments to be discussed are focused on the normal impact of hardened steel spheres onto PMMA target plates, with the intent to examine the dependence of the steel fragmentation on the impact conditions. This section includes the experimental description, experimental results, analysis of the data and the approach to numerical analysis of fragmentation events, and concludes with some numerical analysis of the motion of the debris.

12.2.1. Experimental Configuration and Materials

The experimental configuration for the series of impact fragmentation tests is shown in Figure 12.1. Saboted steel spheres were launched at velocities between about 3 and 5 km/s with a two-stage light-gas gun system. The launch tube diameter was 12 mm. Plastic sabots were separated from the steel spheres through forces produced by a slight gas pressure in the gun range section. Sabot segments are trapped upstream and do not reach the target impact chamber. Velocity of the steel spheres is measured to $\pm 1\%$ accuracy by recording of the time interval during passage between two magnetic coils of known separation. Normal impact occurred in the target chamber at the center of a 75 mm by 75 mm square plastic target plate of thickness between 0.6 and 11 mm. The plastic is PMMA (polymethyl-methacrylate) Rohm and Haas Type II UVA, and has a nominal density of 1186 kg/m^3.

In all experiments a steel sphere 6.35 mm (1/4 in.) in diameter was used. The measured mass was 1.027 ± 0.001 grams. The steel was AISI E52100

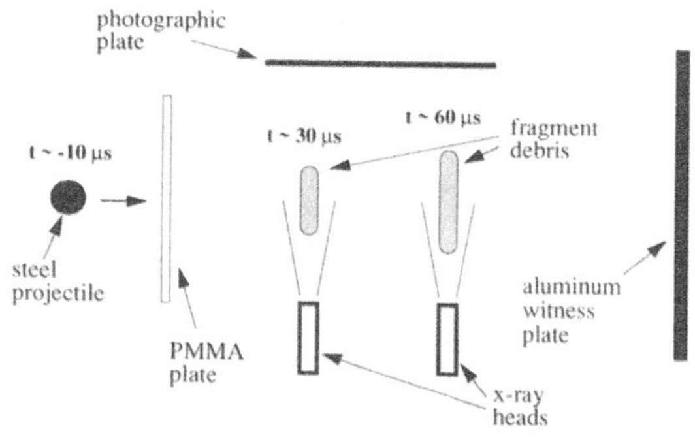

Figure 12.1. Experimental configuration for radiographic and witness plate diagnostic of impact fragmentation experiment. (Timing is representative of a 5 km/s impact.)

high-carbon chromium steel, heat treated to a Rockwell-C hardness of 60 to 67. The density of the steel is 7837 kg/m^3, static yield strength is 2.03 GPa, fracture toughness is 30 to 40 MPa$\sqrt{m}$, and elastic modulus (Young's) is 200 GPa, with a Poisson ratio of 0.29.

Fragment debris is diagnosed at two stations (approximately 150 mm and 300 mm) downstream from the input point. Two 150 keV flash x-ray tubes, placed approximately 400 mm from the line of debris travel, provided orthogonal shadow-graphs of the fragment debris, as shown in Figure 12.1. Appropriate delay times were calculated from the predicted impact velocity and x-ray tubes independently triggered from the second magnetic velocity coil. The x-ray film cassette, using Kodak Direct Exposure film backed by a Quanta Fast Detail screen, was stationed about 100 mm from the centerline of the debris trajectory.

For two experiments (Test 1 and Test 2), the x-ray tubes and film cassette were oriented to obtain an oblique shadow-graph of the fragment debris (the x-ray angle was significantly less than the 90 degree orientation to the line of travel used in the majority of the experiments).

An aluminum target plate 152.4 mm (6 in.) on a side and either 6.35 mm (1/4 in.) or 12.7 mm (1/2 in.) in thickness was placed on axis approximately 400 mm down stream from the PMMA primary target plate to intercept the flux of steel fragments. This witness plate, prepared from 6061-T6 aluminum plate stock, was recovered after each experiment, and provided a post-test passive diagnostic of secondary fragmentation effects.

The configuration parameters for all experiments in the present study are provided in the first two columns of Table 12.1, where h is the PMMA plate thickness and V_i is the impact velocity.

Table 12.1. Experimental impact parameters[a]

Test no.	h (mm)	x_{ps} (mm)	d_s (mm)	V_i (m/s)	Test no.	h (mm)	x_{ps} (mm)	d_s (mm)	V_i (m/s)
1	3.28	[b]	[b]	4460	13	5.37	422	12.7	4430
2	3.28	[b]	[b]	4460	14	5.36	422	12.7	4060
3	1.74	206	6.35	4450	15	5.38	422	12.7	3310
4	1.74	495	6.35	4700	16	5.37	422	12.7	4080
5	3.38	470	12.7	4570	17	3.25	657	12.7	4520
6	3.28	476	12.7	3460	18	4.71	638	12.7	4430
7	3.44	483	12.7	4160	19	5.39	637	12.7	4610
8	1.49	422	6.35	3950	20	4.75	640	12.7	4040
9	1.51	422	6.35	3460	21	4.78	646	12.7	3750
10	0.63	422	6.35	3410	22	0.99	637	12.7	4700
11	0.64	422	6.35	3920	23	11.23	641	12.7	4060
12	0.69	422	9.53	4470	24	9.47	638	12.7	4030

[a] h, primary plate thickness; x_{ps}, plate separation; d_s, secondary plate thickness; V_i, impact velocity.

[b] Oblique angle radiograph—witness plate data were not recorded.

12.2.2. Fragment Debris Experimental Results

A total of 24 experiments of the basic impact configuration described in the previous section were performed in this investigation. The principal experimental variables were the impact velocity, V_i, and the thickness, h, of the primary target plate (PMMA), which defined the amplitude and duration of the impulse transmitted to the steel sphere, and consequently, the intensity of fragmentation of the steel projectile.

An example of the experimental radiographic results is shown in Figure 12.2, where the images for Test 5 are displayed. These images qualitatively illustrate the nature of the fragmentation process observed in all of the experiments in the present study, with the exception of several tests in which parameter extremes were reached. The steel fragments remain well grouped and continue to move along the original trajectory at velocities somewhat less than the initial impact velocity, having been slowed by the impulse delivered to the sphere by the plate. Axial dispersion of the fragment debris is quite limited, with fairly well-defined, nearly planar boundaries forming. Radial dispersion is significant, but still the fragments are contained by rather well-defined limits. These observations hold in general except for the very largest of target thicknesses. Since the PMMA is not recorded in the radiograph, we are assured that only the mass of the original steel sphere is represented in each image.

The foremost objective of the experimental study was to assess, through radiographic diagnostics, the kinematic and structural characteristics of the steel fragment debris produced by the impact of the PMMA plate with the steel sphere.

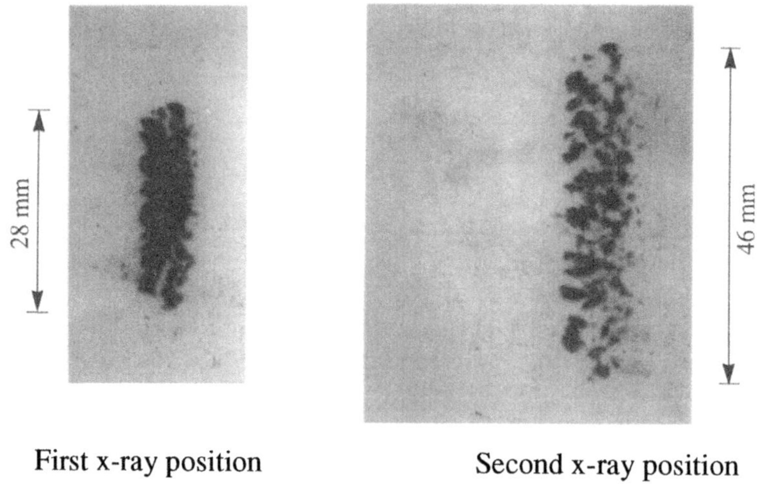

Figure 12.2. Radiographic images from Test 5 illustrating the general fragmentation character observed in the majority of experiments in the present study. (Time interval between images: 33.2 μs.)

Kinematic Properties of the Debris

Several experimental parameters have been extracted from the data and are included in Table 12.2. First is the axial velocity decrease, ΔV, of the fragment debris from the initial impact velocity, V_i. Residual velocity, V_r, of the debris is readily calculated from $\Delta V = V_i - V_r$. The magnitude of ΔV can also be regarded as the change in axial velocity of a stationary sphere due to momentum imparted by the impacting plate. This velocity change is determined from the motion of the geometric center of mass of the debris cloud observed in the radiograph. The most accurate value is calculated by using the separation of the two radiographic images and the times at which the images were made. Corrections are made for magnification and parallax in the radiographic measurement. A detailed table of timing measurements is included in Appendix A of Kipp, et al. (1993a).

As noted earlier, axial dispersion of the fragment debris is small. The axial extent for all of the tests is not more than several sphere diameters at the later radiograph image. The induced radial expansion velocity, V_e, is substantial, however, and varies systematically with initial impact parameters. V_e represents the radial velocity of the outer fringe of the fragment cloud from the centerline and is determined from the change in diameter of the successive radiographic images. Although an errant fragment at the cloud fringe can lead to a degree of subjectivity, the expansion velocity is determined with an accuracy of about ± 20 m/s. In a few cases, at the higher impact velocities, data scatter was somewhat larger. In three experiments (Test 19, Test 23, and Test 24) the fragment debris pattern included a sparse spray of high velocity peripheral fragments. For these tests, radial expansion velocities in Table 12.2 represent the extreme of spray fragments observed on the radiographs.

Table 12.2. Kinematic properties of the debris

Test no.	V_i (m/s)	ΔV (m/s)	V_e (m/s)	V_r (m/s)	Test no.	V_i (m/s)	ΔV (m/s)	V_e (m/s)	V_r (m/s)
1	4460	[a]	[a]	[a]	13	4430	460	290	3970
2	?	[a]	[a]	[a]	14	4060	330	183	3730
3	4450		200		15	3310	240	50	3070
4	4700	150	147	4550	16	4080	330	164	3750
5	4570	270	271	4300	17	4520	195	223	4325
6	3460	200	52	3260	18	4430	345	219	4085
7	4160	250	171	3910	19	4610	410	660[c]	4200
8	3950	130	105	3820	20	4040	295	158	3745
9	3460	170	44	3290	21	3750	270	82	3480
10	3410	50	0	3360	22	4700	85	72	4615
11	3920	60	0	3860	23	4060	785[b]	580[c]	3275
12	4470	90	0	4380	24	4030	680[b]	540[c]	3350

[a] Oblique angle radiograph—parameters not determined.
[b] Velocity of debris cloud front.
[c] Diffused.

In three experiments (Test 10, Test 11, and Test 12) the impulse to the steel ball was insufficient to induce an expansion velocity.

Fragment Size and Number Measurements Through Radiographic Diagnostics and Image Processing

The x-ray image of a cloud of fragment debris produced in a high-velocity impact event produces a shadow-graph of the resulting fragments. The total number of fragments and the average fragment size are important items of data in such an experiment. These data are difficult to extract from radiograph records, however, because of the overlap and shadowing of fragments which occur if sufficient spreading of the fragment debris is not allowed before x-ray imaging. The latter solution is often inconsistent with other objectives of the radiographic diagnostics. For example, preliminary examination of the images shown in Figure 12.2 for Test 5 indicates about 200 particles, with an average fragment size of about 1 mm and a largest fragment size of about 2 mm.

A useful method for extracting fragment number and average fragment size data from radiographs of reasonable tightly packed fragment debris clouds is based on a statistical formula for randomly positioned areas (Johnson and Mehl, 1939). The approach can be conceptualized by imagining coins of various sizes which are tossed and land at random positions on a table of a specified area A_{reg}. After N coins have been thrown, the total area of interest A will be the sum of the areas of the N coins. The area of the table that is covered, however, will be less because of overlapping due to the random placement of the coins. If the area covered is A_{obs}, then the statistical theory leads to a relation between the observed area A_{obs} and the actual area A of the N coins randomly distributed on the region A_{reg},

$$A = A_{obs} \ln(1 - f)^{-1/f}, \tag{12.1}$$

where $f = A_{obs}/A_{reg}$ is the fraction of the region (table) covered by coins. A Taylor expansion of Equation 12.1,

$$A = A_{obs}\left(1 + \frac{f}{2} + \frac{f^2}{3} + \cdots\right), \tag{12.2}$$

reveals that A is always greater than A_{obs} and approaches A_{obs} as f becomes small.

Application of the statistical relation in Eq. (12.1) or Eq. (12.2) to radiographs of fragment debris is reasonably clear. Because of shadowing and overlap of fragments in the relatively dense debris cloud, the fragment area projected on the radiographic image will be less than the projected area of the same fragments if shadowing and overlap does not occur—the latter being the desired property. Thus, image processing of the radiographic record requires two measured properties—the total observed projected fragment area A_{obs}, and an estimate of the area fraction of the debris cloud region masked by the debris fragments. Assuming a random distribution of fragments through the region, Eq. (12.1) provides the required projected fragment area. Note that if f is reasonably small the expression for A is relatively insensitive to inaccuracies in f.

Knowing the total volume of fragmented material within the debris cloud, the number of fragments N and the average fragment size S are related through $V = NS^3$. In addition the total projected area satisfies $A = NS^2$. Solving these relations provides the average fragment size,

$$S = V/A, \tag{12.3}$$

and the total fragment number,

$$N = A^3/V^2. \tag{12.4}$$

The above procedure was used to analyze the majority of the fragment debris radiographs in the present study and extract experimental fragment size and number data. A complete set of image-processed records are included in Appendix B of Kipp, et al. (1993a), and provide a documented record of the radiographic images obtained in the present work. The figures in that appendix also indicate the areas (A_{reg}, A_{obs}) used in the above analysis. A summary of results for all fragment debris radiographs is tabulated in Table 12.3, including both average fragment size and number calculated from Eq. (12.3) and Eq. (12.4).

The fragment size and number data are plotted in Figure 12.3 and Figure

Table 12.3. Fragment size and number results from image-processed radiographs

Test no.	A_{obs} (mm^2)	f	A (mm^2)	N	S (mm)	Test no.	A_{obs} (mm^2)	f	A (mm^2)	N	S (mm)
1	168.4	0.287	198.5	435	0.67	15b	52.1	0.361	64.7	15	2.07
2	108.3	0.069	115.6	86	1.16	16a	93.3	0.519	131.5	127	1.02
3	92.4	0.438	121.5	100	1.10	16b	118.6	0.274	138.5	148	0.97
4a	275.4	0.503	142.6	161	0.94	17a	92.1	0.469	124.3	106	1.08
4b	119.0	0.274	139.1	150	0.96	17b	134.7	0.273	157.4	216	0.85
5a	117.1	0.582	175.5	301	0.76	18a	89.3	0.587	134.6	136	1.00
5b	160.3	0.379	201.5	455	0.67	18b	114.1	0.283	134.2	134	1.00
6a	65.4	0.572	97.0	51	1.38	19a	93.1	0.386	117.6	90	1.14
6b	73.8	0.337	90.0	40	1.49	19b	102.0	0.150	110.7	76	1.21
7a	110.2	0.593	167.0	259	0.80	20a	95.3	0.591	144.2	167	0.93
7b	155.4	0.412	200.3	447	0.67	20b	123.3	0.389	156.2	212	0.86
8a	65.2	0.615	101.3	58	1.32	21a	73.5	0.623	115.0	85	1.17
8b	95.7	0.420	124.1	106	1.08	21b	116.2	0.449	154.3	204	0.87
9a	44.3	0.664	72.8	22	1.84	22a	64.6	0.581	96.7	50	1.39
9b	61.6	0.565	90.7	41	1.47	22b	99.3	0.421	128.8	119	1.04
13a	83.3	0.607	128.2	117	1.04	23a*		0.			
13b	124.3	0.361	154.2	203	0.87	23b	166.2	0.084	173.6	291	0.77
14a	80.0	0.565	117.9	91	1.14	24a	116.2	0.308	138.9	149	0.97
14b*	83.6	0.366	104.1			24b	122.2	0.105	128.0	117	1.05
15a	44.8	0.576	66.7	17	2.01						

* Partial radiograph.

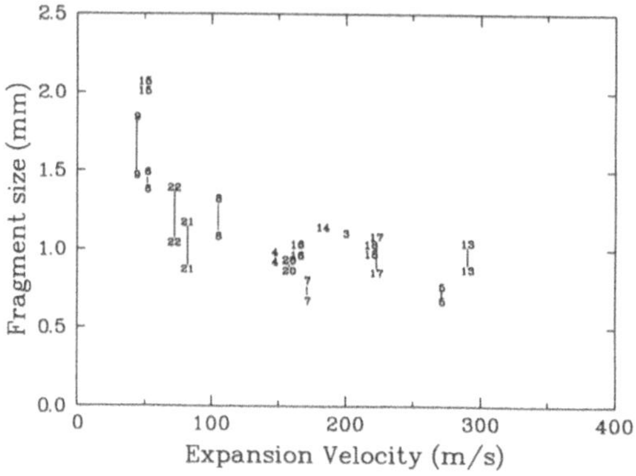

Figure 12.3. Fragment size data vs. radial expansion velocity.

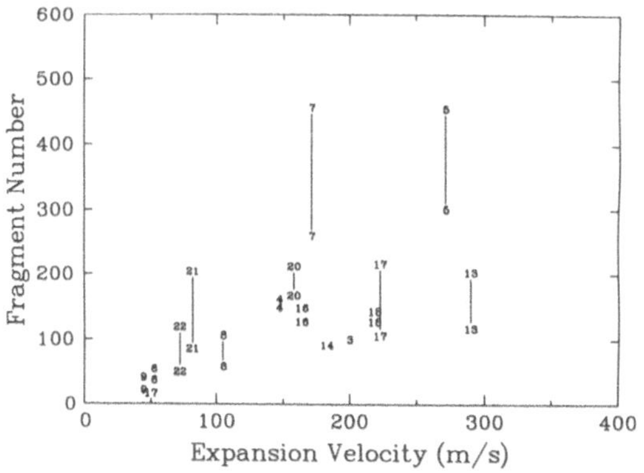

Figure 12.4. Fragment number data vs. radial expansion velocity.

12.4, respectively, as a function of the radial expansion velocity for that experiment provided in Table 12.2. The experiment numbers are used as data symbols in Figure 12.3 and Figure 12.4. For the experiment in which two radiographs of the debris cloud were obtained both points are plotted.

Uncertainties in the present analysis can arise from several sources. The estimate of A_{reg} which encloses the fragment cloud has some subjectivity. This measure affects the value of f, but its influence on the final S and N values is small if f is small. One can readily see other image-processing methods for estimating f but they were not pursued. The image-processing method used to determine A_{obs} requires an edge-selection criterion and further uncertainties can be introduced in this procedure. This uncertainty could be reduced by calibration of the edge-selection method on a radiograph of particles of known size.

The fragment size data are fairly clearly represented in Figure 12.3: the average fragment size is relatively constant above expansion velocities of 150 m/s. The level of scatter probably reflects the uncertainties mentioned above. The fragment number data in Figure 12.4 reflect a similar insensitivity to expansion velocity above 150 m/s. The data for Tests 5 and 7 clearly fall outside of the trend in this plot. The reasons for this behavior are not yet understood.

12.2.3. Analysis of the Steel Debris

It is apparent from the radiograph in Figure 12.2 that the impulse imparted to the steel sphere by the PMMA plate partitions the initial kinetic energy of the steel sphere into kinetic energies of axial translation and radial expansion of the steel fragments, kinetic energy of the PMMA debris, and energy expended in the various dissipative processes active during the impact process. The axial velocity change, ΔV, recorded for the experimental series in Table 12.2, provides a measure of the translational momentum loss by the steel sphere upon impact. These results are plotted against the product of the PMMA plate thickness and the steel sphere impact velocity, hV_i, in Figure 12.5. This latter parameter provides a measure of the impulse delivered to the sphere by the PMMA target. Within experimental scatter, the ΔV data are found to be a single-valued function of the parameter hV_i.

Translational Impulse Analysis

The trend of the data in Figure 12.5 can be reasonably well understood in terms of a relatively basic hydrodynamic description of the sphere and target interaction (e.g., Backman and Goldsmith, 1978). The acceleration of the steel sphere of mass m, is determined from,

$$m\frac{dV}{dt} = \frac{1}{2}\rho V^2 A, \tag{12.5}$$

where $\rho V^2/2$ is the Bernoulli pressure applied by the PMMA target material, of density ρ, under steady flow conditions, and is assumed to apply over the projected geometric area A of the sphere. The velocity V is the equilibrated velocity of the PMMA and the steel sphere. The relatively low target im-

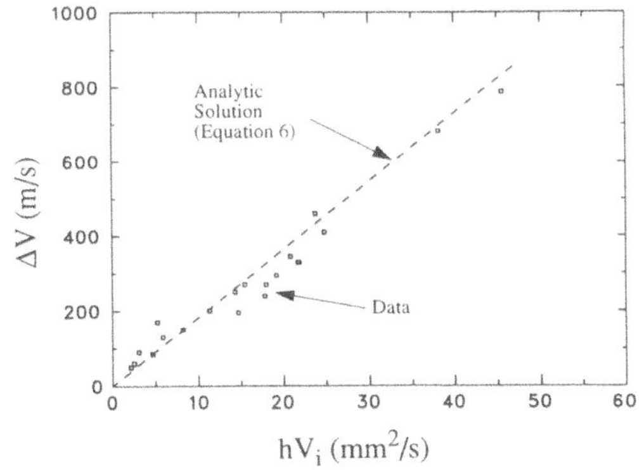

Figure 12.5. Translational velocity reduction data. Dependence on product of target plate thickness and impact velocity.

pedance of PMMA compared with that of the steel projectile permits this approximation. Integration of Eq. (12.5) leads to a predicted exponential change in projectile velocity with plate perforation thickness. For target plate thicknesses h on the order of the sphere diameter, a first order solution of Eq. (12.5) provides the functional relationship of the decrease in velocity with the impact velocity and target thickness,

$$\Delta V = \frac{\rho A}{2m} h V_i. \tag{12.6}$$

Comparison of Equation (12.6) with the measured velocity decrease data, tabulated in Table 12.2, is shown in Figure 12.5, and demonstrates excellent agreement.

Fragment Debris Pattern

In the preponderance of experiments performed, the fragment debris stayed closely grouped or clustered. There was a well-defined region in space within which fragment density was high and a sharp gradient outside of which no fragments were found. This description readily carries over to the characteristics of the impact footprint on the secondary impact plate (see later Sections). In these tests the radial expansion velocity defined previously was relatively easy to establish.

Interesting features are also observed in the radial expansion characteristics of the steel fragment debris following target impact. Several distinct regions of behavior were noted to occur within the parameter range of the present study. First, impulses below which there was no radial expansion of the steel spheres were observed in three experiments performed on target plates approximately 0.6 mm in thickness. The steel sphere remained intact at impact velocities of 3400 and 3900 m/s. A small fragment was spalled off the rear surface of the sphere at a velocity of 4500 m/s.

One of several possible representations of debris formation data is provided in Figure 12.6. Points identify the impact velocity and the target plate thickness, on the vertical and horizontal axes, respectively, the product of which provides an approximate measure of the impulse imparted to the steel sphere. Curves of constant impulse are used to identify boundaries between three regions of behavior in the fragmentation process: Region I–No Fragmentation; Region II–Cluster Fragmentation; and Region III–Diffuse Fragmentation. In Region I, the impulse was insufficient to cause fragmentation. Region II identifies the tests in which fragment debris was clearly clustered in the sense described previously. Within this cluster region, two fairly distinct patterns were observed in the debris cloud; they were most likely a consequence of the target plate thickness relative to sphere diameter. For plate thicknesses less than about 2 mm, distinct spall debris from the back of the sphere leads to some axial divergence and a shallow conical shape to the

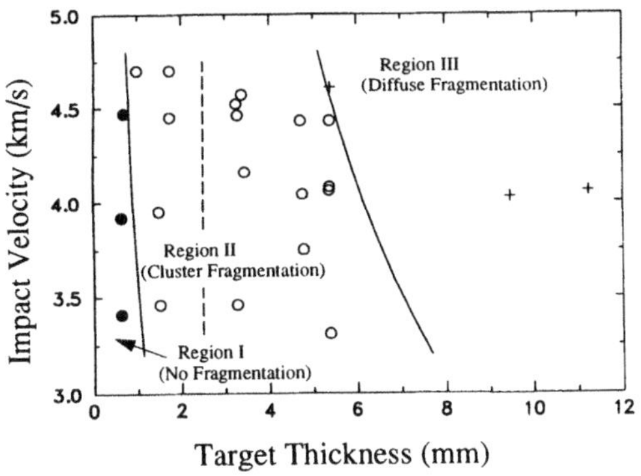

Figure 12.6. Plot of impact velocity vs. target thickness. Different regions of fragment debris characteristics are identified and separated by iso-impulsive curves. The vertical dashed line separates distinct fragment pattern differences in Region II (see text).

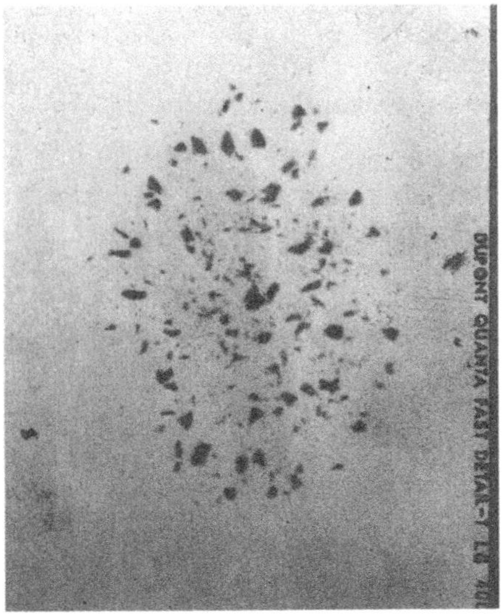

Figure 12.7. Radiograph of steel debris using oblique x-ray diagnostics (Test 2). Note the continuous spatial distribution of the fragments.

debris pattern (for example, Test 4 and Test 8). For plate thicknesses closer to the sphere diameter (3 to 6 mm) a distinct plate or disc shape is observed for the debris pattern.

Also appearing in Region II are two experiments (Test 1 and Test 2) in which an oblique x-ray orientation was used (47 $\pm$ 2 degrees from the shot line) to establish how fragments were distributed through the diameter of the debris disc. The radiograph for Test 2 is shown in Figure 12.7. The debris is moving obliquely away from the point of observation. The radiograph establishes that fragment debris is in fact distributed fairly uniformly through the diameter, and a distinct regular structure in the pattern of the peripheral fragments is observed.

Finally, a clear transition in debris characteristics was observed in several tests in which parameter extremes were achieved (Region III). The impact velocity in Test 19 was in excess of 4600 m/s for the steel sphere on a 5.4 mm plate, and the spheres in Test 23 and Test 24 impacted at slightly lower velocities on plates in excess of 9 mm in thickness. Rather than an abrupt transition in the fragment density, as observed in Region II, there is a more gradual thinning in the density toward the cloud perimeter. The perimeter fragments in Region III have significantly higher expansion velocities (see Table 12.2) than those achieved in Region II. The debris pattern was umbrella shaped with perimeter fragments lagging behind the central cloud.

It is speculated that sufficient impulse has been achieved in the Region III tests to trigger incipient jetting of material from the sphere-plate contact zones. Jetting could account for the excessive radial velocity and trajectory of this part of the fragment debris, however this conjecture has yet to be verified.

Radial Expansion Velocity

For the tests which have Region II behavior for their fragment debris pattern, expansion velocity data from Table 12.2 are plotted as a function of impact velocity, as shown in Figure 12.8. In this graph, tests of nominally the same plate thickness are identified by a common symbol. Although data scatter tends to obscure detailed trends, some observations can be made. First, at similar impact velocities, the expansion velocity increases with increased plate thickness but becomes less sensitive to plate thickness. This trend is most noticeable at the higher impact velocities. Second, there is a critical impact velocity below which fragmentation and subsequent expansion do not occur. This velocity limit is outside of the range of the data for the 0.6 mm plate and is probably around 4000 m/s for a 1.0 mm plate. For

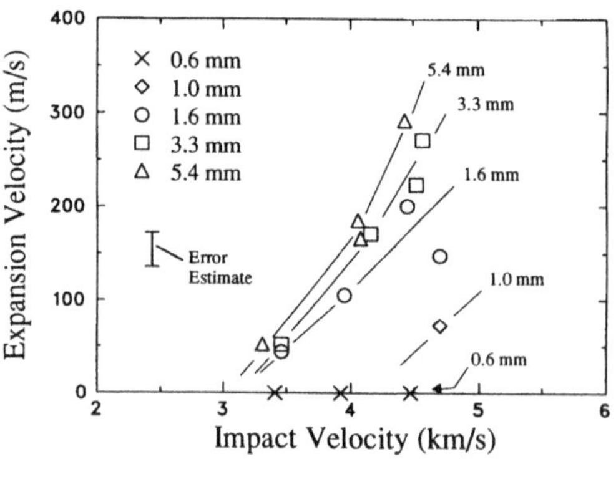

Figure 12.8. Radial expansion velocity of fragment debris. Tests with nominally similar target-plate thicknesses are plotted against the impact velocity.

Figure 12.9. Radial expansion velocity of fragment debris. Tests at nominally similar impact velocities are plotted against target plate thickness.

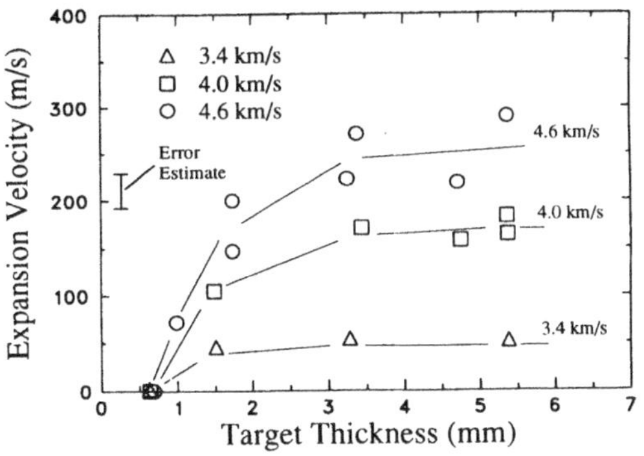

thicker plates (1.6 mm and above) this critical velocity (about 3000 m/s) becomes independent of plate thickness, as found by extrapolating to zero expansion velocity in Figure 12.8.

In Figure 12.9, an alternative representation of the expansion velocity data is shown. Expansion velocity is plotted against plate thicknesses at nominally similar impact velocities. Increased expansion velocity with increasing impact velocity is seen. Flattening of the curves for plate thicknesses above about 2 mm is clearly observed.

Shock Coupling Analysis

The translational impulse imparted to the fragment debris appears to be a consequence of the momentum exchanged during hydrodynamic penetration of the target plate. This conclusion is supported by the nearly linear dependence on plate thickness, in agreement with the analytic expression given in Eq. (12.6) and our numerical computations to be described later. In contrast, the evidence displayed in Figure 12.8 and Figure 12.9 would suggest that radial impulse acquired by the impacting sphere is a consequence of the early shock phase of the interaction and is little affected by the later hydrodynamic penetration phase.

To explore this supposition we will examine a conceptual theory of the shock interaction event and compare implications with the observed trends in the expansion velocity data. Because of the spherical geometry of the impacting projectile, the coupling of shock energy into the projectile is transient and constrained to the central impact surface of the projectile. An analysis of the impact conditions due to Ang (1992) can be used to calculate duration and magnitude of the coupled shock energy.

Referring to the geometry illustrated in Figure 12.10, for a spherical projectile undergoing normal impact at a velocity V_i on a planar surface, the distance of the collision point from the centerline at time t is,

$$a = V_i t \sqrt{\frac{2R}{V_i t} - 1}. \tag{12.7}$$

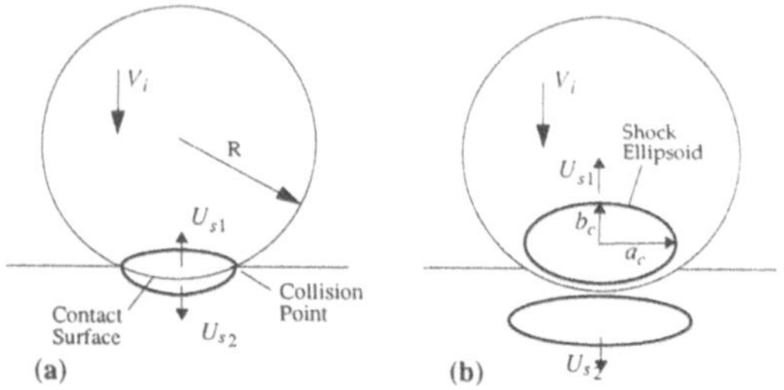

Figure 12.10. Geometry of shock-wave coupling upon normal impact of a spherical particle on a planar surface.

The velocity of the collision point is provided by the derivative of Eq. (12.7),

$$\dot{a} = \frac{V_i\left(\frac{R}{V_i t} - 1\right)}{\sqrt{\frac{2R}{V_i t} - 1}}. \tag{12.8}$$

The collision velocity is initially infinite and decreases with time. As curvature of the impacting sphere at the contact point increases, the collision velocity will continue to decrease (Figure 12.10(a)), ultimately allowing a lateral release wave to propagate inward and quenching the shock coupling phase (Figure 12.10(b)).

A reasonable assumption is that shock waves will couple into both materials over the lateral distance at which $\dot{a}$ reduces to the release wave velocity in the faster material. We will approximate the release velocity by the shock velocity U_{s1} in the faster material. (Release waves are dispersive but U_{s1} is approximately equal to the release velocity at one-half of the Hugoniot particle velocity in a linear shock velocity–particle velocity material.) The time t_1 at which $\dot{a} = U_{s1}$ is calculated from Eq. (12.8) through,

$$U_{s1} = \frac{V_i\left(\frac{R}{V_i t_1} - 1\right)}{\sqrt{\frac{2R}{V_i t_1} - 1}}. \tag{12.9}$$

The corresponding radius at this time is provided by Eq. (12.7),

$$a_c = V_i t_1 \sqrt{\frac{2R}{V_i t_1} - 1}. \tag{12.10}$$

The release time t_2 is in turn calculated from the radius a_c and the release velocity (shock velocity U_{s1}) in the faster material through,

$$t_2 = a_c / U_{s1}. \tag{12.11}$$

The axial thickness b_c of the shock pulse in the projectile (see Figure 12.10(b)) is then determined from,

$$b_c = \tfrac{1}{2}U_{s1}(t_1 + t_2). \tag{12.12}$$

A reasonable estimate of the region of shock coupling into the projectile is an ellipsoidal volume with major axis a_c and minor axis b_c. Hence the coupled shock energy (kinetic and internal) is provided by the volume of material times its energy density,

$$E_s = \frac{4\pi}{3} a_c^2 b_c \frac{p_h^2}{\rho U_{s1}^2}. \tag{12.13}$$

The shock energy provided by Eq. (12.13) is necessarily approximate due to the simplification needed to complete the analytic solution. Also, there is not a clear transition time between the shock and the steady hydro-phases of penetration. Nevertheless, comparisons of Eq. (12.13) with wave-code simulations (see later Sections) of spherical impact on planar surfaces have shown reasonably close agreement (within about 25%).

A portion of the shock energy coupled into the projectile will be dissipated in shock heating. This heat energy is calculated to be about 10% of the shock energy at the present impact amplitudes. Momentum conservation requires that a portion of the shock energy also be converted into kinetic energy associated with axial motion. This energy is also about 10% of the total shock energy. Most of the shock energy will go into plastic distortion of the projectile and radial kinetic energy of fragments if the impact is sufficiently intense.

Carrying through the analysis needed to calculate the shock energy from Eq. (12.13) for a 4600 m/s impact velocity, and reducing this value by the required heat and axial kinetic energies provides an energy available for radial expansion of about 33 J. If we assume that this energy is fully coupled into the uniform radial expansion of a uniform disc of fragment debris (K.E. $= mV_r^2/4$), an expansion velocity of $V_e = 360$ m/s is calculated. Comparison with the appropriate expansion data in Figure 12.11 (see also Figure 12.9) finds the calculated value somewhat higher than the data.

Although approximations in the analytic solution could probably encompass the observed difference between calculation and experiment, it is, nevertheless, worthwhile attempting to account for energy absorbed in the fragmentation process to assess the expected level of influence on the calculated expansion velocity. First, energy to create fragment surface area can be expected to reduce the radial kinetic energy. For a fracture toughness of $K_c = 35$ MPa$\sqrt{m}$, a fracture surface energy from $\Gamma = K_c^2/2E$ is calculated to be about 3830 J/m^2. At an impact velocity of 4600 m/s a total number of about 100 fragments can be calculated, or estimated from the radiographs. For this degree of fragmentation, a fragment surface energy of 2.8 J is calculated. Reducing the 33 J expansion energy by this amount leads to a slightly lower expansion velocity of $V_e = 343$ m/s (see Figure 12.11).

Static tensile tests on the present steel provide a strain to failure of $\varepsilon_f \approx 0.05$ (5%). Assuming a similar strain to failure under the dynamic conditions

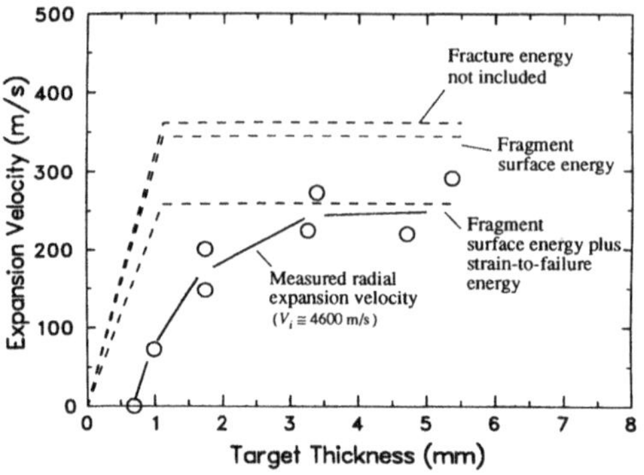

Figure 12.11. Comparison of calculated and measured radial expansion velocities for debris fragments at 4600 m/s impact velocity.

of the impact fracture process, a strain-to-failure energy, $E_f = Y\varepsilon_f$, ($Y =$ 2 GPa is the yield stress) provides an additional fracture energy of 13.1 J. Accounting for both fracture surface energy and strain-to-failure energy further reduces the expansion velocity to $V_e = 258$ m/s. It is interesting to note from the radiographic record for Test 12 that this projectile indicates a plastic distortion of about 5% without fragmentation.

As noted earlier the shock energy is coupled into the spherical projectile in a time $t_1 + t_2$ (see Eqs. (12.9)–(12.12). This time requires that the target plate be thicker than $U_{s2}(t_1 + t_2)/2$ where U_{s2} is the shock velocity in the target material. For an impact velocity of 4600 m/s this thickness is calculated to be approximately 1.1 mm. Thinner target plates would lead to reduced shock energies and radial expansion velocities as is indicated by the linear decrease for thickness below 1.1 mm in Figure 12.11. In contrast, target plates thicker than 1.1 mm would not increase the coupled shock energy and the expansion velocity would be independent of increasing plate thickness. This predicted behavior roughly approximates the observed behavior (Figure 12.11).

Debris Cloud Expansion Analysis

Considering the level of approximations adopted in achieving an analytic solution to the radial expansion characteristics of the Region II fragment debris, the agreement is certainly better than can be expected. Nevertheless, the agreement between trends and rough magnitudes tends to support the premise that the radial expansion and dominant fragmentation characteristics of the debris pattern are established earlier in the interaction by the shock aspect of the impact, and are only weakly affected by the later hydrodynamic loading portion on the projectile.

12.2.4. Target Crater Characteristics

Although a study of the PMMA target crater formation process was not a primary objective in the present investigation, the results were noteworthy

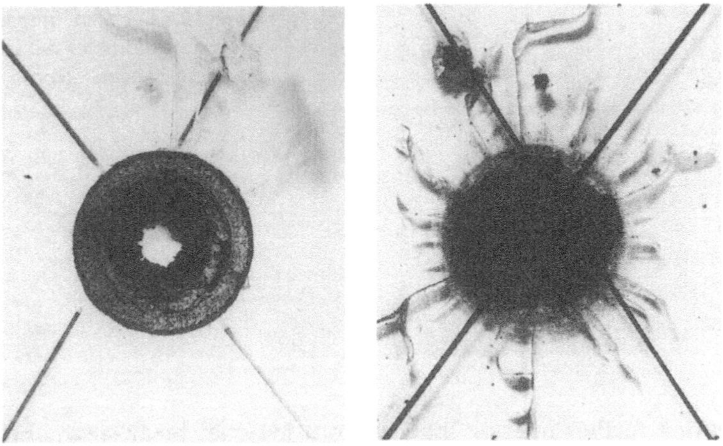

Figure 12.12. Photographs of the impact (left) and exit (right) craters in the PMMA target plate for Test 19. Plate thickness = 5.39 mm and impact velocity = 4.61 km/s.

and warrant some discussion. A number of the target plates were seriously shattered due to the primary or subsequent impacts within the target tank. Consequently, a systematic examination of all the data was not possible. Nevertheless, a sufficient number of plates survived to reveal characteristic trends of the crater geometry.

The target plate from Test 19 remained intact after the test and a photograph of the impact crater region is shown in Figure 12.12. The crater characteristics observed in Figure 12.12 are representative of all of the tests with the exception of those experiments in which very thin plates were used. These latter include tests 10, 11, and 12 which used 0.6 to 0.7 mm thickness plates. In these tests simple punch-through left a hole only slightly over projectile diameter and no collateral crater damage was observed. A transition to the cratering behavior seen in Figure 12.12 occurred with the 1.0 mm plate used in Test 22 and for all thicker plates.

In all tests the physical hole left in the PMMA target plate after the penetration process was never larger than 10 mm in diameter and was usually somewhat less. The hole diameter for Test 19 (Figure 12.12) was about 8 mm. (Recall that the projectile sphere was 6.35 mm in diameter.)

The concentric radial spall region on both the front (impact) and back (exit) of the target plate was the most striking feature of the impact crater. The diameter of the spall crater was slightly larger in front than in back. A sharply-defined circular spall lip was observed on the front crater. The spall lip of the exit crater was more jagged and less uniform. This general character of the spall crater is clearly seen in the photographic results of Test 19 in Figure 12.12. An interior hackle or crazed region in the PMMA was observed to extend a few millimeters beyond the radius of the spall lip—

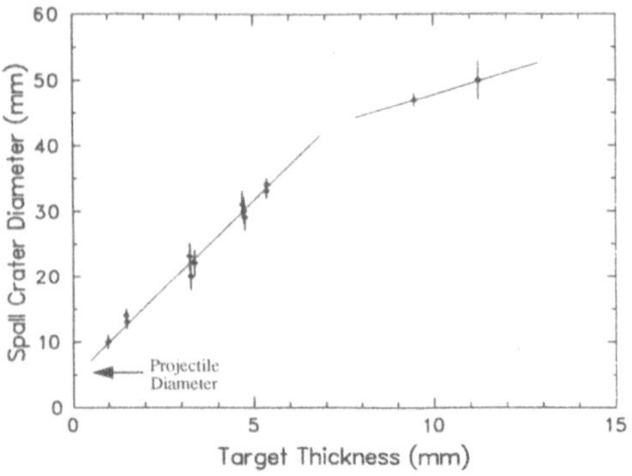

Figure 12.13. Front (impact side) spall-crater diameter data. Impact velocities ranged from 3.3 to 4.7 km/s.

representing further interior spall without material break-away. Finally a collection of 15 to 25 irregular spoke cracks extending radially outward another 10 to 30 mm completed the collateral crater damage effects. Propagation of several of these cracks to the plate edge caused breakup of most of the target plates.

Some systematics were extracted from the target crater data. First, the diameter of the radial front spall zone was found to be relatively insensitive to changes in the impact velocity over the range studied (~3.3 to 4.7 km/s) for target plates of similar thicknesses. Diameters were 20 and 23 mm, respectively, for Tests 6 and 17, and 33 and 34 mm, respectively, for Tests 15 and 19.

On the other hand, a clear dependence of the front spall crater diameter on target thickness was observed. All tests for which a spall crater diameter could be measured are plotted in Figure 12.13. The data include impact velocities over the range of 3.3 to 4.7 km/s. A linear increase in crater diameter with plate thickness up to thicknesses approaching 6 mm was observed. This linear trend is not continued for the two thicker plates.

12.2.5. Summary of Fragmentation Theory and Code Implementation

We have developed the capability to produce fragment size predictions from calculations using wave propagation codes that solve the equations expressing conservation of mass, momentum, and energy for a continuum. Although plots produced from standard wavecode calculations sometimes appear to depict a collection of discrete fragments, the physical mechanisms that control these processes, such as surface tension or real physical heterogeneities and microstructure, are not currently included in the codes. It is not feasible for continuum mechanics wavecodes, except perhaps in one dimension (Kipp and Grady, 1985), to account for the complete, explicit formation of discrete fragments. But extensive work has demonstrated that continuum models of various levels of sophistication can successfully address the damage processes of void and crack growth leading to material failure (e.g., Grady and Kipp, 1989; Johnson and Addessio, 1988).

Previous work (Grady, et al., 1990b) has produced dynamic fragmentation theories that are based on the assumption that strain rate and temperature at the time of failure control subsequent fragmentation. Considerable progress has been made in extracting this information from Lagrangian wavecodes (Grady, et al., 1990b; Johnson, et al., 1990), which are excellent at tracking the history associated with each material element and can easily save the required information at the time of fracture. It is a much more difficult task for an Eulerian wavecode to maintain accurate values of this information because of the degradation accumulated by the repetitive rezoning used in the convection process. However, the extensive deformation associated with these impact and penetration events encourages the use of an Eulerian wave propagation code.

The wavecode that was used for the calculations was the Eulerian finite-difference shock wave propagation code CTH (McGlaun, et al., 1990). This code has a general internal state variable capability which allows information to be saved in variables that are advected with the material as it crosses the cell boundaries. Modifications were made in the code to locate each time and position at which the tensile stress criterion for a single material, rather than a mixed-material cell, was exceeded and material fracture occurred, as defined by the fracture algorithm. All diagonal components of the strain-rate tensor,

$$\dot{\varepsilon}_i = \frac{\partial v_i}{\partial x_i}, \tag{12.14}$$

including the hoop component in cylindrical coordinates,

$$\dot{\varepsilon}_R = \frac{v_R}{R}, \tag{12.15}$$

where v_i and x_i are velocity and position components and R refers to radius, were then calculated and the maximum value was stored as an internal state variable. While the internal state variables provide storage locations for the strain rate and temperature information that must be saved from the calculation, there is a problem with diffusion of these quantities as material motion occurs. For instance, even though a calculation may be performed in which only a single fracture of a material cell takes place during the entire calculation, subsequent material motion may result in spurious values of strain rate appearing in all cells through which the material has passed. Fortunately, these values typically have a very small magnitude except in the region of the mesh containing the bulk of the fractured material. However, diffusion does result in some spreading and loss of localization of the fractured material, so extreme care was taken to assess such effects and ensure that reasonable fragment size distributions were obtained. Since most fractures occurred at high strain rates on the order of 10^3 to 10^6 per second, it was possible to discard significantly lower strain rates as having been produced by diffusion. Mass fraction weighting was used to maintain the proper convected amplitudes of the strain rates and temperatures.

Files containing strain rate and temperature at the time of fracture are saved periodically during the simulation of the impact, and these are examined to determine the extent of the fractured regions as time progressed. When the fracture process is complete, the data are postprocessed outside the wavecode to produce fragment size distributions. Dynamic fragmentation theories predict an average local fragment size at a given strain rate and temperature, and the number of fragments with this average size is determined by the local mass of the material that fractures. The dynamic fragmentation theories that were used to process the strain rate and temperature information have been described in detail elsewhere (Grady, 1988; Grady, et al., 1990a; Grady, et al., 1990b), and they will only be summarized here. Various types of fragmentation mechanisms have been identified, depending on the strain rate and temperature at fracture. The data can also be used to determine the mass distribution of fractured material in the solid, liquid, and vapor phases.

For the present purposes, the average fragment size S will be determined in three different fragmentation regimes. These are:
(1) solid spall dominated by fracture toughness, for which

$$S = \left(\frac{\sqrt{24}K_c}{\rho c\dot{\varepsilon}}\right)^{2/3}, \tag{12.16}$$

(2) solid spall dominated by the flow stress, for which

$$S = \left(\frac{1.2Y}{\rho\dot{\varepsilon}^2}\right)^{1/2}, \tag{12.17}$$

and (3) liquid spall above the melt temperature, for which

$$S = \left(\frac{48\gamma}{\rho\dot{\varepsilon}^2}\right)^{1/3}. \tag{12.18}$$

In these equations, ρ is the density, $\dot{\varepsilon}$ is the strain rate, c is the sound speed, γ is a constant value of the surface tension; the temperature and strain rate dependent yield strength, Y, is given by

$$Y = Y_0\left(1 - \frac{T}{T_m}\right)^n\left(\frac{\dot{\varepsilon}}{\dot{\varepsilon}_0}\right)^m, \tag{12.19}$$

and the temperature dependent fracture toughness, K_c, is given by

$$K_c = K_{c0}\left(1 - \frac{T}{T_m}\right)^{n'} \tag{12.20}$$

where Y_0 is the reference yield strength, K_{c0} is the reference fracture toughness, T is the temperature, T_m is the melt temperature, $\dot{\varepsilon}_0$ is a reference value of the strain rate (one per second), and n, m, and n' are constants. In the solid regime, the transition from fracture toughness to flow stress dominated spall occurs at a strain rate given by

Table 12.4. Fragmentation material properties

	Hard Steel
Density ρ (kg/m^3)	7850
Sound speed c (m/s)	4620
Surface tension γ (N/m)	1.5
Yield strength Y_0 (GPa)	2.0
Fracture toughness K_{c0} (MPa m $^{1/2}$)	40
Melt temperature T_m (K)	1800
n	1
m	0.1
n'	-1

$$\dot{\varepsilon}_t = \sqrt{\frac{0.003\rho c^4 Y^3}{K_c^4}}. \tag{12.21}$$

Appropriate material properties were obtained for the steel sphere for the fragmentation regimes of brittle, ductile, and liquid spall, as shown in Table 12.4. Application of the above formulas allows each point in the strain rate–temperature plane to be mapped into a fragment size for a given material.

The fragmentation theories described above are derived assuming spall induced by uniform volumetric dilatation, so they are most applicable to the prompt fragmentation of the steel sphere induced by the impact. The fracture process is complete within a few microseconds following impact. The strain rate and temperature files generated by each of the calculations that were produced immediately after this time were used for the fragment size predictions.

To obtain the fragment size distribution from the analysis, files are generated by the wavecode calculations that contain the strain rate and temperature at the time of fracture for each cell containing fractured material. There is thus a mass associated with each strain rate–temperature pair, which is just the total mass, m, of the material in the cell. The simplest assumption to make when determining fragment size distributions is that all of the mass in the cell produces equal particles of size S, given by the appropriate one of Eq. (12.16) to (12.18). Thus, letting i be the cell index, where $i = 1, \ldots N$ and N is the total number of cells containing fractured material, the cell data consist of N pairs m_i, S_i where m_i is the mass of material in cell i and S_i is the size of all the particles in the cell. If the data are arranged in order of increasing fragment size, so that $S_i \leq S_{i+1}$, then the cumulative mass of fragments less than or equal to size S_i is

$$M(S_i) = \sum_{j=1}^{i} m_j \tag{12.22}$$

However, as previously described (Grady and Kipp, 1985), statistical considerations indicate that a distribution of fragment sizes should be determined about the mean fragment size calculated for each cell, S_i. The form of the distribution is obtained by assuming that fragments in the mass m_i are Poisson-distributed. This leads to a probability distribution of finding a fragment of mass μ in the cell within a tolerance $d\mu$ given by

$$dP(\mu) = \frac{1}{\mu_a} e^{-\mu/\mu_a}\, d\mu, \tag{12.23}$$

where μ_a is the average, or mean, value of the fragment mass in the cell. Integrating from 0 to μ and multiplying by the total mass m_i of the cell gives the cumulative mass of fragments of mass less than or equal to μ in the cell

$$M(\mu) = m_i[1 - e^{-\mu/\mu_a}]. \tag{12.24}$$

Now, assuming that the mass of the fragment is related to the cube of the fragment size (such as for cubic or spherical particles) and noting that the average mass μ_a corresponds to a fragment having the average size S_i,

$$\frac{\mu}{\mu_a} = \left(\frac{S}{S_i}\right)^3. \tag{12.25}$$

Therefore, the cumulative mass of fragments in cell i having a size less than or equal to S is

$$M(S) = m_i[1 - e^{-(S/S_i)^3}]. \tag{12.26}$$

Finally, the total cumulative mass of all fragments in all cells having a size less than or equal to S is

$$M_T(S) = \sum_{i=1}^{N} m_i[1 - e^{-(S/S_i)^3}]. \tag{12.27}$$

It was found that if the original distribution [Eq. (12.22)] is sharply peaked, a large spread is generated by the statistical relations. However, if the original distribution already contains a large range of fragment sizes, the additional statistical spread is minimal. The statistical distributions do show the addition of a tail at small, possibly aerosol-sized, fragments, but the total additional mass in the aerosol source term so generated is negligible. In application, the cell data is grouped into "bins" of a chosen fragment size increment.

The spectral distribution in size, corresponding to Eq. (12.23), is

$$dP(S) = 3\frac{S^2}{S_i^3} e^{-(S/S_i)^3}\, dS. \tag{12.28}$$

The original distribution of average sizes can be depicted as in Figure 12.14(a) where each bin contains the number of fragments, N_i, of an average fragment size, S_i. These parameters are used in Eq. (12.28) to define the associated statistical distribution for that particular bin, as represented in Figure 12.14(b). All of these statistical distributions are summed to obtain

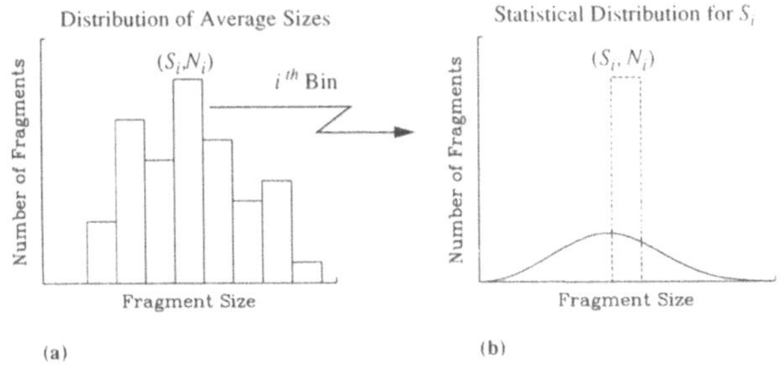

Figure 12.14. Process of transforming the original fragment size distribution to a statistical fragment size distribution.

the final statistical distribution of fragment sizes for the event. Note that although the distribution is exponential in mass, it is exponential in the cube of the size, and the resulting coefficient in Eq. (12.28) contains the square of the size, leading to the character of the statistical distribution in Figure 12.14(b).

12.2.6. Numerical Simulations of Steel Sphere Impact on PMMA Targets

Impact simulations of a steel sphere onto a PMMA target plate were made with a physical space that was partitioned into uniform square cells, with a resolution of 0.05 mm (about 60 cells in the radius of the sphere). The two-dimensional, axisymmetric geometry encompassed a 20 mm radius and a 20 mm axial length, requiring 160,000 computational cells. This radius is sufficiently large that edge effects of the target do not influence the fragmentation behavior of the steel sphere. The steel and PMMA were both represented as low temperature Mie-Gruneisen solids, in which the Hugoniot was described with a linear shock velocity–particle velocity relationship. The properties used in the calculations are listed in Table 12.5. The fragmentation parameters for the steel were listed in Table 12.3. The mass of the sphere used in the calculations was 1.052 g, slightly larger than the experimental mass of 1.027 g.

As an example, consider the case of Test 5, in which a steel sphere impacts a 3.38 mm PMMA target plate at a velocity of 4570 m/s (cf. Table 12.1). The computed sequence of penetration that occurs is plotted in Figure 12.15. Note that the sphere is deformed as it progresses through the plate, with the leading surface undergoing significant flattening. The impact pressure is about 30 GPa, well above the 2 GPa yield strength of the steel. Just after 1 μs, void is beginning to be inserted into the sphere by the code as spall fracture commences. It is apparent from these plots that the PMMA material in front of the sphere has been accelerated to a higher velocity than the exit velocity of the sphere, as is easily confirmed from a one-dimensional pres-

Table 12.5. Material parameters used for calculations

	PMMA	Hard Steel
Density (kg/m^3)	1186	7850
Bulk sound speed (m/s)	2598	4570
Linear shock velocity—particle velocity slope	1.516	1.49
Gruneisen coefficient	0.97	2.17
Yield strength (GPa)	0.2	2.0
Poisson ratio	0.32	0.29
Fracture stress (GPa)	0.15	4.0

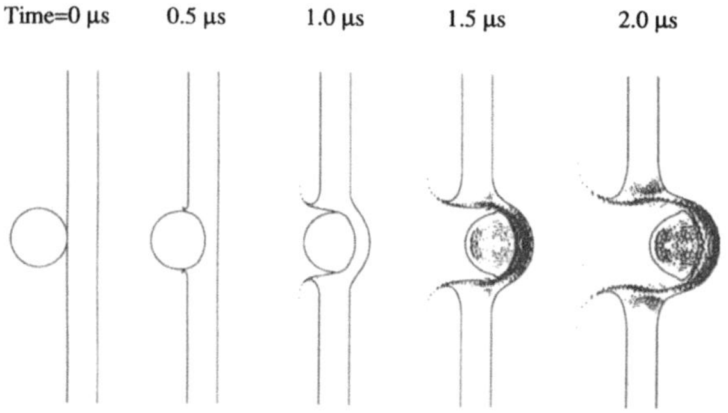

Figure 12.15. Calculated sequence of a 6.35-mm steel sphere, with normal incidence velocity of 4570 m/s, perforating a 3.38-mm PMMA target plate (Test 5).

sure–particle velocity diagram of the impact. Flash laser photographs of similar events also clearly show the dispersed PMMA debris leading the clustered steel debris (Ang, 1992).

Tracer particles, embedded in the sphere to provide point histories of computed variables, indicate that fracture is completed between 1.5 and 2 μs. After this time, the PMMA imparts very little impulse to the steel sphere. This response is illustrated in Figure 12.16, where the minimum principal stresses through the axial diameter of the sphere are plotted. (The first pulse is recorded at the impact surface, and the final pulse at the trailing surface.) The tensile stress is observed to reach a limit of about 4 GPa, then unload as void is added to the local cells to relieve the state of tension. The strain rates going into tension along the axis of the sphere are about 2×10^5 per second. This strain rate occurs in the regions that are driven into tension by release wave interactions. Divergent strain rates near the periphery of the sphere tend to drop below 1×10^5 per second. As the shock pulse transits the axis of the sphere, the amplitude decays from an initial peak of about 30 GPa to less than 10 GPa. The immediate consequence of this decay in impulse is a larger decrease in particle velocity at the leading edge of the sphere than at the trailing edge, so that the trailing material in the sphere has a relative

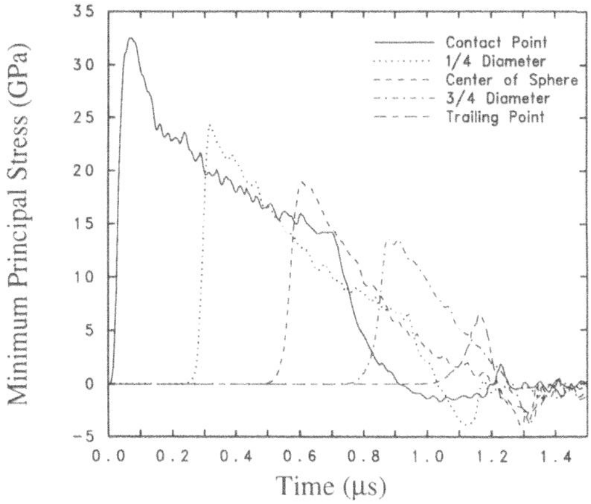

Figure 12.16. Histories of minimum principal stress at points along the axis of the steel sphere.

velocity towards the leading edge of the sphere. This relative velocity of the leading and trailing surfaces of the sphere explains the tendency of the particles to stay clustered in a thin disk, as seen in the radiographs (cf. Figures 12.2 and Figure 12.7). As a consequence of the impulse delivery being completed by about 2 μs, the formation of fragments has also been completed by then, and the fragmentation post-analysis calculations can be made. The debris is basically in free flight by this time, expanding as seen in the radiographs (in Figure 12.2) at much later times (e.g., 30 and 60 μs).

When the steel fragmentation characteristics were computed for this example, it was found that less than 25% of the mass of the steel sphere had fragmented. That is, the fracture stress of 4.0 GPa was exceeded in only a quarter of the volume of the sphere. Examination of the tensile strain rates that occur throughout the sphere indicate a range from about 6×10^4 per second to over 1×10^6 per second. Under these conditions, the expected spall stress would range from 2.2 GPa to 5.6 GPa, based on derived expressions by Grady (1988) for brittle fracture. Recalculating this example with a fracture stress of 2.0 GPa results in about 80% of the sphere fracturing; with a fracture stress of 1.0 GPa, 100% of the sphere fractures. It is clear that fracture based on a constant tensile stress limit does not accurately represent the behavior of this steel, nor of other materials with similar strain-rate dependence. In addition, at a larger strain-to-failure, and less confining pressure, the fracture strength will decrease from that of the spall value.

To further elucidate this issue, it is of interest to evaluate the shock energy coupled into the sphere considered earlier by analytic methods, as defined by Eq. (12.13). In the current example of a 3.38 mm PMMA target plate perforated by a 6.35 mm steel sphere, that analysis indicates the steel sphere acquires approximately 110 J of energy over the duration of the impact process. This calculation is based on an initial impact pressure of about 36 GPa, with a corresponding shock velocity of 5780 m/s and particle velocity of

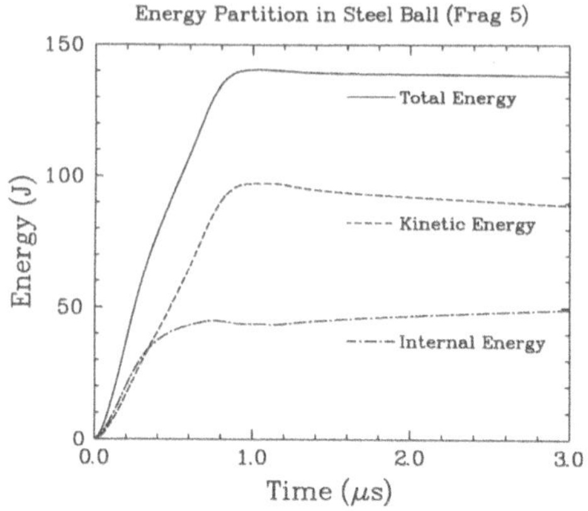

Figure 12.17. Internal, kinetic, and total energy for the steel sphere from the numerical analysis of Test 5.

787 m/s. (In this analysis, t_1 is 0.15 μs, t_2 is 0.34 μs, a_c is 2 mm, and b_c is 1.4 mm.) This energy increase represents the combined kinetic and internal shock heating, and assumes a reverse ballistic impact (i.e. the sphere is initially at rest). One-half of this energy is internal energy, or about 55 J. From the numerical analysis, the individual energy components associated with the steel sphere can be determined. The increases of the kinetic and internal energies as a function of time for the steel sphere are plotted in Figure 12.17. The reference internal energy assigned by the shock wave code to the steel (141 J) has been subtracted from the internal energy so that only changes in the internal energy acquired during the impact are plotted. The total change in energy of the steel sphere during the impact process is about 140 J. The change in internal energy is about 50 J, which corresponds well with the calculated analytic value; the kinetic energy tracks the internal energy until about one-half microsecond, and then the kinetic energy continues to increase to about 90 J. The impact process is dominated by changes in kinetic energy of the sphere, and the majority of this change has occurred in the first microsecond. The internal energy of the sphere is acquired during the first one-half microsecond, and changes little after that. The sum of t_1 and t_2 is about 0.5 μs, supporting the numerical results. This indicates that the primary shock contribution to the internal energy of the sphere has occurred well before the shock has even completed a single transit of the axis of the sphere, which takes over one microsecond (cf. Figure 12.16). The energy does not distinguish between shock heating and plastic deformation heating. The sphere is still decelerating for another one-half microsecond, as steady hydrodynamic forces act on it.

Fragmentation Calculations

As a working basis for this example, the fragmentation process was permitted to proceed using the 1.0 GPa fracture stress limit. The resulting computed steel fragment size distribution is shown in Figure 12.18, labelled "w/o

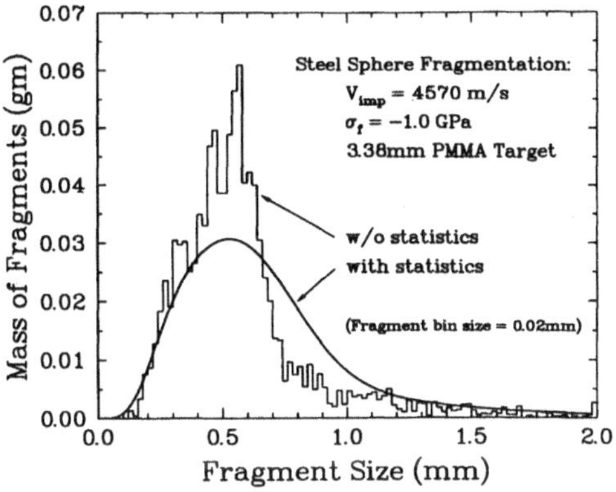

Figure 12.18. Mass distribution of steel fragments vs. fragment size resulting from the impact of a 6.35-mm steel sphere with a 3.38-mm PMMA target plate (Test 5). This distribution represents the sum of the average local fragment size for each steel cell. Both original and Poisson statistics distributions are plotted.

statistics." This distribution sums the masses of fragments in each material cell, where the average fragment size is calculated from the local strain rate, as defined in Eqs. (12.16) to (12.21). The figure plots the total steel fragment mass determined for each fragment size "bin", summed throughout the sphere. Each bin includes a fragment size increment of 0.02 mm. The largest mass of fragments have an average fragment size of about 0.6 mm. The temperature in the steel has increased only about 100 K, so that the majority of the material falls in the brittle, or fracture toughness dominated, region of fragmentation.

As discussed in the previous section, the fragment size calculated for a given strain rate represents the average fragment in an exponential (Poisson) distribution of sizes. The masses associated with each of the fragment sizes shown in Figure 12.18 (curve labelled "w/o statistics") can be modified to account for this statistical spread by employing the concepts expressed in Eqs. (12.22) to (12.27). When these equations are applied to the current example, the mass associated with each particle size results in the distribution labelled "with statistics" in Figure 12.18. The Poisson statistics (in which the same size "bins" were used) tend to broaden the distribution up to about 1.1 mm fragment sizes, and do not significantly affect the upper and lower extremes of sizes. The integral of the mass with size, as described by Eq. (12.27), results in the cumulative mass as a function of fragment size. Figure 12.19 contains both the original calculated cumulative mass distribution and the associated statistical cumulative mass distribution as a function of fragment size. The total mass of steel accounted for up through particle sizes of 2 mm is 1.01 g, or about 96% of the 1.05 g mass of the sphere used in the calculations.

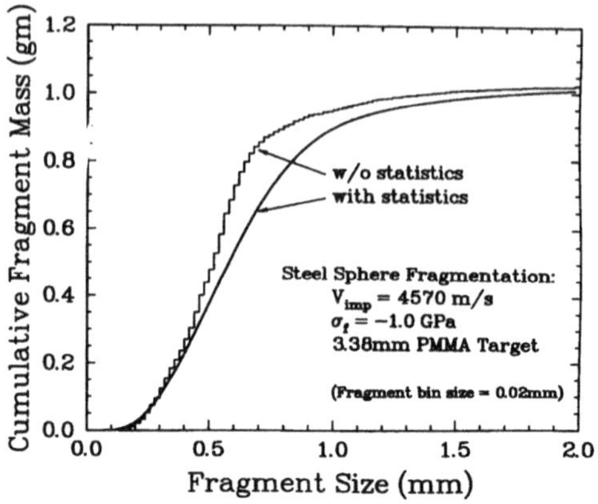

Figure 12.19. Cumulative mass distribution of steel fragments vs. fragment size resulting from the impact of a 6.35 mm steel sphere with a 3.38 mm PMMA target plate (Test 5). This distribution represents the accumulated mass sums of the average local fragment size. Both original and Poisson statistics distributions are plotted.

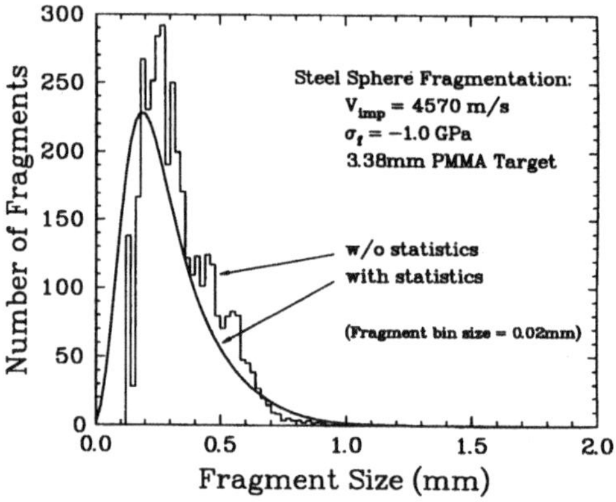

Figure 12.20. Final number distribution of steel fragments vs. fragment size resulting from the impact of a 6.35-mm steel sphere with a 3.38-mm PMMA target plate (Test 5). Both original and Poisson statistics distributions are plotted.

The corresponding distributions that show the number of fragments at each fragment size are plotted in Figure 12.20. Now the broadening effect of the Poisson statistics is apparent for the small size particles. The largest number of particles are of size 0.2 mm; the largest mass of particles are of size 0.6 mm. This prediction compares well with the experimentally observed fragment size of about 0.7–1.1 mm (Figure 12.3) at this impact energy. The cumulative number of particles calculated here is nearly 3800, the majority of which are less than 0.5 mm in size. The average fragment size, based on this number of fragments, is about 0.4 mm (spherical diameter). The average particle size determined from the radiographs for this example (Test 5) was about 0.7 mm, for about 400 particles (Table 12.3). Hence this computed number of fragments is nearly an order of magnitude too large, suggesting that the estimate of strain rate at the time of failure is too large by about a factor of 2 [cf. Eq. (12.16)]. Some effort is required to evaluate the fracture model being used to ensure that fracture occurs at the proper time, since the governing strain rate is directly tied to this aspect of the model. It has also been observed that the strain rates, and consequently the average calculated fragment sizes, depend on the equation of state being used for the materials. Issues relating to the joining of the energy-based and statistical fragmentation theories may also contribute to uncertainty in the fragment number prediction.

Target Thickness Variations

To examine trends of the behavior resulting from target thickness variations, a suite of eight impact calculations was made in which the impact velocity was fixed at 4500 m/s. Variations in PMMA target thickness from 0.76 mm (1/32 inch) to 6.35 mm (1/4 inch) covered most of the range of the experiments, providing a large spread in the amplitude of the impulse imparted to the steel sphere. From these calculations, comparisons can be made with experimental data for the loss in axial momentum, and the expansion velocity of the particles debris cloud.

The loss in axial velocity as a function of target plate thickness is plotted in Figure 12.21, and the numerical data clearly compares well with the experimental data. This figure can be compared with Figure 12.5, noting that the numerical simulations are representing the experimental data in much the same way as the analytic expression. The calculated curve in Figure 12.21 is not quite linear, and tends to have an upward curvature as the PMMA target thickness increases beyond 3 mm.

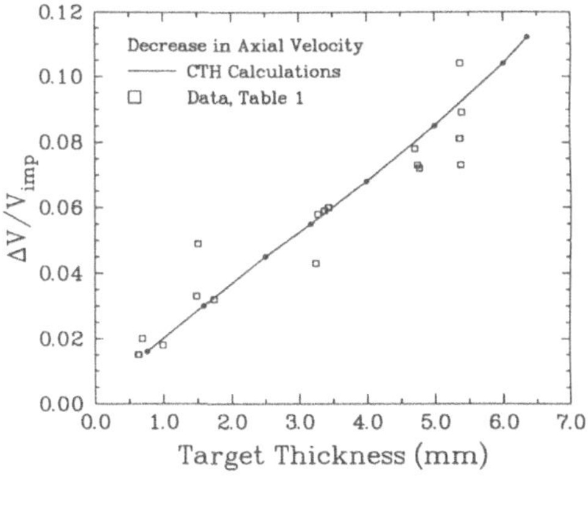

Figure 12.21. Decrease in axial velocity of the steel sphere as a function of PMMA target plate thickness. Data are from Table 12.1.

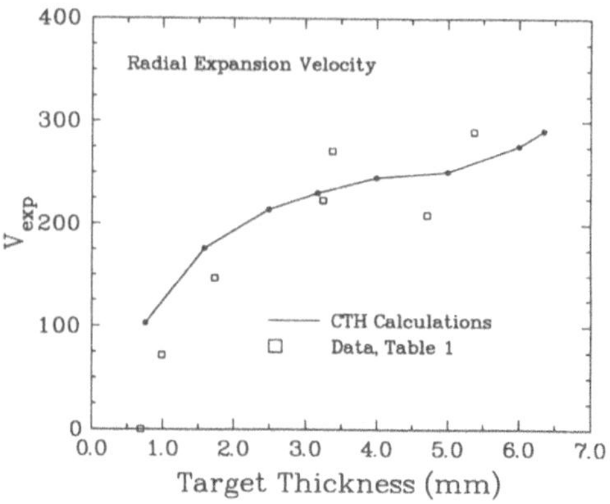

Figure 12.22. Variation of maximum steel particle debris cloud expansion velocity with PMMA target plate thickness. Data are from Table 12.1.

The expansion velocity of the outer debris edge is plotted in Figure 12.22 as a function of PMMA target plate thickness. The computed points were determined by evaluating the maximum radial momentum acquired by the steel during penetration. The computed values generally fall within the range of the experimental data. This figure can be compared to Figure 12.9, in which three curves of constant impact velocity are included. In the present figure, only an impact velocity of 4500 m/s has been tabulated. In principle, the source of the particles in the debris cloud can be determined from velocity histories at points distributed throughout the sphere. Along the initial surface of the sphere, the amplitude of the maximum lateral velocity varies from zero at the leading impact point to a maximum near the equator, returning to zero at the trailing point on the axis. Analysis of the velocity histories on a cross-section at the equator of the sphere indicates a linear increase in velocity from zero on the axis to maximum on the surface, resulting in the distribution of fragments throughout the disk seen in the radiographs. This continuous spatial distribution of particles is in contrast to the commonly observed hollow debris cloud that forms when a target imparts a larger energy to the projectile (e.g., Grady and Passman, 1990; Piekutowski, 1992). We note that as the thickness increases, the calculated expansion velocity begins to increase again. This calculational result may relate to the experimentally observed transition to Region III diffuse fragmentation (Figure 12.6, Table 12.2) in which anomalously large expansion velocities were observed.

An initial attempt was made to quantify the influence of target thickness, or imparted impulse, on the fragmentation characteristics of the steel sphere. For the present calculations with an impact velocity of 4500 m/s, the fragment size distributions for several thicknesses of PMMA targets are compared. The comparison is made on the basis of the statistical fragment size distributions, as illustrated in Figure 12.18. A statistical distribution was determined for each impact case. The results are plotted in Figure 12.23. The least impulse imparted to the sphere is for a PMMA thickness of 0.76 mm,

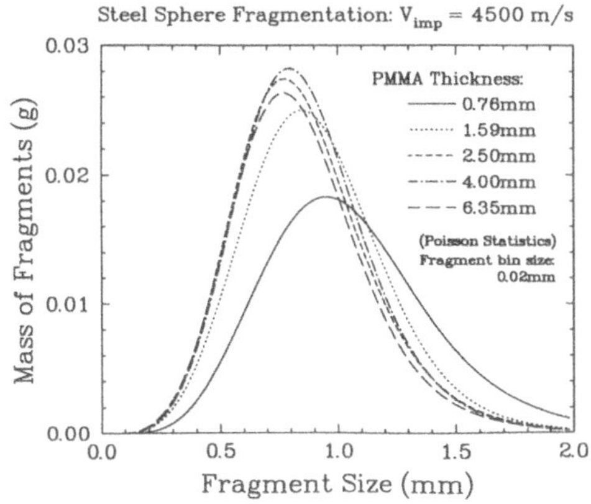

Figure 12.23. Comparison of statistical fragment size distributions for several thicknesses of PMMA targets, and an impact velocity of 4500 m/s.

and the fragment size distribution has the largest average fragment size (about 1 mm) of all the distributions, as should be. However, the remaining distributions are observed to be nearly indistinguishable from one another, suggesting that the sphere is fragmented in the same way, on average, for PMMA target thicknesses from 1.6 mm to 6.35 mm. However, from radiographs of the debris clouds, the qualitative data indicate a trend of continually smaller fragments as the target thickness increases. The implication is that fracture of the steel sphere is not being correctly represented. As a counter, however, the image-processed fragment size data (Figure 12.23) indicates a degree of insensitivity to impact conditions above an expansion velocity of about 100 m/s. Indications of this error were mentioned in previous paragraphs regarding the threshold fracture stress required to obtain complete fracture of the steel sphere. Further work will be necessary to sort out the exact source of the error, and correct it in a satisfactory manner. The comparisons made in Figure 12.23, however, are of the kind desirable for evaluating effectiveness of armor or warheads.

Target Profile Comparison

In some cases, the PMMA targets were recovered intact following an impact experiment. This provided an opportunity to compare the calculated crater and residual hole profiles with that of the recovered target. For example, the cross-section of the PMMA target for Test 19 is sketched in Figure 12.24(a) (also see Figure 12.12). The original steel sphere has been included for reference of scale. This target profile is representative of the recovered targets. The residual hole in the PMMA is only slightly larger than the original diameter of the steel sphere. Both the front and rear faces of the target have been excavated to a radius of nearly five times that of the sphere radius, so that the remaining central section has the appearance of a washer retained in the target. In contrast, the calculated target profile is shown in Figure 12.24(b). The strength material model for the PMMA was elastic–perfectly

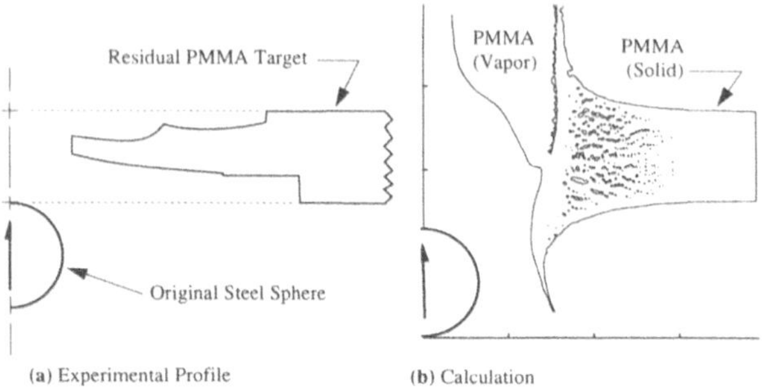

Figure 12.24. Cross-section recovered (a) and calculated (b) comparisons of PMMA target for Test 19 (4610 m/s impact onto 5.385-mm PMMA).

plastic. Although there is a large fractured region in the target, extending to about four and one-half sphere radii, there is no evidence of preferential front or back face surface excavation. The residual target hole radius is over twice the original projectile sphere radius. The front and rear crater lips are characteristic of this material treatment. At this time, the significant differences between the calculated and experimental target profiles have not been explained. Attempts to modify fracture strengths, yield strengths, or equations of state (with and without vaporization) for the PMMA have had only negligible influence on the profile calculated in Figure 12.24(b). It has already been noted that the impulse imparted to the sphere is correct, because the decrease in axial velocity is corroborated with the experiments (cf. Figure 12.21). PMMA is a visco-plastic material, and shock wave propagation and release wave structure are affected by this property, particularly below 6 GPa (Schuler and Nunziato, 1974). This major modification to the equation of state would be a profitable area to explore. In addition, there is also the distinct possibility that PMMA fractures in a much more complex manner than currently modeled.

12.2.7. Secondary Target Damage

As indicated in the experimental configuration shown in Figure 12.1, a secondary target (witness) plate was placed in the line of flow, down stream from the primary impact point, and intercepted the projectile and PMMA target debris. These secondary target plates were prepared from 6061-T6 aluminum plate stock into square plates 15 cm on a side and with several thicknesses (see Table 12.1). Depending on specific experiment parameters (impact velocity, PMMA plate thickness, range to secondary plate), the secondary target plates suffered various degrees of impact damage.

The damage experienced by secondary objects exposed to the high-velocity fragment debris of a primary fragment event represents an effect for which computational and analytical solution methods are in an early stage of development. Consequently, the supplemental data from the present series of experiments are useful for guiding theoretical model development focused on secondary-damage high-velocity impact events.

Experimental Results

The thickness d_s of the aluminum target plates was either 6.35 mm (1/4 in) or 12.7 mm (1/2 in), although one 9.53 mm (3/8 in) plate was used. Secondary plates were positioned behind the primary PMMA plates by a distance x_{ps} which ranged between about 40 and 65 cm (a somewhat shorter distance was used in one early test). These dimensional data are provided in Table 12.1.

Several features of the secondary plate damage are provided in Table 12.6. First, the loss of plate mass caused by the impact and determined by measuring the weight of the plate before and after the test was included. Because of uncertainties in the final disposition of the steel projectiles, and in the difference measurement, an uncertainty of about ± 1.0 g was assigned to these data. Second, in almost all of the tests, whether or not target penetration occurred, the impact debris footprint encompassed a well-defined area on the secondary plate. (Tests 19, 23, and 24 were the exceptions to this observation.) This footprint area was also included in Table 12.6. In those experiments in which the primary fragment debris penetrated the secondary target, removing a substantial mass of aluminum, the projected area of the hole was also measured and recorded in Table 12.6. Typically this area was 60 to 80% of the footprint area for that test.

Table 12.6. Experimental properties of secondary impact damage

Test no.	Footprint area[b] (cm^2)	Mass[c] (g)	Damage mode[d]	Test no.	Footprint area (cm^2)	Mass[c] (g)	Damage mode[d]
1[a]	—	—	—	13	27.2	2.6	spl[e]
2[a]	—	—	—	14	15.9	1.1	prf
3	9.2 (6.7)	15.6	plg	15	8.1 (0.9)	0.5	plg
4	15.6 (13.6)	30.4	plg	16	16.0	4.4	spl
5	33.2	3.1	spl	17	32.3	4.4	prf
6	8.4	7.6	prf	18	30.8	0.1	spl
7	22.8	2.8	prf	19	84.2	2.4	prf[ef]
8	9.4 (7.4)	14.1	plg	20	18.5	5.1	prf[e]
9	5.6 (4.1)	8.6	plg	21	12.1	6.9	prf
10	3.7 (2.0)	8.6	plg	22	10.7 (8.0)	22.7	plg
11	5.7 (2.5)	8.6	plg	23	full plate	1.1	spl[f]
12	7.8 (3.6)	13.6	plg	24	177.0	2.0	spl[f]

[a] Plate was not recovered.
[b] Number in parentheses is hole area where plugging occurred.
[c] Accuracy of 1 g estimated.
[d] plg, plugging; prf, perforation; spl, splash.
[e] Back surface spall.
[f] Diffused footprint (see text).

Secondary target plate damage in the present series of experiments could be generally grouped into three classes. These classes were identified in order of severity as splash, perforation, and plugging.

In the plate damage identified as splash the target plate was not penetrated by the steel projectile debris. There is usually a degree of bulging of the back side of the plate over the area behind the region of impact. Damage to the front of the target plate was perceived as a crater extending over an area comparable to the spread of the fragment debris observed in the radiographic record extrapolated to the point of impact. This crater area (footprint) is tabulated in Table 12.6. The crater cavity appeared to be accommodated as much by the back-surface bulging as by the splash of target aluminum during impact. Only a few grams of target material are removed by the splash process (see Table 12.6). The crater floor appeared chaotic with evidence of many subcraters and deeper channels of varying depths associated with the impact of individual fragments.

The character of plate damage called perforation was similar in appearance to that of splash. The only difference is that impact intensity was sufficient to cause full penetration of one or more individual projectile fragments. These penetrations were typically channels through the target thickness with openings ranging from about one to several millimeters. The minimum mass of steel fragment required to perforate the aluminum witness plate has not been determined empirically, although there have been numerous studies made for other projectile/target material combinations to ascertain the ballistic limit velocity (e.g. Backman and Finnegan, 1984; Cour-Palais, 1987; Herrmann and Wilbeck, 1987; Hohler and Stilp, 1987). A brief numerical study indicates that perforation of a 12.7 mm aluminum target will occur with a single steel particle of 2.5–3 mm diameter, at an impact velocity of 4300 m/s. However, allowing for a second particle impact in a previous crater, two 1.7 mm particles spaced 10 mm apart (about the cloud thickness at impact), will perforate the aluminum target, with about one half the kinetic energy required by a single particle. The average size particles for this experiment was about 0.7 mm (Table 12.3), so there would not be much expectation of two particles of 1.7 mm in the same trajectory, and perforation of the witness plate is unlikely (see Table 12.6).

For plate damage classified as plugging, the collective impulse of the fragment debris was sufficient to remove (plug) a region of the plate leaving a hole comparable to the projected area of the debris cloud. For the tests in which the primary PMMA plate was insufficient to fragment the steel sphere, typical single-particle plugging phenomena were observed. The plug hole diameter was somewhat larger than the projectile diameter and increased with increasing impact velocity (see Table 12.6). When primary projectile fragmentation and debris spread occurred before secondary impact the size of the plug region was significantly larger.

Three impact experiments (Tests 19, 23, and 24) did not readily fall into any of the three classes of secondary plate damage just described. These three tests have already been discussed in terms of the substantially broader dispersion of fragment debris observed on the respective radiographic records. Impact damage on the secondary target also reflected this broad dispersion of debris. Although areas for the impact footprint are tabulated for Tests 19 and 24 in Table 12.6, the radius of these two footprints were not distinct, consisting of numerous independent fragment craters which decreased gradually in impact density with increasing radius. The footprint of Test 23 encompassed the full area of the 15 by 15 cm plate. In each of these three tests a major central crater ($\sim$15 mm in diameter) caused by a large fragment or cluster of fragments was observed. This crater was virtually identical in Tests 23 and 24, causing significant back surface bulging and incipient spall cracking. In Test 19 a central hole approximately 8 mm in diameter was punched through. There is some evidence for the responsible central fragment or fragment cluster in the radiographs for these three tests.

Photographs of four of the secondary target plates are shown in Figure 12.25 and Figure 12.26 to illustrate the various damage modes observed in the present test series. The secondary plates shown were all 12.7 mm in thickness and placed approximately 64 cm behind the PMMA impact plate (see Table 12.1). Velocities for the tests shown ranged from about 4.0 to 4.7 km/s and various thickness PMMA plates were used. The plate for Test 18 in Figure 12.25(a) illustrates the localized splash mode of secondary target damage observed in many of the tests. The plate was not perforated and all fragment impacts are confined (clustered) within a relatively narrow solid angle subtended from the point of impact. In Test 20 slightly different impact conditions led to a smaller area of impact and perforation of several fragments through the plate thickness. Conditions for Test 22, shown in Figure 12.26(a), led to a tight group of the fragment debris and complete plugging

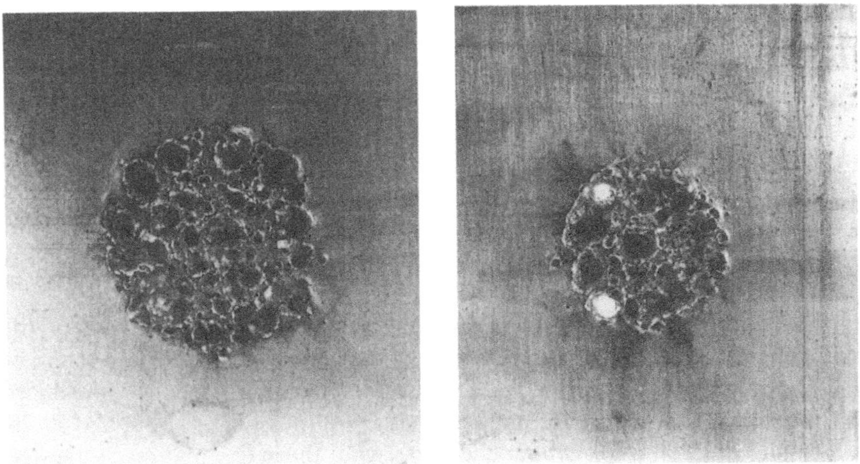

(a) (b)

Figure 12.25. Secondary target damage modes.

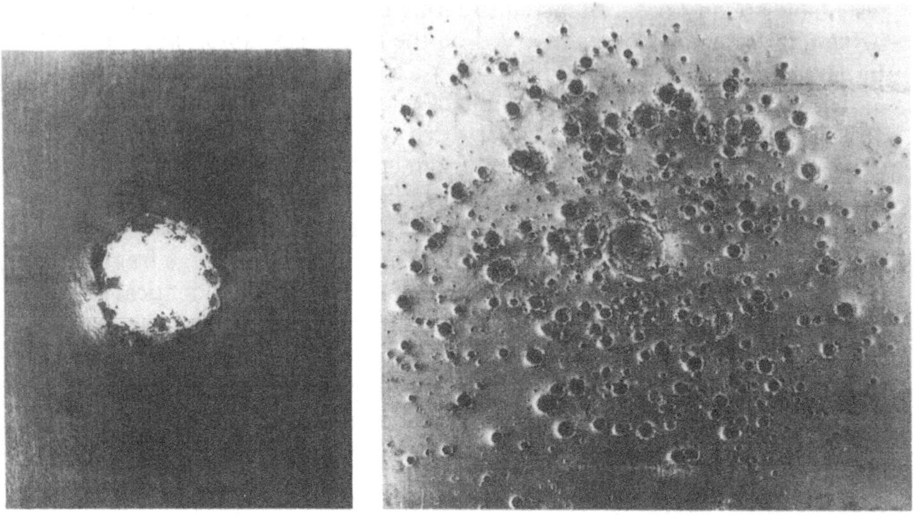

(a) (b)

Figure 12.26. Secondary target damage modes.

damage to a section of the secondary target. The secondary target plate in Figure 12.26(b) for Test 24 illustrates the transition to enhanced debris expansion and the more diffused damage footprint observed in several tests at high velocities and/or with thick target plates.

12.2.8. Numerical Analysis of Debris Cloud Impact on Aluminum Target

An explicit approach to examining the effects of a debris cloud impacting a secondary target is to define each particle in the cloud and compute the excavation of the target resulting from the cumulative cratering of each particle. In this section, impact of the particles in a debris cloud representing those formed in Test 5 is numerically simulated with CTH. This example provides insight into the nature of multiple impacts and the influence of the cloud parameters on the target response. For the present generation of computational hardware, this three-dimensional problem requires significant memory and execution time, and is not yet a routine method of debris-cloud analysis.

The calculations of debris characteristics from the initial impact of the steel sphere with a PMMA target were discussed in a previous section. The size and number distribution characteristics of the debris that was produced in that impact simulation were represented in Figures 12.18, 12.19, and 12.20. In principle, a desirable technique would permit the extrapolation of the steel debris from the initial impact with the PMMA target to the location of the next target—the aluminum witness plate in this example—complete with spatial and size distribution of the particles in the debris cloud. The extrapolation step has not been taken yet; although the expansion velocities are available, the axial dimensions of the debris cloud have not been determined from the simulation, primarily because of a lack of knowledge regarding the particle interactions during the time just after sphere breakup commences. In addition, the total number of particles computed in the example tended to be too large, based on qualitative analyses of the radiographs.

In order to define a tractable problem, the number of particles in the debris cloud was constrained to be about 300, with a particle size distribution that was assumed to be exponential. That is, the number of particles of a given mass was based on an exponential function of the particle mass, normalized so that the total mass of the steel particles summed to the original mass of the steel sphere, M, of 1.05 g. The maximum particle size was less than 2 mm, corresponding to a mass of less than 40 mg for the largest particle. The number of particles of mass m, N_m are distributed according to bin size intervals of Δm, and determined from

$$N_m = \frac{N_0}{\mu} e^{-m/\mu} \Delta m, \tag{12.29}$$

where N_0 is the total number of particles, and μ is the average particle mass, related to the total mass of the original steel sphere by $M = N_0 \mu$. For the present analysis, the average particle mass is about 3.5 mg, and the average particle size (diameter) is about 0.76 mm.

The dimensions of the debris cloud envelope at the location of the aluminum witness plate were determined from the positions and times of the radiographs relative to the position of the witness plate. On this basis, the cloud, assumed to be a circular disk, has a diameter of about 55 mm, and a thickness of about 10 mm.

The impact velocity of the particles onto the witness plate is 4300 m/s, and the outer expansion velocity is about 270 m/s (cf. Table 12.2). The coordinate positions of each particle were assigned with a random number generator: an (x, y, z) position was assigned, and that coordinate was checked to ensure it was interior to the envelope, and then compared against all previous particles for overlap; if no overlap occurred, then the coordinates were retained; otherwise a new set was assigned and rechecked. This procedure was followed for each of the 300 particles until all the particles were contained within the envelope, and no overlap existed. In addition, each particle was assigned an normal impact velocity of 4300 m/s, as well as a radial velocity based on the expansion velocity times the particle position radius normalized to the envelope radius.

In this three-dimensional calculation, the normal impact direction of motion was in the x-coordinate, and the two lateral coordinates were y and z. The physical space of the cloud and 12.7 mm aluminum witness plate was defined by $-2\text{ mm} < x < 28\text{ mm}$, $-37.5\text{ mm} < y < 37.5\text{ mm}$, and $-37.5\text{ mm} < z < 37.5\text{ mm}$, with uniform 0.3 mm cubical cells, resulting

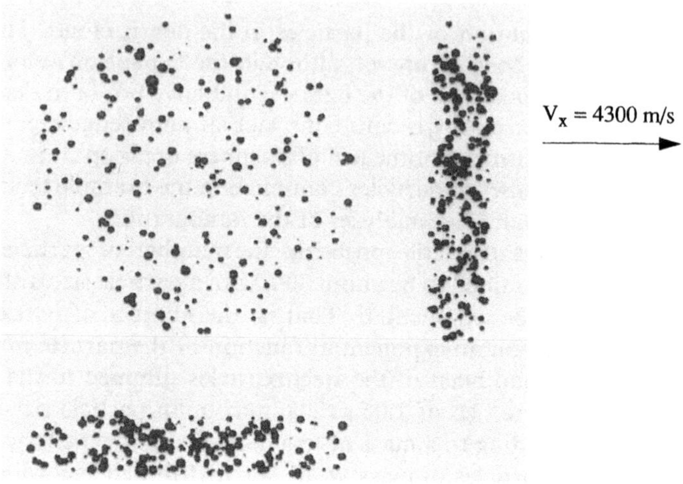

Figure 12.27. Initial conditions of the 297 debris cloud particles.

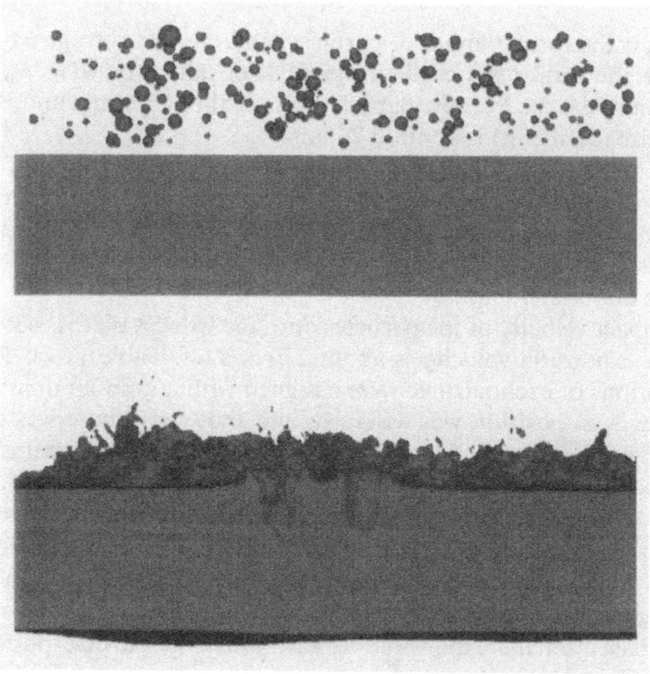

Figure 12.28. Edge view of the aluminum witness plate at 0 and 8 μs.

in about 6.25 million cells in the calculation. Views of the initial conditions are shown in Figure 12.27. One can note that in this simulation there are clear differences from the radiographs in Figure 12.2 and Figure 12.7. Of particular note is the fairly regular distribution of particles on the periphery of the radiograph in Figure 12.7. The resolution of the calculation limits the number of cells to about 2 to define the average particle dimension. Consequently, some of the particles will be shapeless masses impacting the target rather than well-defined spheres. At these velocities, however, the initial shape of the particle is not of primary importance in determining the excavated crater. An edge view of the initial conditions is shown in the upper section of Figure 12.28.

By 8 μs the craters have all been formed and the steel particles exhausted their momentum in the excavation process. An edge view of the target plate at this time is shown in the lower section of Figure 12.28. The jagged crater lip formation is evident in this view. No particles have perforated the aluminum target, but a significant bulge is observed to have formed on the back side of the target. A perspective top view of the craters formed is shown in Figure 12.29. (In this view, the edges have been trimmed from the aluminum plate to include only that material that contains craters.) Although there is clearly significant coupling between craters, extensive contiguous original surface material remains. All these observations are consistent with the recovered witness plate. An estimate of the computed excavation volume of aluminum is 2.75 cm^3. There are many evidences of multiple impacts forming craters, and rare instances of crater connectivity below the surface.

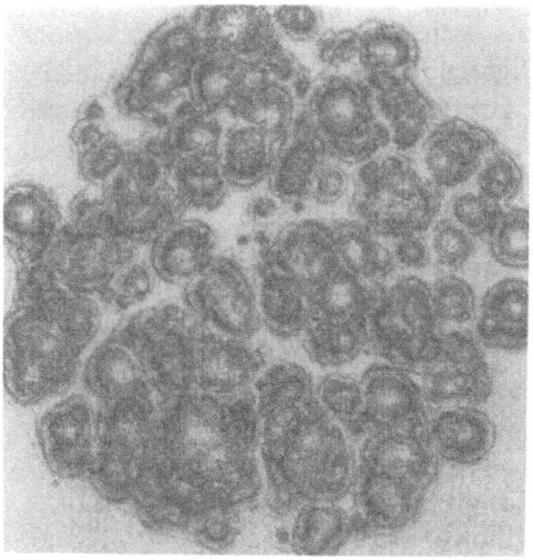

(Right eye)

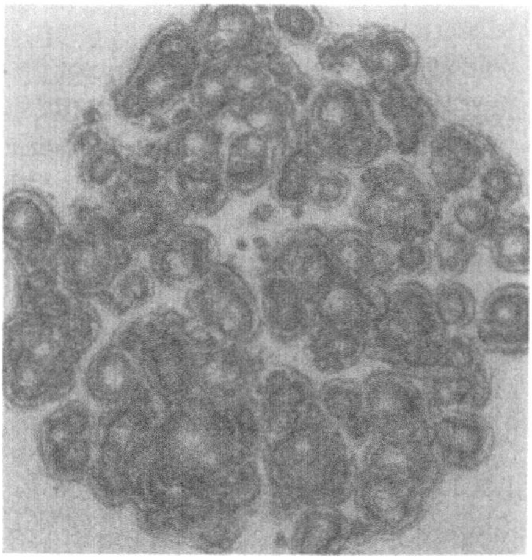

(Left eye)

Figure 12.29. Perspective top view of craters at 8 μs (stereo cross-eyed pair).

12.3. Copper Sphere Impact on Steel Targets

An impact configuration involving two distinct metals for projectile and target was defined with the specific intent of obtaining radiographic diagnostics of the debris cloud shortly after perforation was complete, and to recover the fragments for post-shot analysis. In the present investigation we have undertaken a study of the high-velocity interaction of a copper projectile and a steel plate in the velocity range of 3.5 to 5.0 km/s for the purpose of testing the wave-propagation code. High-resolution radiography has been used to diagnose the crater formation and debris motion within the time frame accessible to computational simulation. Both normal and oblique impact experiments have been performed to test the full three-dimensional capabilities of the wave propagation code. Additional diagnostics have also addressed further details of the fragment size and trajectory properties of the metal debris. The radiographs were successfully obtained. The particles were also successfully captured in foam blocks.

The major effort of this computational wave code analysis was focused on an oblique impact experiment performed at an impact velocity of about 4.5 km/s. Simulated radiographs (projected density plots) were constructed and compared directly with experimental x-ray data.

12.3.1. Experimental Configuration and Materials

The experimental configuration for the impact fragmentation tests is shown in Figure 12.30. Saboted OFHC copper spheres were launched at velocities between about 3 to 5 km/s with a two-stage light-gas gun system. The launch tube diameter was 12 mm. Lexan sabots were separated from the copper spheres through forces produced by a slight backpressure of nitrogen in the gun range section. Sabot segments were trapped upstream and did not reach the target impact chamber. Velocities of the copper spheres were measured to $\pm 2\%$ accuracy by recording of the time interval during passage between two magnetic coils of known separation. Impact of the copper spheres occurred in the target chamber at approximately the center of a 75 mm by

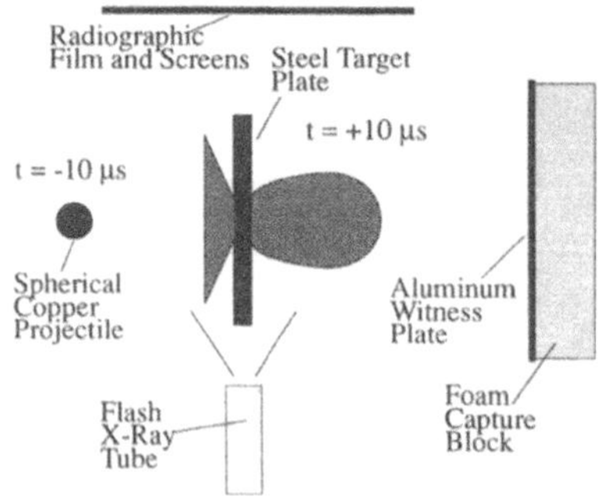

Figure 12.30. Experimental configuration for radiographic and fragment witness diagnostics of copper-on-steel impact fragmentation experiments. (Timing is representative of a 4.5 km/s impact.)

Table 12.7. Experimental conditions

Test no.	Impact Velocity (km/s)	Copper Mass (g)	Copper Diameter (mm)	Plate Thickness (mm)	Plate Inclination (degrees)	X-Ray Times (μs)
Test 1	4.96	1.20	6.36	5.61	0.0	11.6
Test 2	4.52	1.20	6.35	5.60	30.8	11.6
Test 3	3.43	1.20	6.36	5.60	0.0	131.4

75 mm square steel target plate of thickness 5.60 mm. Critical experimental parameters are provided in Table 12.7.

In the reported experiments, spheres of 99.99% pure copper with a 0.2% offset yield strength of about 270 MPa were used for the impacting projectiles. Target plates were prepared from 4340 steel (AMS 6359 plate stock) heat treated to a Rockwell hardness of 41–42. In all experiments a copper sphere 6.35 mm (1/4 in.) in diameter was used. The measured mass was 1.20 ± 0.001 g. The density of the steel is 7850 kg/m^3, yield strength is 2.0 GPa, fracture toughness is 30 to 40 MPa$\sqrt{\text{m}}$, and elastic modulus (Young's) is 200 GPa, with a Poisson ratio of 0.29.

Two impact configurations were tested in the present study. In all tests the steel target plate was supported by a bolt attached to one corner. In Tests 1 and 3 the plates were positioned so that the plane of the plate was vertical (identified as 0.0 degrees of inclination in Table 12.7). Normal impact within ± 1 degree was achieved in these tests. For Test 2 the top of the plate was tilted out of the vertical by 30.8 degrees achieving an oblique impact of the copper sphere on the target plate.

Impact ejecta in Tests 1 and 2 was diagnosed approximately 11.6 μs after impact on the front surface of the target plate. One 150 keV flash x-ray tube, placed approximately 1.5 m from the line of projectile travel, provided orthogonal shadow graphs of the fragment debris, as shown in Figure 12.30. Impact delays were determined from electrical shorting impact switches placed on the front of the target plate. The x-ray film cassette, using Kodak Direct Exposure film backed by a Quanta Fast Detail screen, was stationed about 15 cm from the debris trajectory.

For Tests 1 and 2 the x-ray image was centered on the target plate and the early-time record provided detailed geometry of crater formation, and both front and back surface ejecta debris. In Test 3 a late-time image of back surface fragments was obtained which provided more detailed information on particle number and disposition.

An assembly to witness and capture behind-target fragment debris was also included in the experimental configuration. A block of polyurethane foam (30 cm on a side and 7.5 cm in thickness) with a density of 320 kg/m^3 was positioned 25 cm behind the target plate. A sheet of 30 cm by 30 cm aluminum foil 0.075 mm in thickness was placed on the front of the foam block and provided a witness measurement of the number and distribution of debris fragments. Fragments were captured within the foam block and were interrogated with post-test static radiography.

12.3.2. Experimental Results

Radiography

The radiograph for Test 1 at 11.6 μs corresponding to a normal impact at 4.96 km/s is shown in Figure 12.31. The leading edge of the behind target debris is moving at a velocity of approximately 2.3 km/s, slightly less than half of the initial impact velocity. Velocities in excess of 3 km/s were determined for the highest velocity fragments in the back-splash debris. In contrast, large fragments associated with the lip of the spall ring (both front and back of the target plate), emerging from the peripheral region of the penetration spall crater, indicated an axial velocity component of about 0.4 km/s. At the 11.6 μs time into the motion, fragments appear fully formed although high fragment density precludes resolution of individual particles.

The radiograph resulting from the Test 2 oblique impact experiment at 4.52 km/s is shown in Figure 12.32. The impact debris in this test was also imaged 11.6 μs after impact. The leading edge of the behind-target debris is moving at approximately 1.8 km/s. A slight asymmetry in the envelope of the behind-target ejecta debris due to the oblique impact is noted, and most certainly has more material on the upper side of the cloud than on the lower side, where there are observed to be more spaces between fragments. From the radiograph, it is not possible to discern the interior character of the debris cloud. Details in the backsplash debris from the impact surface asso-

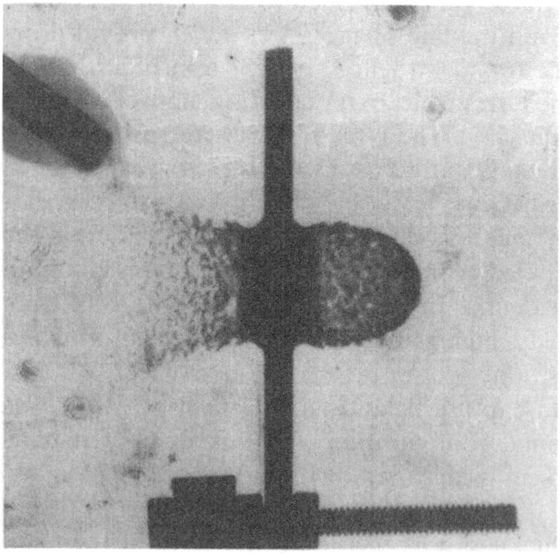

Figure 12.31. Radiographic record of 4.96 km/s normal impact at 11.6 μs (Test 1).

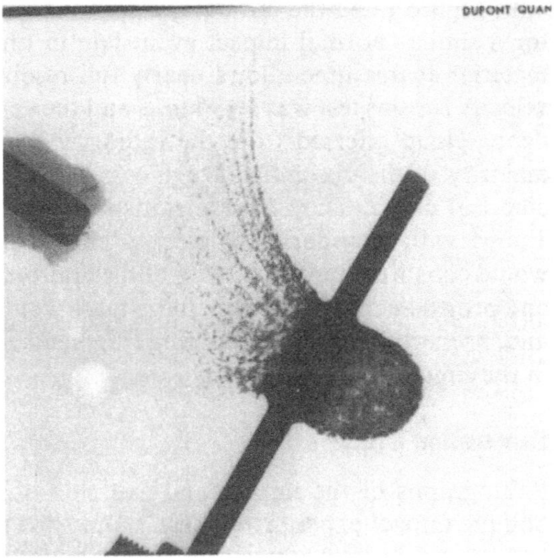

Figure 12.32. Radiographic record of 4.52 km/s oblique impact at 11.6 μs (Test 2).

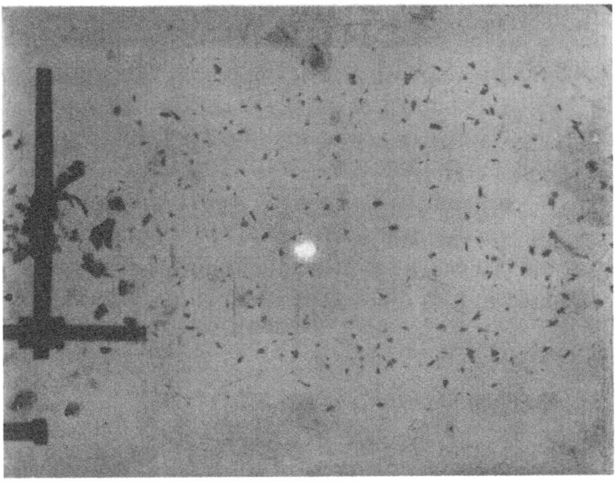

Figure 12.33. Radiographic record of 3.43 km/s normal impact at 131.4 μs (Test 3).

ciated with jetting caused by the spherical impact are particularly apparent. The curved traces of debris on the impact face of the target originate from a radially increasing crater, and are finest at the time of impact, and increase in size as the crater becomes larger. A ray pattern associated with the circumferential instability and breakup of the jet material is also observed. Note the defined spacing of these rays of debris, their circumferential spacing determined by the strain rates of the expanding crater. Leading fragments in the back-splash debris are traveling at velocities near 6 km/s.

In Figure 12.33 the radiograph for Test 3 shows the behind-target debris for a similar normal impact event late in time. The low spatial density of material at this time allows nearly full resolution of all fragments. Impact velocity for this test was 3.43 km/s, and the velocity of the leading edge of the debris cloud inferred from the radiograph is approximately 1.6 km/s. The majority of the observable fragments are uniformly distributed through an elliptical cloud. They have a statistical mean projected area very close to 1 mm^2 with a remarkably narrow size distribution. Thus these particles would constitute only about 2 g of the approximately 13 g of the target ejecta and projectile mass. The remaining mass would be the back splash fragments and, primarily, the large spall ring fragments residing close to the target plate in the Figure 12.33 radiograph.

Penetration Crater

Photographs of the impact and exit side of the penetration crater for the oblique impact experiment (Test 2) are shown in Figure 12.34. Structure of the normal impact crater for Tests 1 and 3 were visually identical to that shown in Figure 12.34. In Test 2 the top of the plate was tilted away from the projectile trajectory to achieve the approximately 30 degree oblique impact configuration The crater shows an approximately circular penetration hole and a spall ring region on both impact and exit sides out to a somewhat larger spall lip radius. Penetration damage on exit and impact sides showed no visual differences. Crater hole and spall lip diameters for the three impact tests are provided in Table 12.8.

Target plates for impact Tests 1 and 2 were also sectioned through the center of the crater and polished for metallographic examination. For the Test 2 specimen the crater was sectioned vertically to expose possible asymmetry in the damage modes due to the oblique impact. The photograph of

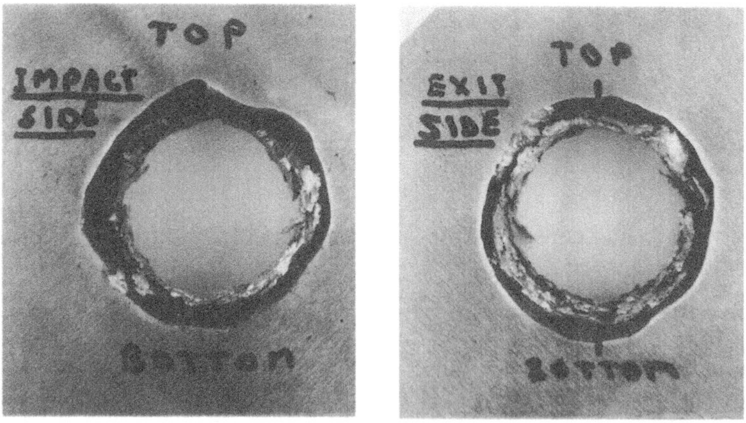

Figure 12.34. Penetration crater for 4.52 km/s oblique impact experiment.

Table 12.8. Experimental parameters

Test no.	Initial Plate Mass (g)	Plate Mass Loss (g)	Hole Diameter (mm)	Crater Lip Diameter (mm)
Test 1	250.9	19.3	19 ± 1	27 ± 2
Test 2	251.7	16.6	18 ± 1	26 ± 2
Test 3	250.0	10.7	15 ± 1	22 ± 2

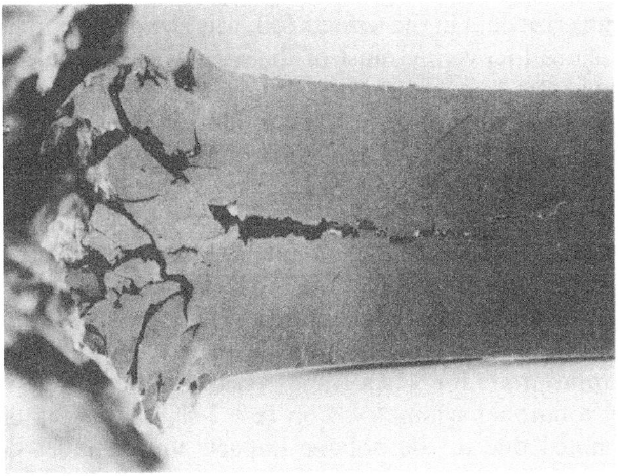

Figure 12.35. Metallographic section through center of penetration crater for 4.96 km/s normal impact experiment.

the Test 1 target plate shown in Figure 12.35 is representative of the results. The extent of the spall ring region is indicated. Close examination revealed extensive shear banding and small-scale cracking. The central spall crack seen in Figure 12.35 circumscribed the penetration crater, extending into the plate about 3 to 10 mm. This central spall crack was not observed in the oblique impact experiment.

Mass loss from the crater region was determined by weighing target plates before and after the impact tests. Results provided in Table 12.8 show a consistent monotone increasing trend with impact velocity.

Witness Plate and Capture Data

The passive methods used to interrogate the behind-target fragment debris were also informative. The foam block and foil positioned behind the target were sufficient to intercept a majority of the high-velocity behind-target debris. The large spall ring fragments which were ejected with a large angle with respect to the projectile trajectory were not captured by this method.

First, the aluminum foil placed at the front of the foam block provided an excellent diagnostic of the fragmentation pattern and intensity. Penetration craters in the foil provided only a rough indication of size, however, since the hole areas were somewhat larger than the corresponding projected fragment area. The number density of fragments was readily determined from the witness foil, however, and provided valuable data for validating computational fragmentation codes since the mass flux at a point in space is readily calculated in the wave-code analysis, and hence couples fragment size and number.

It should be noted that the qualitative fragment size distribution. as inferred from hole size data in the witness foil, was strongly bimodal in character. A broad spread (covering most of the witness foil) of larger fragments (holes) ranging from about 1 mm to about 5 mm in size was observed. In addition an intense spray of very fine particles (hole size between 20 and 200 μm) within about a 10 cm central region was witnessed by the foil. This effect was intense in Tests 1 and 2, but very weak in the lowest velocity Test 3.

A particle number density for fragment hole sizes of approximately 0.5 mm and larger was determined from witness foils from the three tests. These fragment number data are plotted in Figure 12.36. The data are plotted as a function of radius from the line of projectile travel. This representation is appropriate for Tests 1 and 3 because the normal impact conditions imply rotational symmetry. For Test 2 slight skewing of the debris pattern was noted due to the oblique impact, thus number density both vertically upward and downward from the projectile line are plotted in Figure 12.36.

Static radiographs of the foam blocks used to capture the debris ejected from the target plate were taken. Views in the direction of impact are shown in Figure 12.37 and Figure 12.38. Note the distribution of particles through the diameter of the cloud. There is a large contrast in the apparent average fragment size between Figure 12.37 and Figure 12.38. The large impact ve-

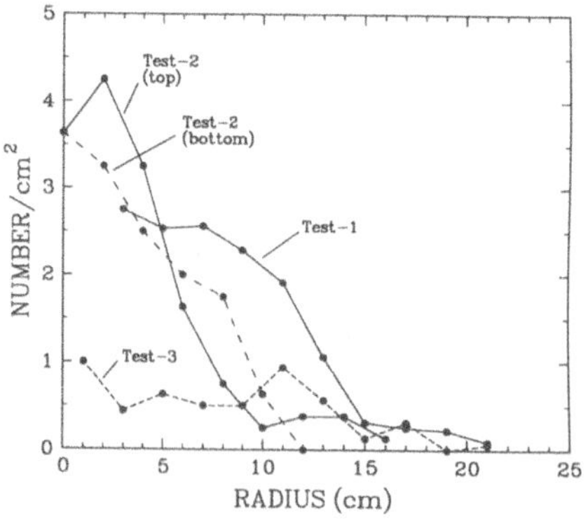

Figure 12.36. Number of fragment density data determined from penetration craters in aluminum witness foils.

Figure 12.37. Static radiograph of foam block used to capture behind-target ejecta fragments: Test 1 (6.3 mm copper sphere impact velocity = 4960 m/s; steel target thickness = 5.61 mm).

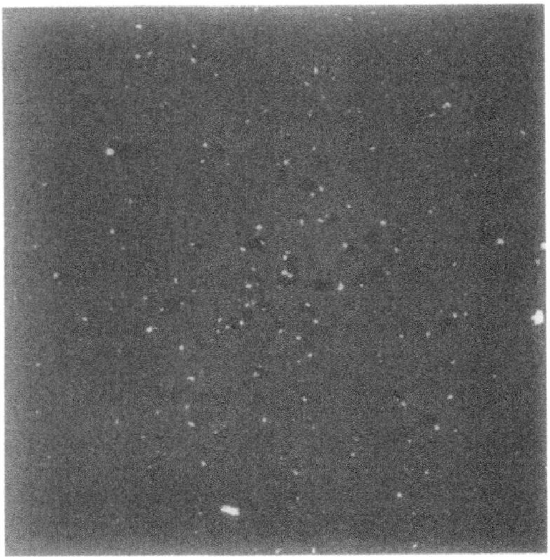

Figure 12.38. Static radiograph of foam block used to capture behind-target ejecta fragments: Test 3 (6.3 mm copper sphere impact velocity = 3430 m/s; steel target thickness = 5.60 mm).

locity of Test 1 (4960 m/s) results in relatively small fragments, and many deep craters in the foam (Figure 12.37). The much lower impact velocity of Test 3 (3430 m/s) results in an apparently larger average fragment size, and considerably fewer deep craters in the foam block (Figure 12.38). The center region of the higher velocity impact is nearly devoid of larger fragment sizes, those appearing in an annular ring about the center.

The radiographs were image processed (See Appendix C of Kipp et al., 1993a) and the fragment areas reduced to digitized data (Figure 12.39). The abscissa is projected area of the fragment. The higher impact velocity has

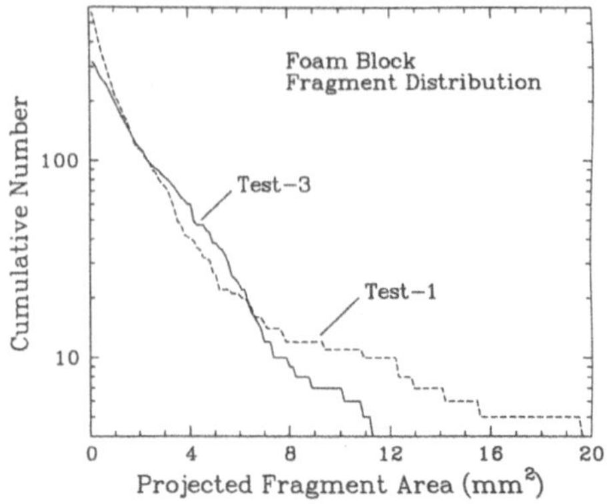

Figure 12.39. Fragment distribution data for normal impact Tests 1 and 3 determined from static radiographs of fragments captured in foam blocks.

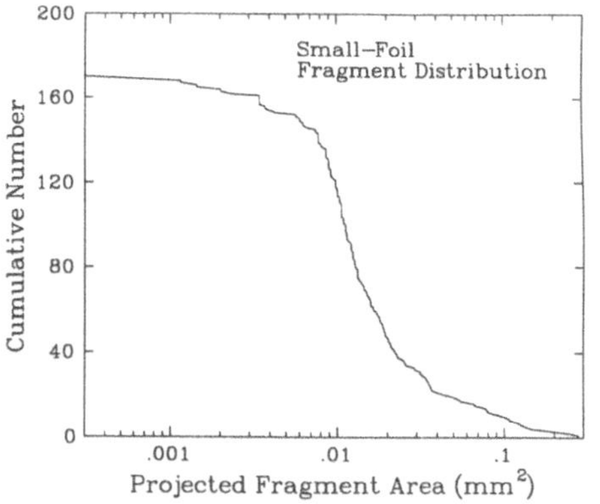

Figure 12.40. Statistical particle size data for fine spray particles determined from aluminum witness foil.

nearly 600 fragments recorded, whereas the lower impact velocity has only about 300 fragments recorded in the foam blocks. There is a crossing of distributions in the cumulative number data shown in Figure 12.39. This is a consequence of approximately equal masses of material being partitioned into two distributions—one with a larger number than the other. The distribution with more fragments has the mass concentrated nearer the small fragment end, with a steeper slope to the large fragment end (assuming exponential distributions). In fact the data are quite linear in over much of the fragment area range, and are represented quite readily with an exponential distribution. The larger chunky fragments are presumed to be pieces of the spall ring that break off and move at relatively slow velocities compared to the rest of the fragments.

At these impact velocities, there is sufficient shock heating locally to cause melt of some of the metal (primarily copper), and the penetration of these fine particles causes a network of fine holes in the aluminum witness foil (Figure 12.40). A representative section of the impact foil from Test 2 was image-processed, and the size distribution data (actually foil hole areas) are provided in Figure 12.40. The dominant hole diameter is about 50 μm approximately consistent with predicted break-up sizes of liquid melt fragments (Grady and Kipp, 1994).

12.3.3. Numerical Analysis of the Copper Sphere Impact

The primary objective of the impact penetration experiments and radiographic diagnostics was to provide spatially resolved data of the crater formation process and impact ejecta motion for the purpose of validating three-dimensional (3-D) computational codes in the regimen of high-velocity material interaction. The Eulerian wave propagation code CTH (McGlaun et al., 1990) was again the computational tool of choice in the present investigation. The computational study focused primarily on the oblique impact experiment (Test 2) which provided the most stringent test of the 3-D features of the wave propagation code.

In the oblique impact geometry that was simulated, there is one plane of symmetry, a plane through the axis of impact, appearing in plan view like the sketch in Figure 12.30. A computational resolution of 0.2 mm was the best that could be obtained, based on available computer storage. This resulted in just over 6 million computational cells, each 0.2 mm cubes. The material properties for the copper and steel are provided in Table 12.9. The $\alpha \leftrightarrow \varepsilon$ solid-solid polymorphic phase transformation was included for the steel, a phase transformation that occurs at about 13 GPa, and has been shown to be an important factor in certain kinds of impact scenarios (Bertholf, et al., 1975). The low- and high-pressure linear shock velocity–particle velocity parameters are included in Table 12.9. The deviatoric behavior was modeled as elastic–perfectly plastic, with the yield strengths indicated in Table 12.9. Fracture in the computational analysis was treated by introducing void into the computational cell when a principal stress exceeded the tensile fracture stress until the stress state in that cell was reduced to zero.

The calculation was initiated with the copper sphere in contact with, and surface tangent to, the steel target plate. The initial velocity vector of all cells within the copper sphere was directed into the target plate 30 degrees from

Table 12.9. Material parameters used for copper sphere impact

	Copper	Hard Steel
Density (kg/m^3)	8920	7850
Bulk sound speed (m/s)	3904	4620
Linear shock velocity—particle velocity slope	1.489	1.74 α-iron
		1.64 ε-iron
Gruneisen coefficient	1.99	1.75
Yield strength (GPa)	0.7	2.0
Poisson ratio	0.351	0.29
Fracture stress (GPa)	0.3	1.2

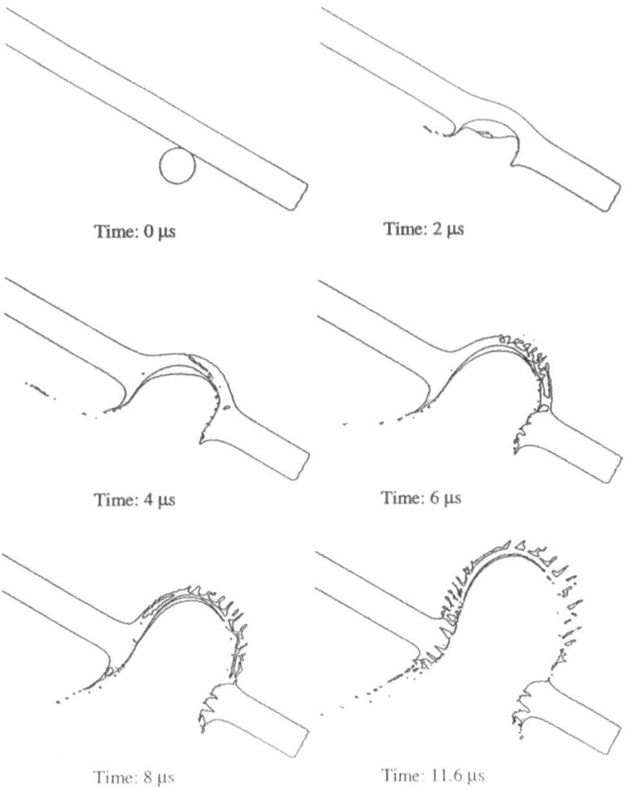

Figure 12.41. Symmetry plane views of the impact of the copper sphere onto an inclined steel target. Times are 0, 2, 4, 6, 8, and 11.6 μs.

the target normal, and with a magnitude of 4.52 km/s. Impact stresses are nearly 120 GPa. A sequence of images at the symmetry plane is recorded in Figure 12.41. In these figures, the sphere is moving from bottom to top. At 2 μs, the crater has begun to form, and the rear of the target is bulging. At 4 μs, the target is nearly perforated, with a spall region visible in the steel layer at the top of the bubble. Although the impact was oblique, already there is evidence of hole symmetry forming that has the appearance of normal impact. The copper is preferentially layered on the steel about the impact axis. By 8 μs, the debris cloud is nearly completely formed. At the final computed time of 11.64 μs, which corresponds to the time of the radiograph, the debris cloud has completely separated from the rear of the steel target, with a copper layer that has remained contiguous over a substantial section of the inside of the debris. The target hole still has the appearance of a normal impact, but the debris cloud is asymmetric, with more material contained on the left than on the right of the cloud. This is more clearly seen in the edge view of the target at 11.64 μs (Figure 12.42) where all the material

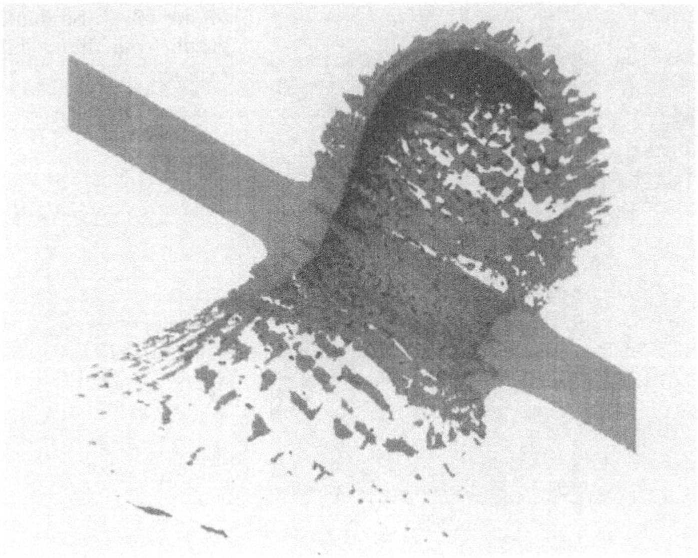

Figure 12.42. Edge view of the target at 11.64 μs (the darker material is copper).

on one side of the symmetry plane has been removed revealing the interior of the crater and the bubble of debris ejecta. Much of the copper (the darker material) is continuing to thin and has not yet undergone instability breakup. On the other hand, all of the ejecta steel has fragmented by this time. The size and shape of the fragments reflect in part the way the material is convected by the code, and the method of representing the data (half volume fractions). The fragment sizes were not computed in this instance. The cross-section (Figure 12.42) also shows the presence of spall rings formed at the front and rear of the target, as well as cracks penetrating the target. Hydrodynamic spreading of the copper occurs over the interior of the expanding steel debris shortly after impact and, late in time, interface separation between the steel and copper is achieved because of the difference in mechanical impedance. Artificial contributions to this separation may also come from the interface tracking algorithm in the code. Recent code modifications eliminate this material separation (Hertel, 1992).

Another representation of the oblique copper-on-steel impact calculation was provided by constructing a projected density normal to the symmetry plane at the final computational time of 11.6 μs. This simulated radiograph was then compared directly with the measured flash x-ray record at the corresponding time for the oblique impact experiment (Figure 12.43). The calculated profile of both the forward and backward ejected debris along with the mass density distribution throughout the debris cloud were in excellent agreement with the measured radiographic data. This nearly perfect overlay of the material boundaries indicates that the material motion in the

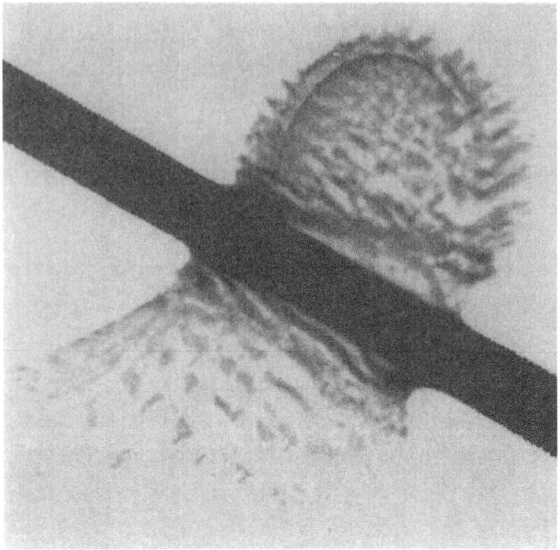

Figure 12.43. Simulated radiograph from the calculation, at 11.6 μs.

impact has been properly computed. The presence of the larger mass of material at the top and left of the debris cloud is consistent with the densities of material observed in the radiograph (Figure 12.32).

The upward directed backsplash jet clearly observed in the x-ray image (Figure 12.32) is also prominent in the numerical simulation (Figure 12.42 and Figure 12.43). Although rapid instability particulation is observed, trajectories and velocities of about 7.5 km/s are consistent with the radiographic record. (Velocity of the observed leading jet fragments in Figure 12.32 was about 6 km/s.) The jet is comprised of both copper and steel material.

A final view of the impact event at 11.6 μs is a stereo (cross-eye) perspective provided by the pair of images in Figure 12.44. Note that the debris bubble is indeed hollow, and that there is copper mixed with steel in the impact-surface ejecta.

Due to entropy production in the primary shockwave in the copper projectile, substantial temperature increases occur. The calculated residual temperatures in the copper (at 11.6 μs) are below incipient melt (~1356 K) for nearly all of the mass (Figure 12.45). Approximately 0.16 g (13%) of the copper remains in the melted state. Only traces of the steel are calculated to remain melted.

At approximately 5 μs into the calculation, breakup of the behind target ejecta initiates (cf. Figure 12.41). Breakup is not a consequence of the tensile fracture model. Rather, it is thought to be a result of kinematic instabilities in the rapidly thinning shell of ejecta probably stimulated by numerical perturbations. Fragmentation of the copper occurs later in time than the steel, consistent with the expected effect of the substantially lower plastic flow stress of the copper on instability breakup. The trend of material to align on

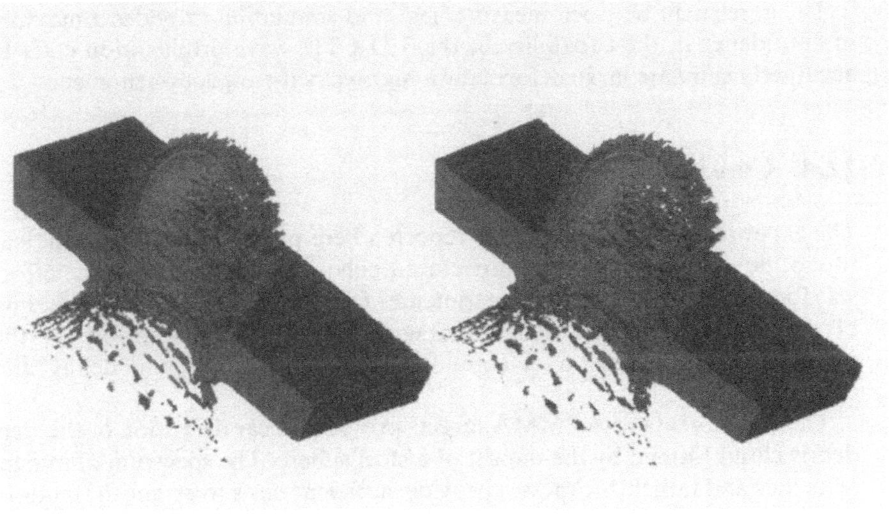

Figure 12.44. Stereo perspective of the oblique impact at 11.6 μs.

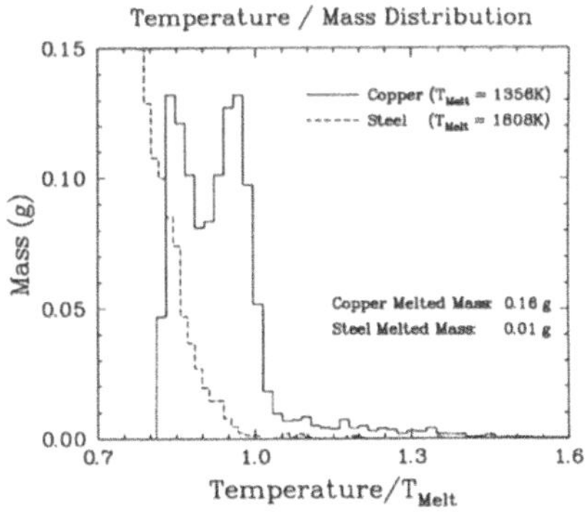

Figure 12.45. Distribution of copper mass with temperature at 11.6 μs.

45° planes is a consequence of preferential material convection in the interface tracking algorithm.

Fractures which initiated at later times (Figure 12.41) in the plane of the 4340 steel target plate, on the other hand, are a consequence of stress states exceeding the tensile fracture criterion. These fractures are consistent in trend with the fractures observed in sectioned target plates used in the impact experiments. The calculated crater hole size of about 17 mm at 11.6 μs is in excellent agreement with the measured hole size (see Table 12.8). Note the dramatic contrast of the crater in this metal target with that of the PMMA target (Figure 12.24).

The agreement between measurement and simulation provides a measure of confidence in the capability of the 3-D CTH wave propagation code to accurately simulate large deformation high-velocity impact phenomena.

12.4. Conclusions

The experiments and calculations reported here provide insight into the nature of debris formed during impact and penetration of a target. The effort was focused on providing an experimental foundation and an understanding of material response in dynamic fragmentation for the purpose of the validation and expansion of computational fragmentation modeling and calculation.

The experiments with PMMA targets provide a clear definition of the steel debris cloud formed by the impact of a steel sphere. The spectrum of impact velocities and target thicknesses provide sufficient basis to establish trends in debris cloud formation. Some aspects of the fragmenting steel sphere data are represented quite well by the calculations, particularly the axial velocity decrease and the radially divergent velocity imparted to the sphere. The former property, the axial velocity decrease, is primarily a function of the hydrodynamics of impact; the radial velocity acquired by the particles of the sphere is more sensitive to the modeling of the fracture, particularly the time at which the material separates. The ability to calculate detailed fragment size distributions is the first step in defining quantitative properties of debris clouds for subsequent interactions. The quantitative comparisons of debris characterization will require some additional refinement of the data from the radiographs. The simulations have also demonstrated that there is additional work required on the fracture model used here to more precisely determine the time of fracture and the condition of the steel at breakup.

High-resolution radiographic images of ejecta motion and fragmentation caused by impact of copper spheres on steel plates at 3 to 5 km/s are a principal result of the present study. The images were obtained at times after impact which are amenable to, and for the purpose of, verification of computational simulation of the impact event. Simulations with the 3-D Eulerian hydrocode CTH were in good agreement with all major features of the radiographic data. The study uncovered minor numerical difficulties for which improvements were implemented over the course of the research.

The remarkably good agreement in the size and character of material breakup in the computational simulation with radiographic data remains incompletely understood. A reasonable understanding of the plastic instability breakup mechanisms of expanding debris bubbles and stretching jets is beginning to emerge (e.g. Grady, 1987; Romero, 1989). Whether instability breakup of material in the computational simulation captures a measure of the correct physical instability mechanisms, or is strictly a numerical instability issue, remains to be worked out.

Passive fragmentation diagnostics were successfully implemented in the present experimental study in terms of witness foils and low-density foam fragment capture devices. Interrogation with static radiography and image processing methods provided statistical size, number density and trajectory data. Implementation of energy-based fragmentation models within the 3-D CTH hydrocode has provided a tool for the analysis of debris breakup in the hypervelocity impact event (Kipp, et al., 1993b).

The extrapolation of the debris characterization to subsequent targets is a step that requires additional technique development. Other tasks remaining which would enhance our understanding of the debris formed in these impact events include:

1. Image processing of fragmentation radiographs to extract statistical fragment size data,
2. Measure spall crater profiles on representative primary target impact plates,
3. Completion of several tests in which recovery of the fragment debris will be accomplished and statistical size analysis performed on the recovered data.

12.5. Acknowledgments

This work was performed at Sandia National Laboratories, supported by the U.S. Department of Energy under contract #DE-AC04-76D00789.

The authors gratefully acknowledge the support of Sandia National Laboratories funding through a Laboratory Directed Research and Development grant for the fiscal years 1990 to 1992. The expert help of R. Moody in the experimental aspect of this work is acknowledged, as is the support of Millie Elrick in the use of CTH. The radiographs were image-processed by David Johnson.

12.6. References

Ang, J.A. Sandia National Laboratories, Personal communication (1992).

Backman, M.E. and Goldsmith, W. The mechanics of penetration of projectiles and targets. *Int. J. Engng. Sci.* **16**, 1–99 (1978).

Backman, M.E. and Finnegan S.A. A phenomenological investigation of the impact of compact projectiles against plates at speeds up to 3 km/s. Proceedings of Eighth International Symposium on Ballistics, October 23–25, 1984, Orlando, Florida.

Bertholf, L.D., Buxton, L.D., Thorne, B.J., Byers, R.K., Stevens, A.L., and Thompson, S.L. Damage in steel plates from hypervelocity impact. II. Numerical results and spall measurement. *J. Appl. Phys.* **46**, 3776–3782 (1975).

Brown, W.K. *J. Astrophys. Astr.* **10**, 89–112 (1989).

Cour-Palais, B.G. Hypervelocity impact in metals, glass, and composites. *Int. J. Impact Engng.* **5**, 221–237 (1987).

Englman, R., Jaeger, Z., and Levi, A. Percolation theoretical treatment of two-dimensional fragmentation in solids. *Phil. Mag. B* **50**, 307–315 (1984).

Glenn, L.A. and Chudnovsky A. Strain energy effects on dynamic fragmentation. *J. Appl. Phys.* **59**, 1379–1380 (1986).

Grady, D.E. Local inertia effects in dynamic fragmentation. *J. Appl. Phys.* **53**, 322–325 (1982).

Grady, D.E. Fragmentation of rapidly expanding jets and sheets. *Int. J. Impact Engng.* **5**, 285–292 (1987).

Grady, D.E. The spall strength of condensed matter. *J. Mech. Phys. Solids* **36**, 353–384 (1988).

Grady, D.E. Particle size statistics in dynamic fragmentation. *J. Appl. Phys.* **68**, 6099–6105 (1990).

Grady, D.E. and Kipp, M.E. Geometric statistics and dynamic fragmentation. *J. Appl. Phys.* **58**, 1210–1222 (1985).

Grady, D.E. and Kipp, M.E. Fragmentation of solids under dynamic loading. In: *Structural Failure* T. Wierzbicki and N. Jones (Eds.), John Wiley & Sons, New York (1989).

Grady, D.E. and Kipp, M.E. Experimental and Computational Simulation of the High-Velocity Impact of Copper Spheres on Steel Plates. *Int. J. Impact Engng.* **15**, 645–660 (1994).

Grady, D.E. and Passman, S.L. Stability and fragmentation of ejecta in hypervelocity impact. *Int. J. Impact Engng.* **10**, 197–212 (1990).

Grady, D.E., Dunn, J.E., Wise, J.L., and Passman, S.L. Analysis of prompt fragmentation. Sandia National Laboratories Report, SAND90-2015 (1990a).

Grady, D.E., Swegle, J.W., and Ang, J.A. Analysis of prompt fragmentation Oct. 89–Sept. 90. SAND91-0483 (1990b).

Herrmann, W. and Wilbeck, J.S. Review of hypervelocity penetration theories. *Int. J. Impact Engng.* **5**, 307–322 (1987).

Hertel, E.S., Jr. A comparison of the CTH Hydrodynamics Code with experimental data. Sandia National Laboratories Report, SAND92-1879 (1992).

Hohler, V. and Stilp, A.J. Hypervelocity impact of rod projectiles with L/D from 1 to 32. *Int. J. Impact Engng.* **5**, 323–331 (1987).

Johnson, W.A. and Mehl, R.F. Reaction kinetics in processes of nucleation and growth. *Trans. A. I. M. M. E.* **135**, 416–458 (1939).

Johnson, J.N. and Addessio, F.L. Tensile plasticity and ductile fracture. *J. Appl. Phys.* **64**, 6699–6712 (1988).

Johnson, G.R, Stryk, R.A., Holmquist, T.J., and Souka, O.A. Recent EPIC Code developments for high velocity impact: 3D element arrangements and 2D fragment distributions. *Int. J. Impact Engng.* **10**, 281–294 (1990).

Kipp, M.E. and Grady, D.E. Dynamic fracture growth and interaction in one dimension. *J. Mech. Phys. Solids* **33**, 399–415 (1985).

Kipp, M.E., Grady, D.E., and Swegle, J.W. Experimental and numerical studies of high-velocity impact fragmentation. Sandia National Laboratories Report SAND93-0773, August (1993a).

Kipp, M.E., Grady, D.E., and Swegle, J.W. Numerical and experimental studies of high-velocity impact fragmentation. *Int. J. Impact Engng.* **14**, 427–438 (1993b).

McGlaun, J.M., Thompson, S.L., and Elrick, M.G. CTH: a three-dimensional shock wave physics code. *Int. J. Impact Engng.* **10**, 351–360 (1990).

Melosh, H.J., Ryan, E.V., and Asphaug, E. Dynamic fragmentation in impacts:

hydrocode simulation of laboratory impacts. *J. Geophys. Res.—Planets*, **97**, No. E9 (1992).

Piekutowski, A.J. Properties of largest fragment produced by hypervelocity impact of aluminum spheres with thin aluminum sheets, AIAA 92-1588, AIAA Space Programs and Technologies Conference, March 24–27, 1992, Huntsville, Alabama.

Romero, L. The instability of rapidly stretching jets. *J. Appl. Phys.* **65**, 3006–3016 (1989).

Schuler, K.W. and Nunziato, J.W. The dynamic mechanical behavior of polymethyl methacrylate. *Rheol. Acta* **13**, 265–273 (1974).

Smith, V. Kaman Sciences Corporation, Personal communication (1989).

CHAPTER 13

Simplified Models of Fracture and Fragmentation

D.R. Curran and L. Seaman

13.1. Introduction

Current engineering models of fracture and fragmentation usually rely almost exclusively on analytical fits to empirical "field data," so it is necessary to generate a new data set for each new application. We will not discuss such models here, even though they have proven very useful in the weapons effects community. Instead, we will concentrate on emerging physics-based models that, at least to some degree, describe the nucleation of damage at weak spots in the material, the growth of the damage with associated material softening, and the ultimate coalescence of the damage to form a fragment size and velocity distribution. The input to such models is based on material properties measured in the laboratory, and the models therefore hold the promise of being much less dependent on expensive (and sometimes unobtainable) field data than are the current engineering models.

In complex structural applications, the models are used in finite element "hydrocodes" to predict the structural response to dynamic loads. The models in the hydrocodes require a description of damage-induced material softening to predict the locations in the structure where time-dependent localized failure will occur, and to describe the resultant time-dependent load-shedding process that affects the response of neighboring elements in the structure. In addition, the models require a description of the fragmentation process to predict hazards and fragment loads on adjacent objects.

Current practice in approaching such problems has heretofore necessarily been somewhat ad hoc. That is, the hydrocode operator will typically try first the simplest failure model available and compare the results with experimental results. If the agreement is poor, the next simplest model will be tried, and so on, until some desired degree of agreement is attained.

Recent and ongoing increases in computing power now make it possible to take a more physically-based approach, the approach that is the subject of this chapter.

13.2. Fracture and Fragmentation Process

When a material element experiences sufficiently high combined tensile and shear stresses or compressive and shear stresses, microscopic cracks, voids, or shear bands nucleate at heterogeneities. Figures 13.1 through 13.3 show some examples.

At higher loading rates, more damage is formed, partly as a result of the higher "viscous" stresses and partly because of the more nearly adiabatic conditions [Curran et al., 1987]. After nucleation, the microscopic damage grows to coalescence.

During both the nucleation and growth phases, the material is "softened" as a result of two related effects. First, the load-bearing area in a material element is reduced, so the remote stresses inferred from the forces on the element faces are less than the local stresses in the intact material. Second, the damage strain rate (due to opening and extending cracks, slipping and extending shear bands, or growing voids) subtracts from the applied total strain rate to reduce the elastic strain rate and the associated stress rate. Modeling these softening effects is discussed in detail by Curran et al. [1987].

Both the nucleation and growth processes are time-dependent and can be described on the average by the evolution of a damage function defined as the microscopic crack, band, or void size distribution function, as shown schematically in Figure 13.4. In Figure 13.4, $N_g(R)$ is the number of cracks per unit volume that have radii greater than R.

As N_g evolves towards more and larger cracks or bands, it reaches a level where coalescence occurs, producing a fragment size distribution that is simi-

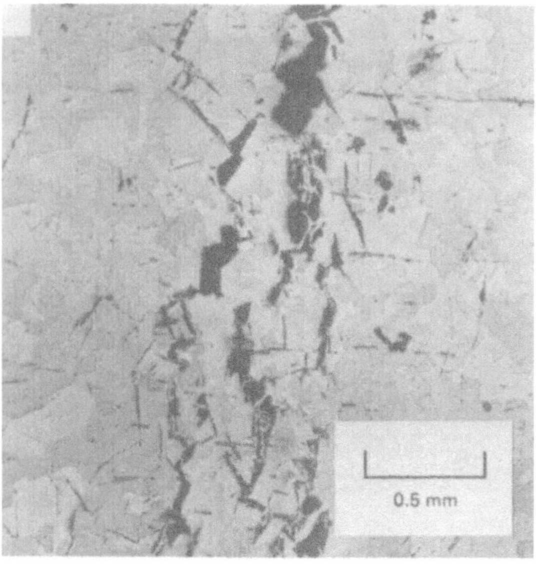

Figure 13.1. Photomicrographs of microcracks in Armco iron.

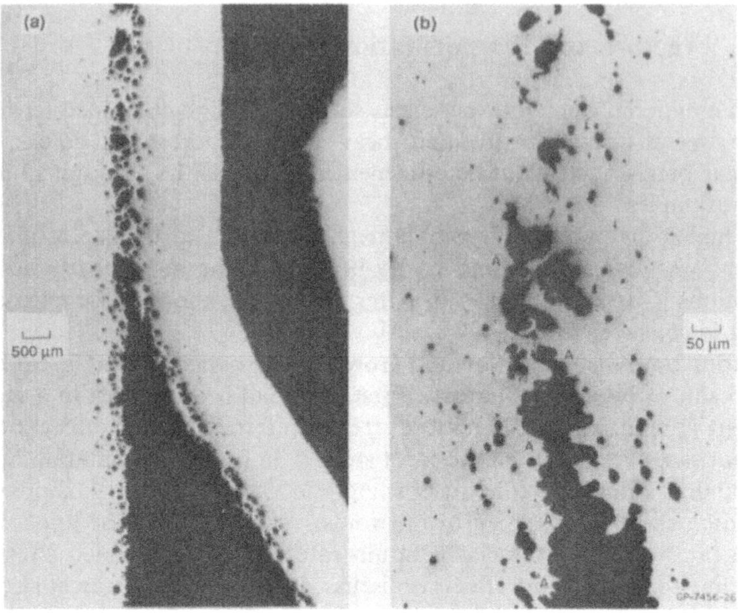

Figure 13.2. Ductile cracks in aluminum. (a) Ductile crack propagation by void coalescence, (b) Tip of ductile crack shown in (a) at higher magnification. Material failure by void coalescence due to necking of the regions separating the voids is apparent near points A.

Figure 13.3. 4340 Steel fragments. Polished and etched cross sections of individual projectile fragments from impact experiment, showing coalesced adiabatic shear zones at their boundaries.

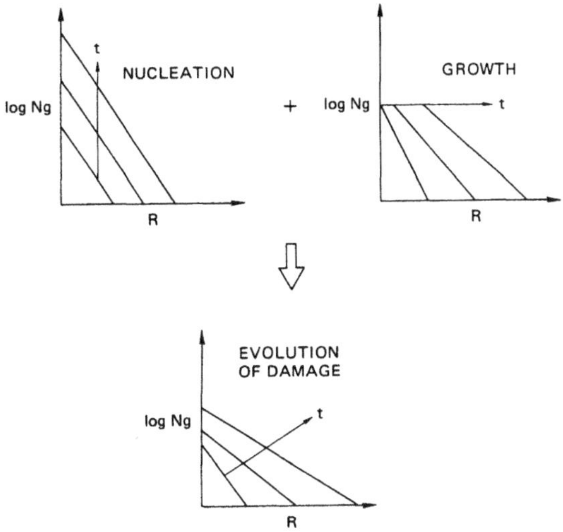

Figure 13.4. Evolution of microscopic damage.

lar to the crack or band size distribution at the onset of coalescence. In general, fragmentation begins as the microdamage begins to coalesce. Coalesced cracks and shear bands tend to form many small, roughly equiaxed fragments, whereas the nearly spherical voids ("cavitation") tend to form "tear surfaces" that produce fewer and larger fragment "petals."

The geometry of the fragmenting object has a strong effect on the fragmentation. If the microscopic damage occurs near free surfaces, as in the case of exploding thin-walled rings or cylinders, then the microscopic cracks, bands, or voids may tend to coalesce with the free surfaces before they coalesce with each other. In that case, the classical analysis of Mott [1947] applies. Figure 13.5 shows a double exposure flash x-ray picture of a fragmenting steel ring, and Figure 13.6 compares the fragment size distribution measured from the x-ray film with the distribution predicted by Mott's model. The measured fragment size distribution was obtained by counting fragments within ranges of thicknesses, such as between 0.28 mm and 0.56 mm; the computed results were organized in a similar manner. We see that Mott's model predicts the main features of the distribution. (Integration of the distribution function of Figure 13.6 from infinity to R yields an N_g function with an approximately exponential form like those in Figure 13.4.)

Mott's model assumes a random distribution of microscopic flaws in the ring wall with an associated distribution of strains-to-failure. The activated flaws are assumed to coalesce with the ring wall surface instantaneously, producing rather slowly diffusing unloading waves that protect adjacent material from further damage. Thus, the fragment size distribution depends on the initial flaw size distribution and the competition between loading rate and the unloading wave speed.

In more complicated geometries, where most of the fragmentation results

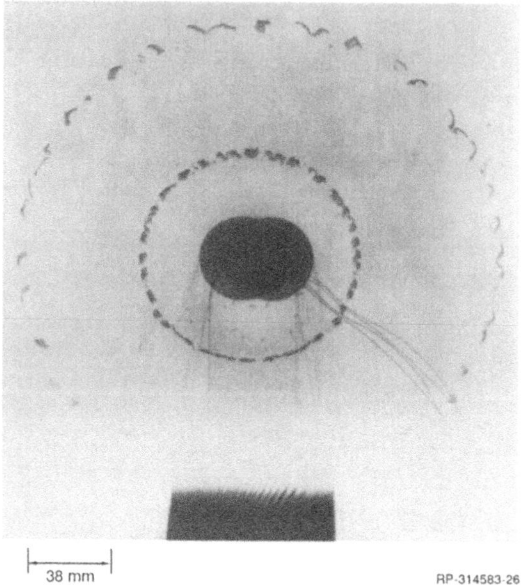

Figure 13.5. Flash x-ray pictures of expanding fragment clouds.

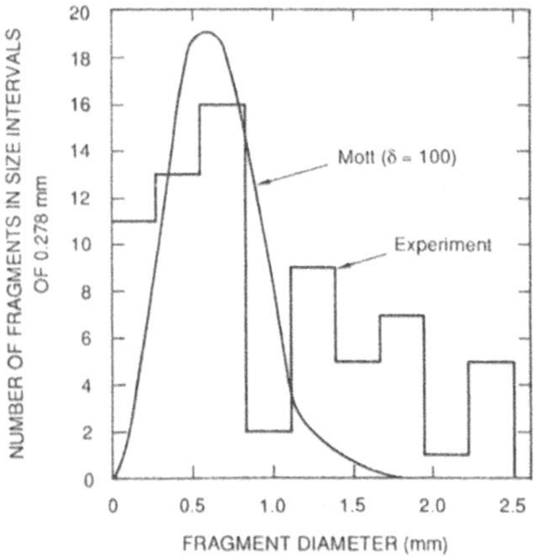

Figure 13.6. Fragment size distribution.

from microscopic cracks or shear bands coalescing with each other rather than with free surfaces, the unloading rate is usually governed not by the unloading wave speed but by the growth (extension) rates of the cracks or bands, because the equilibrium size of the unloaded region around each crack is proportional to the crack or band size, and the unloading wave speeds are normally faster than the crack or band extension speeds.

However, the basic fragmentation process remains that postulated by Mott: the cracks or bands nucleate at weak spots in the material, and the final fragments consist of "unloaded" material around the coalesced cracks or bands.

13.3. Computer Models of the Fracture and Fragmentation Process

13.3.1. Sophisticated Models

During the past 20 years, workers in the failure modeling community have developed a family of material fracture and fragmentation models that describe the evolution of microscopic damage from nucleation to coalescence. A review was written by Curran et al. in 1987, and more recent discussions of constitutive relations for the evolution of microdamage are given in a 1991 compendium by Desai et al. and in a text by Nemat-Nasser and Hori [1993].

The present authors and colleagues have developed a family of such models with the name NAG/FRAG (Nucleation And Growth of damage to FRAGmentation) [Curran et al., 1987]. The family has three main members, SHEAR (for shear banding), BFRACT (for brittle cracking), and DFRACT (for ductile void growth). Each of the above family members has modules of varying complexity to account for cases where, for example, rate dependence or anisotropy is important or can be neglected. However, in all cases the NAG/FRAG models describe in detail the evolution of the damage function shown in Figure 13.4, and they require as input laboratory-generated data for the nucleation rate of inherent microscopic flaws in the material as well as data for post-nucleation growth (extension) rate of the flaws.

A key laboratory experiment is the "controlled fragmenting cylinder (CFC)" test, in which thick-walled cylinders are explosively expanded until stopped by a massive confining cylinder, which freezes a distribution of shear bands in various stages of growth for quantified examination [Curran et al., 1987]. For ductile voids and brittle cracks, on the other hand, plate impact experiments are used [Curran et al., 1987].

Both the nucleation and growth rates are functions of stress and/or strain. Furthermore, the evolving damage is allowed to soften the material and relax the stresses. A more detailed discussion of these processes is given later when we present simplifications of the NAG/FRAG models, and in the Appendix, where we discuss well-posedness and stability questions.

As the cracks or bands intersect each other or specimen boundaries, they form fragments, and the NAG/FRAG models describe this process in an average way to predict fragment size distributions. Figures 13.7 and 13.8 show NAG/FRAG calculations of a naturally fragmenting warhead, and a comparison of the computed and observed fragment size distributions [Zukas et al., 1982]. Figures 13.9 and 13.10 show similar results for debris from a long steel rod impacting a steel target plate at 1550 m/s [Dullum and Seaman, 1985; Erlich et al., 1980].

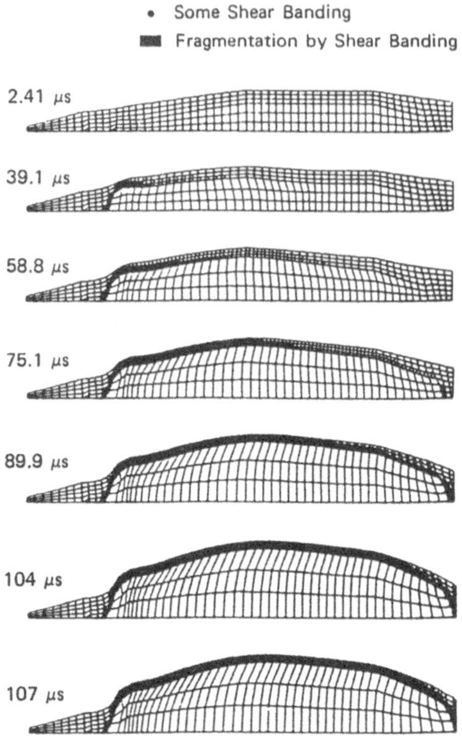

Figure 13.7. Progress of detonation and shear band damage computed in the two-dimensional simulation of fragmenting projectile.

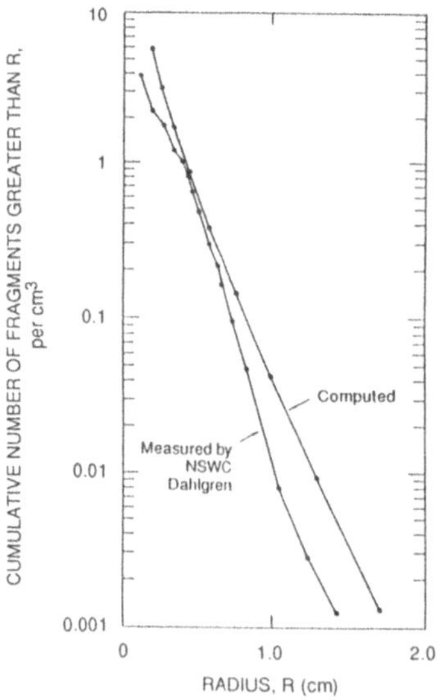

Figure 13.8. Comparison of computed and observed fragment size distribution for sample fragmenting projectile.

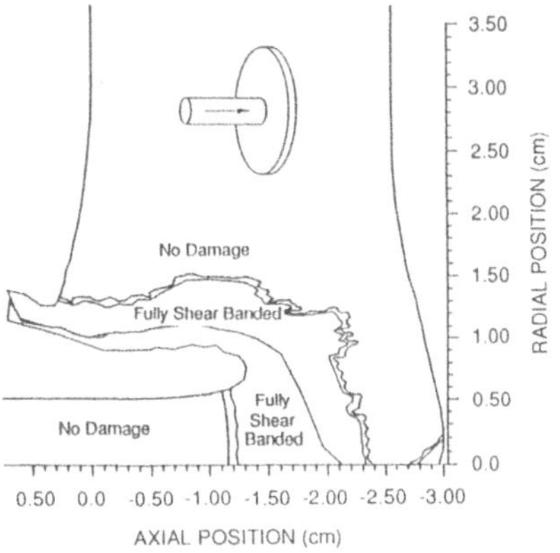

Figure 13.9. Long rod and plate configuration at 28.0 μs for a TDL simulation of a 1550 m/s impact.

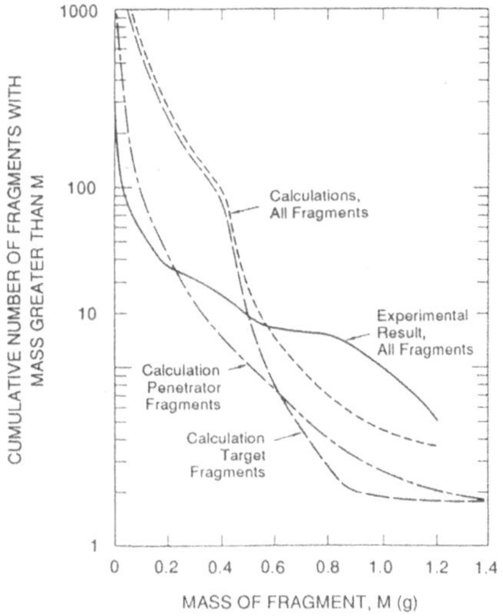

Figure 13.10. Comparison of measured downrange fragment size distribution from experiment shown in Figure S1 and distributions calculated from computer simulation.

An advantage of the NAG/FRAG approach is that the input parameters to the models can be related to the material's microstructure, thereby giving the materials scientist insight for improving material performance. Two disadvantages, however, are the need to perform detailed analysis of a significant number of laboratory material property tests to obtain NAG/FRAG model parameters, and the need to carry along 20 or more damage parameters in hydrocode calculations. For large-scale calculations, simplifications of the NAG/FRAG approach are very useful. In the remainder of this chapter, we discuss several such simplifications.

13.3.2. Simplifications

We will discuss two types of simplification, the first involving the use of simple physical analyses to reduce the scope of the laboratory experimental program required to provide input to sophisticated models and the second involving simplifications of the models themselves. An example of the former is when prior knowledge of the flaw size distribution in virgin material allows the nucleation rate function to be inferred a priori. An example of the latter is when the material is "overdriven" so that the microdamage coalesces quickly to form fragments, and the location of the fragmenting material in the structure is known a priori, so that simplified "end state" fragmentation models are applicable.

In the following paragraphs we discuss such simplifications for the three main microstructural failure modes in homogeneous solids: shear banding, brittle cracking, and ductile void growth.

13.3.2.1. Simplified Shear Banding Models

Analyses to reduce required laboratory data

For fragmenting rounds and back-of-the-armor debris for heavy steel armor, adiabatic shear banding is the main fragmentation mechanism. As described by the NAG/FRAG SHEAR model, the bands nucleate at microscopic heterogeneities with the geometry of penny-shaped or half-penny-shaped cracks and grow (extend) to coalescence [Curran et al., 1987]. As the bands grow in number and size, the average shear stresses across the bands decrease owing to the reduction of the area of intact material (assuming that the bands are thermally softened).

The size of shear bands upon nucleation under dynamic conditions probably depends on strain rate and heat conduction, although some cases may be dominated by geometric softening [Deve et al., 1988; Harren and Asaro, 1989; Meyers et al., 1993]. That is, if the flaw is considered to be penny-shaped with a radius R, adiabatic heating at the band boundary can only occur if the strain concentration at the boundary remains hotter than its surroundings during the nucleation time. This is expressed by the inequality

$$R \leq \sqrt{\frac{2.5k\gamma_{cr}}{\dot{\gamma}}}, \qquad (13.1)$$

where k is the thermal diffusivity, γ_{cr} is the critical strain for localization, and $\dot{\gamma}$ is the strain rate. Thus, the right-hand side of Eq. (13.1) is the minimum size flaw that can be activated. It is also the minimum possible fragment size, corresponding to instantaneous coalescence of the smallest activated flaws.

For the fragmenting munition experiment of Figures 13.7 and 13.8, where the average strain rate was about 2×10^4/s and the critical localization strain is taken to be about 0.2, the above formula gives a minimum activation band length ($2R$) less than 50 μm, in agreement with observations from experiments.

We may also estimate the average band spacing on nucleation. Grady and Kipp [1987] have estimated the average shear band spacing on the basis of an analysis of the competition between thermal diffusion from the heated band and the unloading wave speed. Assuming a minimum work principle, they derived an average band spacing, $2s$, given by the formula

$$2s = 2[(9c^2\rho k)/(\tau_y \alpha^2 \dot{\gamma}^3)]^{1/4} \tag{13.2}$$

where c is the specific heat, ρ is the density, k is the thermal diffusivity, τ_y is the unsoftened shear yielding stress, and α is a linear thermal softening coefficient defined by

$$\tau = \tau_y[1 + \alpha(T_0 - T)], \tag{13.3}$$

where T is the temperature.

Grady and Kipp interpret $2s$ from Eq. (13.2) to be the average fragment size (diameter). However, this can only be true in the special cases where the nucleated bands intersect specimen free boundaries before they coalesce (as in thin-walled, expanding rings or cylinders, when $2s$ is the average circumferential fragment dimension), or when all of the nucleated bands coalesce with neighboring bands. In fact, it is often observed that fragments contain uncoalesced bands. Thus, Eq. (13.2) should tend to predict more numerous and smaller fragments than are observed.

For application to naturally fragmenting munitions or back-of-the-armor debris problems, the complete fragment size distribution is desired, not just the average fragment size. Also, as discussed above, all nucleated bands do not coalesce to form fragments. As shown in Figures 13.8 and 13.10, the NAG/FRAG approach has the capability to take these complexities into account while simulating the fragmentation process. It would be of great value if the NAG/FRAG parameters could be estimated without extensive laboratory experimentation.

Some recent attempts in this direction are described by Grady and Kipp [1985] and Curran et al. [1987]. Grady and Kipp point out that a random Poisson fragmentation process leads to an exponential distribution in fragment size, a distribution often seen in practice [Curran et al., 1987]. Figures 13.8 and 13.10 show this feature as well. In fact, the distribution of Figure 13.8 is well described by the Poisson distribution

$$N_g(R) = N \exp(-R/R_1) \tag{13.4}$$

where $N_g(R)$ is the number of fragments per unit volume with radii greater than R, N is the total fragment concentration, and R_1 is the characteristic size of the distribution. For the data of Figure 13.8, N is about 20 fragments per cubic centimeter, and R_1 is about 0.15 cm.

An interesting feature of this distribution is that R_1 is the average frag-

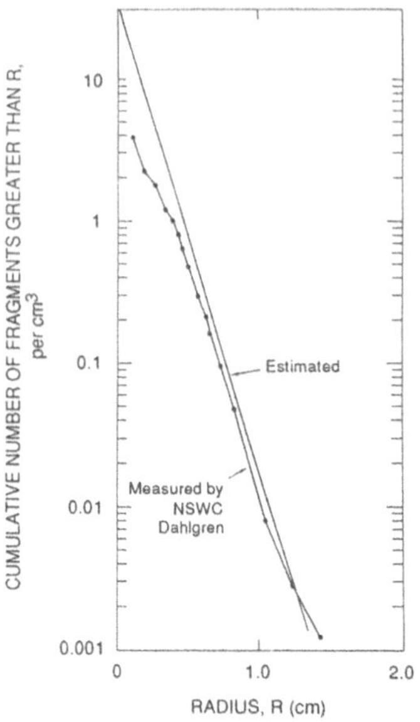

Figure 13.11. Comparison of computed and observed fragment size distribution for sample fragmenting projectile.

ment size, and the average fragment volume (assuming spherical fragments) is $8\pi R_1^3$. If we assume that a rough estimate of the total fragment concentration, N, is given by

$$N \cong 1/8\pi R_1^3 \tag{13.5}$$

then, for $R_1 = 0.15$ cm (the actual average size), $N = 12$ fragments per cubic centimeter.

Figure 13.11 compares the above distribution with the experimental one from Figure 13.8. The agreement is clearly excellent. Thus, if the average fragment size can be estimated accurately, it appears that we can make a remarkably good estimate of the whole distribution for expanding rings and cylinders. More comparisons of this type are needed to explore this possibility.

As discussed above, in other cases the geometry is more complex and the fragmentation process may be more complicated. The material may experience complicated strain histories for which it is not easy to specify an average strain rate during the fragmentation process. An example is armor penetration, where hydrocode calculations predict complicated strain paths. Also, details of the softening process may be needed, in which case the above end-state fragmentation model cannot be used.

However, even in such cases, it may be possible to use a variation of the above simplified approach to reduce the experimental effort required to provide input to the sophisticated NAG/FRAG model. As discussed earlier, the NAG/FRAG models contain formulae for the nucleation and growth rates for the microscopic shear bands. For the nucleation rate, we interpret the Grady/Kipp spacing [Eq. (13.2)] to be the spacing between nucleation sites.

In their 1987 paper, Grady and Kipp also derived a formula for the minimum localization time:

$$t_n = [36\rho^3 c^2 k/\tau_y^3 \alpha^2 \dot{\gamma}]^{1/2} \quad (13.6)$$

Equations (13.5) and (13.6) are combined to obtain the nucleation rate:

$$dN/dt = \beta N/t_n = \beta(0.0013)(\tau_y/\rho)^{9/4}[(\alpha^{5/2})/(c^{5/2}k^{5/4})]\dot{\gamma}^{11/4} \quad (13.7)$$

where β is the number of bands required to make a fragment (3 to 4 for roughly equiaxed fragments).

The above strain rate dependence to the 11/4 power compares well with the dependence to the 5/2 power inferred earlier from prior CFC experiments [Curran et al., 1987]. For the steel properties used above, the multiplier of the strain rate term in the above equation is 3.3×10^{-3}, where the strain rate is measured in inverse seconds.

We can also use Eq. (13.6) to estimate the growth (extension) rates of the shear bands. As noted by Curran et al. [1987], as the shear band front advances it carries a region of enhanced strain whose size is proportional to the band radius, provided that material rate effects and inertial effects can be neglected. If we consider that shear band growth is a successive nucleation process at the band tip, then the growth rate is given by

$$\dot{R} = bR/t_n = bR[(\alpha\tau_y^{3/2})/(6\rho^{3/2}ck^{1/2})]\dot{\gamma}^{1/2}, \quad (13.8)$$

where R is the band radius, and b defines the size of the enhanced strain field at the band tip and is on the order of unity. With the steel properties used above, the multiplier of the strain rate term in (Eq. 13.8) is 1.4×10^4, where the strain rate is measured in inverse seconds. Equation (13.8) has a weaker strain rate dependence than the linear dependence inferred from prior CFC experiments [Curran et al., 1987], but it predicts higher growth rates at the strain rates of interest.

To complete the input to a NAG/FRAG model, the size distribution at nucleation must be specified. For a physically based approach, we choose Eq. (13.4) for the shape of the distribution, with R_1 based on microstructural heterogeneities if possible, with a lower cutoff given by Eq. (13.1).

In short, combining the above analyses with the NAG/FRAG model presents the possibility of providing a detailed hydrocode model of the evolution of damage to fragmentation, but with a significant reduction in the laboratory data required as input. To further investigate this possibility, additional NAG/FRAG calculations using Eqs. (13.7) and (13.8) are needed for comparison with available data from CFC experiments and from range tests.

Simplification of the Models

Grady and Kipp's Eq. (13.2) is an end-state fragmentation model for shear banding, and it often shows good agreement with experimental data. For example, the fragmenting munition of Figures 13.7 and 13.8 is basically an expanding cylinder where the average strain rate was on the order of 2×10^4/s, and thus Eq. (13.2) can be used to estimate the average fragment size. Using a yield strength of 1.5 GPa, an α of $0.0013/K$, and handbook values for the other properties yields an average fragment radius of 0.15 cm. Comparison with Figure 13.7 shows that this is very close to the observed value.

It is also desirable to develop simplified stress relaxation (softening) models for shear banding. In the sophisticated SHEAR model, a multiplane plasticity model similar to that used by Peirce, Asaro, and Needleman [1983], Deve et al. [1988], and Harren and Asaro [1989] is used to allow nucleation, growth, and coalescence of shear bands on any of nine separate planes in the material element. Softening occurs on each plane separately as the number and extent of the bands (with assumed zero shear strength) accumulate and reduce the load-bearing area fraction on the plane. Thus, SHEAR produces a very anisotropic behavior in many cases.

However, in cases where mainly only one plane is active (as in the fragmenting cylinders discussed earlier), simplifications may be possible. One possibility is to define an estimated time-to-failure as the time on the average for a shear band to extend to the next band. Such a time-to-failure can be estimated by dividing s from Eq. (13.2) by $\dot{R}$ from Eq. (13.8). Then, after a threshold condition is exceeded, the shear strength can be relaxed exponentially with a characteristic time given by a multiple of the above time-to-failure.

We have not yet thoroughly explored the above approach to a simplified shear banding model, but we have done so for tensile cracking, as discussed next.

13.3.2.2 Simplified Tensile Fracture and Fragmentation Model

Review of Sophisticated Model

Before discussing simplified models of tensile fracture and fragmentation, we will first briefly review the BFRACT [Curran et al., 1987] member of the NAG/FRAG family. BFRACT is a micromechanical model that has been shown in many applications to describe in detail the evolution of microscopic damage and the development of fragmentation in brittle materials. The model requires detailed input regarding the initial flaw size distribution (nucleation rate function) and the microcrack growth rates, and it produces a detailed description of the damage evolution and the resultant fragment size distribution for each material element (computational cell), just as did the shear band model of Figure 13.8.

Figures 13.12 and 13.13 show some of the results of BFRACT calculations for a single Armco iron material element subjected to tensile strain histories at several constant strain rates. For Armco iron, the tensile microcrack nucleation rate was found by analysis of plate impact tests [Curran et al., 1987] to be

Figure 13.12. BFRACT computations of energy partition for fracturing Armco iron.

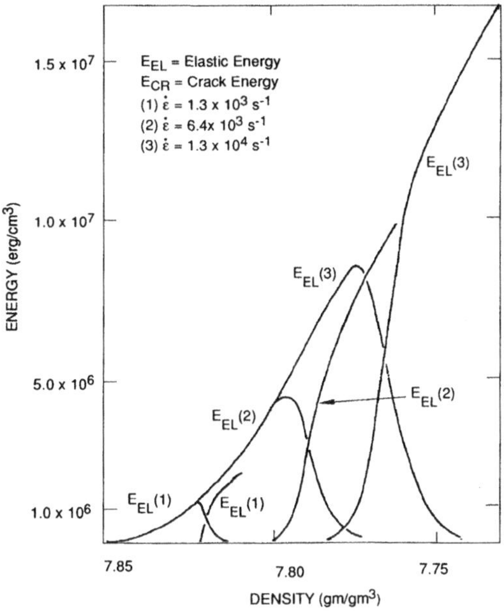

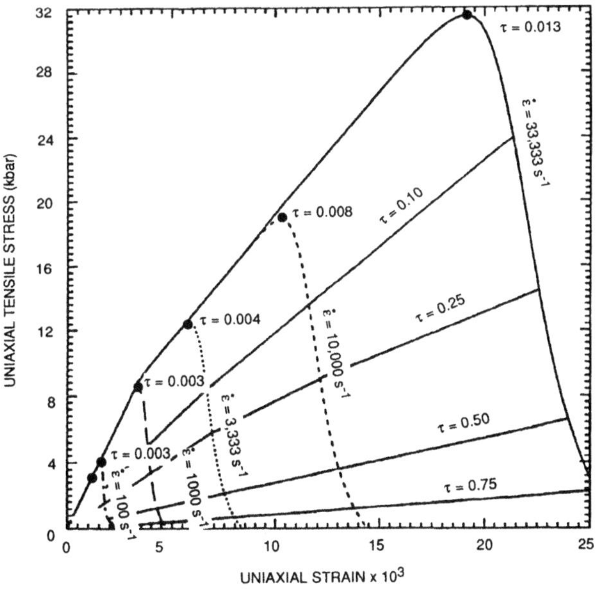

Figure 13.13. BFRACT computations of tensile stress-strain paths for several strain rates in Armco iron.

$$\partial N_t/\partial t = N_{0t} \exp[(\sigma - \sigma_{n0})/\sigma_1] \qquad (13.9a)$$

and the growth rate was found to be a viscous rate until the crack speed reaches the Rayleigh wave speed, C_r

$$\partial R/\partial t = R[(\sigma - \sigma_{g0})/\eta] \qquad (13.9b)$$

where $\sigma_{g0} = K_{1c}R^{-1/2}$; N_{0t}, σ_{n0}, K_{1c}, σ_1, and η in Eq. (13.9) are material constants; and N_t is defined by the observed crack size distribution

$$N_g(R) = N_t \exp(-R/R_1) \qquad (13.10)$$

where the stresses are positive in tension, $N_g(R)$ is the number of cracks per unit volume with sizes greater than R, and R_1 is the characteristic size of the microcrack distribution. For Armco iron, the values of the above parameters are

$$K_{1c} = 3.4 \times 10^7 \text{ dyn/cm}^{3/2}$$

$$N_{ot} = 4.6 \times 10^{12} \text{ no/(cm}^3 \text{ s)}$$

$$\sigma_{n0} = 0.30 \text{ GPa}$$

$$\sigma_1 = 0.46 \text{ GPa}$$

$$\eta = -1667 \text{ dyn-s/cm}^2$$

The initial value of R_1 is $R_{10} = 5 \times 10^{-5}$ cm.

Figure 13.12 shows the energy partition paths in energy-strain space, in which the material element was subjected to purely volumetric tensile strain at the strain rates shown. BFRACT partitions the energies into elastic, plastic, and fracture components:

$$\sigma_{ij} d\varepsilon_{ij} = \sigma_{ij}[d\varepsilon_{ij}^e + d\varepsilon_{ij}^p + d\varepsilon_{ij}^c] \qquad (13.11a)$$

where the crack opening strain is taken from the classical formula for elastic, penny-shaped cracks [Sneddon and Lowengrub, 1969]

$$\varepsilon_{ii}^c = 32(1 - v^2) N_t \sigma R_1^3/E, \qquad (11b)$$

and where E is Young's modulus and v is Poisson's ratio.

In the triaxial loading of Figure 13.12, the plastic strain increment is zero, so only elastic and crack opening strains are present.

Figure 13.12 shows the strong rate-dependence of the fracture and fragmentation energies. We see that higher rates produce higher fracture and fragmentation energies. Furthermore, the energy is first stored elastically, then is fed into work to open and extend the cracks as the number and sizes of cracks increase.

The drop in elastic energy corresponds to stress relaxation as the elastic strain in the intact material (which produces stress) is converted to elastic crack opening strain. Figure 13.13 shows the results of similar calculations (this time under uniaxial tensile strain) in which stress-volume paths are plotted for several high strain rates. The figure also shows values of a damage parameter, τ, which, as discussed later, is a measure in a material element of the ratio of crack opening strain to elastic strain in the intact material ($\tau = 1$ corresponds to complete failure). These calculations illustrate the important fact that the evolution of microdamage begins very early in the loading but the stress does not reach its peak until much later in the process, when a kind of "avalanche" occurs, and the stress drops fairly rapidly thereafter as elastic strain energy in intact material is converted to elastic crack opening energy. This type of behavior is similar to that in dislocation dynamics models of plasticity.

BFRACT thus provides a fairly complete description of tensile fracture and fragmentation. However, it requires extensive input data on crack nucleation and growth. In the remainder of this chapter, we discuss simplifications to BFRACT that are appropriate and useful in many applications.

Analyses to Reduce Required Laboratory Data

For metals, the microcrack nucleation and growth functions obtained from extensive laboratory experiments [Curran et al., 1987] are usually complex enough that a priori derivation of them is a risky business at best. However, for ceramics and other extremely brittle materials, it is perhaps feasible to base estimates of nucleation and growth functions on elastic analyses of nucleation at pre-existing flaws and on growth of penny-shaped cracks in elastic material. In such cases it is then unnecessary to perform laboratory experiments to determine nucleation and growth functions. In the following paragraphs we discuss simplified models in which brittle elastic behavior is assumed and for which only basic mechanical and microstructural properties are required as input.

Simplified Models

Grady and colleagues [for a review, see Grady, 1988] have developed a series of spall and fragmentation models based on a global energy balance approach. These models predict the spall strength, the time-to-failure, and the average fragment size for constant strain rate loading conditions and have proved very valuable in a variety of applications. The energy balance approach has been used for both brittle and ductile fracture and fragmentation [Grady, 1988]. Here, we will briefly summarize the approach for the brittle case.

The approach is based on three main assumptions:

(1) The sum of the elastic strain energy and the nontranslational kinetic energy must equal or exceed the surface energy dissipated in brittle fracture and fragmentation. In most cases, the nontranslational kinetic energy is negligible, so this assumption becomes

$$(\sigma^2/2) \geq (3K_c^2/s), \tag{13.12}$$

where K_c is the fracture toughness and s is the average fragment size (the fragments are assumed to be equisized spheres or cubes of dimension s).

(2) There is a "horizon condition" that specifies

$$s \le 2ct_s, \tag{13.13}$$

where c is the speed of sound and t_s is the time-to-failure, the spall time. Justification for this assumption can perhaps be obtained by looking at Figure 13.13. We see that the onset of microcracking occurs early in the load cycle, avalanches at the peak stress (Grady's σ_s), and then reaches full coalescence at about twice the time at peak stress. Since the fragment size will be roughly equal to the crack size on coalescence, $s/2$ should be about equal to the crack velocity times $2t_s$ (where t_s is the time at peak stress, and we assume that the crack nucleates at a point and propagates radially outward so that the fragment diameter is twice the crack propagation distance). Since the crack velocity in brittle materials is typically about $c/3$, where c is the longitudinal sound speed, we arrive at a formula like Eq. (13.13) with the inequality replaced by an equals sign and the coefficient 2 replaced by 4/3. This agreement with Eq. (13.13) is within the stated approximations of the energy balance approach.

(3) The material is "favorably disposed" (sufficient flaws are present) so that an equals sign can be used in Eq. (13.12).

Then, assuming that the material remains linearly elastic up to the peak stress (which Figure 13.13 shows to be a rough approximation),

$$\sigma_s = \rho c^2 \dot{\varepsilon} t_s \tag{13.14}$$

and one can solve Eqs. (13.12) through (13.14) for σ_s, t_s, and s:

$$\sigma_s = (3\rho c K_c^2 \dot{\varepsilon})^{1/3} \tag{13.15}$$

$$t_s = (1/c)(\sqrt{3}K_c/\rho c \dot{\varepsilon})^{2/3} \tag{13.16}$$

$$s = 2(\sqrt{3}K_c/\rho c \dot{\varepsilon})^{2/3} \tag{13.17}$$

Thus, Eq. (13.17) says that the average fragment size is proportional to the strain rate to the inverse 2/3 power, a result that is consistent with much experimental data for brittle materials.

To get a simplified estimation of the fragment size distribution. we could use the procedure discussed earlier for the shear band-produced fragmentation, i.e., try

$$N_g(R) = (1/8\pi s^3)\exp(-R/s) \tag{13.18}$$

That is, we simply assume a Poisson's distribution of the type given by Eq. (13.11) and let the average fragment size from Grady's model determine both N_t and R_1.

A significant advantage of this simplified approach over the more complex BFRACT model is that no information is required about the flaw size distribution or crack growth rates; only handbook values of elastic modulus and the fracture toughness are required. In practice, two additional parameters are often introduced in the simplified approach: a threshold stress or strain that must be exceeded before fragmentation is allowed, and a characteristic time over which the strain rate is averaged (to prevent spurious dependence on transient excursions in strain rate).

A disadvantage of the above approach is that in many cases of interest the applied strain rate is not constant, a situation leading to ambiguities in finite element code applications as to what strain rate or average strain rate to use in Eq. (13.17). Also, the model, in contrast to the BFRACT model, does not give the history of microdamage evolution and the associated material softening history, information that is sometimes needed to determine deformation localization in structures.

We present here another approach that lies in complexity between the full BFRACT treatment and the simple energy balance approach. In contrast to the energy balance approach, it contains a dependence on microstructural parameters; it also contains a rough description of the softening process.

In this approach we retain the BFRACT dependence on knowledge of the initial flaw size distribution, but we simplify the picture by assuming that the flaws are all pre-existing penny-shaped cracks. Then, for a given flaw size distribution and remotely applied stress, we know which flaws are activated ("nucleated"); they are those with sizes greater than $(K_c/\sigma)^2$, where K_c is the microstructural fracture toughness as before. That is,

$$R_c \geq (K_c/\sigma)^2 \tag{13.19}$$

where R_c is the critical flaw size.

Any initial flaw size distribution is allowed, but for purposes of illustration let us assume the commonly observed distribution

$$N_g(R_c) = N_{tc} \exp(-R_c/R_{c1}) \tag{13.20}$$

The c subscripts refer to the flaws (cracks) that have the potential of growing to coalescence to form fragments.

N_{tc} is the total number of flaws per unit volume. A minimum fragment size can thus be defined; it is

$$s_{\min} = N_{tc}^{-1/3}, \; s_{\min} > R_{c1} \tag{13.21}$$

That is, if all the flaws grew and coalesced, the average fragment size would be the average distance between the flaw centers (and, of course, the distance between flaw centers is assumed to be greater than the initial flaw sizes).

If we now assume that all active cracks will grow to coalescence. the average fragment size will be the spacing between the active crack centers. Thus.

$$s = [N_g(R_c)]^{-1/3} = s_{\min} \exp(R_c/3R_{c1}) = s_{\min} \exp(K_c^2/3\sigma^2 R_{c1}) \tag{13.22}$$

Again referring to the BFRACT example of Figure 13.13, let σ in Eq. (13.22) be the peak stress, σ_s, and let σ_s be given approximately by

$$\sigma_s = \rho c^2 \dot{\varepsilon} t_s \tag{13.23}$$

where t_s is the time at peak stress, as before.

If we again assume that $2t_s$ is given approximately by the time for a crack to propagate the distance $s/2$ at a velocity of $c/3$, then

$$t_s = 3s/4c \tag{13.24}$$

Thus, Eqs. (13.22) through (13.24) can be used to solve for σ_s, t_s, and s, similar to the way that Eqs. (13.15) through (13.17) were used in the energy balance approach. Furthermore, Eq. (13.18) can be used, as before, to approximate the fragment size distribution, since we expect that it will reflect the flaw size distribution of Eq. (13.20).

That is, Eqs. (13.22) through (13.24) can be combined to give

$$[\dot{\varepsilon}/A][(s/s_{\min})\sqrt{\ln(s/s_{\min})}] = 1 \tag{13.25}$$

where $A = 0.8\ K_c/[\rho c s_{\min}\sqrt{(R_{c1})}]$. Eq. (13.25) is plotted in Figure 13.14 and, at strain rates on the order of A, gives a dependence of s on strain rate to about a negative 1/2 power, a relation that is similar to the negative 2/3 power dependence produced by the energy balance approach. However, in the present case, the results depend not only on the modulus and K_c but also on parameters of the initial flaw size distribution, namely, $s_{\min}$ and R_{c1}. The results are dependent on whether there are large numbers of small flaws (small $s_{\min}$ and R_{c1}), or small numbers of large flaws (large $s_{\min}$ and R_{c1}).

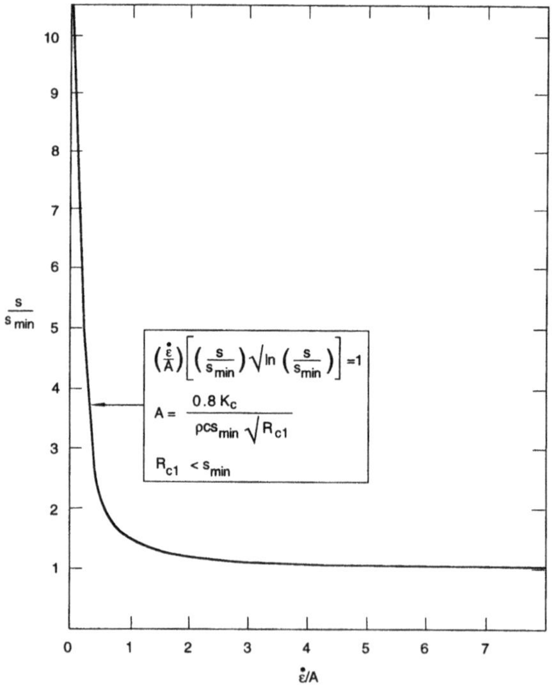

Figure 13.14. Average fragment size as a function of strain rate.

Equation (13.25) represents a modified end-state fragmentation model. We would also like to include a simplified stress-relaxing (softening) algorithm that approximates the sophisticated softening shown in Figures 13.12 and 13.13. A possible way is to allow the stress to rise as though no evolving damage were present to the strain rate dependent value of σ_s given by Eq. (13.23), and thereafter to relax both the elastic moduli and the stresses exponentially with a characteristic time proportional to t_s. That is, after reaching σ_s we relax the stresses according to

$$\sigma = M_0 \Delta\varepsilon \exp(-\Delta t/nt_s) + \sigma_0 \exp(-\Delta t/nt_s) \tag{13.26}$$

where the subscript 0 refers to the value of the parameter at the beginning of a time step Δt, and Eq. (13.26) applies after the BFRACT threshold condition for nucleation of damage is exceeded. The parameters σ, ε, and M represent the stress, strain, and elastic modulus tensors, respectively. The coefficient n is an adjustable constant chosen to fit BFRACT calculations like those of Figures 13.12 and 13.13. Note that in Eq. (13.26) the value of t_s is a function of the current strain rate through Eqs. (13.24) and (13.25). For cases where the strain rate is not constant, the criterion for activating the stress relaxation Eq. (13.26) is that the stress exceed $\sigma_s(\dot{\varepsilon})$.

The damage function τ, whose isovalues are plotted in Figure 13.13, is defined [Curran et al., 1987] as the third moment of the crack size distribution [Eq. (13.11)]:

$$\tau = \int R^3 \, dN = 6N_t R_1^3 \tag{13.27}$$

For tensile cracks in an elastic medium, τ has an especially simple physical meaning: it is proportional to the ratio of the total crack opening strain to the elastic strain in the intact material. That is, Curran et al. [1987] have shown that the crack opening strain due to all the microcracks is [Sneddon and Lowengrub, 1969]

$$\varepsilon_c = 32(1 - \nu^2)N_t R_1^3 \sigma/E = 5.3(1 - \nu^2)\tau\sigma/E \tag{13.28}$$

and σ/E is proportional to the elastic strain ε_e in the intact material. With the assumption of uniaxial strain and elastic response,

$$\sigma/E = (1 - \nu)\varepsilon_e/(1 - 2\nu)(1 + \nu) \tag{13.29}$$

Solving Eqs. (13.28) and (13.29) for τ, and recalling that $\varepsilon = \varepsilon_e + \varepsilon_c$, yields

$$\tau = [6(1 - 2\nu)/32(1 - \nu)^2]\{[E(1 - \nu)(\varepsilon/\sigma)/(1 - 2\nu)(1 + \nu)] - 1\} \tag{13.30}$$

Thus, the evolving damage during the stress relaxation can be obtained from Eq. (13.30).

The simplified model discussed above must, in common with the much more complex BFRACT model, result in well-posed and stable solutions when used in hydrocode applications [Curran and Seaman, 1992]. In the Appendix, we show that both BFRACT and the simplified model represented by Eqs. (13.19) through (13.26) fulfill these well-posedness and stability requirements.

Equations (13.19) through (13.30) represent gross simplifications over the algorithms in BFRACT, which describe the anisotropic behavior due to crack orientations as well as the complex time-dependent softening due to the combined effects of microcrack nucleation, growth, and opening. However, it is possible that for overdriven cases leading to fragmentation of materials that exhibit relatively isotropic microcrack orientations, the simplified model would give similar results.

To test that possibility, we used the simplified model of Eqs. (13.19) through (13.30) to calculate the same family of stress histories at different imposed strain rates in Armco iron as is shown in Figure 13.13 for the BFRACT model. The results are shown in Figure 13.15. Overlaying Figures 13.13 and 13.15 shows that the agreement is rather good for the stress paths, but the damage levels are overpredicted in the simplified model by about a factor of 2. The good stress path agreement was obtained by setting n in Eq. (13.26) to 0.14 and by approximating Eqs. (13.24) and (13.25) by the single expression

$$t_s = 1.13 \times 10^{-4}(\dot{\varepsilon})^{-0.555} \tag{13.31}$$

where the unit of strain rate is s^{-1}.

Equation (13.31) is justified as follows. The impact data that were used

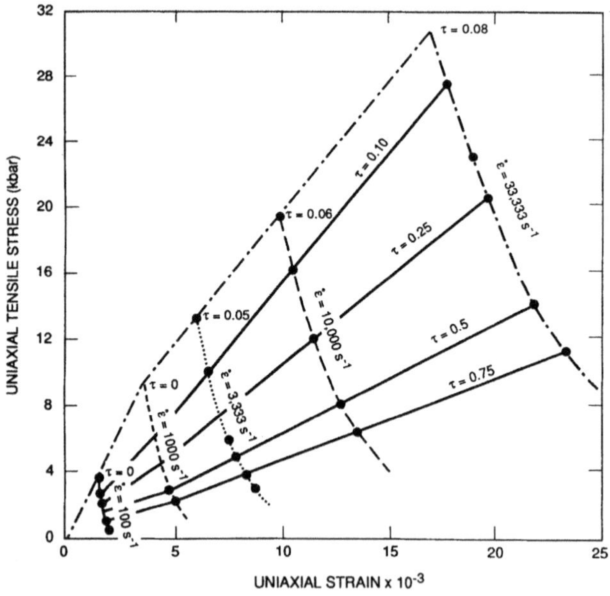

Figure 13.15. Simplified model computations of tensile stress-strain paths for several strain rates in Armco iron.

to calibrate the BFRACT model for Armco iron resulted in a value of $K_c = 3.4 \times 10^7$ dyne/cm$^{3/2}$ [Curran et al., 1987; Shockey et al., 1973]. Also, microcracks smaller than about 40 μm were found to produce negligible crack opening strain and associated softening, and the s_{min} corresponding to an R_{c1} of 40 μm is about 100 μm. These values give a value of $A = 10^4$ s^{-1}. For this value of A, the three highest strain rates in Figure 13.13 have an inverse strain rate dependence of s [from Eq. (13.25)] ranging from about 0.3 to 0.6. The lowest two strain rates have values closer to 0.75. For simplicity, we chose one dependency, and the value of 0.555 of Eq. (13.27) gave the best agreement with the BFRACT calculations.

The simple model of Eqs. (13.19) through (13.30) does not give the "correct" value of the coefficient in Eq. (13.31) but, rather, a value about one-tenth that amount. This is because the cleavage cracks in Armco iron were observed to obey a viscous growth law rather than the constant value of $c/3$ assumed in the above simplified model [Curran et al., 1987; Shockey et al., 1973]. It is true that microcracks larger than about 40 μm were observed to travel at speeds on the order of $c/3$, but the viscous growth leading up to that size causes a significant "delay time." On the other hand, the observed characteristic decay time in Eq. (13.26) ($n = 0.14$) is consistent with the coalescence time for cracks larger than 40 μm.

For Armco iron, therefore, we replace Eq. (13.24) with a time estimated from the viscous growth equation:

$$dR_{c1}/dt = R_{c1} T_1 (\sigma - \sigma_{g0}) \tag{13.32}$$

where σ_{g0} is the K_c-dependent growth threshold stress and T_1 is equal to 0.0006 cm^2/dyne-s [Curran et al., 1987; Shockey et al., 1973]. If we approximate $\sigma - \sigma_{g0} = \rho c^2 \dot{\varepsilon} t$ and integrate R_{c1} in Eq. (13.32) from 0.5 μm to 40 μm (where softening becomes important), we obtain a formula for t_s

$$t_s = 0.87 \times 10^{-4} (\dot{\varepsilon})^{-0.5} \tag{13.33}$$

that is very close to the fitted relation of Eq. (13.31). (The unit of strain rate is again s^{-1}.)

The values of the damage function τ shown in Figure 13.15 were calculated from Eq. (13.30) with Armco iron values of 210 GPa for E and 0.28 for v. The reason for the simplified model's overprediction of damage is that the effect of plastic strain was neglected. Comparison of Figures 13.13 and 13.15 shows that good agreement could be obtained by halving the coefficient in Eq. (13.30).

In summary, it appears that the above simplified elastic fracture model can be "tuned" to give results very similar to those of the sophisticated BFRACT model, even for cases like that of Armco iron where elastic-plastic response and nonelastic crack growth speeds complicate the picture. The untuned simplified model is expected to do a better job on brittle materials like ceramics.

Interestingly, for material properties and flaw sizes typical for brittle ceramics, Eq. (13.25) predicts that for strain rates greater than about $10^4\ s^{-1}$ the average fragment size will tend to be almost equal to the initial flaw spacing and relatively independent of strain rate. That is, for such high rates the initial flaws will all rapidly coalesce to form the minimum fragment size. On the other hand, for lower strain rates the fragment size will have roughly the inverse dependence on the strain rate to the 1/2 power, corresponding to cases where not all the flaws are activated and only the larger ones eventually coalesce to form fragments.

As we noted for the shear banding case discussed earlier, more comparisons of the predictions of Eq. (13.25) with more BFRACT calculations and experimental data are needed to test the usefulness of this simplified approach.

13.3.2.3. Simplified Ductile Fracture and Fragmentation Model

A parallel approach can be taken to construct a simplified version of DFRACT, the NAG/FRAG model for the nucleation, growth, and coalescence of microscopic ductile voids [Curran et al., 1987]. An advantage in this case is that DFRACT is isotropic, and its only stress dependence is on the mean stress. A characteristic time is obtained from an experimentally determined void growth viscosity. The post-peak stress relaxation is associated with this time, and adjustible parameters are calibrated by comparison with single-cell DFRACT calculations over a range of imposed strain rates.

13.4. Conclusions

Physics models of fracture and fragmentation have evolved to the point where they can be used in finite element hydrocodes to make fair-to-good predictions of damage, softening, and final fragment size and velocity distributions in dynamically loaded structures. The models show promise for reducing significantly the amount of experimentation necessary to predict fracture and fragmentation in specific applications. However, the more sophisticated models are demanding in their requirements for computing power. A number of simplifications to such models appear to hold promise for reducing these demands while maintaining reasonable physical accuracy.

13.5. Appendix

Well-Posedness and Stability of the Fracture Models

Constitutive relations that include softening must, when combined with the equations of conservation of mass, momentum, and energy, form an equation set that, when combined with the boundary and initial conditions, is well-posed and yields stable solutions. In this Appendix, we prove that both

BFRACT and the simplified model of Eqs. (13.19) through (13.26) fulfill these requirements. The proof is restricted to equations for uniaxial stress in one space dimension, but we expect that it can be generalized to three spatial dimensions [Curran and Seaman, 1992].

The equations for conservation of mass and momentum are:

$$\partial\varepsilon/\partial t = \partial u/\partial h \tag{13.A.1}$$

$$\rho\partial u/\partial t = \partial\sigma/\partial h \tag{13.A.2}$$

where h is the Lagrangian position coordinate.

The constitutive relation is

$$\partial\varepsilon/\partial t = (\partial\sigma/\partial t)/E + \partial\varepsilon P/\partial t + \partial\varepsilon^c\partial t \tag{13.A.3}$$

where E is Young's modulus, $\partial\varepsilon P/\partial t$ is the plastic strain rate, and $\partial\varepsilon^c/\partial t$ is the elastic crack opening strain rate.

For the plastic strain rate, we choose a commonly used relation:

$$\partial\varepsilon P/\partial t = (\sigma - Y)/E\tau \tag{13.A.4}$$

where τ is a characteristic time based on dislocation dynamics and Y is a constant yield strength. Eq. (13.A.4) was chosen for illustration, but the following stability and uniqueness arguments are not dependent on the specific form of the plasticity relation provided that the relation is dissipative, i.e., plastic deformation requires positive work.

The crack opening strain from the (noninteracting) crack size distribution of Eq. (13.4) is [1]:

$$\varepsilon^c = (\sigma/M)NR_1^3 \tag{13.A.5}$$

where $M = E/[32(1 - \nu^2]$. Thus,

$$\partial\varepsilon^c/\partial t = D(N, R_1)(\partial\sigma/\partial t)/M + F(\sigma, N, R_1)\sigma/M \tag{13.A.6}$$

where $D = NR_1^3$ and $F = (\partial N/\partial t)R_1^3 + 3NR_1^2(\partial R_1/\partial t)$. The specific dependence of F on its variables depends on the specific nucleation and growth relations used for the microcracks. Experimental data usually give $\partial N/\partial t$ and $\partial R_1/\partial t$ to be functions of stress such that the derivatives of F with respect to its variables are all positive. That is, increasing the crack opening strain rate requires an increase in stress and an associated increase in size and number of the cracks. In short, the process is dissipative.

Combining Eqs. (13.A.3) and (13.A.6) yields the constitutive relation

$$\partial\varepsilon/\partial t = (\partial\sigma/\partial t)/M(N, R_1) + G(\sigma, N, R_1) \tag{13.A.7}$$

where M is a "softened" modulus

$$M(N, R_1) = E/[1 + ED(N, R_1)/M] \tag{13.A.8}$$

and

$$G(\sigma, N, R_1) = (\sigma - Y)/E\tau + F(\sigma, N, R_1) \tag{13.A.9}$$

In the description of the simplified model leading to Eq. (13.26) in this chapter, G is given by σ/nt_s and M by $M_0 \exp(-t/nt_s)$, whereas in the BFRACT model the functions are usually much more complicated [Curran et al., 1987]. For example. for Armco iron the BFRACT relations for nucleation and growth rates are

$$\partial N/\partial t = N_{0t} \exp[(\sigma - \sigma_{n0})/\sigma_1] \tag{13.A.10}$$

and

$$\partial R_1/\partial t = R_1 T_1(\sigma - \sigma_{g0}) \tag{13.A.11}$$

where N_{0t}, σ_{n0}, σ_1, T_1, and σ_{g0} are experimentally determined constants.

We next operate on Eqs. (13.A.1) and (13.A.7) by $\partial/\partial t$, and on Eq. (13.A.2) by $\partial/\partial h$, and combine to obtain

$$\partial^2\sigma/\partial t^2 - (M/\rho)\partial^2\sigma/\partial h^2 + A\partial\sigma/\partial t + B \tag{13.A.12}$$

where

$$A = \partial G/\partial\sigma - [(\partial M/\partial N)(\partial N/\partial t) + (\partial M/\partial R_1)(\partial R_1/\partial t)]/M^2 > 0 \tag{13.A.13}$$

and

$$B = (\partial G/\partial N)(\partial N/\partial t) + (\partial G/\partial R_1)(\partial R_1/\partial t) > 0 \tag{13.A.14}$$

A and B are positive because $\partial M/\partial N$ and $\partial M/\partial R_1$ are negative, i.e., the modulus is softened by increasing damage. The other terms in A and B are positive because of the dissipative nature of the model.

Equation (13.A.12) is similar to a class examined by G.B. Whitham [1959, 1974]. Whitham showed that equations like (13.A.12) with $B = 0$ are well-posed and stable if A is positive. This result can be easily extended to Eq. (13.A.12) as follows:

Linearize Eq. (13.A.12) around σ', N', and R_1'. Let $\sigma = \sigma' + \Delta$. Then Eq. (13.A.12) becomes

$$\partial^2\Delta/\partial t^2 - (M'/\rho)\partial^2\Delta/\partial h^2 + A'\partial\Delta/\partial t + B' = 0 \tag{13.A.15}$$

Next, define

$$S = \Delta + (B'/A')t \tag{13.A.16}$$

Combining Eqs. (13.A.15) and (13.A.16) yields

$$\partial^2 S/\partial t^2 - (M'/\rho)\partial^2 S/\partial h^2 + A'\partial S/\partial t = 0 \tag{13.A.17}$$

Equation (13.A.17) is of the form proved well-posed and stable by Whitham. That is, S decays with time if A' is positive. Therefore, $\Delta = S - (B'/A')t$ also decays with time.

In the simplified model leading to Eq. (13.26), the stress relaxation is akin to adding another term like the plastic strain rate of Eq. (13.A.4). In fact, since the simplified model had no plasticity, it is equivalent to reinterpreting the plasticity term as a crack opening term and setting F equal to zero. On the other hand, since the simplified model softens the modulus, an equivalent D function is implicitly retained.

Thus, the simplified model can be considered a special case of BFRACT and will pose no problems involving loss of well-posedness or stability.

13.6. References

Curran, D.R., Seaman, L., and Shockey, D.A. *Physics Rep.* **147**(5&6) (1987).

Curran, D.R. and Seaman, L. In *Shock Compression of Condensed Matter 1991*, S.C. Schmidt, R.D. Dick, J.W. Forbes, and D.G. Tasker, Eds., pp. 395–398, Elsevier Science Publishers B.V. (1992).

Desai, C.S., Krempl, E., Frantziskonis, G., and Saadatmanesh, H. Eds. *Constitutive Relations for Engineering Materials*, ASME Press (1991).

Deve, H., Harren, S., McCullough, C., and Asaro, R.J. *Acta Metall.* **36**(2), 341–365 (1988).

Dullum, O. and Seaman, L. Simulations of Projectile Penetrations with a Shear Banding Model, Poulter Laboratory Technical Report 004-84, SRI International (1985).

Englmen, R., Rivier, N., and Jaeger A. *Phil. Mag.* **B56**, 751 (1987).

Erlich, D.C., Seaman, L., Caligiuri, R.D., and Curran, D.R. SRI International Annual Report to the Ballistic Research Laboratory and the Army Materials and Mechanics Research Center, Contract No. DAAK11-78-C-0115 (1980).

Grady, D.E. and Kipp, M.E. *J. Appl. Phys.* **58**, p. 1210 (1985).

Grady, D.E. and Kipp, M.E. *J. Mech. Phys. Solids* **1**, 95–118 (1987).

Grady, D.E. *J. Mech. Phys. Solids* **36**(3), pp. 353–384 (1988).

Harren, S.V. and Asaro, R.J. *J. Mech. Phys. Solids*, **37**(2), 191–232 (1989).

Meyers, M.A., Vecchio, K.S., and Andrada, U. *Proceedings of Symposium on Shear Bands and Viscoplastic Theories*, University of California at San Diego, La Jolla, CA, September 14–19, 1992, to be published as a separate volume of the *Mechanics of Materials* (1993).

Mott, N.F. *Proc. Roy. Soc. London A* **189**, pp. 300–308 (1947).

Nemat-Nasser, S. and Horii, M. *Micromechanics: Overall Properties of Heterogeneous Materials*, North-Holland, Amsterdam (1993).

Peirce, D., Asaro, R.J., and Needleman, A. *Acta Metall.* **31**, 1951 (1983).

Shockey, D.A., Seaman, L., and Curran, D.R. Final Technical Report to the Air Force Weapons Laboratory, Report No. AFWL-TR-73-12, Contract No. F29601-70-C-0070 (1973).

Sneddon, I.N. and Lowengrub, M. *Crack Problems in the Classical Theory of Elasticity*, John Wiley & Sons, New York (1969).

Whitham, G.B. (1959), *Communications on Pure and Applied Mathematics*, Vol. XII, p. 113.

Whitham, G.B. *Linear and Nonlinear Waves*, John Wiley & Sons, New York (1974).

Zukas, J.A., Nicholas, T., Swift, H.F., Greszczuk, L.B., and Curran, D.R. *Impact Dynamics*, John Wiley & Sons, New York (1982).

CHAPTER 14

A Unified Theory of Flow, Hot Spots, and Fragmentation with an Application to Explosive Sensitivity

J.K. Dienes

14.1. Introduction

Practical knowledge of material failure and its control goes back to chipping in the stone age, forging in the bronze, and annealing of blown glass by the Phoenicians. The theory, however, is not nearly as systematic as in physics or as quantitative as in biology. Analysis of plastic flow began in France in the nineteenth century and a theory of brittle fracture was put forth in England early in the twentieth, but important issues such as brittle-ductile transition and combined plastic and brittle behavior are only now beginning to be understood, with dislocation behavior at crack tips being one of several central issues. Since crack-tip behavior is very sensitive to impurities, grain structure, temperature, strain rate, and many other factors, the possibility of an exact, predictive theory for complex solids is no more promising than for sociology, where so many different complex considerations interact. Geometric nonlinearities, anisotropy, and phase changes present additional complications.

Still, it now seems possible to formulate a unified theory that reflects the most important phenomona. The theory presented here utilizes concepts from continuum mechanics to define stress, strain, and rotation rigorously for arbitrarily large deformations, as well as their tensor rates. This provides a theoretical basis for superimposing strain rates—an old idea, but one that was originally put forth without justification as an *ansatz* for combining elastic and plastic behavior. The current approach allows us to show the validity of the superposition principle for large deformations and for types of material behavior other than plastic and linear-elastic. This allows us to treat microscopic fracture, macroscopic plastic flow and geometric nonlinearities in a unified way. In addition, we can treat compressive and tensile failure separately in ways consistent with observations, thereby accounting for the difference in compressive and tensile strength, which may be very large in brittle materials. Though for some applications it may not be necessary to have such a general approach, it is useful to have a unified theory that places various kinds of deformation and failure into a single conceptual framework.

The theory has a statistical aspect that allows us to relate fracture at the microscopic scale to large-scale deformation. The need to connect behavior on different scales arose originally in the oil-shale program, when we needed to know the permeability of rubblized rock. Connected cracks were considered to be the main source of permeability, and we were able to develop the necessary theory based on statistical concepts (Dienes, 1982a). Another application concerns explosive sensitivity, where we are also faced with a need to understand microscopic behavior because the formation and coalescence of hot spots can be responsible for accidental explosions caused by low-energy incidents such as XDT. In this connection it is noted that microscopic conduction and melting may strongly influence the release of explosive energy, and explain the long delay times observed in explosions resulting from low-energy accidents. They are accounted for in an extension of the theory reported in Section 14.6. The experimental basis for this kind of behavior has been discussed by Bowden and Yoffee (1952), and new theoretical insights have been provided by Frey (1981). Explosive hot spots are interesting examples of localization with reactive heating (Dienes, 1982b).

The current approach differs considerably from the use of flow laws, in which an attempt is made to account for the behavior of brittle materials by modifying plasticity theory as discussed, for example, by Sandler and Baron (1985). That approach fails, however, to account for microscopic behavior, instabilities, or for the strain-rate and size effects observed in brittle materials such as rock, glass, ceramics, and grains of explosive. An early attempt by Drucker and Prager (1952) to account for dilatancy by relating it to the pressure dependence of the flow stress has been shown to be in conflict with experimental evidence by Spitzig and Richmond (1984). The premise for the flow-law approaches is that materials are microscopically *stable* and that strain due to cracking is analogous to strain due to plastic flow. It is clear that such approaches represent only a first attempt to develop constitutive relations—they ignore the evidence for *unstable* behavior given by crack bursting, acoustic emission, stepping of stress-strain curves, strain-softening, and many other kinds of erratic behavior that are observed in the brittle and brittle-ductile regimes. In order to make flow laws tractable, isotropy is usually assumed, though it is well known that this can be only a rough approximation. Isotropic hardening fails to account for the hysteretic behavior observed in cyclic plastic flow, or for the formation of parallel sets of cracks.

Whereas the flow-law approach modifies the treatment of plastic strain rate to account for failure, the current approach considers fracture to reduce the elastic moduli and adds a strain rate due to crack growth. This approach is sometimes criticized for using too much computer time and being too complex, but the increase in computer time is only twofold, and the complexity is not significantly greater than in modern damage theories. For this investment we get a representation of the behavior at the microscopic level that can account for phenomena such as permeability, fragmentation, shear banding, and XDT that cannot be analyzed by existing methods. It cannot be claimed that the treatment of microscopic behavior in this chapter is especially realistic, since there are many simplifying assumptions. However, the theory is capable of generalization in many different ways as the need arises. Some of the generalizations that are in progress are discussed in the final section.

14.2. Theory of Finite Deformation, Polar Decomposition, and Polar Rates

The following approach draws on classical continuum mechanics, making it possible to characterize arbitrarily large geometric nonlinearities. The approach differs from classical continuum mechanics in making extensive use of polar decomposition, polar rates, and in the treatment of materials as ensembles with random defects.

The deformation of a continuum at any point is characterized by the deformation gradient

$$\mathbf{F} = \left(\frac{\partial x_i}{\partial X_j}\right) \tag{14.1}$$

where $x_i(X_j, t)$ denotes the coordinates at time t of a point with the initial coordinates X_i. The polar-decomposition theorem makes it possible to write the matrix $\mathbf{F}$ as the product of a positive-definite matrix and an orthogonal matrix in two different ways:

$$\mathbf{F} = \mathbf{VR} = \mathbf{R}\overline{\mathbf{V}}. \tag{14.2}$$

Geometrically, this means that the deformation of an infinitesimal line element connecting two points in a continuum at time $t = 0$ can be represented by a rotation $\mathbf{R}$ followed by a spatial stretch $\mathbf{V}$, or by a different (rotated) stretch $\overline{\mathbf{V}}$ followed by the same rotation $\mathbf{R}$. If the material contains defects, we consider the medium in the analysis to be an ensemble average of different samples. The mean motion can then be considered as continuous, though members of the ensemble may exhibit discontinuous character in the neighborhood of defects. The mathematical aspects of the continuum theory are discussed by Chadwick (1976) and implications for materials with defects have been analyzed by Dienes (1987).

Since constitutive relations usually involve rates of matrix quantities, it is important to have them defined in a generally valid manner. (Many of the matrix variables are tensors, in the physical sense, but many are not, as discussed by Dienes (1987, Appendix C)). The following approach is useful and can be considered valid for arbitrarily large strains and rotations. The velocity gradient $\mathbf{G}$ is related to the deformation gradient $\mathbf{F}$ through

$$\mathbf{G} = \left(\frac{\partial u_i}{\partial x_j}\right) = \left(\frac{\partial \dot{x}_i}{\partial X_j}\right)\left(\frac{\partial X_j}{\partial x_k}\right) = \dot{\mathbf{F}}\mathbf{F}^{-1}, \tag{14.3}$$

with the factorization an application of the chain rule. In this expression u_i denotes the velocity

$$u_i(X_j, t) = \frac{\partial x_i(X_j, t)}{\partial t} = \dot{x}_i(X_j, t); \tag{14.4}$$

that is, the rate of change of position of a point fixed in the material having initial coordinates X_j. Now, any matrix can be represented as the sum of symmetric and antisymmetric parts, so we can write

$$\mathbf{G} = \mathbf{W} + \mathbf{D} \tag{14.5}$$

where **W** is antisymmetric and is usually called the vorticity, and **D** is symmetric and is called the stretching by Truesdell and Toupin (1960). It is claimed by those and other authors (Eringen, 1967, for example) that **W** has a physical interpretation as the rate of rotation of the principal axes of stretching, **D**. That this is not true can be seen by means of a counter-example pointed out by P. Blewett (1992). In simple shear, the example investigated by Dienes (1979), the vorticity is not zero. However, the flow is steady and the stretching is constant, so the principal axes of stretching are fixed in space. Thus, they do not rotate. The apparent paradox can be resolved by restating the theorem: it should say that material fibers lying on the principal axes of the stretching tensor **D** have an instantaneous rate of rotation given by **W**. It will be shown below that this is different from the rate **Ω** based on a fixed reference system.

By combining Eqs. (14.1), (14.3) and (14.5) it can be shown that

$$\mathbf{D} + \mathbf{W} = \dot{\mathbf{V}}\mathbf{V}^{-1} + \mathbf{V}\boldsymbol{\Omega}\mathbf{V}^{-1} \tag{14.6}$$

where

$$\boldsymbol{\Omega} = \dot{\mathbf{R}}\mathbf{R}^{-1} \tag{14.7}$$

is an antisymmetric matrix representing the rate of material rotation. (This is not the **R** of Truesdell and Toupin, who treat rotation using an infinitesimal theory). By premultiplying and postmultiplying Eq. (14.6) by **V**, and combining the result with its transpose, it can be shown that

$$\dot{\mathbf{B}} + \mathbf{B}\boldsymbol{\Omega} - \boldsymbol{\Omega}\mathbf{B} = 2\mathbf{V}\mathbf{D}\mathbf{V} \tag{14.8}$$

where

$$\mathbf{B} = \mathbf{V}^2 = \mathbf{F}\mathbf{F}^T. \tag{14.9}$$

It is useful to work with the Signorini strain

$$\boldsymbol{\varepsilon} = \tfrac{1}{2}(\mathbf{B} - \mathbf{I}). \tag{14.10}$$

This spatial strain (based on the spatial stretch **V**) is related to the stretching **D** by

$$\dot{\boldsymbol{\varepsilon}} + \boldsymbol{\varepsilon}\boldsymbol{\Omega} - \boldsymbol{\Omega}\boldsymbol{\varepsilon} \equiv \hat{\boldsymbol{\varepsilon}} = \mathbf{V}\mathbf{D}\mathbf{V} \tag{14.11}$$

which follows readily from Eqs. (14.8) and (14.10). The quantity on the left side of Eq. (14.11) I call the polar strain rate. Though the stretching **D** is called the strain rate in some of the early works on continuum mechanics, this is somewhat misleading since it is not the rate of any well-defined, history-independent quantity. On the other hand, the left side of Eq. (14.11) is the rate of change of the Signorini strain involving a rate (albeit a special kind, the polar rate) appropriate to spatial tensors. For small deformations, **V** is approximately the identity matrix **I** and, consequently, the strain rate $\hat{\varepsilon}$ can be approximated by **D**.

Matrices of the form

$$\dot{\mathbf{A}} + \mathbf{A\Omega} - \mathbf{\Omega A} = \hat{\mathbf{A}} \tag{14.12}$$

represent the rate of change of **A** in axes having a fixed orientation, accounting for the effects of material rotation on the components of **A**, and may be called polar rates since $\mathbf{\Omega}$ arises from polar decomposition. (Calling this an Eulerian rate seems to me quite misleading, though common practice. This is the rotation in fixed axes—Euler certainly did not have this in mind when formulating the equations of motion.)

The preceding formal treatment suggests that we may attach a physical (or, more precisely, a kinematic) interpretation to the relation of quantities with bars and those without—in fact, the bars are introduced as a suggestive notation to represent the transformation from fixed axes x_i to polar axes $\overline{x_i}$, of which Eq. (14.2) involves the first example—and is in general

$$\overline{\mathbf{A}} = \mathbf{R}^T\mathbf{A}\mathbf{R}; \tag{14.13}$$

this is a rotation by the amount **R**. Coordinates in polar axes $\overline{x_i}$ are related to those in fixed axes x_i by

$$\overline{x_i} = R_{ji}x_j. \tag{14.14}$$

It is straightforward to show that

$$\dot{\overline{\mathbf{A}}} = \mathbf{R}^T\hat{\mathbf{A}}\mathbf{R}. \tag{14.15}$$

Hence, the polar rate $\hat{\mathbf{A}}$ is related to the rate in polar axes $\dot{\overline{\mathbf{A}}}$ by a rotation. The rate in polar axes is an ordinary time derivative, but it has to be emphasized that this ordinary derivative has to be intrepreted in rotating (polar) axes. This is complex and confusing because orthogonal axes can become extremely skewed, making it difficult to visualize the rotation and strain. Nevertheless, polar decomposition provides an exact, though somewhat mystifying, means for computing this kind of behavior. Examples are given by Dienes (1987) wherein it is also shown that polar derivatives have many of the properties of ordinary derivatives. The Signorini strain of Eq. (14.10) is related to the polar strain by

$$\bar{\varepsilon} = \mathbf{R}^T\varepsilon\mathbf{R}. \tag{14.16}$$

The strain that is called a polar strain here is usually called the Green strain in continuum mechanics. The above equation shows that the Green strain and Signorini strain are the same physical quantity, but are specified in different axes. Thus, they are not competing measures, as is often implied in the literature of continuum mechanics, but different representations of the same measure. It is straightforward to show that

$$\dot{\boldsymbol{\varepsilon}} = \overline{\mathbf{V}}\overline{\mathbf{D}}\overline{\mathbf{V}} \tag{14.17}$$

with the dot denoting an ordinary derivative, providing a simple (and, I believe, new) relation between the stretching **D** and the rate of change of Green strain. This is sometimes called a Lagrangean rate, but the axes are not fixed in the material—thus, this is a misleading terminology. The rotation **R** is not fixed in the material. *It is best understood as the matrix that takes* **F** *into a quantity* **V** *that can be considered a stretch because it is positive definite*. It does not seem possible to assign any more direct physical interpretation to **R** and the rotated axes given in Eq. (14.14). If the material is rigid, **R** is, of course, the rotation. The rate of change of rotation $\boldsymbol{\Omega}$ is more easily understood—more precisely, we can visualize its vector equivalent $\boldsymbol{\omega}$, discussed below.

The preceding relations have been used by Dienes (1975) to show that the vorticity is related to the rate of fiber rotation by

$$\boldsymbol{\omega} = \mathbf{w} + (\mathbf{I}\ \mathrm{tr}\ \mathbf{V} - \mathbf{V})^{-1}\mathbf{z} \tag{14.18}$$

where ω, **w**, and **z** are vectors related to antisymmetric matrices by

$$\Omega_{ik} = e_{ijk}\omega_j \tag{14.19a}$$

$$W_{ik} = e_{ijk}w_j \tag{14.19b}$$

$$Z_{ik} = e_{ijk}z_j \tag{14.19c}$$

with

$$\mathbf{Z} = \mathbf{DV} - \mathbf{VD}. \tag{14.20}$$

Thus, the vorticity, **w**, is quite different from the rate of material rotation, $\boldsymbol{\omega}$, in general, though they may be equal in special cases. When a line in a material is considered at time $t = 0$, then $\mathbf{V} = \mathbf{I}$, but as time proceeds **V** does not remain the unit matrix and the term added to the vorticity in Eq. (14.18) does not remain zero. It may be surprising to students of classical hydrodynamics that a perfect fluid with zero vorticity can have a finite rate of rotation, since texts usually state that vorticity is a measure of rate of rotation. Still, it is easy to visualize that in an ideal vortex this cannot be the case. The vorticity in an ideal vortex is zero (except at the center), but materials in adjacent circles rotate about the center at angular velocities that vary inversely with distance. Thus, it takes longer for a point at a large radius to complete one revolution than for a point at a small radius. A line joining two points on adjacent circles will stretch and rotate. The line will have no initial angular velocity, but eventually it attains the local angular velocity about the vortex center, and its normal will asymptotically align with the radius vector.

Explicit solutions for the cases of pure shear, the ideal vortex, and pure torsion of a cylinder have been determined by Dienes (1987), but in general the equations have to be integrated numerically to determine $\mathbf{\Omega}$. Sometimes $\mathbf{R}$ is called a "rigid body rotation," but this is a misnomer, since the deformed axes are skewed, becoming almost parallel in extreme cases. The interpretation of $\mathbf{R}$ that seems to me the most useful is just that given by Eq. (14.2a); it modifies $\mathbf{F}$ by just the amount required to make it positive definite and, hence, allows us to define a positive-definite stretch.

Given a constitutive law for the Cauchy stress $\boldsymbol{\sigma}$ having the form

$$\dot{\boldsymbol{\sigma}} = \boldsymbol{\varphi}(\boldsymbol{\sigma}, \mathbf{D}) \tag{14.21}$$

in the absence of rotation (obtained, say, from triaxial testing) it is possible to construct the general form for a rotating material, as is necessary when formulating algorithms for numerical integration when the deformation is large. A form of this kind arises in the analysis of elastic-plastic, Maxwell, or kinematic-hardening solids. It was shown by Dienes (1979) that in the presence of rotation this should have the form

$$\hat{\boldsymbol{\sigma}} = \boldsymbol{\varphi}(\boldsymbol{\sigma}, \mathbf{D}). \tag{14.22}$$

This is different from the form proposed by Zaremba, Jaumann and Noll (see Dienes, 1979), though they may nearly coincide when the deformation is small. When materials are anisotropic, deeper considerations are necessary, as illustrated in an important exploratory paper by Zheng (1992).

14.3. Superposition of Strain Rates

Though superposition of strain rates dates back to Reuss (1930), who used it to characterize elastic-plastic behavior, no physical motivation was given, nor adumbrated, nor was it suggested that the principle could be extended to more complex situations in mechanics where processes other than plastic flow are involved. In this section a motivation for the superposition principle that permits extension to fragmentation and the behavior of porous materials is described. Generalization to cover polycrystalline and composite materials is straightforward in principle, but challenging in detail. The section concludes with examples illustrating how the method can be used to obtain some results that are standard in macroscopic plasticity and viscoelasticity.

Though the title of this section contains the phrase "strain rates," the details of the analysis involve primarily the stretching $\mathbf{D}$. However, in view of Eq. (14.11), *if* it can be shown that the effect of simultaneous processes, denoted by an index α, can be represented by

$$\mathbf{D} = \sum_{\alpha} \mathbf{D}^{\alpha} \tag{14.23}$$

with $\mathbf{D}^{\alpha}$ denoting the stretching due to the αth process, then

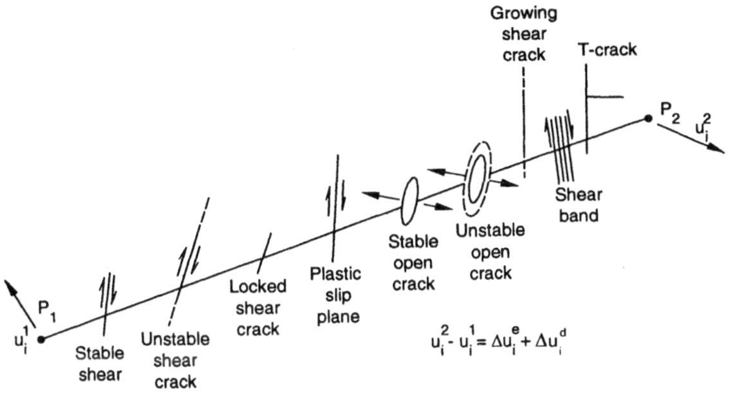

Figure 14.1. Superposition of strain rates: an illustration of how the effects of various defects can combine to influence the velocity increment between two points P_1 and P_2.

$$\hat{\boldsymbol{\varepsilon}} = \mathbf{V}\left(\sum_\alpha \mathbf{D}^\alpha\right)\mathbf{V} = \sum_\alpha \hat{\boldsymbol{\varepsilon}}^\alpha \tag{14.24}$$

where $\hat{\boldsymbol{\varepsilon}}^\alpha$ is the Signorini strain rate due to the αth process. Thus, superposition of stretchings and superposition of strain rates are equivalent. Of course, $\mathbf{V}$ denotes the total stretch, and cannot be separated into contributions from various processes in a corresponding way.

The premise of this section is that deformation is often the consequence of numerous independent physical processes that can be superimposed to obtain an overall deformation. This is illustrated in Fig. 14.1, which illustrates how the difference in velocity between two points $\mathbf{P}_1$ and $\mathbf{P}_2$ can be written as a sum over the individual elements and defects separating the points as

$$\Delta u_i = \sum \Delta u_i^c + \Delta u_i^d, \tag{14.25}$$

where the first sum is taken over the continuous regions between defects, and the second sum is taken over the defects between continuous regions. The continuous regions can be considered to have a general behavior, involving thermoelastic and possibly plastic behavior characterized by rather general thermodynamic laws. In addition, they can be anisotropic and may involve twinning or various phases. The defects are currently considered to be either open or closed. (In real materials defects may be partially open and partially closed, but the theoretical treatment of such flaws is still in the development stage.) The successful application of the superposition method depends on the adequacy of the constitutive laws characterizing the various kinds of defects and their statistics. The current treatment outlines some of the main issues, but many subtle aspects remain to be considered in detail elsewhere.

Now, we turn from an explicit to a statistical treatment and consider the

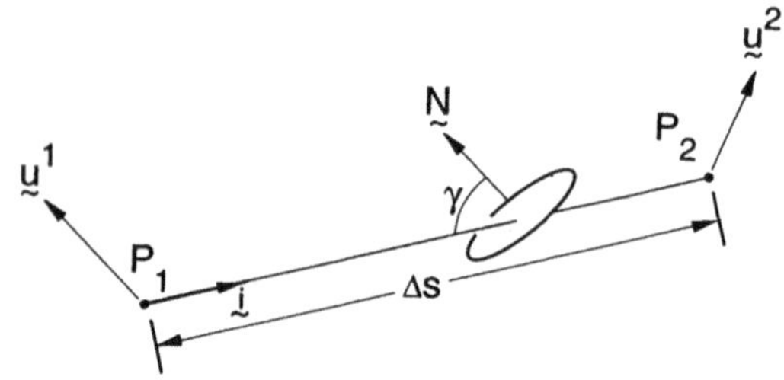

Figure 14.2. The contribution of a defect to the difference in velocity between P_1 and P_2.

defects to be separated into groups of individual type, size and orientation. The number of defects of each type per unit length is denoted by

$$\tilde{L}^\alpha = 1/\tilde{\lambda}^\alpha \tag{14.26}$$

where $\tilde{\lambda}^\alpha$ is the mean free path in the reference state for a particular kind of flaw. This makes it possible to replace the summation over individual defects in Eq. (14.25) by a sum over sets of defects involving the spatial frequency $\tilde{L}^\alpha$, so that

$$\sum \Delta u_i^d = \Delta S \sum \tilde{L}^\alpha \delta u_i^\alpha \tag{14.27}$$

represents the contribution δu_i^α of the individual defects to the velocity difference Δu_i between points $\mathbf{P}_1$ and $\mathbf{P}_2$, as illustrated in Fig. 14.2.

It is very convenient to take the flaws as circular, and likely to be a good approximation in many situations. This may be adduced with an analogy. In Jeans' (1940) development of the kinetic theory of gases, molecules such as O_2 were considered spherical for the purpose of computing collisions, although such an approximation is clearly far from precise (this is not brought out in the text). Nevertheless, the resulting theory was very successful in explaining the behavior of gases, with diffusion, conduction, and viscosity coefficients computed to an accuracy typically on the order of 1%. In the current theory of solid defects, approximating their shape as circular greatly simplifies the statistical arguments, as well as the behavior of individual defects. (Recall, in this connection, that an exact theory of circular cracks was developed in the 1950's when Sack, Segedin, Sneddon, and Keer were able to solve the governing boundary-value problems, as discussed by Dienes, 1989.) The number of defects of each type per unit length is equal to the projected area per unit volume, as illustrated in Fig. 14.3. A set of circular defects with radii in the range $(c, c + \Delta c)$ and orientations occupying a spherical cap of size $\Delta\Psi$ on the unit sphere, as illustrated in Fig. 14.4, is said to be a defect set. Each set has an orientation specified by the normal vector having the usual polar coordinates ϕ and θ, so that the element of area on the unit sphere is

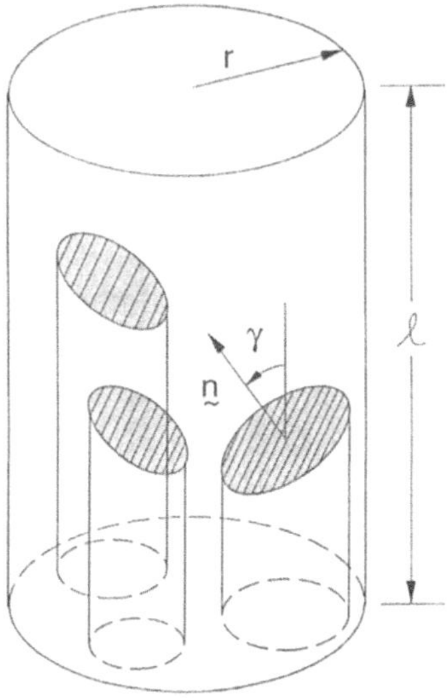

Figure 14.3. Illustration of a control volume containing a random assortment of circular defects used in estimating the number of defects per unit length $\tilde{L} = \pi La$ when the defects all have the same area a and a number density L per 2π, per unit volume.

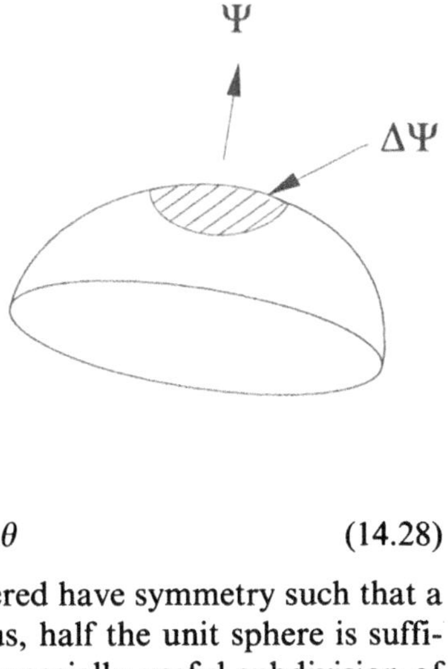

Figure 14.4. Hemisphere defining crack orientation. The cap of area $\mathbf{\Delta\Psi}$ and center $\mathbf{\Psi}$ defines one crack set.

$$\Delta\Psi = \sin\theta\Delta\phi\Delta\theta \tag{14.28}$$

as illustrated in Fig. 14.4. The defects considered have symmetry such that a reversal of 180° leaves them unchanged. Thus, half the unit sphere is sufficient to characterize defect orientation. An especially useful subdivision of half the unit sphere is illustrated in Fig. 14.5, with each subdivision having equal area. With these restrictions, the number of defects per unit length in each crack set can be written

$$\tilde{L}^\alpha = -\pi c^2 L_c^\alpha \Delta c \Delta\Psi \cos\gamma \tag{14.29}$$

where L^α is a continuous function of c, ϕ, and θ representing the number of defects per unit volume exceeding c in radius (hence the negative sign above) per unit solid angle, and the subscript denotes differentiation. The angle between $\overline{\mathbf{P}_1\mathbf{P}_2}$ and the initial crack normal $\mathbf{N}$ is denoted by γ, i.e.,

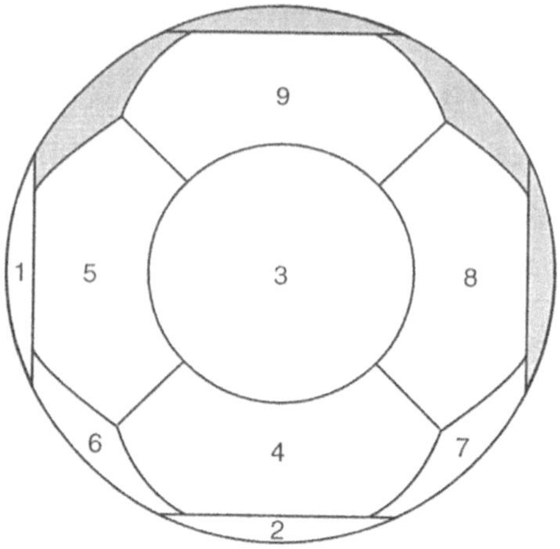

Figure 14.5. Illustration of how a unit hemisphere can be subdivided into elements defining flaw orientation.

$$\cos \gamma = N_k i_k. \tag{14.30}$$

where i_k denotes the unit vector in the direction $\overline{\mathbf{P}_1\mathbf{P}_2}$ in initial coordinates, so that

$$\Delta S i_k = \Delta X_k \tag{14.31}$$

with ΔX_k the projection of the initial element of arc ΔS in the initial reference direction. Upon dividing Eq. (14.27) by Δx_j and considering the mean (in the sense of an ensemble average) deformation to be continuous, we find that the velocity gradient can be written

$$u^d_{i,j} = -F^{-1}_{kj} \sum \pi c^2 L_c \delta u_i N_k \Delta c \Delta \Psi, \tag{14.32}$$

in the limit of small ΔS, where F^{-1}_{kj} denotes an element of $\mathbf{F}^{-1}$. This relation can be put into spatial axes by means of the relation

$$\mathbf{n} = \mathbf{F}^{-T}\mathbf{N}(\mathbf{N}\mathbf{V}^{-2}\mathbf{N})^{-1/2} \tag{14.33}$$

between the current crack normal $\mathbf{n}$ and the initial normal $\mathbf{N}$. (This expression, introduced as Eq. A.66 of Dienes (1987), involves the shorthand superscript "$-T$" implying that matrix inversion and transposition commute.) The result of Eq. (14.32) can be written

$$u^d_{i,j} = -\sum \nu n_j \pi c^2 L_c \delta u_i \Delta c \Delta \Psi \tag{14.34}$$

where

$$\nu = (\mathbf{N}\mathbf{V}^{-2}\mathbf{N})^{1/2}. \tag{14.35}$$

and the summation is taken over defect orientation, size and type. Thus, the velocity gradient associated with each defect set is

$$u^{\alpha}_{i,j} = \nu n_j \pi c^2 \Delta L \delta u_i \tag{14.36}$$

where

$$\Delta L = -L_c \Delta c \Delta \Psi \tag{14.37}$$

represents the (positive) number of defects in each set. All the variables in the sum of Eq. (14.32) depend on the set chosen, but the α superscripts are omitted there, and elsewhere occasionally, to avoid a notational jungle.

Contributions to the velocity gradient of Eq. (14.34) can be divided into symmetric and antisymmetric parts, as indicated by Eq. (14.5). Then, applying the equation to each set,

$$d^d_{ij} = \sum_\alpha d^\alpha_{ij} \tag{14.38}$$

and

$$w^d_{ij} = \sum_\alpha w^\alpha_{ij}. \tag{14.39}$$

Eq. (14.38) is not quite equivalent to the relation postulated in Eq. (14.23) since it does not include the contribution of the continuous regions between detects expressed in Eq. (14.25) but it is obvious that they can be characterized in the same way. Thus, the proposed superposition of velocity gradients implies the superposition of stretchings, which in turn implies the superposition of strain rates as postulated in Eq. (14.24). It also implies superposition of vorticities.

This principle also implies superposition of energy rates, a conclusion that can be derived by operating on Eq. (14.37) with σ_{ij}, and writing

$$\dot{e}^\alpha = \sigma_{ij} d^\alpha_{ij} = a^\alpha \dot{e} \tag{14.40}$$

for the rate of change of internal energy per unit volume associated with the αth deformation mechanism. This energy may go into elastic deformation, and thereby represent a reversible process; into heating, and thereby represent an irreversible process; into formation of new surfaces; and into other microscopic processes. The energy balance is written

$$\dot{e} = \sum_\alpha \dot{e}^\alpha \tag{14.41}$$

with the summation extended over all mechanisms that contribute to the deformation, and not just defects. Each deformation mechanism involves a stretching that can be expressed with a constitutive law of the form

$$\mathbf{D}^\alpha = \mathbf{f}^\alpha(\boldsymbol{\sigma}, \hat{\boldsymbol{\sigma}}, \mathbf{D}, \mathbf{V}) = a^\alpha \mathbf{D} \tag{14.42}$$

and it may possibly involve other variables of the process. This determines the a^α. This implies that

$$\mathbf{W}^\alpha = a^\alpha \mathbf{W}, \tag{14.43}$$

for only if this relation holds can the constitutive laws embodied in Eq.

(14.34) be made consistent with Eq. (14.23). This equation is not a necessary ingredient for calculating deformation, but once the deformation is known, it can be used to interpret the motion and determine the amount of vorticity associated with each type of deformation. It will be shown in the next paragraph that for plasticity the velocity gradient due to dislocations is symmetric, and hence the associated vorticity is zero.

The theory is illustrated below by showing how Eq. (14.34) applies in the case of dislocation motion to obtain the flow laws of ideal plasticity, of kinematic hardening, and of PBXs (Plastic Bonded Explosives), which can be characterized as viscoelastic composites. The defects are taken to be circular dislocation loops in order to retain the geometry assumed in the derivation above, but this is not really an essential feature of the illustration. The average slip velocity due to the passage of dislocation loops is

$$\delta u_i^d = \frac{2b}{d} \mathrm{v}_i \tag{14.44}$$

where b is the appropriate lattice constant, d is the distance between dislocation loops, and v_i is the dislocation velocity. Also assume that this velocity is proportional to the tangential component of traction on the plane of the dislocation loop, T_i' (and is not affected by the normal component of traction), so that

$$\mathrm{v}_i = q_0 T_i', \qquad T_i' = (\delta_{ij} - n_i n_j)\sigma_{jk} n_k. \tag{14.45}$$

Then,

$$\delta u_i^d = q T_i' \tag{14.46}$$

with q a new constant. If this expression is used in Eq. (14.34) and the dislocation loops are all assumed to have the same radius, c, and in addition the slip planes are assumed to be isotropically and homogeneously distributed, then the summation reduces to an integration over half the unit sphere,

$$u_{i,j}^d = -\sigma_{jk} \int \nu n_j \pi c^2 L q (\delta_{ij} - n_i n_j) n_k \, d\psi. \tag{14.47}$$

For this illustration ν (Eq. 14.35) is assumed equal to unity, but its variation may make a significant contribution for large strains when **V** is not well approximated by **I**. This makes it possible to make use of the straightforward results

$$\int n_i n_j \, d\psi = \frac{2\pi}{3} \delta_{ij} \tag{14.48}$$

$$\int n_i n_j n_k n_l \, d\psi = \frac{2\pi}{15} (\delta_{ij}\delta_{kl} + \delta_{ik}\delta_{jl} + \delta_{il}\delta_{jk}). \tag{14.49}$$

Then,

$$u_{i,j}^d = \frac{2\pi^2}{15} c^2 L q \sigma_{ij}' \tag{14.50}$$

where

$$\sigma'_{ij} = \sigma_{ij} - \tfrac{1}{3}\sigma_{kk}\delta_{ij} \tag{14.51}$$

is the deviator stress. This tensor appears in plasticity theory as the consequence of an assumption that the flow behavior depends only on the deviator and not on the mean stress. Here, the appearance of the deviator arises from the assumption that the tangential component of traction controls the dislocation velocity. Since σ'_{ij} is symmetric, Eq. (14.50) shows that there is no vorticity induced by the dislocation motion.

The flow law of Eq. (14.50) can also be written in the simpler form

$$d^{d'}_{ij} = \lambda\sigma'_{ij} \tag{14.52}$$

where λ involves a number of features of the microstructure, and the prime on the "d" emphasizes that it is a deviator quantity ($d^{d'}_{ii} = 0$). But, alternatively, λ can be determined from the macroscopic observation that the deviator stresses satisfy the von Mises criterion

$$\sigma'_{ij}\sigma'_{ij} = 2Y^2 \tag{14.53}$$

for plastic flow (with Y the flow stress in simple shear). This is done by applying Eq. (14.23) to the case of combined elastic and plastic flow. Then λ can be found by operating on both sides of the equation with σ'_{ij} and noting that $\sigma'_{ij}\dot{\sigma}'_{ij}=0$ (more on this following Eq. (14.90)), resulting in the expression

$$\lambda = \sqrt{\sigma'_{ij}d'_{ij}/2Y^2}. \tag{14.54}$$

This is sufficient to determine λ in macrosopic calculations, showing that detailed knowledge of the microstructure is not necessary for this simple, isotropic, small-deformation, flow. However, realistic extension to more complex behavior generally requires some knowledge of the microstructure.

If, following Prager (1955), a "back stress" $\boldsymbol{\alpha}$ is introduced to account for hardening, then Eqs. (14.45a,b) can be replaced by a modified set

$$\mathrm{v}_i = q_0 T'_i, \qquad T'_i = (\delta_{ij} - n_i n_j)\tilde{\sigma}_{jk} n_k \tag{14.55a,b}$$

in which the stress acting on a dislocation loop is replaced by a reduced stress

$$\tilde{\boldsymbol{\sigma}} = \boldsymbol{\sigma} - \boldsymbol{\alpha}. \tag{14.56}$$

In kinematic hardening, macroscopic considerations of hardening and hysteresis suggest that we describe this effect with the relation

$$\hat{\boldsymbol{\alpha}} = b\mathbf{D}^p \tag{14.57}$$

with b the hardening modulus. Microscopically, the single crystal in the grain interior may be considered to be soft as a result of mobile dislocations, while the grain boundary is considered to be hard as the result of dislocation pileups and disclinations. As a result, the grain-boundary region carries a significant part of the loading, shielding the interior which, as a result, is subject to the reduced stress $\tilde{\boldsymbol{\sigma}}$. The concept of grain-boundary shielding is not part of the usual argument for kinematic hardening, which is based, rather, on macroscopic observation of bilinear behavior in plastic flow. It is consistent, however, with the interaction of dislocations and grain boundaries described by Petch (1954) to account for the effect of grain size on strength. Many authors suggest that the back stress is caused by the effect of impurities on dislocation motion but, insofar as I am aware, the suggestion has not been quantified. This subject merits additional study.

In propellants and PBX explosives. hard grains of energetic materials are held together by a binder that usually has very complex rate-dependent behavior. To illustrate the use of Statistical Crack Mechanics (SCM) for such materials we consider the binder to be represented by an incompressible (isochoric is a better term) viscous fluid with viscosity μ in a channel of width h. The viscous behavior represents the most compliant aspect of the binder response and, hence, can be considered to idealize its contribution in the first approximation. Then the contribution of the binder to the velocity difference is given by Eq. (14.27) with the relative velocity across the binder given by

$$\delta u_i = \frac{h}{\mu} T_i' \tag{14.58}$$

where, as in Eq. 14.45, T_i' represents the tangential component of traction. We consider the binder to consist of thin disks of material with volume $\pi c^2 h$. Then, taking L^b to be the number density of binder disks, the volume fraction of binder is

$$f^b = \pi c^2 h L^b. \tag{14.59}$$

Following the same procedure as used in obtaining Eq. (14.28) and illustrated in Fig. 14.3,

$$\tilde{L}^b = \pi c^2 L^b \cos \gamma \tag{14.60}$$

is the average number of binder disks per unit length. Then, following the procedure outined for the example of kinematic hardening,

$$u_{i,j}^b = \frac{4\pi}{15} \frac{f_b}{\mu} \sigma_{ij}' \tag{14.61}$$

is the contribution of binder to the velocity gradient. Note that the divergence of this contribution is identically zero.

14.4. Reduced Moduli

Just as we analyze elastic and plastic strain rates in the flow of ductile materials, we need to study the effect of microfracturing in brittle materials such as rocks, ceramics and explosive grains. In this section a calculation by Dienes (1978, 1980, 1981, 1982a,b, 1983a, 1989), of the reduction in elastic moduli due to a distribution of microcracks is summarized, as well as the added strain rate due to microcrack growth. A strain rate due to microcrack growth occurs only when the stress level is above a critical value. The 1989 derivation considers finite-deformation effects, which include crack rotation, but this leads to a variety of geometric complications that may not be warranted in view of our limited knowledge of the physics of microfracturing. Thus, only the simplest version of these results is presented here. In many practical problems involving large deformation most of the strain is plastic or viscoelastic, and the elastic and microcrack strains are small. Thus, the formulation presented here may be adequate for many large deformation problems, though the fracture analysis is based on results from linear theory.

The compliance of the matrix material is denoted by $\mathbf{C}^m$, and superscripts $(\ )^o$ and $(\ )^s$ are used to denote the added compliance due to open cracks and shear cracks. Then the net compliance can be written

$$\mathbf{C} = \mathbf{C}^m + \mathbf{C}^o + \mathbf{C}^s. \tag{14.62}$$

The compliance is given in terms of microscopic quantities by Dienes and Margolin (1980); for open cracks

$$C^o_{ijkl} = \beta \sum c^o_{ijkl} F(\theta, \phi) \Delta\Psi, \tag{14.63}$$

and for closed (shear) cracks

$$C^s_{ijkl} = \beta \sum c^s_{ijkl} F(\theta, \phi) \Delta\Psi, \tag{14.64}$$

where the summation is taken over discretized crack orientations. The quantities c_{ijkl} are the fabric tensors: for open cracks

$$c^o_{ijkl} = b_{ijkl} - \nu_e a_{ijkl} \tag{14.65}$$

where

$$a_{ijkl} = n_i n_j n_k n_l \tag{14.66}$$

and

$$b_{ijkl} = \tfrac{1}{2}(\delta_{il} n_j n_k + \delta_{jl} n_i n_k + \delta_{ik} n_j n_l + \delta_{jk} n_i n_l). \tag{14.67}$$

This result makes use of the exact solution for the penny-shaped crack (a term "coined" by the British), which assumes an isotropic elastic matrix material—hence the use of ν_e to represent Poisson's ratio. For shear (closed) cracks the corresponding result is

$$c^s_{ijkl} = b_{ijkl} - 2a_{ijkl}. \tag{14.68}$$

Cracks are treated as open in the numerical model when the normal component of traction is positive (tensile), and otherwise as closed. Crack orientation may vary significantly when the material deformation is large, and then the normal has to be calculated from Eq. (14.33). In some cases it is sufficient to use the initial value $\mathbf{N}$ for $\mathbf{n}$. The effective crack volumes are

$$F^o = \int_0^\infty n^o(c, t) c^3 \, dc \tag{14.69}$$

for open cracks and

$$F^s = \int_0^\infty n^s(c, t)c^3 \, dc \tag{14.70}$$

for closed (shear) cracks. (These are not real volumes, but quantities that come out of the analysis and have an interpretation as volumes circumscribed on the cracks.) The constant

$$\beta = \frac{8}{3\mu}\frac{1 - ve}{2 - ve} \tag{14.71}$$

arises from analytic solutions for the open and closed penny-shaped cracks, as discussed by Dienes (1989).

In addition to changing the compliance, cracks can contribute to the strain rate by growing when the stresses are high enough to cause instability, causing the crack volumes of Eqs. (14.69) and (14.70) to change. The criterion for instability of open cracks is

$$c > \frac{\pi}{2}\frac{2 - ve}{1 - ve}\frac{\gamma\mu}{A - veB^2/2}, \tag{14.72}$$

where

$$A = \sigma_{ij}n_j\sigma_{ik}n_k, \qquad B = \sigma_{ij}n_in_j. \tag{14.73a,b}$$

For closed cracks the criterion is

$$c > \frac{\pi}{2}\frac{2 - ve}{1 - ve}\frac{\gamma\mu}{(\sigma_s - \tau)^2}, \qquad \sigma_s = \sqrt{A - B^2}, \qquad \tau = -\bar{\mu}B \tag{14.74a,b,c}$$

accounting for the effect of friction through the coefficient of friction $\bar{\mu}$. The quantity σ_s represents the tangential component of traction on a crack. The stretching due to growth of open cracks is

$$d_{ij}^o = \dot{C}_{ijkl}^o\sigma_{kl} \tag{14.75}$$

and for closed cracks

$$d_{ij}^s = \dot{C}_{ijkl}^s\sigma_{kl}. \tag{14.76}$$

If cracks grew without coalescence the statistical problem would be trivial but, in fact, crack intersections are a vital part of material failure. If cracks grew without coalescence the largest crack would be the most unstable, and it would grow and separate the mass very rapidly while the others are unaffected. Though this can happen in some circumstances, in this article we consider only the alternative, fragmentation, in which failure is the result of microcrack growth and coalescence, and no single crack dominates. This is consistent with the observation that the stress-strain curves for brittle materials often show a gradual reduction of compliance as the stress increases. Furthermore, we observe crack networks with complex structure in micrographs of damaged material. This growth and coalescence of cracks results, ultimately, in fragmentation if the loading is of sufficient duration and intensity. The view of fragmentation taken here is that crack growth can be investigated in detail using known methods, whereas the study of fragments per se does not involve any well-defined physical processes. The statistical theory occupies the remainder of this section.

In this simplest version of crack statistics it is assumed that crack growth terminates when the crack edge, which is approximately circular, intersects α_c other cracks. It is shown by Dienes (1978) that this behavior is characterized by a Liouville equation

$$\dot{L} + L_c = -\dot{M} \tag{14.77}$$

where $L(c, \theta, \phi, t)$ denotes the number of cracks with radius exceeding c that have not coalesced at time t, and have a normal characterized by θ and ϕ. Growth does not affect orientation. $M(c, \theta, \phi, t)$ represents the number density of inactive cracks. In this model cracks are divided into active and inactive sets. Those capable of further growth are called active, and have distribution L; those that cannot grow further as a result of intersections are called inactive and have distribution M. It is assumed in this formulation that cracks remain plane. In some microfracturing studies such as Nemat-Nasser and Obata (1988) cracks are assumed to develop wings as they grow under compression, and this is observed in the experiments of Nemat-Nasser and Horii (1982), but these are not observed in micrographs or in the field. On the other hand, clear evidence for planar growth may be seen in the experiments reported by Howe, Gibbons and Webber (1985). An explanation for the absence of wings is given by Kalthoff and Winkler (1987), who observe that under dynamic conditions cracks tend to grow in their own planes, whereas at low strain rates there is a tendency for wings to form. Another key to the difference appears in the paper by Horii and Nemat-Nasser (1986) who note that under near hydrostatic conditions cracks tend to grow in their own plane, whereas when the stress state is strongly biaxial (a large difference in principal strains) there is a tendency for wing cracks to develop. The circumstances that favor high strain rates also tend to favor nearly hydrostatic stress states and, hence, cracks growing in their own plane.

The simplest assumption for coalescence is the depletion law

$$\dot{M} = kL \tag{14.78}$$

and an analysis of the mean free path given by Dienes (1989) leads to the expression

$$k = 2\pi^4 \dot{c} L_0 \bar{c}^2 / \alpha_c \tag{14.79}$$

for the coalescence constant, assuming an exponential initial distribution,

$$L = L_0 e^{-c/\bar{c}}. \tag{14.80}$$

In Eq. (14.79) $\dot{c}$ denotes the crack speed and α_c the number of intersections needed to terminate crack growth. The Liouville equation obtained by combining Eqs. (14.77) and (14.78) has a particularly simple solution when the initial distribution is exponential, which, fortunately, is what is usually observed. The solution is piecewise continuous, with a discontinuity in slope at $c = \dot{c}t$. For $c > \dot{c}t$:

$$L = L_0 e^{-(c-\dot{c}t)/\bar{c}-kt}, \tag{14.81a}$$

$$M = L_0 \frac{k}{\bar{\beta}}(e^{\bar{\beta}t} - 1)e^{-c/\bar{c}}, \tag{14.82b}$$

while for $c < \dot{c}t$:

$$L = 0 \tag{14.83a}$$

$$M = \frac{L_0}{\bar{\beta}}\left(\frac{\dot{c}}{\bar{c}}e^{-kc/\dot{c}} - ke^{-c/\bar{c}} - \bar{\beta}e^{-kt}\right) \tag{14.84b}$$

where

$$\bar{\beta} = \dot{c}/\bar{c} - k. \tag{14.85}$$

It is straightforward to show that the total number of cracks remains constant in this solution. A more general expression for k accounts for the initial anisotropy induced by bedding cracks, Eq. 3.29 of Dienes (1987a), and the effect of increasing cross section is discussed by Dienes (1987b). It is not difficult to modify the theory to account for variable crack speed $\dot{c}$ by introducing total crack growth as a new independent variable replacing the time, t.

14.5. Synthesis of a Unified Constitutive Law

A constitutive law for combined plasticity and microcracking is considered here, with the plasticity represented by kinematic hardening and the microcracking represented by the variable modulus theory of the previous section. The combination of these processes is justified by the superposition principle of Section 14.3. The effect of plasticity could be represented, in principle, by an assumed defect structure, as in Johnson (1992), but it is simpler to treat it as part of the matrix behavior, and our current understanding of dislocations does not justify complex modelling in SCM at this time (though the issue remains open). Computer algorithms could be developed to account for simultaneous microcracking and dislocation motion, but the current craze for large-scale computers is absorbing most of the available programming resources.

The premise for kinematic hardening is that the yield surface translates at a rate proportional to the plastic stretching so that

$$f(\tilde{\boldsymbol{\sigma}}') = \tilde{\sigma}'_{ij}\tilde{\sigma}'_{ij} - 2Y^2 = 0 \tag{14.86}$$

where f represents the plastic potential and $\tilde{\boldsymbol{\sigma}}$ is defined by (14.56) The back stress, $\boldsymbol{\alpha}$, is governed by the hardening law of Eq. (14.57). The linear hypoelastic constitutive law is an adequate approximation for most purposes, since the elastic strain is small, (the main contribution usually being the plastic strain):

$$d^e_{ij} = C_{ijkl}\hat{\sigma}_{kl} \qquad \text{or} \qquad \hat{\boldsymbol{\sigma}} = \mathbf{C}^{-1} : \mathbf{D}^e. \tag{14.87}$$

In numerical calculations the elastic stretching can be obtained from the superposition principle by

$$\mathbf{D}^e = \mathbf{D} - \mathbf{D}^p - \mathbf{D}^{oth} \tag{14.88}$$

where $\mathbf{D}^{oth}$ denotes the stretching due to other processes (particularly, crack growth here, but there are other possibilities such as nonlinearities from the equation of state). To close the set of constitutive equations we write for the plastic stretching

$$d^p_{ij} = \lambda \frac{\partial f}{\partial \sigma'_{ij}} = \lambda \tilde{\sigma}'_{ij}. \tag{14.89}$$

Substitution of Eqs. (14.87) and (14.89) into Eq. (14.88) followed by multiplication by σ'_{ij}, contraction, and use of the relation

$$\tilde{\sigma}'_{ij}\hat{\tilde{\sigma}}'_{ij} = 0, \tag{14.90}$$

leads to an explicit representation for the Lagrangean multiplier

$$\lambda = \frac{\mathrm{tr}\{\tilde{\boldsymbol{\sigma}}' : \mathbf{C}^{-1} : (\mathbf{D} - \mathbf{D}^{oth})\}}{2bY^2 + \mathrm{tr}(\tilde{\boldsymbol{\sigma}}' : \mathbf{C}^{-1} : \tilde{\boldsymbol{\sigma}}')}. \tag{14.91}$$

It should be noted that Eq. (14.90) is equivalent to $\tilde{\sigma}'_{ij}\dot{\tilde{\sigma}}'_{ij} = 0$ as a result of symmetry considerations. Eqs. (14.87) and (14.12) can be combined to produce the constitutive relation

$$\dot{\boldsymbol{\sigma}} = \mathbf{C}^{-1} : (\mathbf{D} - \mathbf{D}^{oth} - \lambda\tilde{\boldsymbol{\sigma}}') + \boldsymbol{\Omega}\boldsymbol{\sigma}' - \boldsymbol{\sigma}'\boldsymbol{\Omega}. \tag{14.92}$$

This expression can be used to update the stresses in a finite-difference scheme.

14.6. Application to XDT

The discovery that mild mechanical shocks could lead to violent explosions in solid propellants was of great concern to rocket designers and users. This anomalous behavior was observed in experiments at Hercules by Jensen, Blommer, and Brown (1981) and at Lawrence Livermore National Laboratory by Green et al. (1981). They found that impact of propellant cylinders onto a steel plate at rather low speeds could produce explosions that were at least as violent as those caused by high-speed impact, which are initiated by relatively strong shock waves and referred to as SDT (Shock to Detonation Transition). Since the cause was unknown, the detonation process was called XDT. Two features of the low-speed impacts were particularly remarkable. One was that the high-pressure air shocks measured adjacent to the impact were produced at very late times, on the order of 40 microseconds, whereas SDT occurred within 2 microseconds for high-speed impacts. The second remarkable feature was that the low-speed shots were not reproducible, the high pressures being generated in only about one out of five cases.

At Los Alamos we proposed Dienes (1982b) that this anomalous behavior could be explained by the ignition of hot spots. The irregular size and density of microscopic flaws would explain the erratic character of the results. Many mechanisms for hot-spot formation are known, but order-of-magnitude analyses Dienes (1983b) suggested that frictional forces in shear cracks are the most vigorous source of heating. Interfacial heating of cracks is quite different from the void-collapse mechanisms most often considered. Its occurence is difficult to confirm since nucleation sites for shear cracks, which contain no void volume, are not readily observable and care is required to dispay even relatively long shear cracks. Though many collegues have maintained that shear cracks do not exist, they are clearly visible in the experiments of Howe et al. (1985), Chaudhry (1992), and Elban et al. (1993). Shear cracks in reactive materials are of particular interest since they can be more readily observed than in nonreactive solids, and because of their potential role in deflagrations and detonations. They are especially visible because of the reactive products they leave and because of their low melting point. Relatively large volumes are involved in the frictional heating in shear cracks compared to crack-tip heating, which involves a relatively small volume but is sometimes proposed as an ignition mechanism. The importance of the size effect in ignition was demonstrated by Frank-Kamenetskii (1942) when he solved the governing equations in closed form. His analysis concerned, however, the steady state heat equation, while we are concerned with unsteady, erratic processes in XDT. Especially vigorous heating occurs during crack growth, but some heating can occur in stable cracks as a result of interfacial sliding. As a result of this complexity we are forced to numerical solution of the governing equations. These equations govern only very thin zones adjacent to the cracks, and must be solved with a very fine grid, typically with zones well under a micron in thickness. Fortunately, it seems adequate to solve the equations in one dimension, since the heated regions are thin compared to the crack length.

In the first calculations Dienes (1982b) assumed that when the temperature reached the melting point the frictional heating would decrease so as to just maintain the temperature at the melting point. It seemed probable at the time, however, that the melting point was raised by several mechanisms involving the local pressure. Since then I have revised this scenario, which required some assumptions that did not hold up under closer scrutiny. It now appears that the temperature can rise well above the melting point in a very thin boundary layer. Though viscous heating in liquids is usually negligible, in the thin boundary layer of a rapidly shearing crack this effect can be significant. In the series of impact tests on 105 mm shells performed by Howe et al. (1985), cracking at low speeds was observed in the TNT and violent explosions occurred at high speeds. At the higher impact speeds (on the order of 600 m/s), Howe infers (1993) that temperatures well above melting must have occurred, since TNT can be safely maintained at the melting point for days. It is likely that the high temperatures were the result of viscous heating that occurred as the result of interfacial sliding in the cracks, which have grown to centimeters in length at ignition time. There is some evidence of melting in micrographs of TNT that was recovered. Some examples of crack temperature histories are shown in Figure 14.6.

In these calculations the frictional heating per unit area is given by

$$\dot{q} = \bar{\mu}\tau v \tag{14.93}$$

where τ is the tangential component of traction, $\bar{\mu}$ is the coefficient of friction, and v is the sliding velocity. The heating acts as a boundary condition for the heat equation

$$\rho C_v \dot{T} = kT_{xx} + \rho\dot{Q}. \tag{14.94}$$

The heating per unit mass, $\dot{Q}$, is given by the sum of chemical and mechanical source terms as

$$\dot{Q} = Q_r v e^{-E/kT} + \dot{\varepsilon}[(1-f)Y + f\mu\dot{\varepsilon}] \tag{14.95}$$

The first term on the right accounts for chemical reactions and the second for viscoplastic deformation, with Y the flow stress in shear of the solid phase;

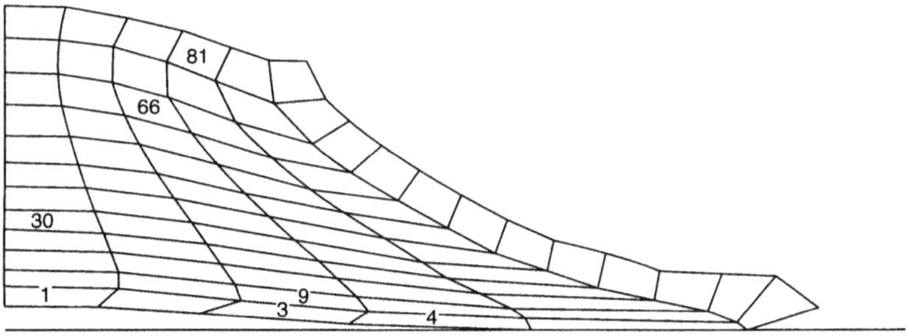

Figure 14.6a. Deformation of an explosive cylinder at 30 μs as computed by SCRAM, showing the finite-difference grid. This represents the shotgun test in which cylinders of sample material are fired against a steel target to determine explosive and propellant sensitivity. In these calculations the target is considered rigid. The numbers in certain cells represent the cell numbers, and the crack temperatures for the numbered cells are plotted in the succeeding figures. The temperatures are plotted for cracks oriented so that they bisect the coordinate axes of the figure.

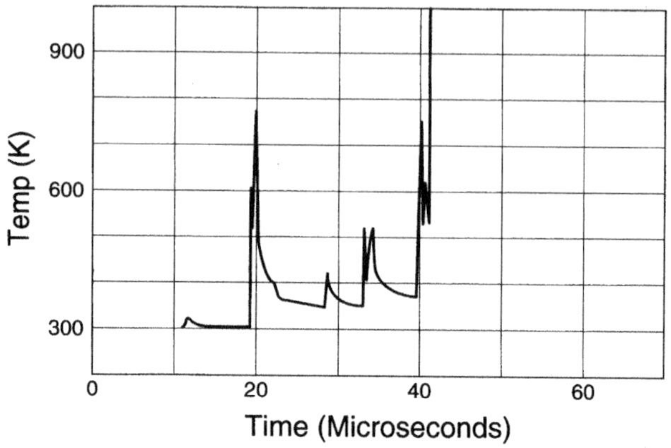

Figure 14.6b. The surface temperature of cracks that bisect the coordinate axes of Fig. 14.6a is shown for cracks in cell 1. Ignition occurs at approximately 42 μs. Melting of HMX occurs at 519 K, and its effect on the temperatures can be seen in inhibiting both the rise and fall of temperature. At ignition the temperature rises to a very large value in a very short time, less than a single time step, 10 ns in this calculation. Including depletion terms in the chemistry or changing the time step does not significantly modify the rapid increase in temperature.

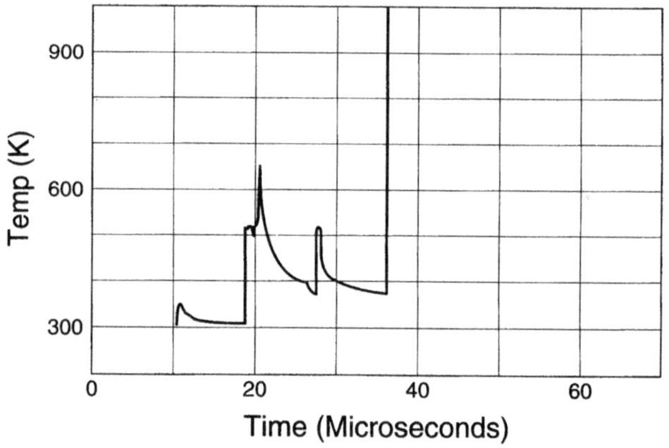

Figure 14.6c. This history for cell 3 illustrates the resistance to temperature changes offered by the latent heat of melting, as well as the increase in temperature above the melting point due to the high strain rate in the molten boundary layer. The rapid increase in temperature occurs during periods of crack growth, and is followed by slow decay as heat is conducted away from the crack surface.

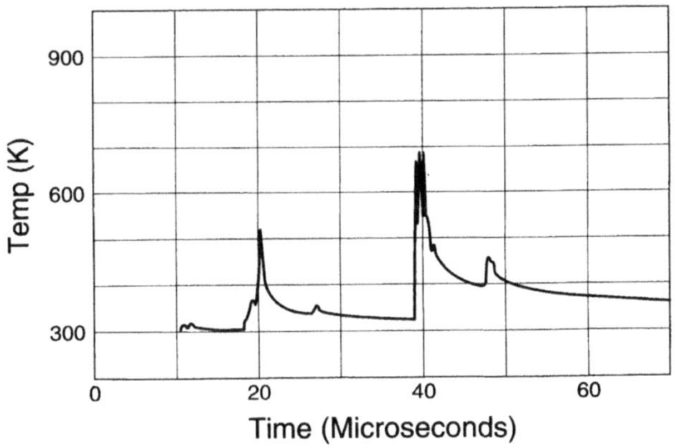

Figure 14.6d. The temperature changes in cell 4 are similar in character to those in cell 3, but in this case ignition does not occur.

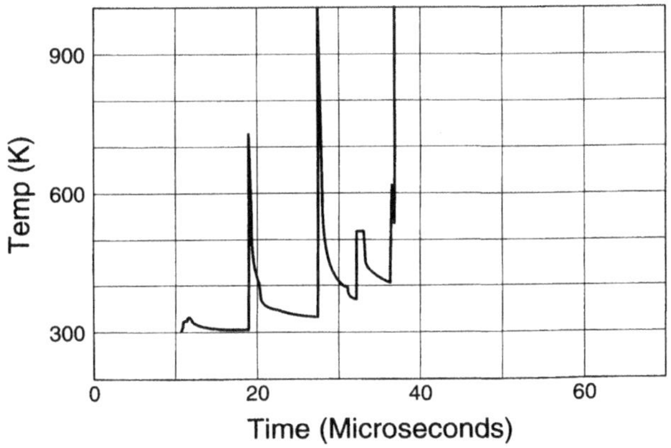

Figure 14.6e. Cell 9 is adjacent to and above cell 3, and ignition occurs at roughly the same time, though details of the temperature spikes are somewhat different.

μ, the viscosity of the melt; $\dot{\varepsilon}$, the strain rate; and f, the fraction of the phase transition completed. Only at or above the melting point does the second term contribute to the heating. The boundary condition at the crack surface is taken as

$$-kT_x = \dot{q}. \tag{14.96}$$

The sliding velocity of Eq. (14.93) used to compute $\dot{q}$ is not necessarily the crack velocity, since the interfacial material may be melting or molten. We take the sliding velocity

$$v = (1 - f)v_c \tag{14.97}$$

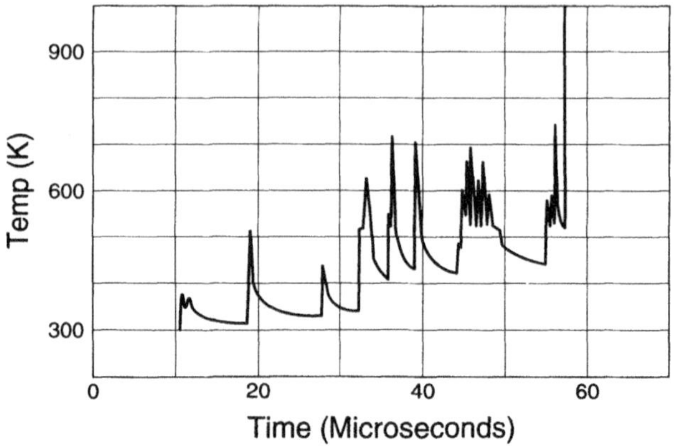

Figure 14.6f. The temperature in cell 30 shows a secular increase interrupted by occasional spikes, culminating in ignition at 57 μs.

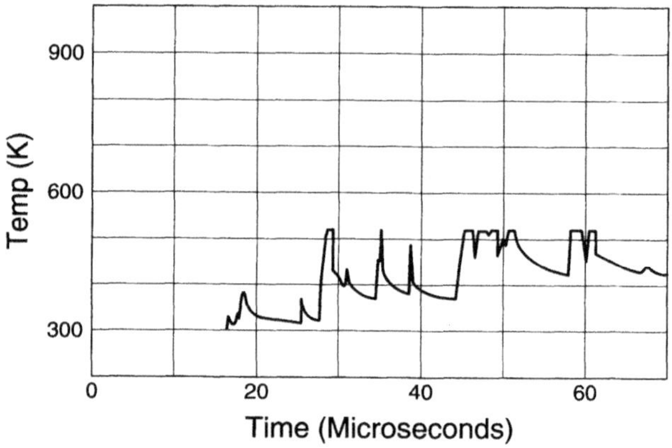

Figure 14.6g. The temperature spikes in cell 66 never rise above the melting point, the heating being limited by the relatively low pressures far from the impact plane.

to be the fraction of the total velocity v_c that is associated with sliding, and the remainder is associated with local shearing in the melting and molten material adjacent to the crack. Thus,

$$\dot{\varepsilon} = fv_c/l \tag{14.98}$$

where l is the width of the region below the melting point.

The ignition calculation has been embedded in a finite-difference code we call SCRAM (Statistical CRAck Mechanics) that uses the calculation method described in the body of this report. This is a variant of the original ALE code developed by Hirt, Ruppel and Amsden, and revised in the SALE code by Amsden, Ruppel, and Hirt (1980), which is used in current calculations.

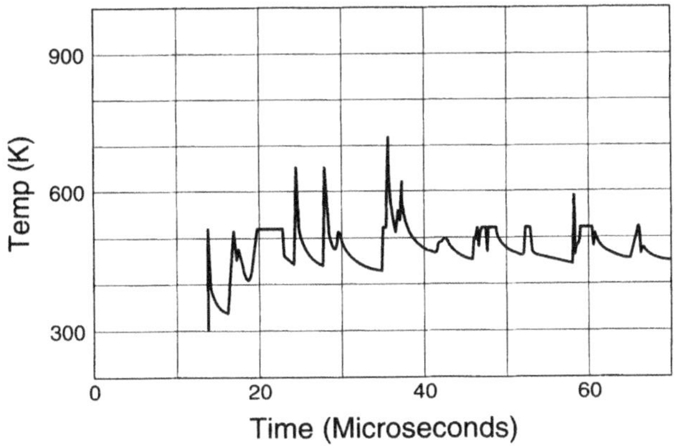

Figure 14.6h. In cell 81 it can be seen that the temperature can spend relatively long time at the melting point without ever igniting.

This is reported by Dienes in the Theory of Deformation, Part III. The crack velocity is computed as the central velocity of a penny-shaped shear crack

$$v_c = \frac{4(1-v)}{\pi\mu}(\dot{\tau}c + \tau\dot{c}) \tag{14.99}$$

with τ the absolute value of the tangential component of traction. Of course, $\dot{c}$ is zero if the crack is stable. The mechanical constants for propellants and explosives are not well known, so the values used in these calculations, given in the table below are only best guesses. These values are representative of solid propellants using HMX as a major ingredient, except that in some cases the values are those of HMX. The values are in many cases based on the Los Alamos Handbook, Gibbs and Popolato (1981). It is unfortunate that the complete set of constants is not available for any single material. This is a consequence of the historical fact that a complete set of properties could not have been used in practical calculations until modern computer methods for solid mechanics became available. Now, there is a critical need for more data on the mechanical properties of reactive materials.

The results of SCRAM calculations are given in Fig. 14.6, which shows the deformantion at 30 μs and the temperature history on shear cracks in several computational cells that are of particular interest. The target was considered rigid in these calculations. More recently we have incorporated the SCM algorithm into PRONTO (Taylor and Flanagan, 1987), a finite element code that has several features not available in SALE—a general multimaterial capability, a generator that allows for setting up very complex geometries in two and three dimensions, and slide lines between materials. Results of PRONTO calculations are shown in Figs. 14.7–14.9. The iron target is not shown in the figure, since it contributes very little to the deformation. Fig. 14.7 shows the variation in density following impact, Fig. 14.8 shows crack

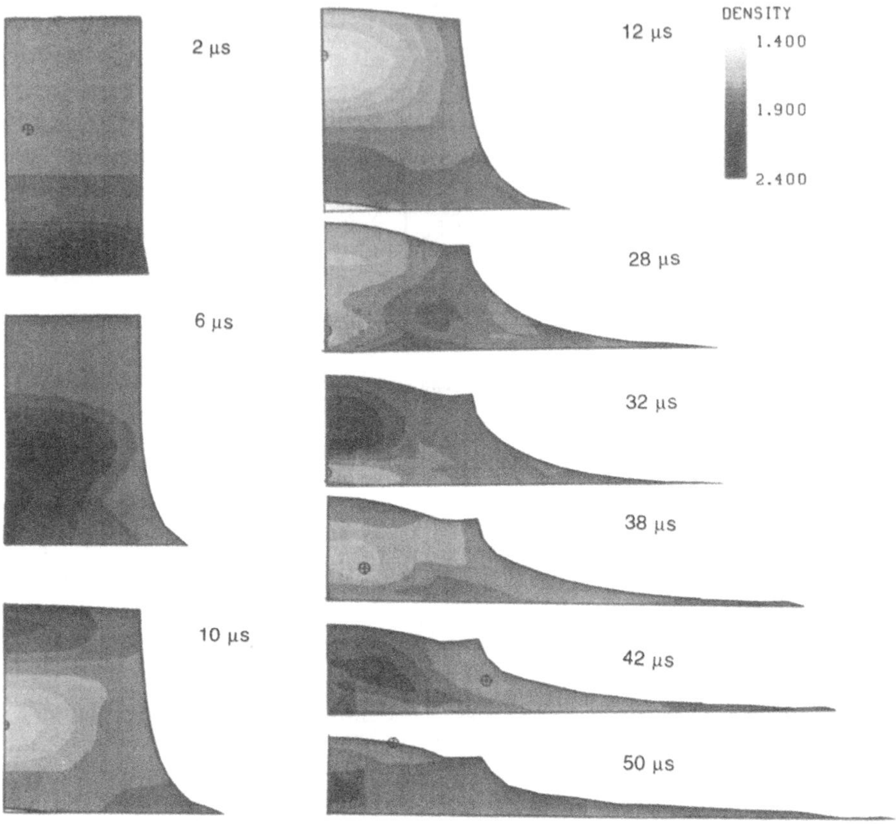

Figure 14.7. Impact of an axisymmetric high-explosive cylinder against a steel cylinder at 500 m/s. The initial density was 1.8 g/cc and the other properties are given in the table. These calculations were performed using PRONTO with the cylinder resolved into a 6 × 12 finite-element grid with equal sides. Note that the general shape at 30 μs computed with the SCRAM code agrees well with the shapes at 28 and 32 μs shown above. The shock wave at early times is not well resolved because of the small number of zones, but there is no difficulty, and only small expense, in running with higher resolution. These calculations are done as part of a study to explore the feasibility of analyzing brittle fracture and explaining XDT, and no claim, is made concerning their quantitative validity. However, the main features of the shotgun test are modelled adequately insofar as they are known.

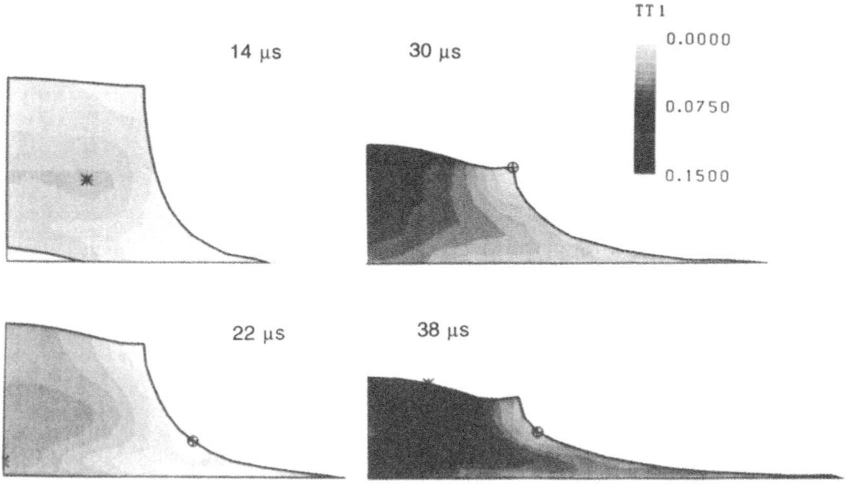

Figure 14.8. Cracks bisecting the coordinate axes are represented in this figure. Their radius is denoted by TT1. The initial radius of 100 μm has increased almost 15-fold at 38 μs. Their deformation is generally the result of shear occurring in the impact process, though the cracks can open when the normal component of traction is tensile. Frictional heating of these cracks is responsible for the high temperatures and occasional ignition illustrated in Fig. 14.6. The thesis of this work is that such cracks are responsible for XDT and other explosion phenomena that occur in explosives and propellants at late times.

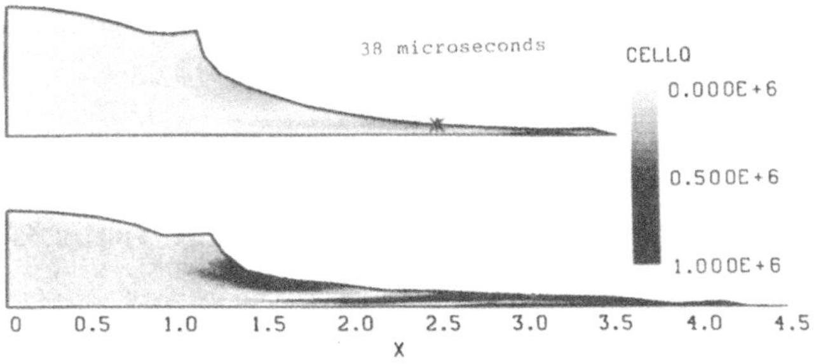

Figure 14.9. Plots of CELLQ, the chemically released energy per unit mass due to the cracks in set TT1. The energy is concentrated in the neighborhood of shear cracks. The maximum of 215 J/kg at 50 μs is much less than the (roughly) 4 MJ/kg potentially available in explosions, but is concentrated in hot spots that may be adequate to initiate total reaction. This quantity depends strongly on the number and size of cracks initially in the explosive.

growth, and Fig. 14.9 shows the chemical heating, which is the integrated value of the first term on the right of Eq. (14.95). The transition from ignition to burn is currently under study.

Table of Material Constants (cgs units, except as noted)

AMU, shear modulus	1.18e9
AMUB, static coefficient of friction	0.7
AMUBD, dynamic coefficient of friction	0.3
ANU, Poisson's ratio	0.4924
BKH, strain hardening	0.1e9
CBARZ, initial radius of cracks	0.0100
CD, speed of growth of open cracks	6.e3
CDS, speed of growth of closed cracks	3.e3
CV, specific heat	8.0e6
GAMMA, surface energy for fracture toughness	1.e5
GRU, Gruneisen's ratio	1.5
S, Slope of $U_s - U_s$ line in HOM Equation of State	1.4
SCRN, number of cracks per cc per 2π	1.0
YZ, flow stress in pure shear	1.5e9
AK, conductivity	1.e-3
E, energy barrier in Arrhenius kinetics, cal/mole	52700
Q, Heat of Reaction, cal/g	500
TZERO, initial temperature	300
Z, pre-exponential factor in Arrhenius kinetics	5e19
TMLT, melting point, degrees Kelvin	519
VISCM, viscosity	0.0726
HTMLT, latent heat of melting	0.979e9

14.7. Prospects for Further Developments and Applications

The analysis described here relates deformation and failure processes on the macroscopic and mesoscopic scales. Though it might appear interesting to attempt another level of unification by accounting for phenomena on the atomic scale, this is not promising in connection with the current effort for several reasons. First, because numerical calculations would require a time step much smaller than those used in applications. In addition, the behavior on the atomic scale is so sensitive to impurities and their interaction with grain boundaries that results for a few thousand atoms would not be useful for ordinary materials. Though one might hope for molecular dynamics to provide insights into how fracture toughness, hardening modulus, and other material constants depend or processes, it is not clear how such calculations could account for the effects of grain structure and grain history (which dominate the strength, (Petch, 1955)), even if the behavior of the 10^{16} atoms required to simulate the strength of a few dozen grain were accounted for. At a more fundamental level, the atomic force models in common use do not reflect many of the realities of binding on the atomic level, though in special cases progress is being made. Atomic scale simulations might be useful to develop some of the constants used in mesoscale calculations, but this has not yet been demonstrated.

The approach described here provides a way to account for microstructure and, hence, for phenomena that are not accessible to macroscopic calculations, such as permeability resulting from crack networks and explosive hot spots. Analysis of geophysical processes such as earthquakes, volcanoes, impact cratering, and triggering of avalanches provides another application of such theory. A drawback of the method is that many of the constants required are not normally known for specific materials. At SRI a parallel effort by Seaman, Curran, and Murri (1985) and many others has involved experiments capable of generating the data necessary for their modelling effort. Ideally, theory, numerical calculations, material characterization and dynamic experiments should go together, but most studies have been too fragmented to allow this kind of modelling to reach a useful level of sophistication. The urgency of the cold war was frequently used to justify the absence of long-term planning. Perhaps the new world situation will allow for a more coherent expenditure of research funds.

Many processes require additional study before credible predictions can be made. Seven examples are cited here. Dilatancy needs to be investigated, a process that depends on the microscopic generation of voids under compression but eludes quantitative description. Strain-rate effects due to the finite speed of crack growth are predicted by the approach in this article, but better data are needed and the results should be tested. Size effects are well-known to civil engineers but have been inadequately characterized in constitutive modelling. Much data is based on tests of very small samples and thereby ignores the effect of large defects. Phase changes and other high-pressure effects need to be accounted for. The coalescence model described here assumes a constant mean free path, but it seems clear that in many situations the mean free path will decrease in the course of fragmentation as crack size increases. A theoretical approach to this process has been described by Dienes (1985) but has not been implemenented in any code. Many practical materials are composites, and failure of such materials has not been sufficiently characterized. The brittle-ductile transition is mentioned in this chapter but it depends on so many effects that we can consider it far from adequately described.

Many scientists with whom I have discussed the analysis described here suggest that it should be tested experimentally, but this does not seem promising, per se, for several reasons. First, the material properties and microstructure are not well enough known for any single material to carry out promising calculations. Though, in principle, the appropriate measurements could be made, a major effort would be required which no existing laboratory is equipped to undertake. Second, even if such an effort were carried out successfully for one material, it still would not prove much about other materials—there are too many variations. If the effort were not successful or, more likely, were ambiguous, it would not prove the underlying concepts wrong, but merely that there are additional effects that have not been accounted for.

More promising is to apply this method in special projects where there is enough interest to establish the required material properties. They can be determined from stress-strain curves and other standard tests, or back-fitted by comparison with macrocopic experiments (exploding-wire, impact, etc.), or estimated by analogy with similar materials. It should be emphasized that the properties of interest are standard in mechanics, and can often be estimated from handbook data. For most projects we do not require all the parameters since only a few dominate the processes of interest. In many fragmentation studies we do not normally need a high standard of prediction —all that is necessary is an understanding of the main effects—still a considerable challenge. The general approach presented here is useful in allowing analysts to put various processes into context. What is needed now is a more complete theory and better numerical modelling. The success of the theory will depend on its usefulness in a variety of case histories. It is unlikely that any single theory of materials can predict all possible kinds of failure, but a unified approach helps us to relate and interpret many observations that are otherwise disconnected. In addition, the current approach allows us to understand better the variability of material properties that results from variability in defects, and this may lead us to more effective quality control at the process level. This is the probably the most important application.

14.8. References

Amsden, A.A., Ruppel, H.M., and Hirt, C.W. SALE: A Simplified ALE Computer Program for Fluid Flow at All Speeds, Los Alamos National Laboratory Report LA-8095 (1980).

Blewett, P. Private Communication (1992).

Bowden, F.P. and Yoffe, Y.D. *Initiation and Growth of Explosion in Liquids and Solids*, Cambridge University Press (1952).

Chadwick, P. *Continuum Mechanics*, George Allen and Unwin Ltd., London (1976).

Chaudhri, M.M. Photographic evidence for ignition by friction in a deflagrating explosive single crystal, *J. Phys. D: Appl. Phys.* **25** (1992).

Dienes, J.K. A Statistical Theory of Fragmentation. In: Y.S. Kim, ed. Proc. 19th U.S. Rock Mechanics Symposium, Stateline Nevada (1978).

Dienes, J.K. On the Analysis of Rotation and Stress Rate in Deforming Bodies, *Acta Mechanica*, **32**, pp. 217–232 (1979).

Dienes, J.K. and Margolin, L.G. A Computational Approach to Rock Fragmentation, in D.A. Summers, Ed., The State of the Art in Rock Mechanics, Proc. 21st U.S. National Symposium on Rock Mechanics. Rolla, Missouri (1980).

Dienes, J.K. On the Effect of Anisotropy in Explosive Fragmentation, in H.H. Einstein, ed., Rock Mechanics from Research to Application, Proc. 22nd U.S. Symposium on Rock Mechanics, M.I.T., Cambridge, MA (1981).

Dienes, J.K. Permeability, Percolation and Statistical Crack Mechanics, in R.E. Goodman and F.E. Heuze, Issues in Rock Mechanics, Proc. 23rd Symposium on Rock Mechanics, Berkeley, CA (1982a).

Dienes, J.K. A Frictional Hot-Spot Theory for Propellant Sensitivity, in Proc. 2nd JANNAF Propulsion Systems Hazards Meeting, China Lake, Cal. April 19–21 (1982b).

Dienes, J.K. Statistical Crack Mechanics, in J.-P. Boehler, Ed., Failure Criteria of Structured Media, Proc. Colloque International du CNRS no. 351, Villard-de-Lans, France (1983a).

Dienes, J.K. Frictional Hot-Spots and Statistical Crack Mechanics, in Proc., 1983 Annual Meeting of the Materials Research Society, J.H. Crawford, Jr., Y. Chen and W.A. Sibley, eds. Boston, MA. (1983b).

Dienes, J.K. A Statistical Theory of Fragmentation Processes, *Mechanics of Materials*, **4**, 325–335 (1987a).

Dienes, J.K. Theory of Deformation, Part I, Kinematics, Los Alamos National Laboratory Report, LA-11063-MS, Vol. I (1987b).

Dienes, J.K. Theory of Deformation, Part II, Physical Theory, Los Alamos Report, LA-11063-MS, Vol. II (1989).

Dienes, J.K. Theory of Deformation, Part III, The SCRAM code, in Preparation.

Drucker, D.C. and Prager W. Soil Mechanics and Plastic Analysis or Limit Design, *Quart. Appl. Math* **10** (1952).

Elban, W.L., Sandusky, H.W., Beard, B.C., and Glancy, B.C. Investigation of the Origin of Hot Spots in Deformed Crystals: Final Report on Ammonium Perchlorate Studies, Naval Surface Warfare Center Report NSWCDD/TR-92/206 (1993).

Eringen, A.C. *Mechanics of Continua*. John Wiley & Sons, New York.

Frank-Kamenetskii, A.A. On the Mathematical Theory of Thermal Explosions, *Acta Physicochimica* URSS, Vol. XVI, No. 5–6, pp. 357–361 (1942).

Frey, R.B. The Initiation of Explosive Charges by Rapid Shear, Proceedings of the Seventh Symposium on Detonation, Annapolis, MD (1981).

Gibbs, T.R. and Popolato, A. LASL Explosive Property Data, University of California Press (1981).

Green, L.G., James, E., Lee, E.L., Chambers, E.S., Tarver, C.M., Westmorland, C., Weston, A.M., and Brown, B. Delayed Detonation in Propellants from Low Velocity Impact, in Proc. 7th Symposium on Detonation, J.M. Short, Ed., Annapolis, Md. (1981).

Horii, H. and Nemat-Nasser, S. Brittle failure in compression: splitting, faulting and brittle-dactile transition, *Phil. Trans. Royal. Society London*, **319**, pp. 337–374 (1986).

Howe, P. Private Communication (1993).

Howe, P.M., Gibbons, G. Jr., and Webber P.E. An Experimental Investigation of the Role of Shear in Initiation of Detonation by Impact, Proc. Eighth Symposium (International) on Detonation, Albuquerque, NM (1985).

Jeans, J.H. *An Introduction to the Kinetic Theory of Gases*, Cambridge (1940).

Jensen, R. C., Blommer, E.J., and Brown, B. An Instrumented Shotgun Facility to Study Impact Initiated Explosive Reactions, in Proc. 7th Symposium on Detonation, J.M. Short, Ed., Annapolis, Md. (1981).

Johnson, J.N. Micromechanical Considerations in Shock Compression of Solids, in *High Pressure Shock Wave Compression of Solids*, J.P. Asay and M. Shahinpoor, Eds., Springer-Verlag, New York (1992).

Kalthoff, J.F. and Winkler, S. Failure Mode Transition at High Rates of Shear Loading in Impact Loading and Dynamic Behavior of Materials, Vol. I, C.Y. Chiem, H.D. Kunze and L.W. Meyer, Eds., Informationsgesselschaft, Verlag (1987).

Nemat-Nasser, S. On Finite Deformation Elasto-Plasticity, *Int. J. Solids Structures*, **18**, 857–872 (1982).

Nemat-Nasser, S., and Horii, H. Compression-induced Nonplanar Crack Extension with Application to Splitting, Exfoliation and Rockburst, *J. Geophys. Res.*, **87**, 6805–6821 (1982).

Nemat-Nasser, S., and Obata, M. A Microcrack Model of Dilatancy in Brittle Materials, *JAM*, **110** (1988).

Petch, N.J. The Fracture of Metals, in *Progress in Metal Physics*, B. Chalmers and R. King, Eds., Interscience Publishers (1954).

Prager, W. The Theory of Plasticity: A Survey of Recent Achievements, *Proc. Inst. Mech. Engn.* **169**, 41–57 (1955).

Reuss, A. Berucksichtigung der elastischen Formanderung in der plastizitats Theorie, *Zeits. ang. Math. Mech.*, **10**, 266–269 (1930).

Sandler, I. and Baron, M. Numerical Models for Dynamic Loading, in *Mechanics of Geomaterials*, Edited by Z. Bazant, John Wiley & Sons, Ltd. (1985).

Seaman, L., Curran, D.R., and Murri, W.J. A Continuum Model for Dynamic Tensile Microfracture and Fragmentation, *JAM* **52**, 593 (1985).

Spitzig, W.A. and Richmond, O. The Effect of Pressure on the Flow Stress of Metals, *Acta Metall*, **32**, 457–463 (1984).

Taylor, L.M. and Flanagan, D.P. PRONTO 2D A Two-Dimensional Transient Solid Dynamics Program, Sandia Report SAND86-0594, (March 1987).

Truesdell, C. and Toupin, R.A. The Classical Field Theories, in *Handbuch der Physik*, Vol. III. (1960).

Zheng, Q.S. On the generalization of constitutive laws from their arotational forms, *Acta Mechanica* **91**, 97–105 (1992).

CHAPTER 15

Constitutive Modelling of Spall Fracture

John Eftis

15.1. Physical Features of Spall Fracture of Polycrystalline Materials Induced by Plate Impact

Spall fracture (spallation) is a particular kind of dynamic fracture that results from the tension produced by the interaction of propagating waves of rarefaction. Such waves can be caused by the impact of a projectile (flyer) onto the surface of a structure (target) [1], the detonation of a high explosive in direct contact with the target [2], or by sudden deposition of an intense pulse of energy on the target surface [3]. More specifically, in the case of plate impact, the initial compressive stress wave travelling across the target reflects back at the free surface as a propagating decompression wave. A similar process, oppositely directed, occurs in the thinner flyer plate. The superposition of the reflected decompression wave fronts, if sufficient intensity and time duration, can cause partial or complete separation of the material along a plane perpendicular to the direction of the traveling wave fronts, cf. Figure 15.1, [4–6]. Spallation as a particular manifestation of dynamic fracture and fragmentation in general, has been the subject of considerable investigation, particularly at the Stanford Research Institute [7–13], and at the Sandia National Laboratories, [14–18]. Review articles on spall and dynamic fracture by Meyers and Aimone [19], and by Curran, Seaman and Shockey [20], provide access to much of the extensive literature on the subject.

A plate impact spall experiment consists essentially of a thin flyer plate acting as a high speed projectile that strikes the thicker target plate with an impact velocity that ranges usually between 100 to 400 m/s. The transit time of the stress waves that travel across the plates is of the order of 10^{-7}–10^{-6} s, with corresponding strain rates varying between 10^4 and 10^6 s^{-1}. The impact produces mean stress in the 5–50 kbar low- to moderate-pressure shock range. (An explosive type of impact that can cause mean stress ranging from 100 to 500 kbar provides an illustration of high pressure shocks). Because of the rapidity of the spall event the deformation-damage process takes place under approximately adiabatic conditions. In as much as the deforma-

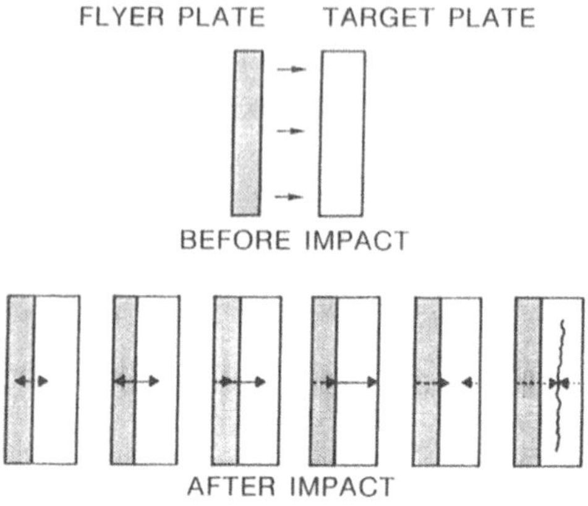

Figure 15.1. Schematic of spall fracture induced by plate impact.

tion that precedes fracture is of small order, temperature variations and thermomechanical coupling induced by the compressive phase of the impact will not be large. The initial porosity in polycrystalline metals is of the order of 10^{-3}, thus the development of crushing type damage as a result of the compressive phase of the wave propagation should be negligible because of the small initial porosity, and also because of the relatively moderate levels of the imposed compressive mean stress. Therefore, for low to moderate impact stress range, the effects of adiabatic heating, mechanical changes of the material, and changes because of thermomechanical coupling, should not be significantly important features of the deformation-material degradation process.

When describing mechanical behavior of polycrystalline materials for ordinary engineering applications, the presence of microvoids within the polycrystal structure is not significant, and is given no consideration. However in circumstances where material degradation occurs, either because of high intensity stress loading, or because of advanced deformation, the porosity of the material becomes an important feature of the mechanical response and cannot be overlooked. Microvoids in polycrystalline metals can vary in size from 10^{-2}–10^{-5} cm., with an average density of the order of 10^6 per cm^3 and an average initial void volume fraction between 10^{-3}–10^{-4}. Post-fracture photomicrographic studies have shown that spallation represents the end result of an accumulation of microdamage that takes place during the tensile phase of the stress wave loading. Figure 15.2 is an illustration of microdamage that consists of somewhat disoriented microcracks of different sizes in one micrograph, and of microvoids of varying size and configuration in the other. The microvoids characterize material failure that is ductile. The microdamage process associated with spall fracture consists mainly of microvoids or microcracks that initially nucleate, enlarge and then join together to form macroscopic surfaces of separation within the material body.

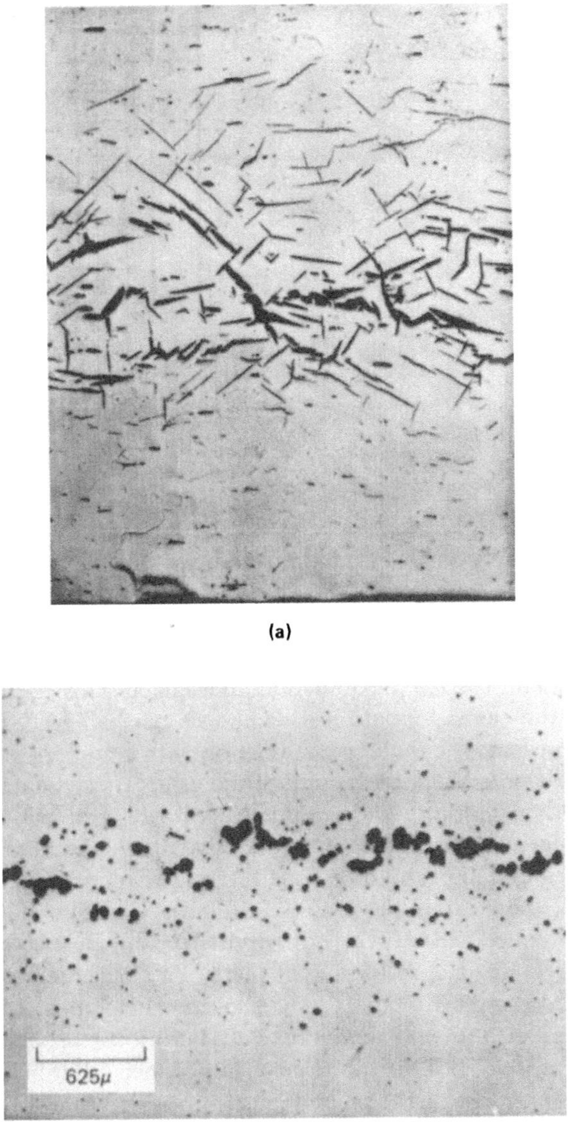

(a)

(b)

Figure 15.2. Photomicrograph of microdamage caused by impact.

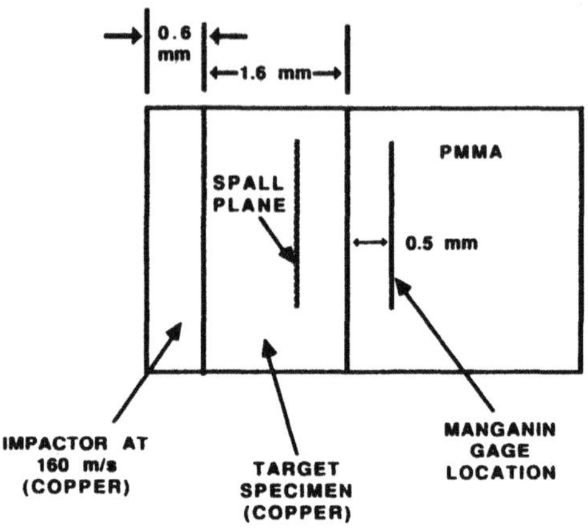

Figure 15.3. Flyer-target-backup plate configuration showing embedded stress gage.

The most common type of plate impact laboratory test utilizes flyer and target plates of equal diameter. The attractiveness of this test configuration for both normal and oblique impact, stems from the occurrence of uniaxial strain across the thickness of the central region of the target for the short time during which the spall phenomenon takes place. This test configuration also provides the easiest situation for analysis of spall fracture. Using an appropriate constitutive model in conjunction with a numerical wave propagation code, it is possible at each point of the target to calculate the velocity, stress, rate of deformation, void volume fraction, as well as other variables at successive increments of time. However, given the extreme physical circumstances that attend high strain rate spall fracture experiments, the measurement of any of these quantities is quite difficult. By means of a laser interferometer technique (VISAR, [21]), it is possible to measure the velocity of points of the rear surface of the target plate. The measurement of stress is made possible by means of a stress gage, embedded in a low impedance buffer material (such as polymethylmethacrylate) located behind the target as shown by Figure 15.3. Due to the large impedance difference between the target and buffer, most of the incoming compressive wave is reflected back into the target upon reaching the target-buffer interface. That portion of the stress wave transmitted across the interface and into the buffer can be recorded by the embedded manganin wire gage. A third type of measurement that can provide comparative data is post-test microscopic examination of the target plate, from which it is possible to obtain information giving the distribution of the void volume fraction across the thickness of the damaged specimen [20]. The results from each of these measurement techniques provide data that can be used for comparison with the calculations provided by the constitutive-damage modeling of spall fracture.

15.2. Elastic-Viscoplastic-Damage Constitutive Model

For shock compression of solids wherein the strength of the shock greatly exceeds the maximum shear stress that the material can withstand, the shear stress in the compressed material is small relative to the mean stress, and neglect of the relatively small shear allows the material response to be modeled as an inviscid fluid, i.e. hydrodynamically [22]. At the much lower pressures encountered in plate-impact-spall experiments, the deviatoric component of the overall stress becomes considerably more significant, making a hydrodynamic treatment of the solid inappropriate. Here the recoverable, nonrecoverable and rate-dependent aspects of the material behavior must be considered. If, furthermore, material failure is to be included, an elastic-viscoplastic-damage type of constitutive theory is best employed to effectively model dynamic failure processes such as spall fracture.

The constitutive model described below is appropriate for the physical circumstances discussed in Section 15.1. Thus, for impact shock in the low to moderate 5–50 kbar stress range, temperature sensitivity need not be included in the constitutive model, nor is kinematic strain hardening important because of the absence of load cycling. The polycrystalline solid, which is represented mathematically as a continuous medium, is actually permeated with microvoids. The microvoid volume fraction $\xi = (V - V_s)/V$, where V is the volume of a material microelement and V_s is the volume of the solid constituent, is taken as a scalar-valued continuous point function of position $\mathbf{x}$ and time t, $\xi = \xi(\mathbf{x}, t)$, as with the other field variables. In so doing, a micro-porous crystalline solid can be treated mathematically as a homogenized continuous medium, where ξ specifies a continuous damage measure at each point. (The description of spall damage by a continuous measure of damage appears to have first been introduced by Davison and Stevens [14].)

15.2.1. Rate of Elastic Deformation

The constitutive equations are expressed in rate form, where the rate of deformation tensor is assumed to have additive elastic and rate-dependent plastic components

$$\mathbf{D} = \mathbf{D}^e + \mathbf{D}^p. \tag{15.1}$$

The rate-independent elastic deformation is of small order in strain, allowing the rate of elastic deformation to be expressed as a linear function of an objective stress-rate $\check{\mathbf{T}}$ of the Cauchy stress tensor $\mathbf{T}$,

$$\mathbf{D}^e = \frac{1}{2\bar{\mu}}\left[\check{\mathbf{T}} - \frac{\bar{\nu}}{1+\bar{\nu}}(\mathrm{tr}\,\check{\mathbf{T}})\mathbf{1}\right]. \tag{15.2}$$

Because of the presence of microvoids the macroscopic properties of the elastic solid can change as the void volume fraction changes. Utilizing a

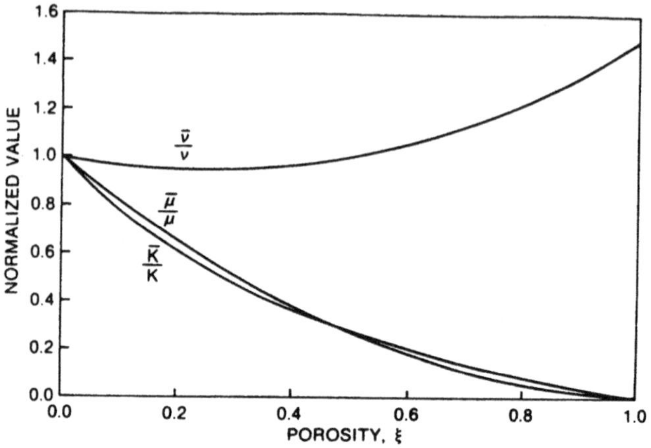

Figure 15.4. Degradation of elastic properties with increasing void volume (damage).

hollow sphere model for the elastic solid with microvoids proposed by Mackenzie [23], the elastic shear and bulk moduli can be expressed by the relations, Johnson [24],

$$\bar{\mu}(\xi) = \mu(1-\xi)\left(1 - \frac{6K + 12\mu}{9K + 8\mu}\xi\right), \tag{15.3}$$

$$\bar{K}(\xi) = \frac{4\mu K(1-\xi)}{4\mu + 3K\xi}, \qquad \bar{\nu} = \frac{1}{2}\frac{3\bar{K} - 2\bar{\mu}}{3\bar{K} + \bar{\mu}}, \tag{15.4}$$

where μ and K are the elastic shear and bulk moduli values for the ideal material without microvoid defects. The strength reduction or softening effect on the elastic moduli caused by an increase of the void volume fraction is illustrated by Figure 15.4.

15.2.2. Rate of Viscoplastic Deformation

Among the various constitutive theories for rate-dependent plastic deformation appropriate for high rates of deformation, the viscoplastic theory proposed by Perzyna [25, 26] appears well suited for modeling dynamic fracture. The rate of inelastic deformation has the form

$$\begin{aligned} \mathbf{D}^p &= \frac{\gamma}{\phi}\Phi(\hat{F})\frac{\partial f}{\partial \mathbf{T}}, \qquad && \hat{F} > 0, \\ \mathbf{D}^p &= 0, && \hat{F} \leq 0. \end{aligned} \tag{15.5}$$

The function Φ is a material functional of the yield function

$$\hat{F} = \frac{f}{\kappa} - 1, \tag{15.6}$$

and the material parameter

$$\gamma = \frac{\gamma_0}{\kappa_0} \tag{15.7}$$

represents the reciprocal of the average (macroscopic) viscosity of the material, where γ_0 is a viscosity coefficient, and κ_0 is related to the quasi-static uniaxial yield stress by the relation $\kappa_0 = (1/\sqrt{3})T_y$. The function ϕ serves as a rate of deformation control function that has for its argument the ratio of deformation invariants $[(I_2'/I_2^s) - 1]$, and is defined such that for quasi static deformation $I_2' = I_2^s$, $\phi(0) = 0$, while for $I_2' < I_2^s$, $\phi(\ldots\ldots) \equiv 0$. Quasi-static deformation here signifies the strain rate below which no significant rate effect of the viscoplastic material is discernable by the usual means of laboratory testing at room temperatures. The rate of deformation invariants are defined as

$$I_2' = |(II_{\mathbf{D}'})^{1/2}| \tag{15.8}$$

where $|\ldots\ldots|$ denotes absolute value and

$$II_{\mathbf{D}'} = -\tfrac{1}{2}\operatorname{tr}(\mathbf{D}' \cdot \mathbf{D}'), \qquad \mathbf{D}' = \mathbf{D} - \tfrac{1}{3}(\operatorname{tr}\mathbf{D})\mathbf{1}. \tag{15.9}$$

The presence of microvoids introduces compressibility into the otherwise incompressible plastic deformation. Thus the yield function f has the form [27–29]

$$f(J_1, J_2', \xi) = J_2' + n\xi J_1^2, \tag{15.10}$$

where

$$\begin{aligned} J_1 &= I_{\mathbf{T}} = \operatorname{tr}\mathbf{T}, \\ J_2' &= -II_{\mathbf{T}'} = \tfrac{1}{2}\operatorname{tr}(\mathbf{T}' \cdot \mathbf{T}'), \\ \mathbf{T}' &= \mathbf{T} - \tfrac{1}{3}(\operatorname{tr}\mathbf{T})\mathbf{1}. \end{aligned} \tag{15.11}$$

The material parameter n proportions the relative effects of the dilatational and deviatoric contributions to the yield process. The isotropic hardening, and the concomitant softening of the material as the microvoid volume increases, is expressed through the yield function

$$\kappa = \kappa(\varepsilon^p, \xi) = [q + (\kappa_0 - q)e^{-\beta\varepsilon^p}]^2[1 - n_1\xi^{1/2}]^2, \tag{15.12}$$

which assumes a nonlinear strain hardening that saturates with strain [30]. The material parameters q and β characterize the isotropic hardening, n_1 is identified below, and ε^p is the equivalent plastic strain invariant

$$\varepsilon^p = \int_0^t 2/\sqrt{3}|(II_{\mathbf{D}^p})^{1/2}|\, dt' \tag{15.13}$$

$$II_{\mathbf{D}} = \tfrac{1}{2}[(\operatorname{tr}\mathbf{D})^2 - \operatorname{tr}(\mathbf{D} \cdot \mathbf{D})]. \tag{15.14}$$

The functional Φ and the rate control function ϕ are taken as power functions of their arguments

$$\Phi(\hat{F}) = \left(\frac{f(J_1, J_2', \xi)}{\kappa(\varepsilon^p, \xi)} - 1\right)^{m_1}, \qquad \Phi\left(\frac{I_2'}{I_2^s} - 1\right) = \left(\frac{I_2'}{I_2^s} - 1\right)^m, \quad (15.15)$$

where m_1 and m are material parameters. With these choices for the various functions, the viscoplastic rate of deformation for the solid with microvoids can be expressed explicitly as

$$\mathbf{D}^p = \frac{\gamma_0}{\left(\frac{I_2'}{I_2^s} - 1\right)^m}\left[\frac{J_2' + n\xi J_1^2}{[q + (\kappa_0 - q)e^{-\beta\varepsilon^p}]^2[1 - n_1\xi^{1/2}]^2} - 1\right]^{m_1}\frac{1}{\kappa_0}(2n\xi J_1\mathbf{1} + \mathbf{T}'). \quad (15.16)$$

By means of (15.5) it follows that

$$\Phi(\hat{F}) = \frac{1}{\gamma}[\mathrm{tr}(\mathbf{D}^p \cdot \mathbf{D}^p)]^{1/2}\Phi\left[\mathrm{tr}\left(\frac{\partial f}{\partial \mathbf{T}} \cdot \frac{\partial f}{\partial \mathbf{T}}\right)\right]^{-1/2} \quad (15.17)$$

Inserting the expressions appearing in (15.16) and making use of the relation

$$\frac{\partial f}{\partial \mathbf{T}} = 2n\xi J_1\mathbf{1} + \mathbf{T}' \quad (15.18)$$

produces the following yield condition for the hardening viscoplastic solid with microvoids

$$J_2' + n\xi J_1^2 - [q + (\kappa_0 - q)e^{-\beta\varepsilon^p}]^2[1 - n_1\xi^{1/2}]^2$$
$$\times\left\{1 + \left[\frac{1}{\gamma_0}[\mathrm{tr}(\mathbf{D}^p \cdot \mathbf{D}^p)]^{1/2}\left(\frac{I_2'}{I_2^s} - 1\right)^m \kappa_0(12n^2\xi^2 J_1^2 + 2J_2')^{-1/2}\right]^{1/m_1}\right\} = 0, \quad (15.19)$$

which illustrates the combined influence of the isotropic hardening, the material softening and compressibility associated with microvoid growth, and the rate of deformation, on the dynamic yield process.

15.2.3. Local Criterion for Fracture

Information obtained from micrographs taken from cross-sections of spalled plates indicates that ductile spallation is the macroscopic terminal state of a microprocess involving microvoid generation and growth [19, 20]. Accordingly a local criterion for ductile fracture is most appropriate, and can be defined as the attainment of a critical value for the void volume fraction at any point of the body (cf. Perzyna [26, 31]). At each such point, a microelement of material experiences separation as the rate of inelastic deformation increases dramatically, so that $\mathbf{D}^p \to \infty$ as $\xi \to \xi_F$. For this to occur we have in conjunction with expressions (15.12) and (15.16) the corresponding condition

$$\kappa(\varepsilon^P, \xi)_{\xi=\xi_F} = [q + (\kappa_0 - q)e^{-\beta\varepsilon^P}]^2[1 - n_1\xi_F^{1/2}]^2 = 0, \qquad (15.20)$$

which identifies the material parameter $n_1 = 1/(\xi_F)^{1/2}$.

15.2.4. Evolution Equation for the Void Volume Fraction

The material degradation, which is represented by increase of the microvoid volume fraction from the initial value ξ_0, can occur from any one of the following microdamage processes: (i) diffusion, (ii) nucleation of new voids, and (iii) void growth and coalescence. However, at non-elevated temperatures and high rates of deformation, damage associated with the diffusion of microvoids is not known to be significant [32, 33]. Therefore nucleation of new voids, in addition to growth of existing ones, represent the essential micromechanisms operative in producing ductile spall damage. Because of the high rates of strain and the very small characteristic time to fracture associated with plate-impact spall experiments, the thermally activated mechanism for microvoid nucleation is considered the most important. It can be modeled by the form suggested by Seaman et al. [9, 10], subsequently modified by Perzyna [26] to further include the effects of void interaction,

$$\left(\frac{d\xi}{dt}\right)_n = \frac{h(\xi)}{1-\xi}\left[\exp\left(\frac{m_2|\sigma - \sigma_N|}{k\theta}\right) - 1\right], \qquad \sigma < \sigma_N \qquad (15.21)$$

where $m_2|\sigma - \sigma_N|/k\theta$ is the nucleation activation energy, $\sigma = (1/3)J_1$ is the mean stress, (where compressive stress is taken as positive), σ_N is the threshold mean stress for nucleation, θ is the temperature, k is the Boltzmann constant, and m_2 is a material parameter. The material function $h(\xi) > 0$ is introduced to give some account of the possible effect of void interaction on the nucleation process.

The rate of increase of the volume fraction of the microvoids calculated by Eftis and Nemes [34] makes use of the hollow sphere model, where the microvoids are replaced by a random distribution of spherical voids of arbitrary size embedded in a viscoplastic material that strain hardens nonlinearly (cf. Figure 15.5), and is given by the following set of relations

$$\left(\frac{d\xi}{dt}\right)_g = \frac{1}{\eta}g(\xi)F(\xi, \xi_0)|\sigma - \sigma_N|, \qquad \sigma < \sigma_N, \qquad (15.22)$$

where

$$F(\xi, \xi_0) = \frac{\sqrt{3}}{2}\xi\left(\frac{1-\xi}{1-\xi_0}\right)^{2/3}\left[\left(\frac{\xi}{\xi_0}\right)^{2/3} - \xi\right]^{-1}, \qquad (15.23)$$

$$\sigma_G = -\frac{1}{\sqrt{3}}(1-\xi)\ln\left(\frac{1}{\xi}\right)[2q + (\kappa_0 + q)F_1(\xi, \xi_0)], \qquad (15.24)$$

and

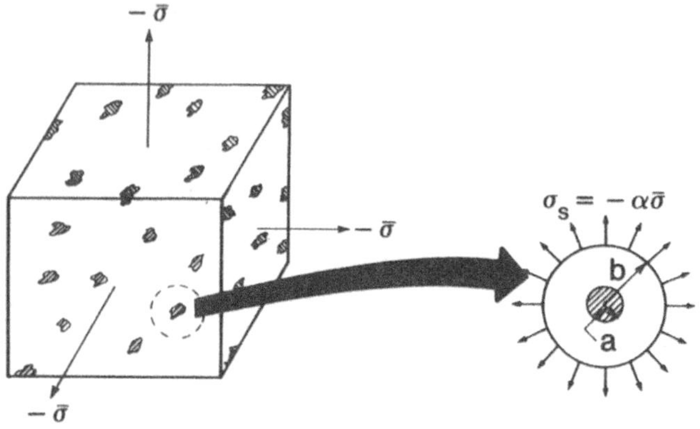

Figure 15.5. Element of the solid with microvoids replaced by idealized spherical voids.

$$F_1(\xi, \xi_0) = \exp\left[\frac{2}{3}\beta\frac{(\xi_0 - \xi)}{\xi(1 - \xi_0)}\left(\frac{1 - \xi}{1 - \xi_0}\right)^{-2/3}\left(\frac{\xi_0}{\xi}\right)^{-2/3}\right] + \exp\left[\frac{2}{3}\beta\frac{(\xi_0 - \xi)}{(1 - \xi_0)}\left(\frac{1 - \xi}{1 - \xi_0}\right)^{-2/3}\right]. \qquad (15.25)$$

These results generalize earlier calculations by Johnson [24] and Perzyna [26], based on the work of Carroll and Holt [35], for the collapse of voids in porous ductile solids. In these expressions σ_G represents the microvoid growth threshold stress, which is shown to depend upon the initial and current state of the porosity, as well as on the material yield and strain hardening, and η is the microviscosity parameter associated with the microvoid growth. The material function $g(\xi) > 0$ accounts for the effects of void interaction during void growth. Since the functions $h(\xi)$ and $g(\xi)$ serve to hasten the void nucleation and growth rates, they can be associated with the void coalescence that takes place at the terminal stage of the void growth process. On the basis of observed void volume distributions in spalled plates the material function $g(\xi)$ was chosen to have the exponential form $g(\xi) = e^{\alpha\xi}$, where α is a material parameter, while the function $h(\xi)$ was assumed to have a constant value (cf. Ref [36]).

Thus, the instantaneous value of the void volume fraction $\xi(\mathbf{x}, t)$ appearing in the constitutive relations (15.3), (15.4) and (15.6), can be determined by time integration of the rate equation

$$\frac{d\xi}{dt} = \frac{\partial\xi}{\partial t} + \mathbf{v}\cdot\text{grad}(\xi) = \frac{h(\xi)}{1 - \xi}\left[\exp\left(\frac{m_2|\sigma - \sigma_N|}{k\theta}\right) - 1\right] + \frac{1}{\eta}g(\xi)F(\xi, \xi_0)|\sigma - \sigma_G|. \qquad (15.26)$$

Table 15.1. OFHC copper material parameters

$\mu = 4.84 \times 10^4$ MPa	$n = 0.25$
$K = 10.49 \times 10^4$ MPa	$\xi_0 = 3 \times 10^{-4}$
$q = 125$ MPa	$\xi_F = 0.32$
$\beta = 6.14$	$n_1 = 1.77$
$\rho_R = 8.93$ gm/cm^3	$\sigma_N = -500$ MPa (tension)
$I_2^s = 7 \times 10^4$ s^{-1}	$m_2 = 2.025 \times 10^{-23}$ cm^3
$\kappa_0 = 9.31$ MPa	$h = 70.37$
$\gamma_0 = 337$ s^{-1}	$\alpha = 20$
$m = \frac{1}{2}$	$\eta = 10$ Poise
$m_1 = 2$	$k\theta = 4.05 \times 10^{-21}$ J

Table 15.2. 6061-T6 aluminum material parameters

$\mu = 2.77 \times 10^4$ MPa	$I_2^s = .1$ s^{-1}
$K = 7.42 \times 10^4$ MPa	$\kappa_0 = 173$ MPa
$q = 219$ MPa	$\gamma_0 = 5.99 \times 10^5$ s^{-1}
$\beta = 14.3$	$m = .25$
$\rho_R = 2.703$ gm/cm^3	$m_1 = 1.55$

15.2.5. Material Parameter Values

For the plate impact-spall fracture simulations that are described below, oxygen-free high purity copper was chosen for the plate material because of the experimental data from high strain rate tests, and from spall fracture studies, that are available in the literature. Pressure-shear wave calculations without material damage for 6061-T6 aluminum are also shown and discussed in connection with oblique plate impact and spall. Discussion of the evaluation and assignment of the material parameter values and material functions for the OFHC copper and the 6061-T6 aluminum can be found in References [30, 36, 37, 38]. Accordingly, only a tabulated summary of values need be provided, and are shown as Tables 15.1 and 15.2

15.3. Simulation of Plate Impact Spall Fracture

15.3.1. Introduction

The constitutive-damage theory presented above can be employed to simulate plate-impact spall fracture experiments in circumstances where the material damage process is essentially ductile. Three experimental configurations have been studied. These include both normal and low angle oblique impact of rectangular plates of equal size, and normal impact of circular plates with different size diameters. In the absence of thermal effects, the system of governing equations that apply for each include the equations of motion

$$\mathbf{x} = \chi(\mathbf{X}, t), \tag{15.27}$$

equations for the balance of mass and of linear momentum

$$\frac{\rho_R}{\rho} = \det \mathbf{F} = \det(\partial \mathbf{x}/\partial \mathbf{X}), \tag{15.28}$$

$$\operatorname{div} \mathbf{T} = \rho \frac{d\mathbf{v}}{dt} = \rho\left(\frac{\partial \mathbf{v}}{\partial t} + \mathbf{v} \cdot \operatorname{grad} \mathbf{v}\right), \tag{15.29}$$

the constitutive rate equation

$$\frac{1}{2\bar{\mu}}\left[\check{\mathbf{T}} - \frac{\bar{\nu}}{1+\bar{\nu}}(\operatorname{tr} \check{\mathbf{T}})\mathbf{1}\right] = \mathbf{D} - \frac{\gamma}{\phi}\Phi(\hat{F})\frac{\partial f}{\partial \mathbf{T}}, \tag{15.30}$$

and the damage evolution equation (15.26). In these equations $\mathbf{x}$ is the position in the current configuration occupied by a material point having the position $\mathbf{X}$ in the reference configuration, ρ, ρ_R are the mass densities in the current and reference configurations, respectively, and $\mathbf{F}$ is the deformation gradient tensor. Detailed component forms for these equations appropriate to each of the experimental configurations can be found in References [37–41].

For both rectangular plate configurations all field variables are functions of one space variable along the thickness direction of the target plate, whereas for the circular plate geometry they are functions of the two space variables along the radial and thickness directions. The spin tensor $2\mathbf{W} = (\mathbf{L} - \mathbf{L}^T)$, $\mathbf{L} = \operatorname{grad} \mathbf{v}$, signifying local rigid body rotation, has zero value only for normal impact of the rectangular plates. For this configuration the stress-rate $\check{\mathbf{T}}$ reduces accordingly to the material time derivative. For the other two impact configurations rigid body rotations do occur, however the computational procedure for calculating stress-rate used in the PRONTO computer code circumvents the need for utilizing any one of the numerous definitions that have been proposed for an objective time derivative of stress, cf. References [41–43]. The numerical calculations for both normal and oblique impact of rectangular plates were performed using the finite-difference wave propagation code Wondy [44]; calculations for normal impact of the circular plates required use of the transient finite-element computer code Pronto [45]. It is noted that the numerical computational scheme cannot actually make use of the equations of motion (15.27) specifying the motion of points of the target plate since they are not explicitly known at the outset. The calculational procedure begins instead with the given velocities of the flyer and target and an assumed velocity gradient across the flyer–target interface, enabling numerical calculation of the velocity gradient and rate of deformation tensors $\mathbf{L}$ and $\mathbf{D}$ for points initially along the target surface at the first time step.

15.3.2. Normal Impact of Rectangular Plates

The first simulation of a plate-impact spall fracture test by the viscoplastic-damage constitutive model describes the effect of the impact of rectangular

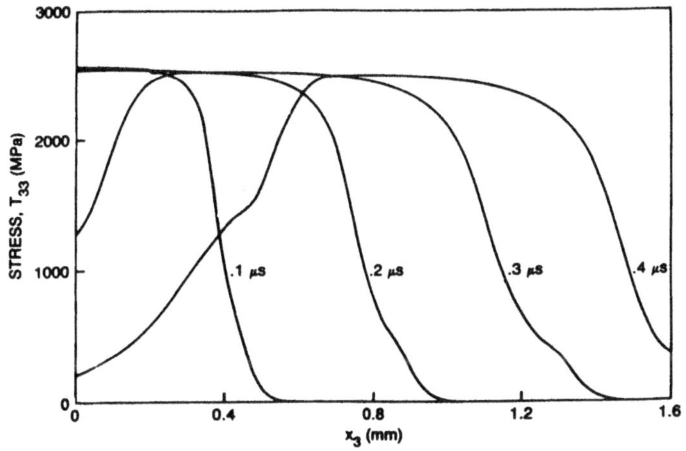

Figure 15.6. Computed compressive stress wave profiles for 160 m/s impact.

copper plates of equal size, where the flyer plate traveling at 160 m/s strikes the target with zero angle of incidence. The flyer and target have plate thicknesses of 0.6 mm and 1.6 mm, respectively. The impact produces a compressive stress of 2,600 MPa (26 kbar) that propagates as a compressive wave across the target thickness. Calculated stress wave profiles are shown in Fig. 15.6. The initially steep shock wave front rapidly evolves into an elastic precursor wave, followed by the slower moving viscoplastic wave front which exhibits wave front dispersion due to the rate and strain hardening properties of the target material. The faster moving wave of elastic unloading can be seen to be catching up with the slower viscoplastic wave of loading. The compressive stress waves in both the target and flyer become tensile waves that are reflected from the rear surface of the plates, traveling in opposite directions to superimpose and produce the material damage leading to the spall fracture of the target. This is illustrated in part by Fig. 15.7, where the superimposed tensile stress profiles across the target are shown for the time interval during which material damage has initiated in the region of maximum tensile stress, and has caused the stress to relax. The corresponding calculated and measured [20] distributions of the void volume fraction are shown by Fig. 15.8, which also shows the localization of the damage, thereby determining the location of the spall fracture plane. The calculated stress history at this location is shown by Fig. 15.9, which clearly shows the arrival of the elastic precursor and viscoplastic compressive wave fronts, followed by unloading and the arrival of the reflected tensile wave and its subsequent relaxation as the microvoid damage process ensues.

In Figure 15.10 the measured [9] and calculated stress histories at a gage embedded in a buffer plate of polymeric material located behind the target plate are shown, cf. Fig. 15.3. The profiles are qualitatively similar, with

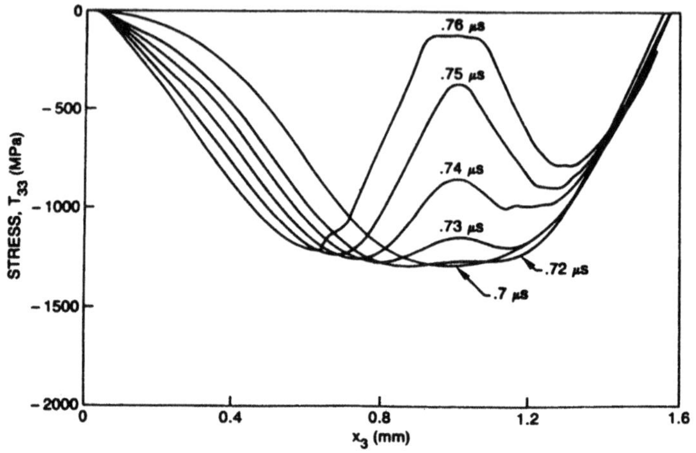

Figure 15.7. Computed tensile wave profile illustrating the effect of damage accumulation (tensile stress is shown negative).

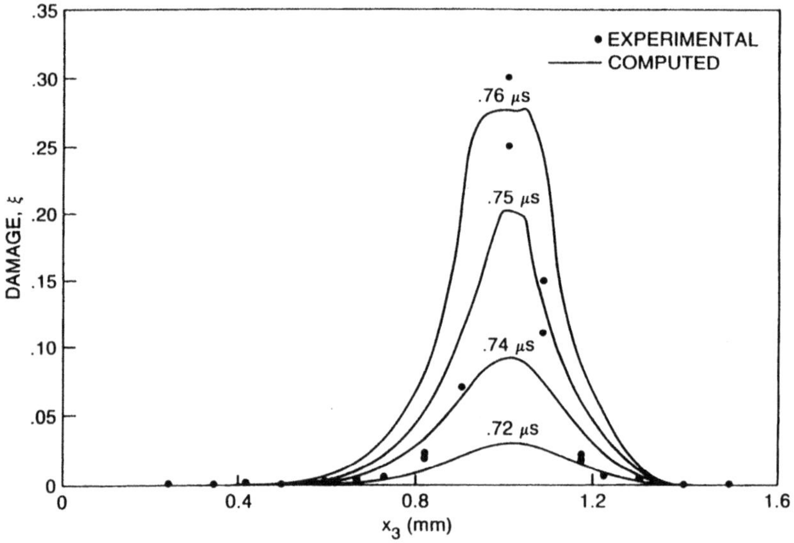

Figure 15.8. Comparison of computed and measured void volume fraction across the thickness of the target plate.

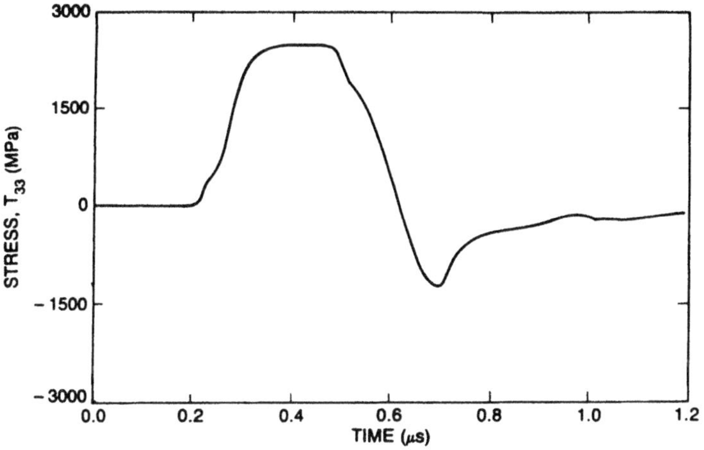

Figure 15.9. Computed stress history at $x_3 = 0.975$ mm (spall plane).

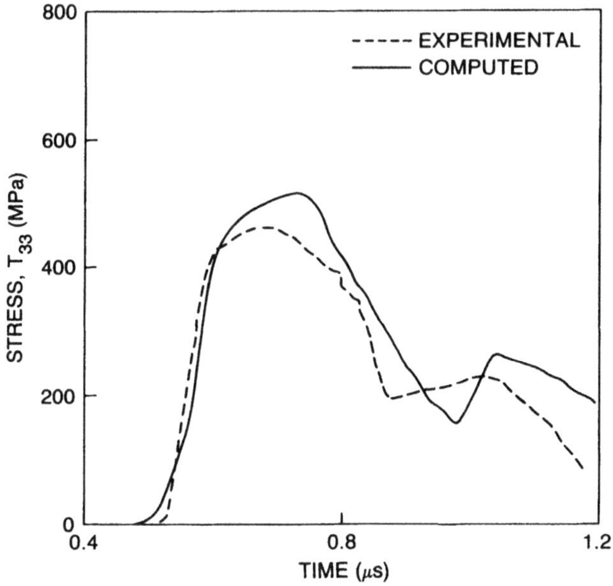

Figure 15.10. Comparison of computed and measured stress histories at stress gage location for 160 m/s impact.

close agreement between the time of arrival of the stress wave fronts. The second peak of the stress-time curve, referred to as the 'spall signal', requires some explanation. As the initial compressive stress wave passes through the target, part of the wave is reflected back at the target-buffer interface as a rarefaction wave combining with the reflected rarefaction wave coming from the flyer, thereby possibly causing spall damage within the target. The remaining part continues on through the buffer resulting in the first compressive peak stress recorded by the stress gage. If spall damage occurs within the target, a second compressive wave caused by recompression of the material adjacent to the spalled surface as the tensile stress there is relaxed, travels to the right from the spall to the buffer surface. Part of this wave is reflected back into the target as a rarefaction wave, while the remaining part propagates through the buffer causing the second stress peak at the stress gage. Thus its identification as the spall signal. The amplitude of the second peak stress is related to the degree of the tensile stress relaxation in the target, and thus to the extent of the damage that has taken place. In a second experiment [9] using this type of plate backup configuration, a 1.14 mm-thick flyer impacts with a 3.18 mm-thick target traveling with a velocity of 129 m/s. The record of the gage in the PMMA buffer is shown in Fig. 15.11, along with the computed stress history. Again there is reasonably good qualitative agreement between the predicted and measured stress-time profiles.

Numerical simulation of this kind of experiment requires the modeling of

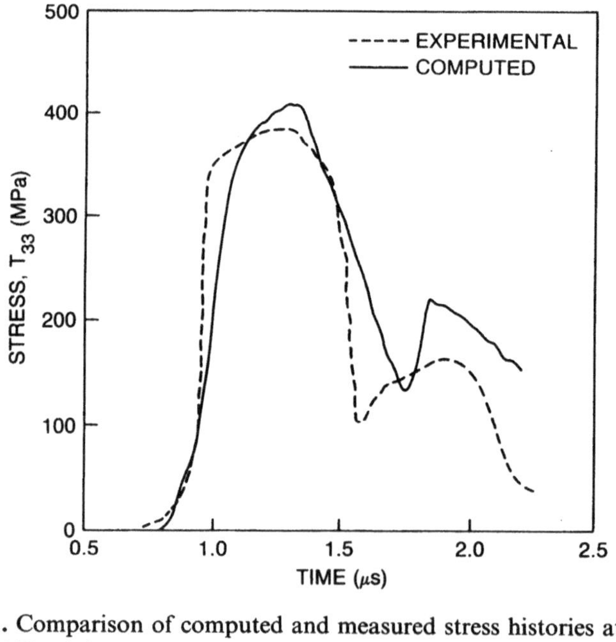

Figure 15.11. Comparison of computed and measured stress histories at stress gage location for 129 m/s impact.

the compressive behavior of the buffer material in addition to the target and flyer. During the course of the experiment the polymeric buffer material experiences a compressive stress wave that can be described by means the pressure-density relation [46]

$$\sigma = \frac{K_0[1 - \rho_R/\rho]}{[1 - s(1 - \rho_R/\rho)]^2}, \tag{15.31}$$

where K_0 represents the bulk modulus in the reference state and s is a material constant. This relation is derived from the empirical observation that for many materials the velocity of a shock wave front is linearly related to the velocity of material particles immediately behind the shock front for pressures ranging from 10^2–10^5 kbar [22]. The material parameter values for the PMMA buffer material used in the calculations are $\rho_R = 1.185$ g/cm^3, $s = 2.467$, and $K_0 = 8.15 \times 10^3$ MPa.

Plate impact experiments performed by Rajendran, Grove and Bless [47] on OFHC copper utilize a configuration where the target plate has a free rear surface, which allows the measurement of the rear surface velocity by means of an interferometer (VISAR). For this experiment a 9-mm thick target experiences impact by a 2-mm thick flyer plate traveling at a velocity of 185 m/s. The calculated compressive stress wave profiles across the target are shown by Fig. 15.12. Because of the greater thickness of the target, thereby requiring a longer transit time, the characteristic features of the elastic-viscoplastic stress wave propagation are here more fully developed than the wave propagation shown in Fig. 15.6. The stress shock front, initially at 3,000 MPa (30 kbars) shows greater dispersion, and the faster moving elastic

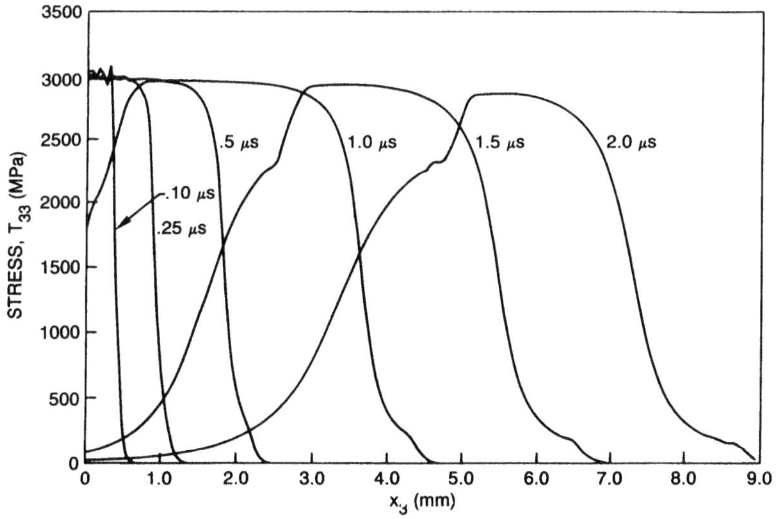

Figure 15.12. Computed compressive stress wave profiles for 185 m/s impact.

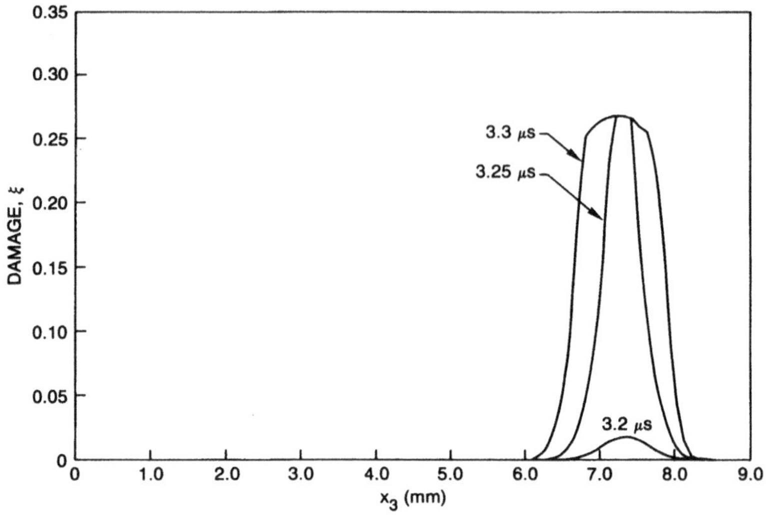

Figure 15.13. Computed void volume fraction across the thickness of the target plate for 185 m/s impact.

precursor wave moves further out ahead of the rounding viscoplastic wave front as the wave propagation progresses. The catch-up by the faster elastic wave of unloading with the wave front for the viscoplastic loading is also more clearly observable. The calculated localization and evolution of damage within the target is illustrated by Fig. 15.13, where because of the greater target-flyer thickness ratio, causes the location of the spall plane to be closer to the rear surface of the target. Figure 15.14 provides a comparison between the calculated target rear surface velocity-time profile and the measured profile obtain by VISAR. The calculated profile shows qualitative similarity with the measured profile, including appearance of a spall signal, that is, of a second velocity peak due to the arrival of a second compressive wave that is associated with the appearance of a fracture surface within the target.

15.3.3. Elastic Acoustical Tensor and Plane Wave Velocity In Damaged Softened Material

A number of materials such as concrete, rock, and dense soils show a decrease in strength during progressive straining after the peak stress is reached. The possibility of plane wave propagation for materials exhibiting such softening behavior has been the subject of recent investigation. Valanis [48] has discussed constitutive equations for softening elastic materials that may not necessarily be governed by hyperbolic type partial differential equations, leading to physical results that are unacceptable. Rice [49] has investigated the localization of plastic deformation for elastic-plastic solids and has

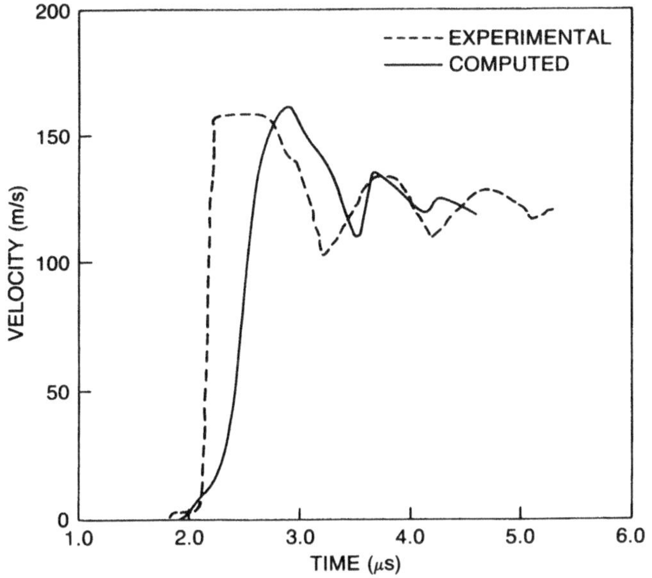

Figure 15.14. Comparison of computed and measured rear-surface velocity histories for 185 m/s impact.

linked material response for small wave-like disturbances to the square of the plane wave propagation speed. Small disturbances may remain stable, or may experience 'divergence' growth, or 'flutter' growth depending on whether the square of the wave speed is real and positive, real and negative, or complex, respectively. In as much as stress wave propagation induced by high speed impact in polycrystalline materials with microvoids can cause material softening and damage localization leading to spallation, a demonstration that the wave speeds will be real valued and positive, and that the character of the governing partial differential equations will remain hyperbolic throughout the process of material softening, damage localization and fracture, is of some interest and value.

Referring to the plate impact configuration discussed in Section 15.3.2, the wave propagation across the target thickness along the x_3 coordinate direction is governed by a system of three quasilinear, first-order, hyperbolic partial differential equations provided by the equation of motion (15.29) and the constitutive rate equation (15.30). In reduced component form they appear as

$$\rho\left(\frac{\partial v_3}{\partial t} + \frac{\partial v_3}{\partial x_3} v_3\right) - \frac{\partial T_{33}}{\partial x_3} = 0 \tag{15.32}$$

$$\frac{1}{2\bar{\mu}}\left[\frac{dT_{11}}{dt} - \frac{\bar{v}}{1+\bar{v}}\frac{dJ_1}{dt}\right] + \frac{\gamma_0}{\phi}\Phi(\hat{F})\frac{1}{\kappa_0}(2n\xi J_1 + T'_{11}) = 0 \tag{15.33}$$

$$\frac{1}{2\bar{\mu}}\left[\frac{dT_{33}}{dt} - \frac{\bar{\nu}}{1+\bar{\nu}}\frac{dJ_1}{dt}\right] - \frac{\partial v_3}{\partial x_3} + \frac{\gamma_0}{\phi}\Phi(\hat{F})\frac{1}{\kappa_0}(2n\xi J_1 + T'_{33}) = 0 \quad (15.34)$$

When coupled with the void volume evolution equation (15.26), this system of equations determine the four unknowns for the stress T_{33}, $T_{11} = T_{22}$, the velocity v_3, and the void volume fraction ξ. The characteristics for this system initially are straight lines with slope $c = dx_3/dt$ given by $c(\xi) = [1/\rho\langle\bar{K}(\xi) + 4/3\bar{\mu}(\xi)\rangle]^{1/2}$, which represents the speed of the wave propagating along the x_3 direction. They do not remain as straight lines, however, because of the dependence of the elastic parameters on the void volume fraction, the values of which change when the mean tensile stress thresholds for void nucleation and void growth are exceeded and material softening and damage follow. Nevertheless the values of the elastic constants $\bar{K}$ and $\bar{\mu}$ remain positive throughout the softening stage of the material damage process. For copper material, with a critical void volume fraction $\xi_F \approx 0.3$, expressions (15.3) and (15.4) show that the elastic moduli are reduced in value by about one-third to one-half as the void volume increases to critical value, cf. Fig. 15.4. With the void fraction ranging from initial value to critical value, $\xi_0 \leq \xi \leq \xi_F$, correspondingly $K > \bar{K}(\xi) > 0$ and $\mu > \bar{\mu}(\xi) > 0$, and the wave speed $c(\xi)$ remains positive throughout the damage process.

This may also be seen by examination of the elastic acoustical tensor $\mathbf{Q}$ associated with propagation of small plane wave disturbances from the natural state in isotropic materials with microvoids. For plane waves with wave number $\mathbf{n}$, the displacement

$$\mathbf{u} = \mathbf{A}\exp[i(2\pi/\lambda)(\mathbf{x}\cdot\mathbf{n} - ct)], \quad (15.35)$$

and the wave speed are the eigenvectors and eigenvalues of the characteristic value problem

$$[\mathbf{Q} - (\rho c^2)\mathbf{1}]\mathbf{A} = 0, \quad (15.36)$$

where the acoustical tensor is given by the symmetric form

$$Q_{kl}(\mathbf{n}, \xi) = [\bar{K}(\xi) + \tfrac{1}{3}\bar{\mu}(\xi)]n_k n_l + \bar{\mu}(\xi)\delta_{kl}. \quad (15.37)$$

Since $\bar{K}(\xi)$ and $\bar{\mu}(\xi)$ remain positive throughout the damage process, the symmetric acoustic tensor is positive definite and the eigenvalue (ρc^2) is real and positive valued, signifying real, positive elastic wave speed $c(\xi)$ for the softened material. These conclusions are also apparent from the spall signals appearing in the measured and calculated wave histories for compressive stress and rear surface velocity shown in Figures 15.10, 15.11, and 15.14. The spall signal provides graphic illustration of the fact that throughout the entire time that the material undergoes softening, ending ultimately in fracture, wave propagation continues through the damaged material, implying that the governing set of hyperbolic partial differential equations (15.32)–(15.33) describing the elastic-viscoplastic wave motion remain hyperbolic throughout.

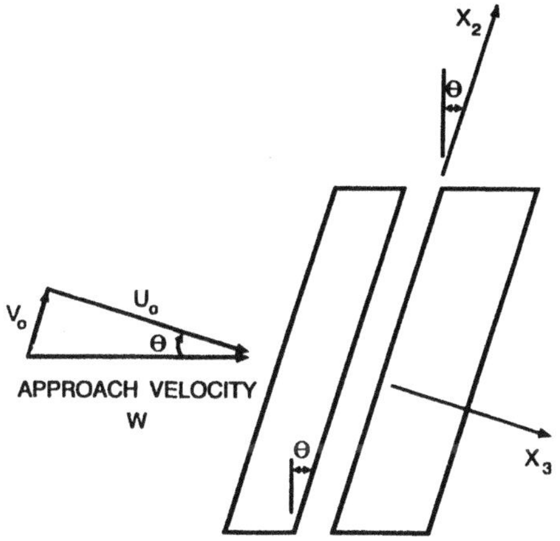

Figure 15.15. Schematic of oblique flyer-target configuration.

15.3.4. Oblique Impact of Rectangular Plates

An oblique impact of rectangular plates can be realized when the flyer and target plates remain parallel and are inclined relative to the direction of the flyer velocity. The experimental configuration is shown schematically by Fig. 15.15. This arrangement imparts velocity components that are normal and transverse to the plane of impact, thereby creating shear and normal stresses that propagate along the thickness direction of the target plate [50]. Thus far, oblique plate impact experiments have been used principally for the study of plastic response of metals at strain-rates above those attainable by the Hopkinson bar technique, with little, if any work reported on its use for spall fracture studies under combined normal and shear stress [51–53]. The constitutive-damage model has been used to simulate low-angle oblique plate-impact experiments that would create material damage and spall fracture. With small angle of obliquity the mean stress that develops in the target will be approximately an order of magnitude greater than the shear stress, with the large tensile mean stress primarily responsible for the spall damage. The analytical details of the constitutive modelization and analysis appropriately specialized to pressure-shear impact loading can be found in References [38, 39].

Oblique impact of plates produces elastic and plastic longitudinal and transverse waves. The elastic longitudinal wave propagating with wave speed c_{EL} is followed by the slower longitudinal plastic wave travelling at wave speed c_{PL}. These waves in turn precede the slower elastic and plastic shear waves c_{ES} and c_{PS}, where $c_{EL} > c_{PL} > c_{ES} > c_{PS}$. The wave characteristics are shown schematically in Fig. 15.16, where the elastic and plastic longitudinal waves and the elastic shear wave are labeled by E, P, and S, respectively. The plastic shear wave is not shown, and to further simplify the diagram the waves are drawn as if propagating at constant velocity, although in actuality

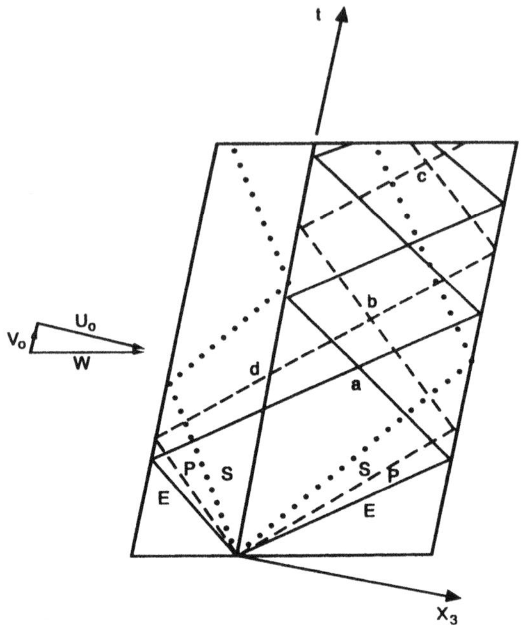

Figure 15.16. Schematic distance–time plot showing wave paths.

they may not. The elastic wave speeds will change through change in value of the elastic constants when the stress thresholds for void volume increase are exceeded. The plastic wave speeds are affected by the strain hardening of the material and, conceivably, by damage accumulation as well. However for purposes of illustration, one can construct the diagram with the wave speeds as constants without introducing serious qualitative error. The superposing of the oppositely directed elastic and plastic waves of rarefaction which occur at the times labeled as points (*a*) and (*b*) in Fig. 15.16, create the tensile mean stress causing material damage and spallation across a plane parallel to the faces of the plate. Point (*d*) corresponds to the time at which both reflected elastic and plastic rarefaction waves from the flyer free surface have crossed the flyer-target interface. Thereafter the normal stress across the plate reduces to zero, and from this time onward the shear induced by the interface friction between the plates stemming from the obliquity of the impact cannot develop, causing slip between the plates.

15.3.4.1. Pressure-Shear Waves in 6061-T6 Aluminum Without Material Damage

Because of the availability of experimental data for oblique impact of 6061-T6 aluminum plates, calculations are shown first for the normal and shear stress wave profiles and for the normal and transverse target rear surface velocity histories assuming no material damage. The experimental simulation considers impact of a 3.2 mm flyer onto a 3.2 mm target with normal and

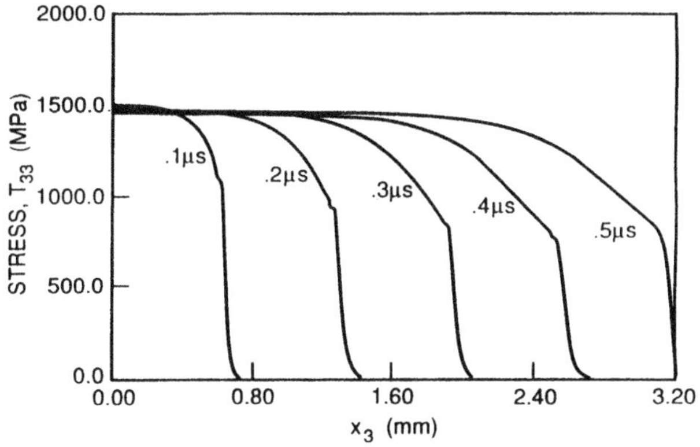

Figure 15.17. Computed normal stress-wave profiles for impact of 6061-T6 aluminum.

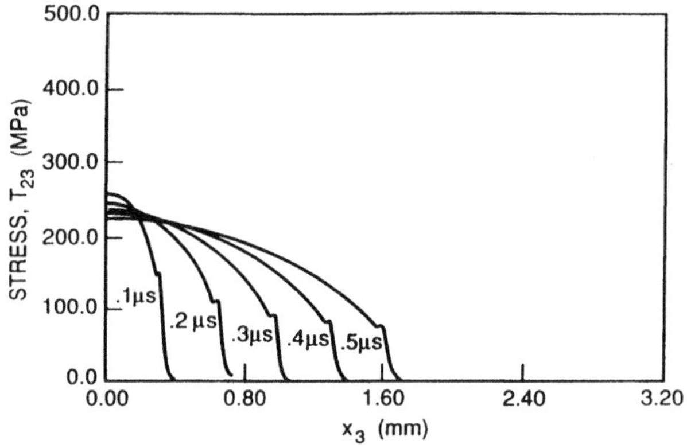

Figure 15.18. Computed shear stress-wave profiles for impact of 6061-T6 aluminum.

tangential impact velocity components $U_0 = 196.7$ m/s and $V_0 = 98.35$ m/s, respectively. The calculated normal compressive and shear wave profiles for the T_{33} and T_{23} stress components propagating across the plate thickness are shown in Figures 15.17 and 15.18. The elastic precursor waves propagating at the longitudinal and shear wave speeds are visible at 0.1 μs, and show decay as the wave fronts move across the plate. Since yield stress is strain-rate dependent, this decay may be due to the reduction of the strain rate as the wave front moves further away from the impact surface. The calculated shear stress wave travels at roughly one-half the speed of the longitudinal wave. Both of the slower moving plastic wave fronts that follow show considerable wave front dispersion owing to the inelastic behavior of the material. The small but continuous reduction of the shear stress calculated at the front face of the target plate is unexpected and difficult to interpret, in view

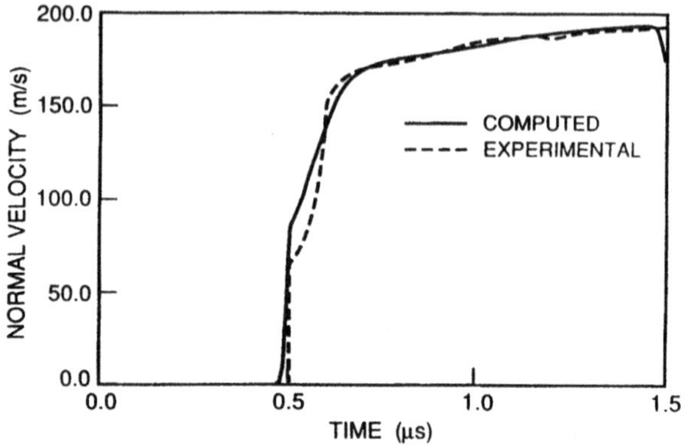

Figure 15.19. Normal rear surface velocity history for impact of 6061-T6 aluminum.

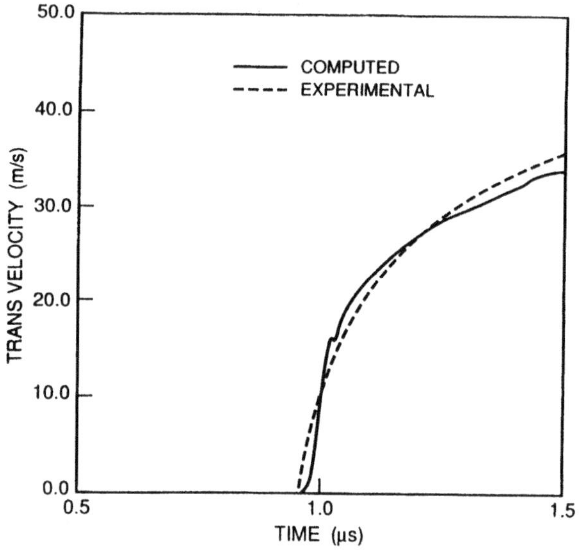

Figure 15.20. Transverse rear surface velocity history for impact of 6061-T6 aluminum.

of fact that the calculated normal stress on this surface is maintained at its initial impact level throughout this same time period.

Predicted normal and transverse velocity histories of the rear surface of the target are shown in Figures 15.19 and 15.20, respectively, along with the measured values reported in Reference [54]. Although the calculated elastic wave front in Fig. 15.19 is higher than the measured value, the remaining features of the predicted, normal and transverse velocity histories agree quite well with the experimentally measured histories.

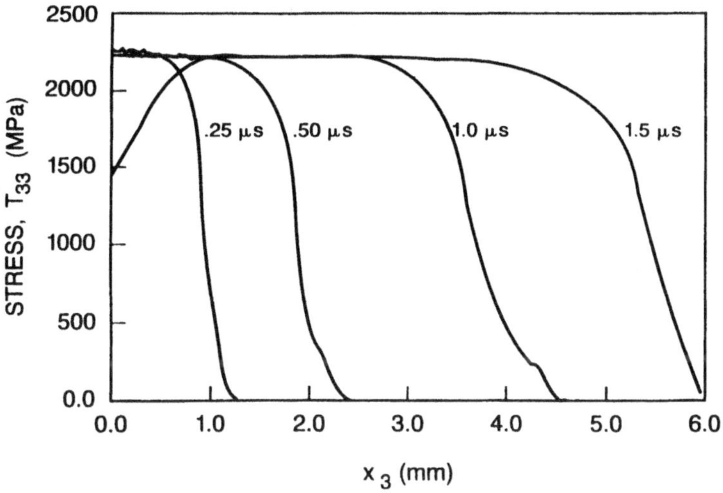

Figure 15.21. Computed normal compressive stress wave profile.

15.3.4.2. Oblique Impact of OFHC Copper with Material Damage and Spallation

The first experimental simulation has a flyer 3 mm in thickness impacting a target plate 6 mm thick, with normal and tangential velocities $U_0 = 140$ m/s and $V_0 = 70$ m/s, respectively. This corresponds to an angle of incidence of 26.6 degrees, which is the angle used in the oblique impact experiments reported by Kim and Clifton [52]. Computed normal compressive stress wave profiles are shown in Fig. 15.21. As with aluminum, a decaying elastic precursor wave is visible. The wave profiles shown are prior to the onset of damage which begins at approximately 2.65 μs after impact as shown by Fig. 15.22. We note that comparison of these calculated compressive stress wave profiles with the profiles calculated when the impact is not oblique, that is, with $U_0 = W = 140$ m/s, $V_0 = 0$, (cf. Fig. 15.6), show no significant differences. The damage localization shown in Fig. 15.22 indicates that the spall fracture will occur along a plane located near the center of the target plate.

The calculated shear stress wave shown in Fig. 15.23, traveling at approximately one-half the speed of the longitudinal wave, is comparable to the shear stress wave calculated for the aluminum prior to the slip condition. Between 0.2 μs and 1.0 μs there is a relaxation of the shear at the front face of the target, and between 1.0 and 1.5 μs unloading of the stress wave begins. Thereafter the slip condition begins at some instant between 1.5 and 2.0 μs, where the reduction of normal stress associated with arrival of the longitudinal rarefaction wave correspondingly reduces the friction induced shear stress to zero as shown by the stress profiles at 2.0 and 2.5 μs. The oscilla-

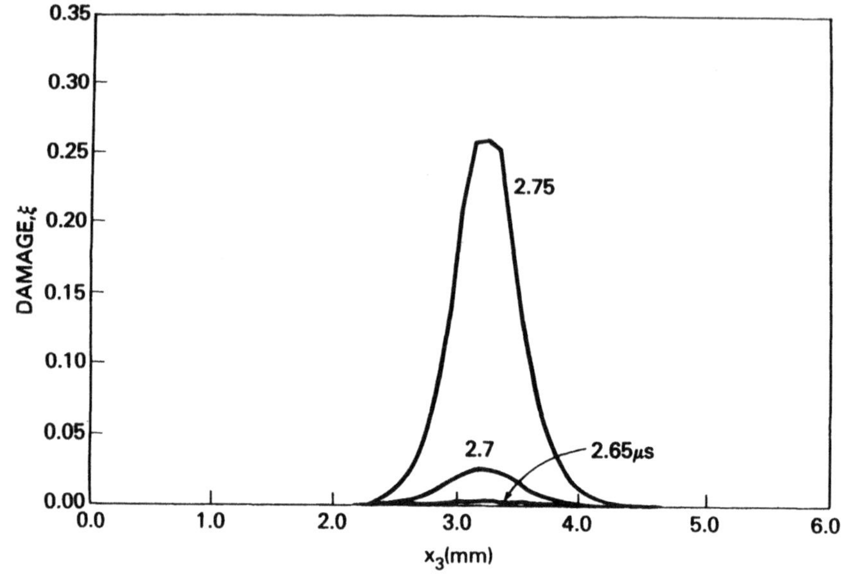

Figure 15.22. Computed void volume fraction across the, thickness of the target plate.

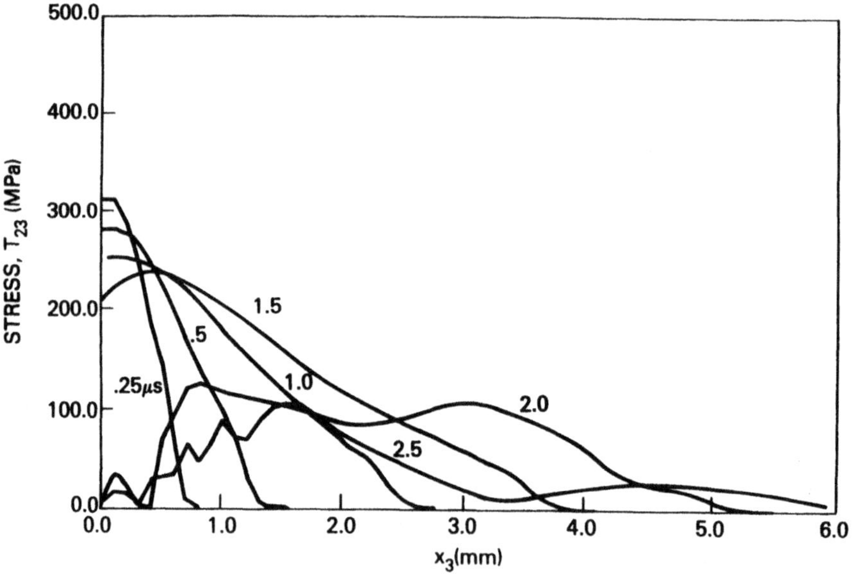

Figure 15.23. Computed shear stress wave profiles.

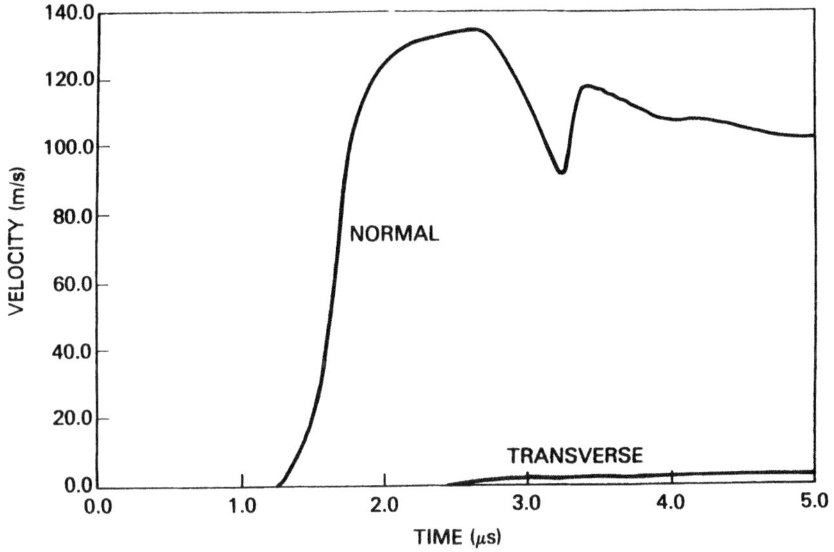

Figure 15.24. Calculated rear surface velocities for $U_0 = 140$ m/s and $V_0 = 70$ m/s.

tions appearing at the left end of these profiles are believed to be numerical in origin. The predicted normal and transverse velocity histories at the rear surface of the target plate for this simulation are illustrated by Fig. 15.24. The normal velocity component which equals 140 m/s at the impact surface shows little attenuation, rising to approximately 135 m/s at the rear surface. The characteristic spall signal is evident at 3.1 μs. The decrease of the transverse velocity from 70 m/s at the impact surface to approximately 3 m/s at the rear surface can be explained as follows. In a second simulation the impact velocity components are reduced to $U_0 = 40$ m/s and $V_0 = 20$ m/s, insufficient to produce spallation. This may be seen by the corresponding rear surface velocity components shown by Fig. 15.25, where there is no spall signal in the normal velocity profile. The transverse velocity is shown by Figures 15.26(a) to 15.26(c). Figure 15.26(a), which is for the smaller impact velocity (producing no spall damage) shows arrival of the elastic precursor shear wave at approximately 2.6 μs, followed by the slower plastic shear wave. The decrease in transverse velocity, beginning at approximately 4.3 μs, reflects the propagation of the slip condition at the front face which occurred just prior to 2.0 μs. Figure 15.26(b) is an enlargement of the transverse velocity-time curve shown in Fig. 15.24, corresponding to the higher velocity of impact (producing spall damage). The leading elastic shear wave is followed by a plateau. However there is no substantial increase in the transverse velocity associated with the arrival of the plastic shear wave. It seems plausible, therefore, that the lack of a substantial plastic wave can be attributed to the damage accumulation which would prevent the plastic shear from being

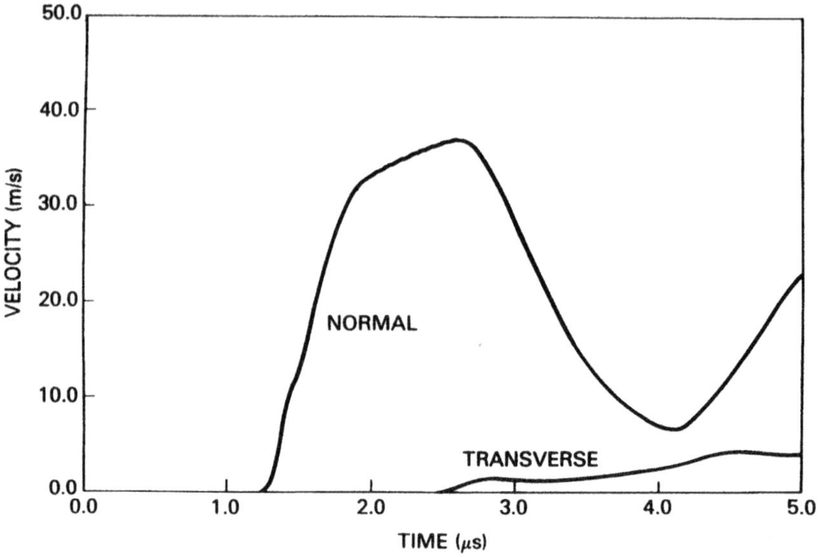

Figure 15.25. Calculated rear surface velocities for $U_0 = 40$ m/s and $V_0 = 20$ m/s.

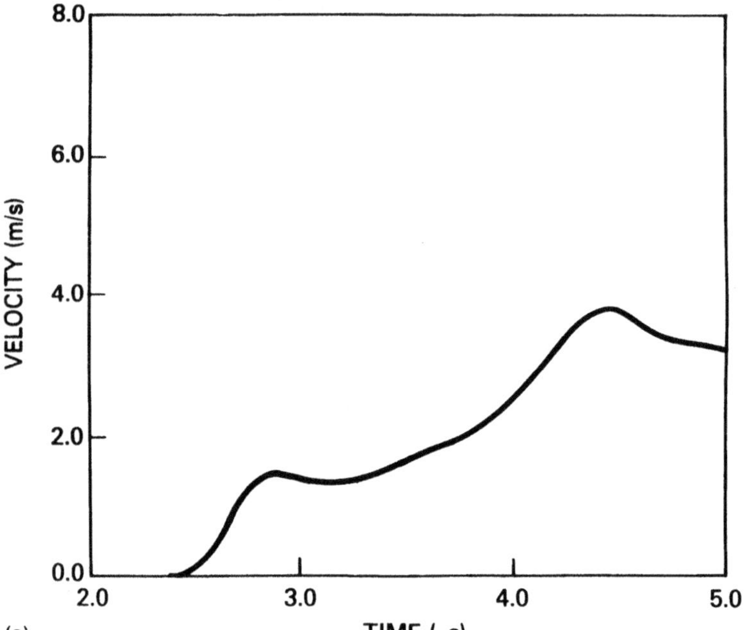

(a)

Figure 15.26. Calculated rear surface transverse velocities (a) $U_0 = 40$ m/s, $V_0 = 20$ m/s, (b) $U_0 = 140$ m/s, $V_0 = 70$ m/s, (c) $U_0 = 140$ m/s, $V_0 = 70$ m/s with damage suppressed.

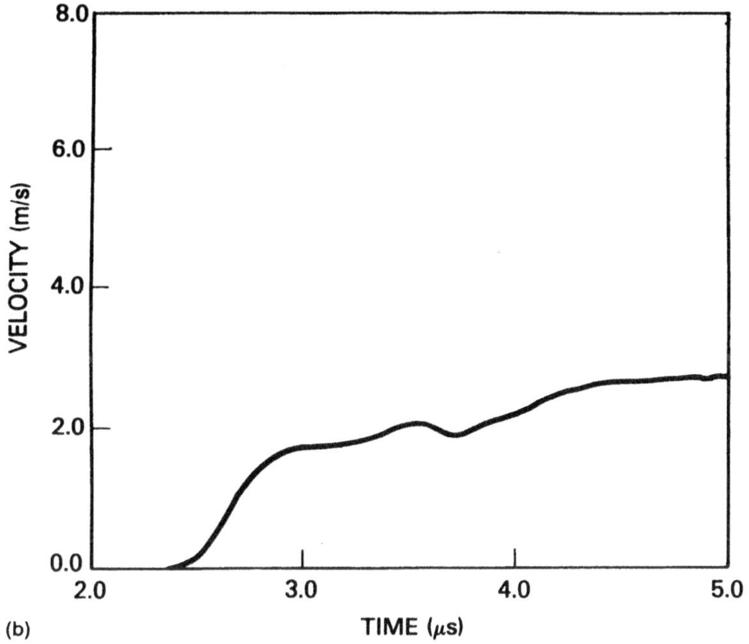
8.0
6.0
4.0
2.0
0.0
VELOCITY (m/s)
2.0
3.0
4.0
5.0
TIME (μs)

(b)

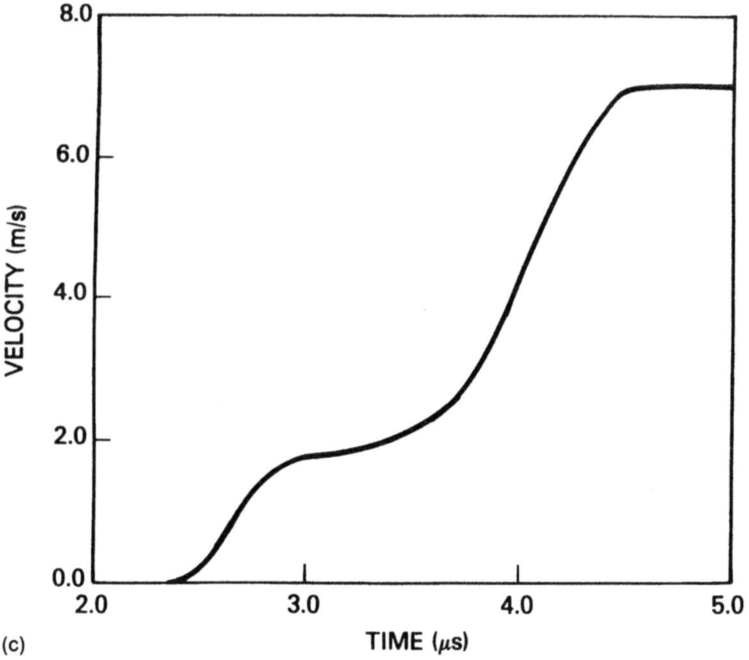
8.0
6.0
4.0
2.0
0.0
VELOCITY (m/s)
2.0
3.0
4.0
5.0
TIME (μs)

(c)

transmitted after the elastic shear had propagated through the material prior to onset of microvoid growth. This appears to be confirmed by the rear surface transverse velocity history shown in Fig. 15.26(c). The physical conditions for Fig. 15.26(c) are the same as for 15.26(b) (high impact velocity and spall damage), with the exception that the generation of damage following the impact has been suppressed. Without growth of the microvoid volume the transverse velocity profile of Fig. 15.26(c) now shows arrival of the elastic shear wave followed by a plastic wave front, similar overall in form to the velocity profile of Fig. 15.26(a) for which there was no damage because of the low velocity of impact.

15.3.5. Normal Impact of Circular Plates With Unequal Diameters

Non-oblique impact of circular plates that have different diameters generates multi-dimensional axisymmetric stress/strain, consisting of three normal components and one shear component that vary in the radial direction, as well as along the plate thickness direction. Because of the edge effect caused by the circumference of the smaller flyer plate, non-planar waves originating from the circumferential edge complicate the pattern of wave deformation. The schematic Fig. 15.27 illustrates the initial plane compressive wave fronts propagating across each plate, and the non-planar wave starting from the circular edge of the flyer. The latter wave traveling radially towards the center of the target reduces the region that is initially subjected only to high mean tensile stress. However if the plates are relatively thin, that is, if their diameters are ten to fifteen times as large as their thickness, a central region of the target will briefly experience large tensile mean stress arising from superposition of the reflected planar longitudinal stress waves prior to arrival of the radial wave. When the impact velocity of the flyer is sufficiently large the induced tensile mean stress in the target reach intensities high enough to cause spall fracture. The plate dimensions for the flyer are radius $R_1 = 20$ mm, thickness $L_1 = 3$ mm. For the target, radius $R_2 = 40$ mm and thickness $L_2 = 6$ mm, cf. Fig. 15.28. The numerical simulations were carried out for a total time of 5 μs after impact, sufficient for several wave propagations across the target thickness. Because of the smaller diameter of the flyer,

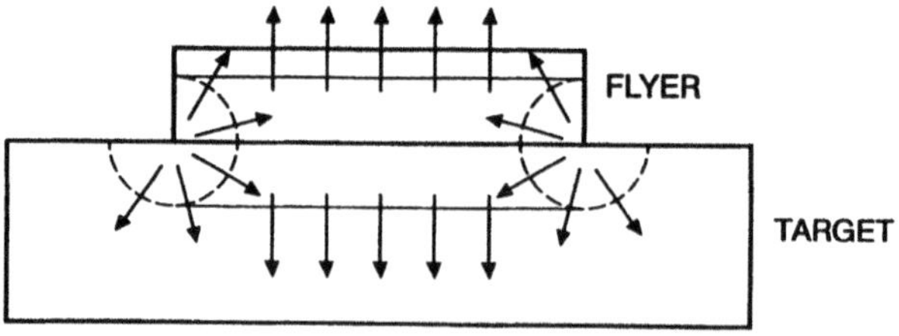

Figure 15.27. Impact of a circular flyer with a larger diameter target.

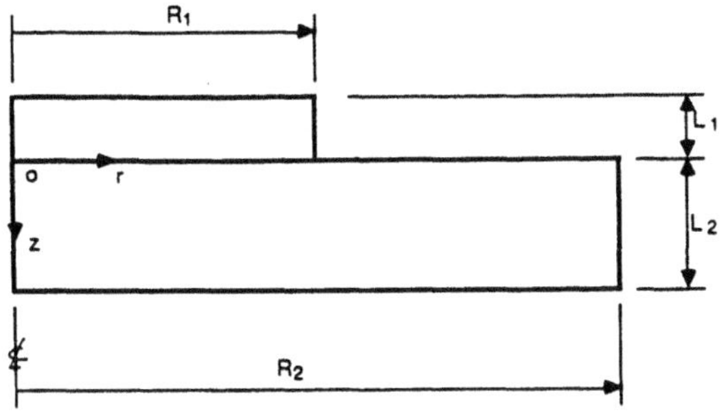

Figure 15.28. Flyer and target plate dimensions.

spall initiates only over the central region of the target. The post-spall behavior of the partially fragmented target was modeled by carrying the numerical calculations to a post-impact time of 50 μs. The calculations required use of the transient finite-element computer code PRONTO, and were carried out on a Cray X-MP computer. Detailed presentation of the various equations, the boundary and initial conditions, etc., as well as discussion of the numerical methods employed, can be found in References [40, 41].

15.3.5.1. Variation of Impact Velocity

The effects of increasing the flyer impact velocity in three steps from 100 m/s to 350 m/s, are shown graphically by Figures 15.29(a)–(c), which are the plate configurations 5 μs after impact. (For illustrative purposes the computed deformations have been magnified by a factor of three.) From these test simulations it was possible to estimate the damage threshold impact velocity at approximately 50 m/s for OFHC copper. This was determined by the first significant increase of the computed maximum void volume fraction over the average initial value. This estimate is in approximate agreement with experimental data and with a previous computational estimate for copper plates having the same flyer-target thickness ratio [55, 56]. It is noted that for these calculations fracture within the target plate was not simulated by severing the finite element grid lines. Accordingly, the extensive material damage appears as excessive elongation of the mesh elements. The corresponding contours of the void volume fraction across the target plate thickness are shown by Figures 15.30(a)–(c). Both sets of figures illustrate the pronounced effect of increasing the flyer impact velocity. At 100 m/s the void volume of the target along midplane shows an increase, although well below the critical value at 0.32. Correspondingly, material softening accompanies the void growth with elongation of the mesh elements across the midplane. At 350 m/s impact the rear spalled section of the target has sufficient momentum to cause further widening of the opening in the plate.

With the impact velocity at 350 m/s Fig. 15.31(a) displays the time varia-

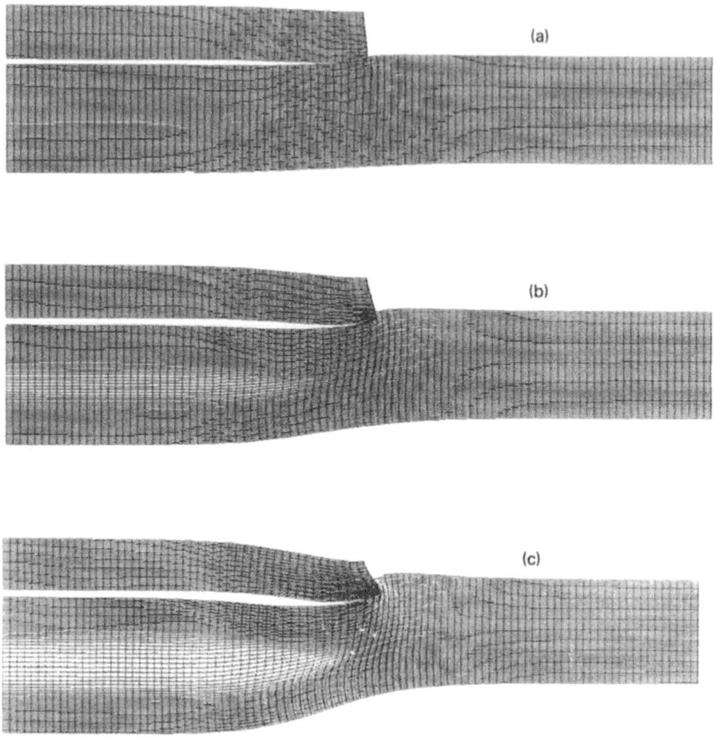

Figure 15.29. Deformed geometry at 5 μs post-impact for impact velocity at (a) 100 m/s, (b) 200 m/s, (c) 350 m/s.

tion of the location of the contour for the T_{zz} normal stress component at 6000 MPa compression. From 1.0 to 1.5 μs the reflected tensile stress wave from the back surface of the flyer travels forward causing forward shift of the location of the compressive stress contour. Simultaneously the inward movement of the vertical edge of the contour reflects the interaction with the non-planar radial wave from the circumferential edge of the flyer, reducing the diameter of the region of the target experiencing very high mean stress. This would seem to indicate why the diameter of the spalled surface initially is smaller than the diameter of the flyer. However as Fig. 15.29 graphically illustrates, as the velocity of impact increases the correspondingly larger momentum of the spalled section of the target causes the diameter of the area of separation to widen.

Figure 15.31(b) provides a similar plot for contours of the shear stress T_{rz} at 150 MPa at 0.5, 1.0 and 1.5 μs following impact at 350 m/s. The stress contours delineate the propagation of the non-planar shear wave from the edge of the flyer, traveling radially in toward the center and across the target thickness.

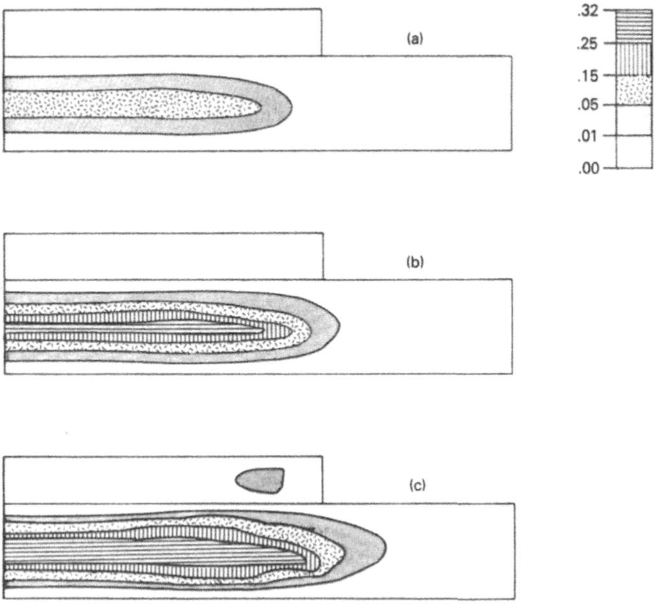

Figure 15.30. Contours for void volume fraction for impact velocity at (a) 100 m/s, (b) 200 m/s, (c) 350 m/s.

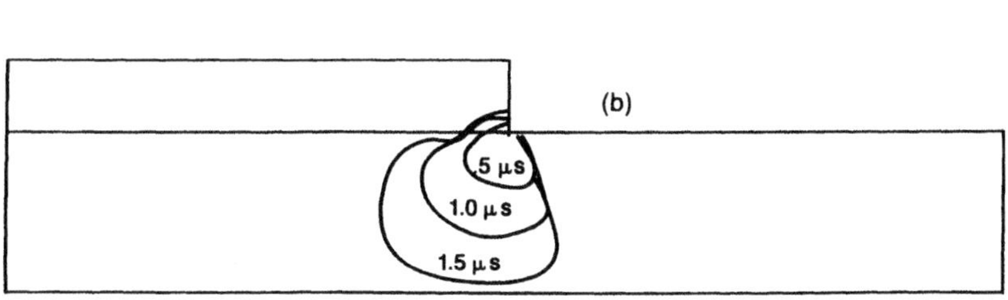

Figure 15.31. Time variation of stress contours after impact at 350 m/s velocity (a) $T_{zz} = 6000$ MPa, (b) $T_{rz} = 150$ MPa.

15.3.5.2. Variation of the Microviscosity Material Parameter

The material viscosity parameter $1/\gamma$ appearing in constitutive equation (15.5) for the viscoplastic rate of deformation is a measure of the viscosity of the polycrystalline material in the bulk (macroscopic) sense, as distinguished from the microviscosity material parameter η which is related to the void growth micro-damage process. Material parameter studies have shown that the parameter η has a major influence on the behavior of the damaged material [34, 36]. It is apparent from equation (15.22) that larger values for the microviscosity has the effect of reducing the rate of microvoid growth, while equation (15.12) indicates an associated strengthening of the material strain hardening as the void volume fraction is reduced.

The viscosity of polycrystalline metals subject to high strain rates at room temperature usually range between 10^3–10^5 P (where 1 Poise = 10^{-7} MPa-s),

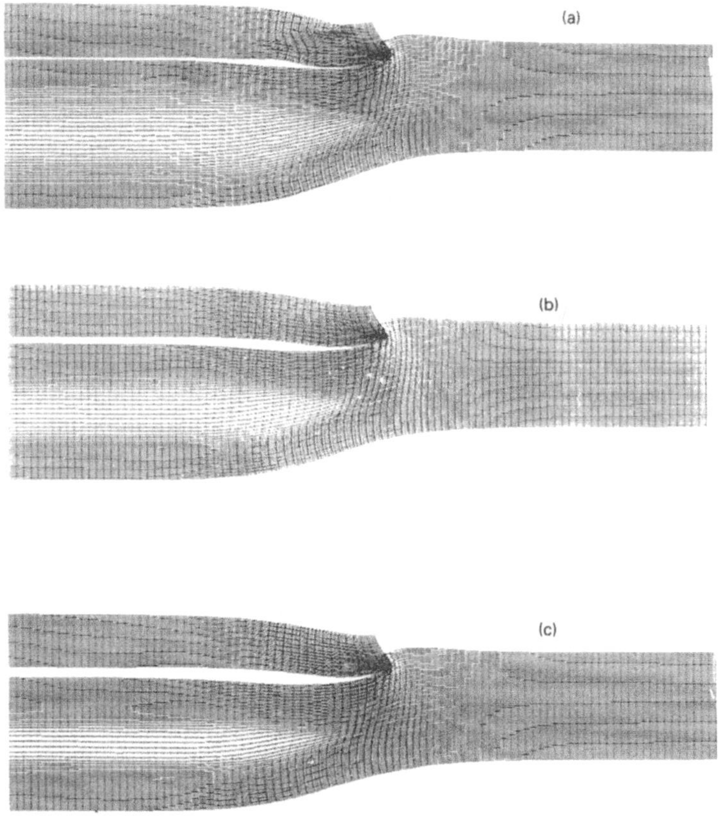

Figure 15.32. Deformed geometry at 5 μs post-impact for microviscosity parameter η at (a) 10 P, (b) 120 P, (c) 500 P. (Deformations have been magnified 3-fold for easier visualization.)

falling to as low as 10^{-2} P at melting temperature [57]. In the numerical simulations of plate impact spall fracture tests for copper at room temperature, the effect of temperature change induced by inelastic deformation was not considered and the microviscosity parameter value $\eta = 10$ P, suggested by Johnson [24], was assumed for copper, giving good correlation between predicted and experimental spall fracture data. This low value adopted for η is attributed to the highly localized heating effects that develop at and around the plastically expanding microvoid walls, creating local 'hot spots' that can rise to a substantial fraction of the melting temperature. To gain some idea of the relative difference between values for the bulk (macroscopic) viscosity $1/\gamma$ and the microviscosity η, it is noted that for copper $1/\gamma = \kappa_0/\gamma_0 = 2.77 \times 10^5$ P, while $\eta = 10^1$ P. Thus $1/\gamma \approx 10^4$ η.

With the flyer impact velocity at 350 m/s, several numerical simulations were performed with the parameter η assigned values of 10 P, 120 P, and 500 P. The corresponding plate configurations and the void volume fraction contours at 5 μs post-impact are shown by Figures 15.32 and 15.33, respectively, illustrating the controlling influence of this material parameter on the development of microvoid growth and material damage. As the microviscosity is increased the extent of damage that can develop within the material after a 5 μs interval is reduced considerably, while the material, which has correspondingly smaller microvoid volume, experiences greater strain hardening.

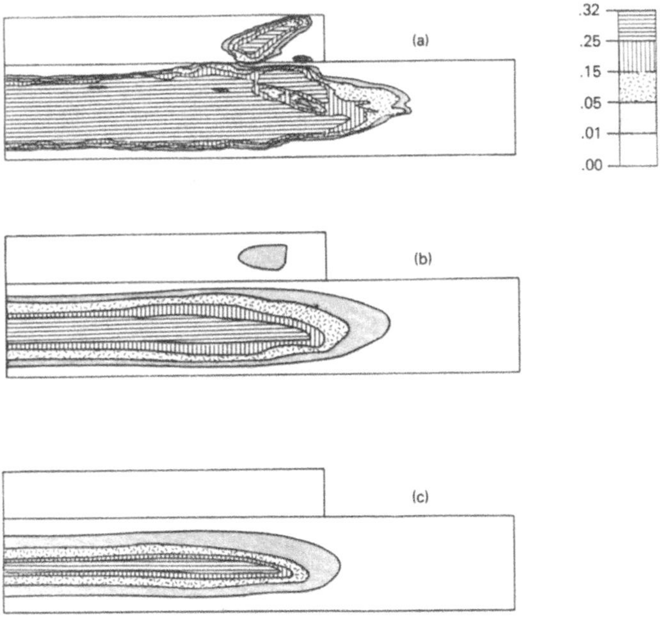

Figure 15.33. Contours for void volume fraction for microviscosity parameter η at (a) 10 P, (b) 120 P, (c) 500 P.

15.3.5.3. Post-spall Behavior

The calculated deformation of the plates at 5, 15, 25, and 50 μs after impact by the flyer traveling at 350 m/s are shown by Figures 15.34(a)–15.34(d). F6r these figures the deformation has not been magnified as in Figures 15.29 and 15.32. Also the element deletion capability available in the PRONTO computer code was utilized, where upon reaching a prescribed void volume fraction the stresses in the finite element are set to zero over several time steps and the element is deleted from further calculations. It should be noted that as the critical value of the microvoid volume is approached in the limit, that is, when $\xi = \xi_F$, signifying the onset of separation of the material, it is clear from expressions (15.20) and (15.16) that the rate of inelastic deformation $\mathbf{D}^p \to \infty$. Therefore, numerically, the condition $\xi = \xi_F$ can be approached but never actually reached in the computational simulation. The prescribed void volume fraction for element deletion for the numerical simulation shown here was chosen therefore to be 0.25, slightly below the critical value of $\xi_F = 0.32$.

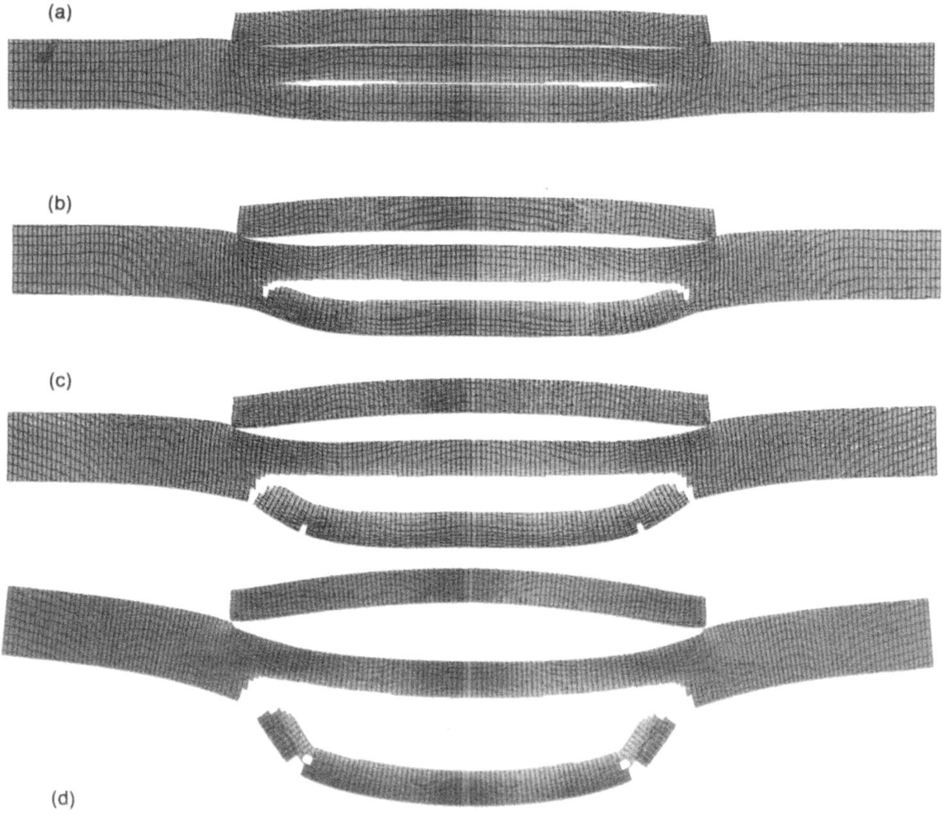

Figure 15.34. Deformed geometry for a 3.5 mm flyer with 350 m/s impact velocity at (a) 5 μs, (b) 15 μs, (c) 25 μs, (d) 50 μs post-impact.

At 5 μs spall has developed along the mid-plane of the target, with the rear portion of the spalled segment moving out and away from the remainder of the plate with momentum sufficient to cause it to continue its outward motion. This leads to fracture at the edges of the spall seen at 15 μs. The progressive deletion of elements along the edge of the spall simulates a shearing of the spalled material away from the remainder of the target. By 25 μs the spalled section of target plate has detached and has developed additional fractures near the detached edges along the outer highly curved tensile edge. At 50 μs, these edge fractures have advanced almost entirely across the fragment thickness. Calculations at later times show only minor changes from the configuration at 50 μs, indicating that no further fracturing of the segment is to be expected. Figure 15.35(a) displays the void volume distribution at 50 μs where, as expected, the highest values are adjacent to the spall plane. The equivalent plastic strain distribution across the plates appearing in Figure 15.35(b) indicate that the regions of highest plastic strains are those regions where the edge effects are present due to the larger shear deformation that occurs. In particular, the outer edge of the flyer along the flyer-target

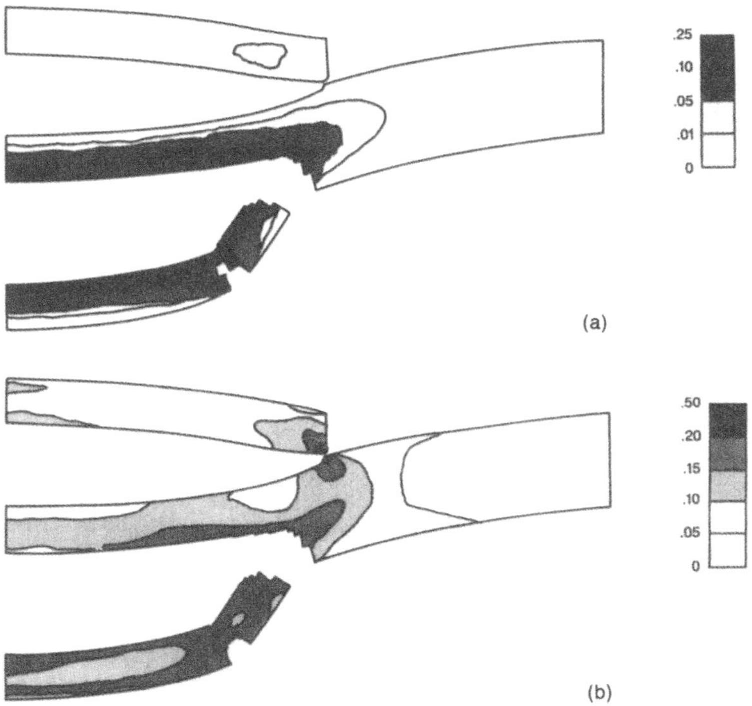

Figure 15.35. Contours due to 350 m/s impact velocity at 50 μs post-impact (a) Void volume fraction, (b) Equivalent plastic strain.

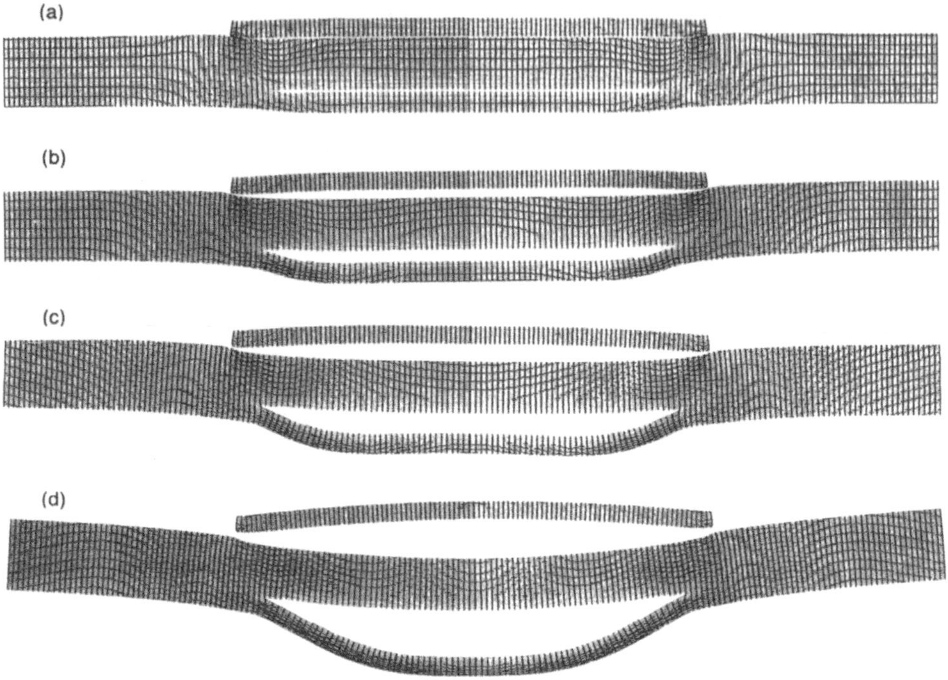

Figure 15.36. Deformed geometry for a 1.5 mm flyer with 350 m/s impact velocity at (a) 5 μs, (b) 15 μs, (c) 25 μs, (d) 50 μs post-impact.

interface, and the edges of the spall, are regions of high shear deformation that are reflected in the higher values of the equivalent plastic strain.

15.3.5.4. Change of Plate Thickness Ratio

For the test simulation illustrated by Figure 15.34, the flyer plate thickness is one-half the thickness of the target, causing the spall plane to appear at the mid-plane. Reducing the flyer thickness to one-quarter the target thickness, that is, with $L_1 = 1.5$ mm and $L_2 = 6.00$ mm, the expected location of the spall will be further back from the impact face at three-fourths the distance to the rear surface. With the flyer impact velocity at 350 m/s, Figures 15.36(a)–15.36(d) show the plate configurations at 5, 15, 25, and 50 μs after impact. Because of the thinner flyer the reflected tensile stress wave from the rear surface of the flyer enters the target sooner, superimposing with the reflected tensile stress wave from the rear surface of the target at a location closer to the rear surface. The calculated spall shown in Figure 15.36 appears at the anticipated location, however the comparatively thinner rear segment of the spalled portion of the target, having less momentum than the spalled segment of the previous simulation, is insufficient to cause fragmentation.

15.4. Other Constitutive-Damage Models for Spall Fracture

15.4.1. Introduction

Early investigations of spall fracture date back to the work of Hopkinson [58, 59], who observed that detonation of an explosive in contact with a steel plate produced not only a depression on the front surface of the plate, but the removal of a scab of metal from the rear surface. Hopkinson correctly deduced that the high tension needed to cause such a fracture was the result of a reflection of the original pressure wave from the rear surface. Hopkinson's experiments also revealed that when using a smaller quantity of explosive such that no scab would be evident, internal cracks appear nevertheless upon sectioning of the plate specimen. Interest in spallation and other dynamic fracture phenomena was renewed by the second world war, and continued into the next decade with the work of Rinehart [60, 61, 2], who proposed the first quantitative spall criterion. It predicts that spall fracture will occur whenever the normal stress carried by the reflected tensile wave is equal to or greater than a critical spall stress that is characteristic of the material. The insufficiency of so simple a criterion became evident, both from the work of Rinehart, and from the tests of Smith [55] and Blincow and Keller [62], when it was shown that different values for the critical spall stress are obtained for different thicknesses of the same plate material.

Guided by these findings Butcher et al. [4] postulated that not only is the magnitude of the tensile stress important in determining spallation, the duration of the tensile wave pulse must also be considered. Thus an empirical relation was introduced which includes both the tensile stress magnitude and the time duration of the pulse such that spall occurs whenever the quantity $[P \exp(h\sigma)] > k$, where P is the pulse duration, σ is the tensile stress and h and k are material parameters. In order to preclude the possibility of predicting spall in circumstances of low stress of long duration, Tuler and Butcher [5] subsequently proposed a criterion whereby spallation occurs at a critical value for the stress impulse

$$K = \int_0^t (\sigma - \sigma_0)^\lambda \, dt, \tag{15.38}$$

where σ_0 and λ are material parameters.

One of the principal shortcomings of the spall criteria described above resides in the fact that they do not reflect the influence that the microdamage within the material can have on the shape, and thus on the magnitude and time duration, of the tensile stress pulse. In order to account for this important effect it is necessary to bring some measure of microdamage into the description of the mechanical behavior of the material. Following in the wake of relatively recent developments in constitutive theory, and in the constitutive modeling of inelastic material behavior and material damage, there has emerged the background of information that enables the description of spall phenomena as the terminal phase of material response to specific dynamic loading processes.

Within the last two decades more than a few constitutive-type theories have been developed for the purpose of modeling spall fracture and fragmentation. In keeping with the constitutive-microdamage type of model to describe impact induced spall fracture discussed in Sections 2 and 3, only those theories similar in general structure are outlined below. Among those that are not included is the constitutive model proposed by Steinberg et al. [63, 64], incorporating the damage model for spall proposed by Cochran and Banner [65]. Although successful in calculating shock wave profiles including the effect of spall in metals, this constitutive model does not explicitly include microdamage processes as an integral part of the constitutive-damage formulation. Johnson [24] presents an analysis of the development of microvoids produced by tensile mean stress, extending the theoretical model of void collapse in a porous ductile material under pressure developed by Caroll and Holt [35]. The results are used to model the ductile spall of copper plate caused by impact. However a general constitutive formulation that includes the microdamage development as an integral part of a constitutive-damage theory is not given. The work of Grady [18] involves development of analytical expressions for the spall strength and fragmentation properties of strong solids, using energy balance principles together with concepts from fracture mechanics such as the fracture toughness. This work does not attempt to describe spall phenomena in terms of a microstructural damage process that is incorporated into a constitutive model that can describe material behavior and failure under dynamic loading.

15.4.2. Stanford Research Institute

The spall research activity of Curran, Seaman, Shockey and colleagues at SRI, cf. References [7–13], has been instrumental in the development of appropriate models for nucleation and growth of microvoids and microcracks in damaged solids. Systematic statistical measurements of microvoid volume and microcrack size distributions for different degrees of spall damage were obtained for ductile and brittle materials, and used to establish nucleation and growth rate expressions. These expressions have been used to model spallation in ductile and brittle materials for a variety of loading conditions.

The ductile fracture model is formulated on the basis of observations of ductile spall fracture of soft aluminum and copper plates. The measured microvoid volume distributions of the nearly spherical microvoids have a form that can be approximated by the relation

$$N(R) = N_0 \exp(-R/R_N), \tag{15.39}$$

where N is the cumulative number of voids per unit volume with radii larger than R, N_0 is the total number per unit volume, and R_N is a distribution parameter. The total void volume is obtained by integrating over the entire distribution (assumed to be spherically shaped)

$$V_V = \int_0^{N_0} N(R)\, dN = \frac{4\pi}{3}\int_0^{\infty} R^3 \frac{dN}{dR}\, dR = 8\pi N_0 R_N^3. \qquad (15.40)$$

Nucleation occurs in the model by adding new voids with a size distribution given by (15.39), at a rate of nucleation given by

$$\frac{dN}{dt} = C_1 \exp\left[\frac{\sigma - \sigma_N}{\sigma_1}\right], \qquad \sigma > \sigma_N \qquad (15.41)$$

where $\sigma = (1/3)J_1$, tensile stress is positive, σ_N is the nucleation threshold mean stress, and C_1 the nucleation rate at the threshold stress and σ_1 are material parameters. Microvoid growth is specified by an equation expressing the change in void radius caused by the tensile mean stress

$$\frac{dR}{dt} = \frac{1}{4\eta}(\sigma - \sigma_G)R, \qquad \sigma > \sigma_G \qquad (15.42)$$

where σ_G is the growth threshold stress, and η is the microviscosity parameter. The total change in void volume is the sum of the contributions associated with nucleation and growth. By means of expressions (15.40)–(15.42) it can be shown that after the time interval Δt the corresponding increment of the microvoid volume has the value

$$V_V = V_{V0} \exp\left[\frac{3}{4\eta}(\sigma - \sigma_G)\Delta t\right] + 8\pi C_1 R_N^3 \exp\left[\frac{\sigma - \sigma_N}{\sigma_1}\right]\Delta t. \qquad (15.43)$$

The constitutive equations utilized in the model are separated into volumetric and deviatoric parts. The non-linear Mie-Gruneisen pressure–volume relation, modified to account for the presence of voids, is adopted for the volumetric portion and has the form

$$\sigma = \frac{\rho}{\rho_s}\left[K\left(\frac{\rho_s}{\rho_0} - 1\right) + \rho_s \Gamma E\right]. \qquad (15.44)$$

Here ρ is the density of the porous material. ρ_s is the solid density with ρ_0 as its initial value, K is the bulk modulus of the solid, Γ is the Gruneisen ratio, and E is the internal energy of the solid.

For the deviatoric deformation the infinitesimal strain tensor is decomposed additively into elastic and plastic parts

$$d\boldsymbol{\Sigma}' = d\boldsymbol{\Sigma}'^e + d\boldsymbol{\Sigma}'^p. \qquad (15.45)$$

The incremental linear elastic constitutive relation

$$d\boldsymbol{\Sigma}'^e = \frac{1}{2\bar{\mu}}\mathbf{T}' \qquad (15.46)$$

includes the reduced shear modulus

$$\bar{\mu} = \mu\left\{1 - \rho V_V 15\left(\frac{1-\nu}{7-5\nu}\right)\right\}, \tag{15.47}$$

where V_V is the void volume given by (15.43), μ is the elastic shear modulus for the non-porous solid and ν is the Poisson ratio. The plastic deformation is taken to be rate-independent, non-hardening and volume preserving. Therefore $d\Sigma'^p = d\Sigma^p$. The yield stress $\bar{Y}$ is reduced because of the voids such that

$$\bar{Y} = Y(1 - 4\rho V_V), \tag{15.48}$$

where Y is the yield stress for the non-porous solid. The incremental plastic constitutive equation coupled with a von-Mises yield criterion can be expressed in the form

$$\begin{aligned} d\Sigma^p &= \frac{3}{2}\frac{d\overline{\varepsilon^p}}{\bar{\sigma}}\mathbf{T}', &\quad \bar{\sigma} &= \bar{Y} \\ d\Sigma^p &= 0 &\quad \bar{\sigma} &< \bar{Y} \end{aligned} \tag{15.49}$$

where the invariants

$$\bar{\sigma} = \{\tfrac{3}{2}\,\mathrm{tr}(\mathbf{T}'\cdot\mathbf{T}')\}^{1/2} \tag{15.50}$$

$$d\overline{\varepsilon^p} = \{\tfrac{2}{3}\,\mathrm{tr}(d\Sigma^p\cdot d\Sigma^p)\}^{1/2}. \tag{15.51}$$

Application of the above equations are carried forward by considering a volumetric strain increment which is decomposed into solid and void strain increments, $\Delta V = \Delta V_S + \Delta V_V$. Starting with a trial volume strain increment, equations (15.43)–(15.51) are then used in a trial and error iterative procedure as a subroutine in a wave propagation code. The model has been used to describe the evolution of damage and spall fracture in aluminum and copper from stress waves caused by impact, explosion and electron beam.

15.4.3. Sandia National Laboratories

Davison, Stevens and Kipp, [16, 17], have developed a constitutive model for ductile spall fracture that incorporates damage, expressed as the void volume fraction, into a viscoplastic constitutive theory based on dislocation mechanics. The description of the elastic-inelastic behavior centers on the assumption that the deformation gradient tensor $\mathbf{F}$ is composed of part $\mathbf{F}^p$ associated with viscoplastic flow, part $\mathbf{M}$ representing the dilatation resulting from the nucleation and growth of microvoids, and a part $\mathbf{F}^e$ associated with thermoelastic deformation such that

$$\mathbf{F} = \mathbf{F}^e\cdot\mathbf{M}\cdot\mathbf{F}^p. \tag{15.52}$$

This is the Lee decomposition, [66, 67], modified by inclusion of the isotropic tensor $\mathbf{M}$ representing the microvoid damage

$$\mathbf{M} = M\mathbf{1} = \left(\frac{1}{1-\xi}\right)^{1/3} \mathbf{1}, \tag{15.53}$$

where ξ is the void volume fraction at each point of the porous solid.

The constitutive equation for the inelastic deformation is developed in terms of a dislocation description of plasticity. The rate of plastic deformation is assumed to be a linear function of the product of the velocity gradient due to slip, Λ, and the plastic deformation gradient

$$\mathbf{D}^p = \Lambda \cdot \mathbf{F}^p. \tag{15.54}$$

In addition the plastic deformation is assumed to be volume preserving,

$$\det \mathbf{F}^P = 1. \tag{15.55}$$

The shear strain rate, $\dot{\gamma}$, is given by

$$\dot{\gamma} = b p_d d_m v(\tau) \tag{15.56}$$

where $\mathbf{b}$ is the Burgers vector and $b = |\mathbf{b}|$, p_d is the dislocation density, d_m is the fraction of dislocations that are mobile, and $v(\tau)$ is the average velocity of a mobile dislocation at the resolved shear stress τ. The velocity gradient associated with plastic flow on a given slip system is given by $\dot{\gamma}(\mathbf{b} \otimes \mathbf{n})$, where $\mathbf{n}$ is the normal to the slip plane. Since the velocity gradient for slip on all systems is additive, the total velocity gradient from slip on all of the slip planes is obtained by integrating

$$\Lambda = \int_{-\pi/2}^{\pi/2} \int_0^{\pi/2} \int_0^{\pi} b p_d d_m v(\tau)(\mathbf{b} \otimes \mathbf{n}) \sin \phi_2 \, d\phi_1 \, d\phi_2 \, d\phi_3. \tag{15.57}$$

The angles ϕ_k, $k = 1, 2, 3$, orient the glide plane, the Burgers vector, and the normal vector to the reference coordinate frame. The form used for the dislocation velocity has

$$v(\tau) = \begin{cases} 0, & \tau \leq \tau_A \\ \left(\dfrac{\mu}{\rho_R}\right)^{1/2} \operatorname{sgn}(\tau) \dfrac{s^m}{1 + s^m}, & |\tau| > \tau_A \end{cases} \tag{15.58}$$

where $s = (|\tau| - \tau_A)/\tau^*$, μ is the shear modulus, ρ_R is the material density in the reference configuration, m and τ^* are material constants and τ_A is given by

$$\tau_A = \tau_{A0}\left[1 + \left(\frac{\gamma}{\gamma_0}\right)^{1/p}\right], \tag{15.59}$$

with τ_{A0}, γ_0, and p as additional material constants. The total number of dislocations and the fraction of mobile dislocations are determined by the evolution equations

$$\frac{d}{dt}(p_d) = n^*(p_d^* - p_d)p_d d_m |v(\tau)| \tag{15.60}$$

$$\frac{d}{dt}(d_m) = -b^* p_d (d_m - d_m^*) d_m v(\tau), \tag{15.61}$$

where n^*, b^*, p_d^*, and d_m^* are material constants.

The recoverable part of the deformation is described by the linear constitutive equation for the thermoelastic solid assuming small deformation.

$$\mathbf{T} = [(\bar{K} - \tfrac{2}{3}\bar{\mu})I_{\Sigma^e} - 3\alpha\bar{K}(\theta - \theta_R)]\mathbf{1} + 2\mu\mathbf{\Sigma}^e, \tag{15.62}$$

where the infinitesimal strain tensor $\mathbf{\Sigma}^e = 2(\mathbf{F}^e - \mathbf{1})$, $\mathbf{I}_{\Sigma^e} = \operatorname{tr} \mathbf{\Sigma}^e$, θ is the current temperature, θ_R is the reference temperature, and α is the thermal expansion coefficient. $\bar{K}$ and $\bar{\mu}$ are the elastic bulk and shear moduli for the solid with microvoids,

$$\bar{K} = K_0\left[1 - \frac{3}{2}\frac{(1 - \nu_0)}{(1 - 2\nu_0)}\xi\right] \tag{15.63}$$

$$\bar{\mu} = \mu_0\left[1 - \frac{15}{7}\frac{(1 - \nu_0)}{(1 - 2\nu_0)}\xi\right], \tag{15.64}$$

where K_0, μ_0 and ν_0 are the moduli and the Poisson ratio for the non-porous solid.

The evolution equation for the microvoid volume fraction is given by

$$\frac{d\xi}{dt} = \left\{f_n \bar{V}_0 + 3f_g\left(\frac{\xi}{1 - \xi}\right)\right\}(1 - \xi)^2, \tag{15.65}$$

where f_n and f_g are functions of the mean stress σ, and $\bar{V}_0$ is the minimum void volume fraction. The forms used for f_n and f_g at the reference temperature are

$$f_n = C_N\left\{\exp\left[\frac{\sigma - \sigma_N + |\sigma - \sigma_N|}{2\sigma_1}\right] - 1\right\} \tag{15.66}$$

$$f_g = \frac{1}{2\sigma_1} C_G\{2\sigma + |\sigma - \sigma_G| - |\sigma + \sigma_G|\}. \tag{15.67}$$

where σ_N and σ_G are the void nucleation and growth threshold stresses, and C_N, C_G, σ_1 are material constants.

The above equations are specialized to uniaxial strain for wave propagation along the X_3 direction for application to plate impact fracture. The equations of motion (15.27) have the form

$$\begin{aligned} x_1 &= X_1 \\ x_2 &= X_2 \\ x_3 &= f(X_3, t). \end{aligned} \tag{15.68}$$

The corresponding deformation gradient and each of the components of equation (15.52) are given by

$$[\mathbf{F}] = \begin{bmatrix} 1 & & \\ & 1 & \\ & & \lambda_3 \end{bmatrix}, \quad [\mathbf{F}^e] = \begin{bmatrix} \lambda_1^e & & \\ & \lambda_2^e & \\ & & \lambda_3^e \end{bmatrix}, \quad [\mathbf{F}^p] = \begin{bmatrix} \lambda_1^p & & \\ & \lambda_2^p & \\ & & \lambda_3^p \end{bmatrix}, \tag{15.69}$$

where $\partial x_3/\partial X_3 = F_{33} \equiv \lambda_3$. Substitution of equation (15.69) into (15.52) and the plastic incompressibility condition (15.55) gives the set of equations relating the stretch components and the void volume fraction.

$$1 = \lambda_1^e M \lambda_1^p, \qquad \lambda_3 = \lambda_3^e M \lambda_e^p, \qquad 1 = (\lambda_1^p)^2 \lambda_3^p. \tag{15.70}$$

Assuming that slip takes place on the family of planes subjected to the largest resolved shear stress, $\tau = \mu(\lambda_3^e - \lambda_1^e)(1 - \xi)^{-2/3}$, which corresponds to angles $\phi_2 = (1/4)\pi$ and $\phi_3 = (1/2)\pi$. Correspondingly, the vectors $\mathbf{n}$ and $\mathbf{b}$ in (15.57) have the values $\mathbf{n} = 1/\sqrt{2}\,(\cos\phi_1, \sin\phi_1, 1)$, $\mathbf{b} = b/\sqrt{2}\,(-\cos\phi_1, \sin\phi_1, 1)$, and equation (15.57) reduces to

$$[\wedge] = \frac{\dot{\gamma}}{4\pi}\begin{bmatrix} -\pi & & \\ & -\pi & \\ & & 2\pi \end{bmatrix}. \tag{15.71}$$

Expression (15.54) can thereby be written as

$$\begin{bmatrix} D_{11}^p & & \\ & D_{22}^p & \\ & & D_{33}^p \end{bmatrix} = \frac{\dot{\gamma}}{4\pi}\begin{bmatrix} -\pi & & \\ & -\pi & \\ & & 2\pi \end{bmatrix}\begin{bmatrix} \lambda_1^p & & \\ & \lambda_2^p & \\ & & \lambda_3^p \end{bmatrix}. \tag{15.72}$$

Using Equation (15.56) gives for the components of the inelastic rate of deformation

$$\begin{aligned} D_{11}^p &= -\tfrac{1}{4} b p_a d_m v(\tau) \lambda_1^p \\ D_{33}^p &= \tfrac{1}{2} b p_a d_m v(\tau) \lambda_3^p \end{aligned}, \qquad |\tau| > \tau_A. \tag{15.73}$$

The constitutive-damage model, coupled with an adiabatic energy rate balance equation, has been applied to model the material damage and spall fracture of aluminum caused by plate impact, showing very good comparison between the calculated and observed rear surface velocity histories [16].

15.4.4. Dayton Research Institute

Most recently, Rajendran, Grove and Bless [68, 69] have developed a viscoplastic model for high strain-rate deformation and spall fracture in ductile materials. The constitutive-damage model views the material as an aggregate consisting of a matrix of solid material, and microvoids. Prior to void nucleation the aggregate is the fully dense void free matrix material, i.e.;

prior to loading, the material is assumed not to have an initial microvoid distribution.

The strain rate is decomposed into elastic and plastic parts

$$\dot{\mathbf{\Sigma}} = \dot{\mathbf{\Sigma}}^e + \dot{\mathbf{\Sigma}}^p \tag{15.74}$$

where $\mathbf{\Sigma}$ is the infinitesimal strain tensor. The elastic strain-rate is related to the stress-rate through Hooke's law

$$\dot{\mathbf{T}} = \mathbf{C} \cdot (\dot{\mathbf{\Sigma}} - \dot{\mathbf{\Sigma}}^p) \tag{15.75}$$

where $\mathbf{C}$ is the elasticity tensor, the components of which are assumed to be unaffected by the presence of the microvoids. Because of the use of the small strain tensor rigid rotations are presumably very small, and the stress-rate appearing in (15.75) can be interpreted as the usual time rate.

The viscoplastic constitutive relation for the aggregate is given by

$$\dot{\mathbf{\Sigma}}^p = \frac{(1 - \xi) Y_m \dot{\Sigma}_m^p}{\operatorname{tr}\left(\dfrac{\partial F}{\partial \mathbf{T}} \cdot \mathbf{T}\right)} \frac{\partial F}{\partial \mathbf{T}} \tag{15.76}$$

where ξ is the void volume fraction, F is the yield function for the aggregate, $\dot{\Sigma}_m^p$ is the equivalent plastic strain rate invariant and Y_m is the effective stress invariant, both for the matrix, defined by

$$\dot{\Sigma}_m^p = \frac{2}{\sqrt{3}} \left\{ \frac{1}{2} \operatorname{tr}(\dot{\mathbf{\Sigma}}^p \cdot \dot{\mathbf{\Sigma}}^p) \right\}^{1/2}, \tag{15.77}$$

$$Y_m = (3J_2')^{1/2}. \tag{15.78}$$

Note that the subscript m denotes association with the matrix and is not a tensorial subscript. The viscoplastic behavior of the fully dense void free matrix material is described by the Bodner-Partom constitutive model [70],

$$\dot{\mathbf{\Sigma}}^p = D_0 \exp\left[-\left(\frac{Z}{Y_m}\right)^{2n} \left(\frac{n+1}{2n}\right) \right] \frac{\mathbf{T}'}{(J_2')^{1/2}} \tag{15.79}$$

where Z is an internal scalar variable that is interpreted as the isotropic hardening of the matrix, n is a material parameter related to the strain-rate sensitivity, and D_0 is the limiting value of the plastic strain rate that is usually taken at $10^8\ \mathrm{s}^{-1}$ for metals. The constitutive theory does not include a yield condition, implying viscoplastic deformation at any level of stress in the matrix material. Combining (15.79) with (15.77) gives the following expression for the equivalent plastic strain rate in the matrix

$$\dot{\Sigma}_m^p = \frac{2}{\sqrt{3}} D_0 \exp\left[-\left(\frac{Z}{Y_m}\right)^{2n} \left(\frac{n+1}{2n}\right) \right]. \tag{15.80}$$

The hardening variable Z is determined by the rate equation

$$\frac{dZ}{dt} = m(Z_1 - Z)\frac{dW^p}{dt} \tag{15.81}$$

where

$$m = m_0 + m_1 \exp(-\propto W^p), \tag{15.82}$$

Z_1, m_0, m_1, and $\propto$ are material parameters and W^p is the plastic work

$$W^p = \int_0^t \mathrm{tr}(\mathbf{T}' \cdot \dot{\boldsymbol{\Sigma}}^p)\, dt'. \tag{15.83}$$

The yield condition for the aggregate has the form

$$F = [2 + (1 - \xi)^2]J_2' + \left[\frac{1 - (1 - \xi)}{3}\right]J_1^2 - \exp\left[-B\left(\frac{\xi}{1 - \xi}\right)\right]Y_m^2 = 0, \tag{15.84}$$

where B is a material parameter, J_2' is the second invariant of the aggregate stress deviator tensor and J_1 is the first invariant of the aggregate stress tensor. The yield function F exhibits the effects of softening due to the presence of voids. It is to be noted that although the description of the aggregate viscoplastic flow is governed by a yield condition, the description of the viscoplastic deformation of the matrix is not. Thus for stress states insufficient to cause inelastic behavior of the aggregate, it is still possible to have viscoplastic deformation of the matrix.

The development of damage within the aggregate, that is, the increase of the void volume fraction because of nucleation and growth, has for the void nucleation the model suggested by Chu and Needleman [71],

$$\left(\frac{d\xi}{dt}\right)_n = F_1(\dot{Y}_m + \dot{\sigma}) + F_2\dot{\Sigma}_m^p \tag{15.85}$$

where

$$F_1 = \frac{1}{s_1\sqrt{2\pi}} \exp\left[-\left(\frac{Y_m + \sigma - \sigma_N}{2s_1}\right)\right]^2 \tag{15.86}$$

and

$$F_2 = \frac{1}{s_2\sqrt{2\pi}} \exp\left[-\left(\frac{\Sigma_m^p - e_N}{2s_2}\right)\right]^2. \tag{15.87}$$

In these expressions σ is the mean stress, f_1, f_2, s_1, s_2, are material parameters and σ_N and e_N are nucleation threshold values for the mean stress and equivalent strain respectively. The microvoid growth is considered to be the only source for the volumetric deformation of the aggregate. The microvoid growth rate is therefore expressed in terms of the plastic dilatation rate of the aggregate

$$\left(\frac{d\xi}{dt}\right)_g = (1 - \xi)\dot{\Sigma}^p_{kk}. \tag{15.88}$$

The constitutive-damage model in conjunction with the STEALTH one-dimensional finite-difference computer code, has been used to simulate the spall behavior of OFHC copper, showing good comparisons with experimental measurements for the rear surface velocity of the target plate.

15.4.5. Additional Comments

The earlier investigations of spall phenomena characterize spall fracture as a material separation that results when the interaction of tensile waves produce stress in excess of the 'spall threshold' of the material. This kind of a "go–no go" spall criterion, in effect, presupposes no prior damage within the material at the instant the material separation takes place. A more realistic view emerged subsequently (Stanford Research Institute) quantifying the material damage leading to spall in terms of the number density and size distributions of the microvoids and microcracks in the material, which in turn suggested model representations for the nucleation and growth rates of the microdamage processes. This was coupled with the recognition that a constitutive-damage model provides an avenue for linking the microstructural damage features to the description of dynamic fracture processes. The next major step towards a capability for realistically describing spall was the representation of the microdamage as a continuous field measure (Sandia National Laboratory), thereby eliminating the need for detailed consideration of the features of the individual micro-defects such as their number, size, orientation, etc.

The constitutive-damage theory presented in Sections 15.2 and 15.3 above, demonstrates quite convincingly that knowledge of the microvoid volume fraction alone appears to be sufficient to enable construction of a reasonably realistic and accurate model for ductile spall fracture when limited to conditions of low to moderate shock. For application to high shock spall phenomena such as explosive impact, temperature sensitivity must be incorporated into both the constitutive and the micro-damage parts of the model to account for the thermomechanical coupling effects that can be expected to be significant. While the model provides threshold conditions for the onset of material damage, including the condition for local fracture, less well understood is the ability to describe the effects of microvoid interaction and coalescence. However the categorization of coalescence as a microdamage process distinct unto itself, as are the void nucleation and growth processes, is open to question. There is physical evidence that suggests that coalescence is actually microvoid/microcrack nucleation and growth on a smaller scale.

The temporal dependence of the spall damage process, that is, on the time duration of the tensile stress impulse associated with the rear surface reflected tensile wave can be demonstrated in the following manner, Perzyna

[26]. Rewriting the microvoid growth rate equation (15.22)

$$\left(\frac{d\xi}{dt}\right)_g = \frac{1}{\eta} g(\xi) F(\xi, \xi_0) |\sigma - \sigma_G(\xi)| \tag{15.89}$$

where the void growth threshold stress $\sigma_G(\xi)$ is specified by equation (15.24). For the purposes of illustration and simplification, an average constant value for the threshold stress can be assumed, that is, let $\sigma_G(\xi) \equiv \bar{\sigma}_G$ for $\xi_0 \leq \xi \leq \xi_F$. Equation (15.89) can then be written in the form

$$\frac{d\xi}{g(\xi) F(\xi, \xi_0)} = \frac{1}{\eta} [\sigma(t) - \bar{\sigma}_G] \, dt, \tag{15.90}$$

which upon integration gives

$$\int_{\xi_0}^{\xi} \frac{d\zeta}{g(\zeta) F(\zeta, \zeta_0)} = \frac{1}{\eta} \int_0^t [\sigma(z) - \bar{\sigma}_G] \, dz. \tag{15.91}$$

The left side of (15.91) can be designated by the function

$$\psi(\xi, \xi_0) = \int_{\xi_0}^{\xi} \frac{d\zeta}{g(\zeta) F(\zeta, \zeta_0)}, \tag{15.92}$$

so that

$$\psi(\xi, \xi_0) = \frac{1}{\eta} \int_0^t [\sigma(z) - \bar{\sigma}_G] \, dz. \tag{15.93}$$

From this relation it can be seen that the microdamage associated with the void growth process is some function of the initial porosity ξ_0, and the stress impulse divided by the material microviscosity, i.e.,

$$\xi(\xi_0, t) = \phi\left(\xi_0, \frac{1}{\eta} \int_0^t [\sigma(z) - \bar{\sigma}_G] \, dz\right). \tag{15.94}$$

This result illustrates the time duration necessary for microvoid growth, and thus for the spall damage process to take place, and bears similarity to the stress impulse spall criterion, equation (15.38), initially proposed by Tuler and Butcher [5].

15.5. References

1. Curran, D.R. Dynamic fracture, in: Impact Dynamics, 333–366, J. Zukas et al. [eds.], J. Wiley, New York (1982).
2. Rinehart, J.S. and Pearson, J. Behavior of metals under impulsive loads, American Society for Metals, 1954, Dover Pub., New York (1965).
3. Gilath, I. and Eliezer, S. Pulsed laser induced spall, a new method to study dynamic properties, Advances in Plasticity, 393–396 (1989), A.H. Khan and M. Tokuda [eds.], Pergamon Press, Oxford (1989).

4. Butcher, B.M., Barker, L.M., Munson, D.E., and Lundergan, C.D. Influence of stress history on time-dependent spall in metals, *AIAA J.*, **2**, 977–990 (1964).
5. Tuler, F.R. and Butcher, B.M. A criterion for the time dependence of dynamic fracture, *Int. J. Fract.*, **4**, 431–437 (1968).
6. Gilman, J.J. and Tuler, F.R. Dynamic fracture by spallation in metals, *Int. J. Fract.*, **6**, 169–182 (1970).
7. Barbee, T.W., Seaman, L., Crowden, R., and Curran, D.R. Dynamic fracture criteria for ductile and brittle metals, *J. Mater.*, **7**, 393–401 (1972).
8. Seaman, L., Shockey, D.A., and Curran, D.R. The growth law for crack propagation under shock conditions, in: Dynamic Crack Propagation, 629–647, G.H. Sih [ed.], Noordhoff, Leyden (1973).
9. Seaman, L., Curran, D.R., and Shockey, D.A. Computational models for ductile and brittle fracture, *J. Appl. Phys.*, **47**, 4814–4826 (1976).
10. Curran, D.R., Seaman, L., and Shockey, D.A. Dynamic failure in solids, *Physics Today*, 46–55 (January 1977).
11. Shockey, D.A., Seaman, L., and Curran, D.R. Microfailure models and their application to nonlinear dynamic fracture problems, in: Nonlinear and Dynamic Fracture Mechanics, *ASME, Appl. Mech. Div.* **35**, 79–104 (1979).
12. Seaman, L., Curran, D.R., and Murri, W.J. A continuum model for dynamic tensile microfracture and fragmentation, *J. Appl. Mech.*, **52**, 593–600 (1985).
13. Shockey, D.A., Seaman, L., and Curran, D.R. The micro-statistical fracture mechanics approach to dynamic fracture problems, *Int. J. Fracture*, **27**, 145–157 (1985).
14. Davison, L. and Stevens, A.L. Continuum measures of spall damage, *J. Appl. Phys.*, **43**, 944–988 (1972).
15. Davison, L. and Stevens, A.L. Thermomechanical constitution of spalling elastic bodies, *J. Appl. Phys.*, **44**, 668–674 (1973).
16. Davison, L., Stevens, A.L., and Kipp, M.E. Theory of spall damage accumulation in ductile metals, *J. Mech. Phys. Solids*, **25**, 11–28 (1977).
17. Davison, L. and Kipp, M.E. Calculation of spall damage accumulation in ductile metals, in: High Velocity Deformation of Solids, 163–175, K. Kawata and J. Shiori [eds.], Springer-Verlag, New York (1978).
18. Grady, D.E. The spall strength of condensed matter, *J. Mech. Phys. Solids*, **36**, 353–384 (1988).
19. Meyers, M.A. and Aimone, C.T. Dynamic fracture (spalling) of metals, Prog. in *Mater. Sci.*, **28**, 1–96 (1983).
20. Curran, D.R., Seaman, L., and Shockey, D.A. Dynamic failure of solids, Phys. Reports, **147**, 253–388 (1987).
21. Barker, L.M. and Hollenbach, R.E. Laser interferometer for measuring high velocities of any reflecting surface, *J. Appl. Phys.*, **43**, 4669–4674 (1972).
22. Davison, L. and Graham, R.A. Shock compression of solids, *Physics Reports*, **55**, No. 4, 255–379 (1979).
23. Mackenzie, J.H. The elastic constants of a solid containing spherical holes, Proc. Phys. Soc. (London), **B**, **63**, 2–11 (1950).
24. Johnson, J.N. Dynamic fracture and spallation in ductile solids, *J. Appl. Phys.*, **53**, 2812–2825 (1981).
25. Perzyna, P. The constitutive equations for rate sensitive plastic materials, *Quart. Appl. Math.*, **20**, 321–332 (1963).

26. Perzyna, P. Internal state variable description of dynamic fracture of ductile solids, *Ind. J. Solids* Struct., **22**, 797–818 (1986).
27. Shima, S. and Oyane, M. Plasticity for porous solids, *Int. J. Mech. Sci.*, **18**, 285–291 (1976).
28. Hancock, J.W. Plasticity of porous metals, in: Yield, Flow and Fracture of Polycrystals, 69–79, T.N. Baker [ed.], Appl. Science Pub., Essex (1983).
29. Doraivelu, S.M., Gegel, H.L., Gunasekera, J.S., Malas, J.C., Morgan, J., and Thomas, J.S. A new yield function for compressible P/M materials, *Int. J. Mech. Sci.*, **26**, 527–535 (1984).
30. Nemes, J.A., Eftis, J., and Randles, P.W. Viscoplastic constitutive modeling of high strain-rate deformation, material damage, and spall fracture, Trans. *ASME J. Appl. Mech.*, **57**, 282–291 (1990).
31. Perzyna, P. Constitutive modeling of dissipative solids for postcritical behavior and fracture, *ASME J. Engng. Materials Tech.*, **106**, 410–419 (1984).
32. Raj, R. and Ashby, M.F. Intergranular fracture at elevated temperature, *Acta Metallurgica*. **23**, 653–666 (1975).
33. Ashby, M.F. and Verall, R.A. Diffusion-accommodated flow and superplasticity, *Acta Metallurgica*. **21**, 149–163 (1973).
34. Eftis, J. and Nemes, J.A. Evolution equation for the void volume growth rate in a viscoplastic-damage constitutive model, *Int. J. Plasticity*, **7**, 275–293 (1991).
35. Carroll, M.M. and Holt, A.C. Static and dynamic pore collapse relations for ductile solids, *J. Appl. Phys.*, **46**, 1626–1636 (1972).
36. Nemes, J.A. A viscoplastic description of high strain-rate deformation, material damage and spall fracture, DSc. Dissertation, George Washington Univ., Washington, D.C. (1989).
37. Eftis, J., Nemes, J.A., and Randles, P.W. Viscoplastic analysis of plate-impact spallation, *Int. J. Plasticity*, **7**, 15–39 (1991).
38. Nemes, J.A. and Eftis, J. Pressure-shear waves and spall fracture described by a viscoplastic-damage constitutive model, *Int. J. Plasticity*, **8**, 185–207 (1992).
39. Nemes, J.A. and Eftis, J. Low-angle oblique impact spall fracture using viscoplastic constitutive theory, in: Advances in Constitutive Laws for Engineering Materials, Vol. II, 633–639, Int. Acad. Pub., Beijing (1989).
40. Nemes, J.A. and Eftis, J. Several features of a viscoplastic study of plate-impact spallation with multi-dimensional strain, *Computers and Structures*, **38**, 317–328 (1991).
41. Eftis, J. and Nemes, J.A. Modeling of impact-induced spall fracture and post spall behavior of a circular plate, *Int. J. Fracture*, **53**, 301–324 (1992).
42. Flanagan, D.P. and Taylor, L.M. An accurate numerical algorithm for stress integration with finite rotations, *Computer Methods in Appl. Mechanics and Engineering*, **62**, 305–320 (1987).
43. Johnson, G.C. and Bammann, D.J. A discussion of stress rates in finite deformation problems, *Int. J. Solids and Structures*, **20**, 725–757 (1984).
44. Kipp, M.E. and Lawrence, R.J. WONDY-V, A one-dimensional finite difference wave propagation code, SAND 81-0594, Sandia Laboratories, Albuquerque (1982).
45. Taylor, L.M. and Flanagan, D.P. PRONTO 2D, A two dimensional transient solid dynamics program, SAND 86-0594, Sandia Laboratories, Albuquerque (1987).

46. Herrmann, W. Nonlinear stress waves in metals, in: Wave Propagation in Solids, 129–183, ASME, New York (1969).
47. Rajendran, A.M., Grove, D.J., and Bless, S.J. A new yield function based dynamic failure model, in: Shock Waves in Condensed Matter—1987, S.C. Schmidt and N.C. Holmes, eds. Elsevier, Amsterdam (1988).
48. Valanis, K.C. On the uniqueness of solution of the initial value problem in softening materials, *J. Appl. Mech.*, **52**, 649–653 (1985).
49. Rice, J.R. The localization of plastic deformation, in: Theoretical and Appl. Mechanics, 207–220, W.T. Koiter, ed., North Holland Pub. (1976).
50. Abou-Sayed, A.S., Clifton, R.J., and Hermann, L. The oblique-plate impact experiment, *Exp. Mech.*, **16**, 127–133 (1976).
51. Abou-Sayed, A.S. and Clifton, R.J. Analysis of combined pressure-shear waves in an elastic/viscoplastic material. *J. Appl Mech.*, **44**, 70–84 (1977).
52. Kim, K.S. and Clifton, R.J. Pressure-shear impact of 6061-T6 aluminum, *J. Appl. Mech.*, **47**, 11–16 (1980).
53. Klopp, R.W., Clifton, R.J., and Shawki, T.G. Pressure-shear impact and the dynamic viscoplastic response of metals, *Mech. Materials*, **4**, 375–385 (1985).
54. Gilat, A. and Clifton, R.J. Pressure-shear waves in 6061-T6 aluminum and alpha-titanium *J. Mech. Phys. Solids*, **33**, 263–284 (1985).
55. Smith, J.H. Three low pressure spall thresholds in copper, in: Dynamic Behavior of Materials, 264–282, ASTM, Philadelphia (1963).
56. Nemes, J.A. and Eftis, J. Use of viscoplastic constitutive theory for simulating spallation thresholds, in: Shock Waves in Condensed Matter, 369–372, (1989), S.C. Schmidt, et al., [eds.], Elsevier Sci. Pub., Amsterdam (1990).
57. Carroll, M.M., Kim, K.T., and Nesterenko, V.F. The effect of temperature on viscoplastic pore collapse, *J. Appl. Phys.* **59**, 1962–1967 (1986).
58. Hopkinson, B. A method of measuring the pressure produced in the deterioration of high explosives or by the impact of bullets, *Proc. Roy. Soc., London* **213A**, 437–456, (1914).
59. Hopkinson, B. Scientific Papers, Cambridge Univ. Press, London (1910).
60. Rinehart, J.S., Some quantitative data bearing on the scabbing of metals under explosive attack, *J. Appl. Phys.*, **22**, 5, 555–560 (1951).
61. Rinehart, J.S. Scabbing of metals under explosive attack: Multiple scabbing, *J. Appl. Phys.*, **23**, **11**, 1229–1233 (1952).
62. Blincow, D.W. and Keller, D.V. Experiments on the mechanism of spall, in: Dynamic Behavior of Materials, 252–263, American Society for Testing Materials, Philadelphia, PA (1963).
63. Steinberg, D.J. and Sharp, R.W. Interpretation of shock-wave data for beryllium and uranium with an elastic-viscoplastic constitutive model, *J. Appl. Phys.*, **52**, **8**, 5073–5083 (1981).
64. Steinberg, D.J. Constitutive model used in computer simulation of time-resolved, shock-wave data, *Int. J. Impact Engng.*, **5**, 603–611 (1987).
65. Cochran, S. and Banner, D. Spall studies in uranium, *J. Appl. Phys.*, **48**, 7 2729–2737 (1977).
66. Lee, E.H. Finite-strain elastic-plastic theory with application to plane wave analysis, *J. Appl. Phys.*, **38**, 1, 19 (1967).
67. Lee, E.H. Analysis of the propagation of plane elastic-plastic waves at finite-strain, *J. Appl. Mech.*, Trans. ASME, **932** (1967).

68. Rajendran, A.M., Grove, D.J., and Bless, S.J. A new yield function based dynamic failure model, in: Shock Waves in Condensed Matter, 359–362 (1987), S.C. Schmidt and N.C. Holmes [edts.], Elsevier, Amsterdam.
69. Rajendran, A.M., Dietenberger, M.A., and Grove, D.J. A void growth-based failure model to describe spallation, *J. Appl. Phys.*, **65**, 4, 1521–1527 (1989).
70. Bodner, S.R. and Partom, Y. Constitutive equations for elastic-viscoplastic strain-hardening materials, *J. Appl. Mech.*, **42**, 385–389 (1975).
71. Chu, C.C. and Needleman, A. Void nucleation effects in biaxially stretched sheets, *J. Eng. Mater. Tech.*, **102**, 249–256 (1980).

CHAPTER 16

A Thermodynamically Consistent Description of Dynamic Continuum Damage

M.W. Lewis and H.L. Schreyer

16.1. Introduction

Continuum damage mechanics is the art of formulating constitutive equations that reflect the loss of stiffness and strength due to the growth of voids and microcracks. Just as plasticity is an idealization of the formation of slip bands, constitutive equations based on damage mechanics can be considered as an attempt to capture the essense of microstructural fracture—but yet result in a constitutive relation amenable to analytical and numerical analysis. In fact, idealized plasticity and damage occur simultaneously, so separating the two effects is merely an attempt to describe essential features. In general, ductile materials are represented by the theory of plasticity, whereas damage mechanics describes brittle materials. Here, the primary interest is in the dynamic response of brittle materials so the framework of continuum damage mechanics is considered to be most appropriate. However, extensions to include both plasticity and damage are possible (Stumvoll and Swoboda, 1993).

Recently, significant research effort has been placed on the representation of ceramic materials under high strain rate (Johnson and Holmquist, 1992; Kipp, Grady and Wise, 1992). As indicated by Shockey et al. (1990) in connection with penetration resistance of confined ceramics, the phenomena are complex and good constitutive models are necessary if the number of parameters to be determined from experimental studies is to be limited. In reality, most brittle damage is directional (Costin, 1983) so many approaches involve, in some sense, fracture mechanics. Rajendran (1982) provides a damage model based on microcrack initiation when Griffith's criterion is met for the largest characteristic microcrack. All smaller microcracks propagate at rates required to maintain an assumed exponential distribution of microcracks with random orientation. Taylor et al. (1986) developed a model based on work by Budiansky and O'Connell (1976) and Kipp and Grady (1979) involving a combination of tensile damage and plasticity in compression. Inconsistencies in the relationship among isotropic elastic moduli and the connection of strain rate to damage have been removed by Thorne (1990) but there still remain some theoretical limitations. The work by Dienes (1983) was used as a basis for a dynamic formulation by Addessio and Johnson (1989) in which micromechanical arguments were used to develop a fairly simple isotropic model suitable for numerical analysis. This work serves as a basis for the current paper.

Although computers allow the use of constitutive equations much more complicated than would be feasible for obtaining analytical solutions, there is still a severe limitation on the complexity that can be accommodated. It is interesting to note that only slight modifications to the von Mises, Prager-Drucker, and Mohr-Coulomb models are actually used for many problems in plasticity although these models represent a significant number of simplifying assumptions. The problem in damage mechanics can perhaps be stated as one of finding the corresponding simple forms for constitutive equations that reflect the essence of the effects of microcracks without being too complicated for numerical analysis.

The present work builds on many of the references cited above, but with a slightly different twist. The assumptions are made that (i) the strain tensor can be decomposed into the sum of elastic and crack strains, and (ii) the Helmholtz free energy consists of the sum of the stored elastic energy and surface energy created by the formation of cracks. Crack strain and average crack length are chosen to be the internal variables which characterize the damaged configuration. Evolution equations for crack strain and average crack length are based on the micromechanical considerations of Dienes (1983). A formulation consistent with thermodynamics is developed in a manner analogous to the theory of plasticity. The dissipation expression provides the appropriate function for use in the definition of a damage surface which is similar to that derived by Addessio and Johnson (1989). However, following Margolin (1983), several cases must be defined according to the signs of the principal values of stress so the damage surface is actually the greatest lower bound of several surfaces. The evolution equations are activated only when state variables lie on the damage surface. When the state variables lie inside the damage surface, additional damage is assumed not to occur. To maintain an upper limit on the speed of average crack growth, the state variables are also allowed to lie outside the damage surface—in which case rate dependence is exhibited through an overstress approach.

Although not necessary, the material is assumed initially to be isotropic. Then, with the growth of microcracks, such a material becomes anisotropic in a manner similar to that of plasticity. However, there is no need to determine the effective elastic moduli unless unloading is an important aspect of an engineering problem. The damage surface that results from microstructural and averaging arguments is of a shape consistent with the form often used in plasticity with a smooth tensile cutoff. Therefore, it is believed that characteristic material features are represented by the derived constitutive relation, which is sufficiently simple to be suitable for efficient use in numerical analyses.

16.2. Basic Formulation

To be specific, suppose that deformations are infinitesimal so no differentiation among various stress and strain tensors is required. The stress, $\boldsymbol{\sigma}$, is taken to be a linear map of the strain, $\mathbf{e}$:

$$\boldsymbol{\sigma} = \mathbf{E}^0 : \mathbf{e} \tag{16.1}$$

The material is assumed to be isotropic initially so the initial form of the elasticity tensor is

$$\mathbf{E}^0 = 3B\mathbf{P}^s + 2G\mathbf{P}^d \tag{16.2}$$

in which the bulk and shear moduli are B and G, respectively. The spherical and deviatoric projection operators are:

$$\mathbf{P}^s = \tfrac{1}{3}\mathbf{i} \otimes \mathbf{i} \qquad \mathbf{P}^d = \mathbf{I} - \mathbf{P}^s \tag{16.3}$$

The second and symmetric fourth order identity tensors are denoted by $\mathbf{i}$ and $\mathbf{I}$, respectively.

As damage occurs, the elasticity tensor changes to $\mathbf{E}$, which might not be isotropic, and the stress-strain relation becomes

$$\boldsymbol{\sigma} = \mathbf{E} : \mathbf{e} \tag{16.4}$$

Suppose that damage can be characterized with a set of variables denoted collectively as D, which may be a combination of scalars, second-order tensors and fourth-order tensors. A specific selection will be made later based on microstructural arguments. Suppose U, the specific internal energy, is assumed to be a function of the strain, the set of damage variables, D, and the entropy, s. The absolute temperature, T, is defined to be the variable conjugate to entropy. Therefore, T and the other conjugate variables are defined as follows:

$$T = \frac{\partial U}{\partial s} \qquad Y_e = \rho\frac{\partial U}{\partial \mathbf{e}} \qquad Y_D = \rho\frac{\partial U}{\partial D} \tag{16.5}$$

in which ρ is the mass density. It is convenient to define the following product as the dissipation rate:

$$D_s = -Y_D\dot{D} \tag{16.6}$$

If r denotes the specific internal heat source, $\mathbf{q}$ the heat flux, and $(\;)\nabla$ the gradient operator, then the local forms of the first law and the second law (Clausius-Duhem inequality) of thermodynamics are:

$$\begin{gathered} p\dot{U} - \rho r + \mathbf{q}\cdot\nabla - \boldsymbol{\sigma} : \dot{\mathbf{e}} = 0 \\ \rho T\dot{s} \geq \rho\dot{\mathbf{U}} - \boldsymbol{\sigma} : \dot{\mathbf{e}} + \frac{1}{T}\mathbf{q}\cdot(T\nabla) \end{gathered} \tag{16.7}$$

With the use of (16.5) and (16.6), (16.7) becomes:

$$\mathbf{Y}_e = \rho\boldsymbol{\sigma}$$
$$\rho T\dot{s} - D_s - \rho r + \mathbf{q}\cdot\nabla = 0 \tag{16.8}$$
$$D_s \geq \frac{1}{T}\mathbf{q}\cdot(T\nabla)$$

In the absence of thermal effects ($r = 0$ and $\mathbf{q} = 0$), the last two equations reduce to

$$\begin{aligned} \rho T\dot{s} &= D_s \\ D_s &\geq 0 \end{aligned} \tag{16.9}$$

The Helmholtz free energy, which involves temperature and not entropy, is a Legendre transformation of the internal energy with respect to the conjugate pair (s, T):

$$\Psi(T; \mathbf{e}, D) = U(s; e, D) - sT \tag{16.10}$$

It follows that

$$s = -\frac{\partial\Psi}{\partial T} \qquad Y_e = \rho\frac{\partial\Psi}{\partial\mathbf{e}} \qquad Y_D = \rho\frac{\partial\Psi}{\partial D} \tag{16.11}$$

For isothermal processes, it is more convenient to use the Helmholtz free energy instead of the internal energy. The form of the energy must be consistent with the stress-strain relation.

Consider damage evolution laws of the form

$$\dot{D} = -\dot{\omega}d(\mathbf{e}, D) \tag{16.12}$$

in which ω is a monotonically increasing variable used to parameterize the evolution equations and d is a set of functions unspecified at this stage. The value of the parameter, ω, might be used to characterize the extent of damage. When the evolution equations are substituted in the dissipation rate, the result is

$$D_s = \dot{\omega}Y_D d \tag{16.13}$$

To ensure that the dissipation rate is always positive, define a damage function

$$\Psi_D = Y_D d - \Psi_0 \tag{16.14}$$

in which Ψ_0 is a positive constant. If Ψ_D is negative, the evolution equations are not evoked, i.e., additional damage is not allowed to occur. When $\Psi_D = 0$, the evolution equations are invoked and the dissipation rate becomes

$$D_s = \dot{\omega}\Psi_0 \tag{16.15}$$

Because Ψ_0 and $\dot{\omega}$ are positive or zero, the dissipation inequality and the second law of thermodynamics are automatically satisfied. Furthermore, the damage evolution equations have been parameterized in terms of the dissipation rate since $\dot{\omega}$ is proportional to D_s. The relation for $\dot{\omega}$ is obtained from the "consistency" condition that

$$\dot{\Psi}_D = 0 \tag{16.16}$$

One method for introducing rate dependence is to allow for an overstress model in which the coefficient $\dot{\omega}$ in the evolution equations is replaced with a material variable, γ, times the distance from the damage surface. For example, the evolution equation for the damage variables would then be

$$\dot{D} = -\gamma\langle\Psi_D\rangle d(\mathbf{e}, D) \tag{16.17}$$

in which $\langle\ \rangle$ denotes the Macauley bracket. Then the dissipation rate becomes

$$D_s = \gamma\langle\Psi_D\rangle Y_D d \tag{16.18}$$

which must be positive.

16.3. Strains Due to Cracks

The development of this section builds on the work of Dienes (1983) and Addessio and Johnson (1989). One essential aspect of the approach is that microcracks are always assumed to exist and that a crack strain field can be associated with these microcracks. The average crack length $\bar{c}$ is an essential parameter so $\bar{c}$ and the crack strain are taken to be the damage parameters for the model.

The total crack strain $\mathbf{e}^c$ is assumed to be composed of a part $\mathbf{e}^t$ due to tension (mode I) and a part $\mathbf{e}^s$ due to shear (modes II and III):

$$\mathbf{e}^c = \mathbf{e}^t + \mathbf{e}^s \tag{16.19}$$

These strains will develop only if appropriate tensile and shear components of stress exist. Let $\mathbf{t}^n$ denote the traction vector on a surface with normal $\mathbf{n}$, and let t_n^n denote the normal component of this vector:

$$\mathbf{t}^n = \boldsymbol{\sigma}\cdot\mathbf{n} \qquad t_n^n = \mathbf{t}^n\cdot\mathbf{n} \tag{16.20}$$

The magnitude of the shear component of traction is

$$t_s^n = [\mathbf{t}^n\cdot\mathbf{t}^n - (t_n^n)^2]^{1/2} \tag{16.21}$$

If G and ν denote the shear modulus and Poisson's ratio, respectively, then the farfield tensile crack strain in a body of volume V due to the presence of a penny-shaped crack with unit normal $\mathbf{n}$ and radius c is (Dienes, 1983)

$$\mathbf{e}^t = \frac{8c^3(1-\nu)}{3VG}\langle t_n^n\rangle\mathbf{n}\otimes\mathbf{n} \tag{16.22}$$

in which the Macauley bracket $\langle\ \rangle$ is used to indicate that this strain is present only if a tensile traction component exists in the direction of the normal to the crack face, and then the crack strain is proportional to the normal component of traction. Similarly, the crack strain due to shear is given by

$$\mathbf{e}^s = \frac{8c^3(1-\nu)}{3VG(3-\nu)} H[t_s^n - \mu\langle -t_n^n \rangle]\{\mathbf{t}^n \otimes \mathbf{n} + \mathbf{n} \otimes \mathbf{t}^n - 2t_n^n \mathbf{n} \otimes \mathbf{n}\} \tag{16.23}$$

Here, the Heaviside function $H[\ ]$ indicates that shear crack strain is activated only if the shear traction t_s^n is sufficiently great to overcome the coefficient of Coulomb friction coefficient μ times the compressive traction $\langle -t_n^n \rangle$. Then the quantity in the brackets $\{\ \}$ implies that the crack strain due to shear is proportional to the shear traction.

In their model, Addessio and Johnson (1989) neglected the frictional resistive force which is equivalent to assuming $\mu = 0$. The assumption may be satisfactory for many problems involving the penetration of ceramic armor because for high strain rates and high pressures there may be frictional melting at crack surfaces in which case μ would be zero. However, micrographic evidence of such melting is not available. As Dienes (1983) notes, the presence of the frictional mechanism provides the hysteresis expected in such materials.

For the purpose of providing a relatively simple theory, we follow the approach of Addessio and Johnson and also ignore the Coulomb limitation although Coulomb friction plays a role in the definition of a damage function to be given later.

To obtain the crack strain due to an ensemble of cracks, the formulation for crack strain due to a single crack must be multiplied by a probability density function $p(c, \mathbf{n})$ and the product integrated over all crack lengths and orientations. The probability density function is assumed to be isotropic in $\mathbf{n}$ and exponential in c:

$$p(c, \mathbf{n}) = \frac{N_0}{\bar{c}} \exp\left(-\frac{c}{\bar{c}}\right) \tag{16.24}$$

in which N_0 denotes the initial crack distribution and $\bar{c}$ the average crack size. N_0 is proportional to the number of cracks per unit volume and has dimension of the reciprocal of volume. The integration over crack lengths is trivial since

$$\int_0^\infty c^3 \exp\left(-\frac{c}{\bar{c}}\right) dc = 6\bar{c}^3 \tag{16.25}$$

It follows, with $\mu = 0$, that

$$\begin{aligned} \int_0^\infty \mathbf{e}^t p(c, \mathbf{n})\, dc &= \frac{16N_0\bar{c}^3(1-\nu)}{VG} \langle t_n^n \rangle \mathbf{n} \otimes \mathbf{n} \\ \int_0^\infty \mathbf{e}^s p(c, \mathbf{n})\, dc &= \frac{16N_0\bar{c}^3(1-\nu)}{VG(3-\nu)} \{\mathbf{t}^n \otimes \mathbf{n} + \mathbf{n} \otimes \mathbf{t}^n - 2t_n^n \mathbf{n} \otimes \mathbf{n}\} \end{aligned} \tag{16.26}$$

The integration of any function of a unit vector over all possible orientations of that unit vector can be transformed into an equivalent integral involving spherical coordinates (Addessio and Johnson, 1989):

$$\int_{\forall|\mathbf{n}|=1} f(\mathbf{n})\, d\mathbf{n} = \int_{\theta=0}^{2\pi} \int_{\phi=0}^{\pi} f[\mathbf{n}(\theta, \phi)] \sin\phi \, d\phi \, d\theta \tag{16.27}$$

in which θ $(0 \leq \theta \leq 2\pi)$ denotes the azimuthal angle, ϕ $(0 \leq \phi \leq \pi)$ the declination angle and $\mathbf{n}$ the outer unit normal on a unit sphere.

Let $\mathbf{p}_i$ denote the principal basis and λ_i the eigenset for the stress tensor, i.e.,

$$\boldsymbol{\sigma} = \lambda_1 \mathbf{p}_1 \otimes \mathbf{p}_1 + \lambda_2 \mathbf{p}_2 \otimes \mathbf{p}_2 + \lambda_3 \mathbf{p}_3 \otimes \mathbf{p}_3 \tag{16.28}$$

If the spherical coordinates θ and ϕ are defined with respect to this basis, then the outer unit normal becomes

$$\mathbf{n} = \cos\theta \sin\phi \mathbf{p}_1 + \sin\theta \sin\phi \mathbf{p}_2 + \cos\phi \mathbf{p}_3 \tag{16.29}$$

Now, if $\mathbf{n}$ is also interpreted as the unit normal to a crack, the traction vector and the normal component of the traction vector are

$$\begin{aligned} \mathbf{t}^n &= \lambda_1 \cos\theta \sin\phi \mathbf{p}_1 + \lambda_2 \sin\theta \sin\phi \mathbf{p}_2 + \lambda_3 \cos\phi \mathbf{p}_3 \\ t_n^n &= [\lambda_1 \cos^2\theta + \lambda_2 \sin^2\theta] \sin^2\phi + \lambda_3 \cos^2\phi \end{aligned} \tag{16.30}$$

The integral over all orientations for the crack shear strain can be determined explicitly. When integrated over the volume, the result is the total contribution $\mathbf{e}_{tot}^s$.

$$\mathbf{e}_{tot}^s \equiv \int_V \int_{\forall|\mathbf{n}|=1} \left\{ \int_0^\infty \mathbf{e}^s p(c, \mathbf{n})\, dc \right\} d\mathbf{n}\, dV = \frac{128\pi N_0 \bar{c}^3 (1-\nu)}{5G(3-\nu)} \boldsymbol{\sigma}^d \tag{16.31}$$

in which $\boldsymbol{\sigma}^d$ denotes the stress deviator.

The determination of the total contribution to the tensile crack strain $\mathbf{e}_{tot}^t$ is more complicated because of the term in the Macauley bracket. All possible situations can be separated into four cases (Margolin, 1983): Case 1 is defined to be the situation when all eigenvalues of the stress tensor are positive, Case 2 is associated with two positive eigenvalues and Case 3 with one positive eigenvalue. Case 4 is the trivial situation when all eigenvalues of the stress tensor are negative and no tensile crack strain is possible so $\mathbf{e}_{tot}^t = 0$. The details of the first three cases are given next.

Case 1 (all $\lambda_i \geq 0$):

When all eigenvalues are positive, the integration can also be carried out analogously to the case for crack shear strain. The result is the remarkably simple expression

$$\mathbf{e}_{tot}^t = \frac{64\pi N_0 \bar{c}^3 (1-\nu)}{15G} [2\boldsymbol{\sigma} + (\operatorname{tr}\boldsymbol{\sigma})\mathbf{i}] \tag{16.32}$$

However, it is also possible for the eigenvalues of the stress tensor to have mixed signs. The result is the following two cases which are much more difficult to investigate. Addessio and Johnson (1989) did not incorporate these additional cases with the result that anomalous results can be obtained with their approach for certain stress paths.

Case 2 ($\lambda_1 \leq 0, \lambda_2 \leq 0, \lambda_3 > 0$):

Now the integration over all possible orientations is more difficult because whether t_n'' is positive or not depends on the relative magnitudes of the eigenvalues and the values of the spherical polar angles. It is convenient to define the variables

$$z_1 = \frac{-\lambda_1}{\lambda_3} \qquad z_2 = \frac{-\lambda_2}{\lambda_3} \tag{16.33}$$

and the angle

$$\alpha = \tan^{-1}\left\{\frac{1}{z_1 \cos^2\theta + z_2 \sin^2\theta}\right\}^{1/2} \tag{16.34}$$

The argument for the integrand in the expression for the total crack strain is positive provided $0 \leq \phi \leq \alpha$ or $\pi - \alpha \leq \phi \leq \pi$. If numerical integration is used the result is

$$\begin{gathered} \mathbf{e}_{tot}^{t} = \frac{64 N_0 \bar{c}^3 (1 - \nu)}{5G} \lambda_3 \mathbf{S} \\ \mathbf{S} = [S_1 \mathbf{p}_1 \otimes \mathbf{p}_1 + S_2 \mathbf{p}_2 \otimes \mathbf{p}_2 + S_3 \mathbf{p}_3 \otimes \mathbf{p}_3] \end{gathered} \tag{16.35}$$

where the tensor, $\mathbf{S}$, has the same principal directions as the stress, $\boldsymbol{\sigma}$. The principal values S_1, S_2, and S_3 are given as functions of z_1 and z_2 in Fig. 16.1.

Instead of using a numerical integration each time this particular situation arises, the curves were fit with the following analytical expressions:

$$\begin{aligned} S_1 &= \frac{1}{(1 + a z_1)(1 + b z_1)^{1/2}(1 + b z_2)^{1/2}} \\ S_2 &= \frac{1}{(1 + a z_2)(1 + b z_2)^{1/2}(1 + b z_2)^{1/2}} \\ S_3 &= \frac{1}{(1 + c z_1)^{1/2}(1 + c z_2)^{1/2}} \end{aligned} \tag{16.36}$$

For $a = 1.283$, $b = 1.421$ and $c = 0.763$, the quality of the fit is also shown in Fig. 16.1.

Case 3 ($\lambda_1 \geq 0, \lambda_2 \geq 0, \lambda_3 < 0$):

For this situation, at least one of λ_1 and λ_2 must be greater than zero or there is no contribution to the crack strain. With the same definition of

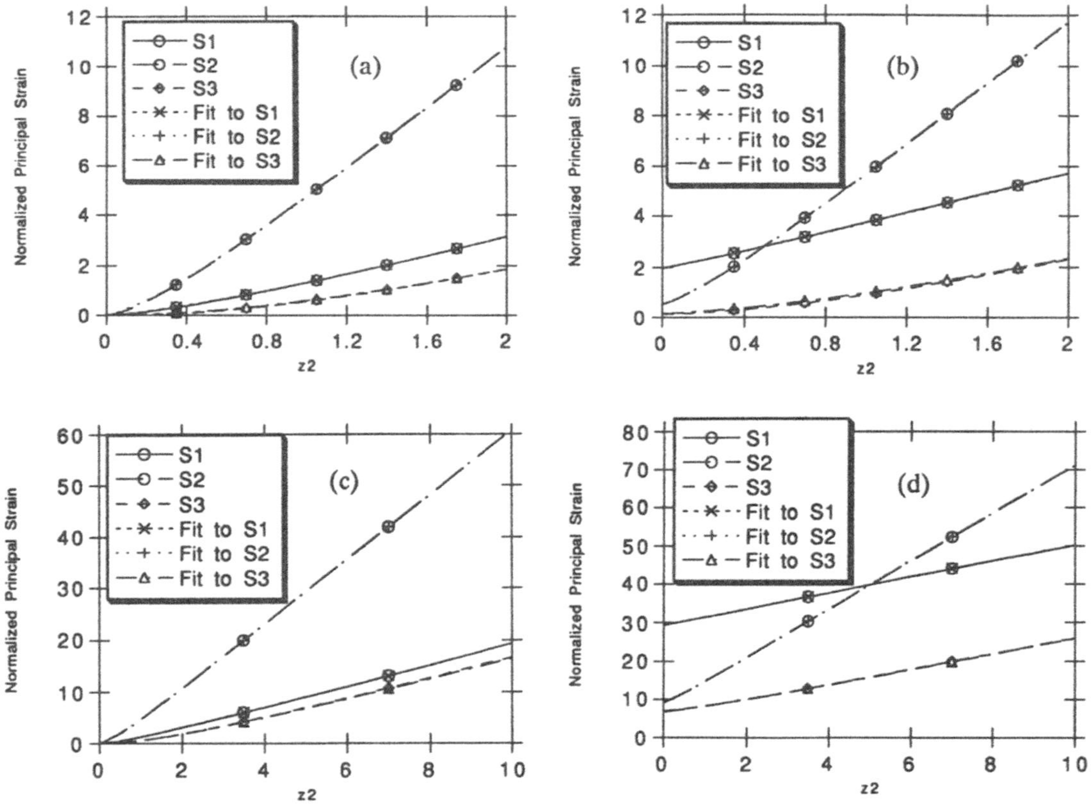

Figure 16.1. Numerically integrated normalized principal crack-opening strains and analytical fits for Case 2. (a) $z_1 = 0$ and $0 < z_2 < 2$, (b) $z_1 = 0.5$ and $0 < z_2 < 2$, (c) $z_1 = 0$ and $0 < z_2 < 10$, (d) $z_1 = 5.0$ and $0 < z_2 < 10$.

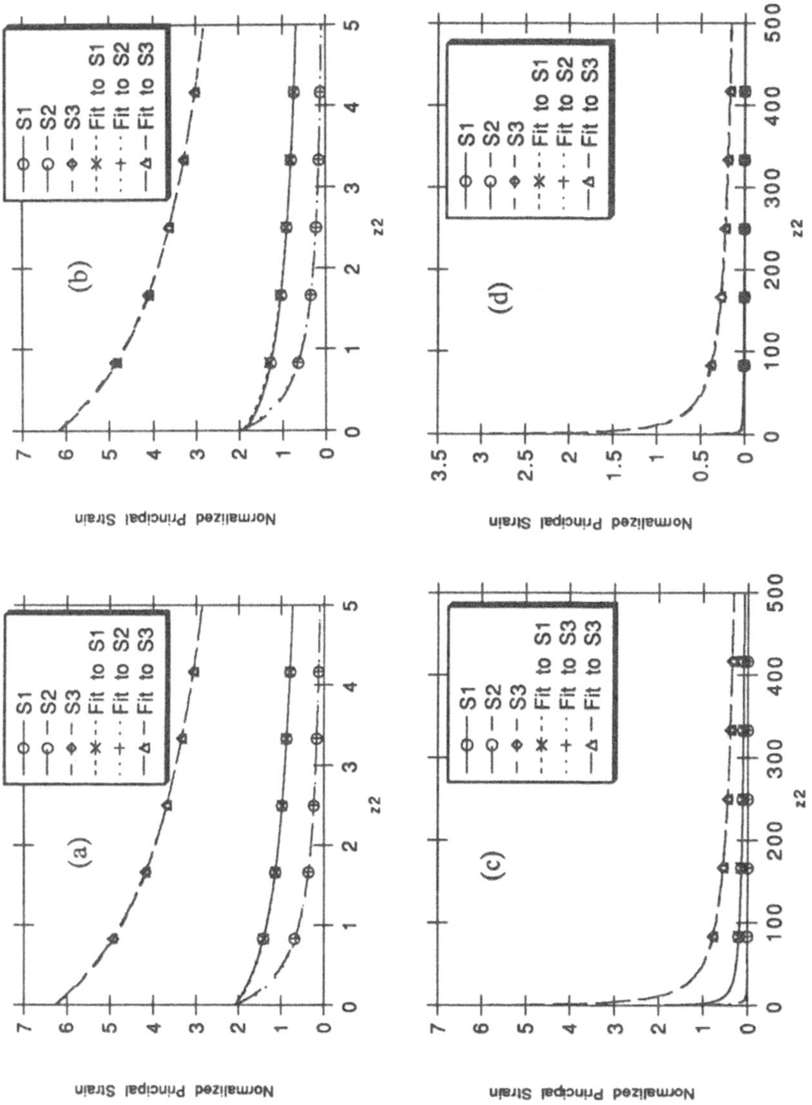

Figure 16.2. Numerically integrated normalized principal crack-opening strains and analytical fits for Case 3. (a) $z_1 = 0$ and $0 < z_2 < 5$, (b) $z_1 = 0.04$ and $0 < z_2 < 5$, (c) $z_1 = 0$ and $0 < z_2 < 500$, (d) $z_1 = 4.0$ and $0 < z_2 < 500$.

variables used for Case 2 given in (16.33) and (16.34) a positive integrand exists when $\alpha \leq \phi \leq \pi - \alpha$. Now the total strain due to tensile cracking is:

$$\mathbf{e}_{tot}^{t} = \frac{-64\pi N_0 \bar{c}^3(1-\nu)}{15G} \lambda_3 \mathbf{S} \tag{16.37}$$

Again, analytical fits are proposed for computational convenience. Numerical quadrature and the fits provided by the following analytical expressions are shown in Fig. 16.2:

$$\begin{aligned} S_1 &= 3z_1 + z_2 - 1 + \frac{1}{(1+az_1)^{3/2}(1+bz_2)^{1/2}} \\ S_2 &= 3z_2 + z_1 - 1 + \frac{1}{(1+az_2)^{3/2}(1+bz_1)^{1/2}} \\ S_3 &= z_1 + z_2 - 3 + \frac{3}{(1+az_1)^{1/2}(1+cz_2)^{1/2}} \end{aligned} \tag{16.38}$$

in which $a = 1.403$, $b = 1.470$, and $c = 0.500$. More details are given by Lewis (1991).

16.4. Damage Model: General Formulation for Crack Strains

The results of the previous section suggest the following simple formulation for damage based on crack strain. Suppose the cracks are isotropically distributed throughout a medium and that the strain in the uncracked part of the medium is called the elastic strain, $\mathbf{e}^e$. The total strain, $\mathbf{e}$, is the sum of the elastic strain and the crack strain, $\mathbf{e}^c$:

$$\mathbf{e} = \mathbf{e}^e + \mathbf{e}^c \tag{16.39}$$

The Helmholtz free energy is assumed to be equal to the sum of the stored energy density and the surface energy created by cracks:

$$\begin{aligned} \Psi &= S_E + C_E \\ S_E &= \tfrac{1}{2}\mathbf{e}^e : \mathbf{E}^0 : \mathbf{e}^e = \tfrac{1}{2}(\mathbf{e} - \mathbf{e}^c) : \mathbf{E}^0 : (\mathbf{e} - \mathbf{e}^c) \\ C_E &= \tfrac{1}{2} b \bar{c}^2 \end{aligned} \tag{16.40}$$

in which $\mathbf{E}^0$ denotes the elasticity tensor of the uncracked medium and b is the surface energy per unit surface area. If the material is originally isotropic, then $\mathbf{E}^0$ is isotropic although the cracked material will not be isotropic.

Suppose the crack strain $\mathbf{e}^c$ and average crack length $\bar{c}$ are chosen to be the damage variables, i.e., the set D becomes

$$D \equiv \{\mathbf{e}^c, \bar{c}\} \tag{16.41}$$

Then the conjugate variables are

$$
\begin{aligned}
\mathbf{Y}_{ec} &\equiv \frac{\partial \Psi}{\partial \mathbf{e}^c} = -\mathbf{E} : (\mathbf{e} - \mathbf{e}^c) = -\boldsymbol{\sigma} \\
Y_{\bar{c}} &= \frac{\partial \Psi}{\partial \bar{c}} = b\bar{c}
\end{aligned}
\tag{16.42}
$$

Each of the expressions for crack strain from the previous section is of the form

$$\mathbf{e}^c = \omega \mathbf{m} \tag{16.43}$$

where $\omega = \bar{c}^3/c_r^3$, c_r is a reference crack length used to make ω dimensionless, and $\mathbf{m}$ depends only on $\boldsymbol{\sigma}$. Assume that the time rate of change of stress is not significant in the evolution of crack strain, i.e.,

$$\dot{\boldsymbol{\sigma}} = \mathbf{E}^0 : \dot{\mathbf{e}}^e = \mathbf{E}^0 : (\dot{\mathbf{e}} - \dot{\mathbf{e}}^c) \approx 0 \tag{16.44}$$

This assumption is consistent with the concept of a damage surface since the stress normally does not change significantly when the damage surface criterion is being met. Then the following evolution equation holds:

$$\dot{\mathbf{e}}^c = \dot{\omega} \mathbf{m} \tag{16.45}$$

where the specific form for $\mathbf{m}$ depends on the specific stress case under consideration. Similarly, the evolution equation for the average crack length is

$$\dot{\bar{c}} = \dot{\omega} m \qquad m = \frac{c_r^3}{3\bar{c}^2} \tag{16.46}$$

which is independent of the stress case. The dissipation rate becomes

$$
\begin{aligned}
D_s &= -\mathbf{Y}_{ec} : \dot{\mathbf{e}}^c - Y_{\bar{c}} \dot{\bar{c}} \\
&= \dot{\omega}[\boldsymbol{\sigma} : \mathbf{m} - b\bar{c}m]
\end{aligned}
\tag{16.47}
$$

Consequently, the damage function of (16.14) reduces to

$$\Psi_D = \boldsymbol{\sigma} : \mathbf{m} - b\bar{c}m - \Psi_{DO} \tag{16.48}$$

where Ψ_{DO} must be positive to provide positive dissipation as given in (16.15). This formulation essentially uses dissipation as the independent parameter describing the damage process, but (16.46) shows that the average crack length could also be used to parameterize the process.

If $\Psi_D < 0$, cracking is assumed not to occur. While damage is occurring, $\Psi_D = 0$. Then (16.46) and (16.48) imply

$$\boldsymbol{\sigma} : \mathbf{m} = b\frac{c_r^3}{3\bar{c}} + \Psi_{DO} \tag{16.49}$$

The term $\boldsymbol{\sigma} : \mathbf{m}$ must be positive and can be defined as the square of an "effective stress." From both a mathematical and a physical viewpoint it is necessary to predict initial crack propagation at a finite stress so the initial value of $\bar{c}$ is chosen to be some positive value $\bar{c}_0$ which can be interpreted as the average crack length of the original material. The model predicts softening because $\bar{c}$ can only increase and the effective stress must therefore decrease. Such a feature does not necessarily reflect failure because of the constraint provided by the compatibility conditions (Neilsen and Schreyer, 1993).

While damage is occurring, the consistency condition

$$\dot{\Psi}_D = 0 \tag{16.50}$$

yields an expression for $\dot{\omega}$ and hence for $\dot{\bar{c}}$ from (16.46). If the predicted value for $\dot{\bar{c}}$ is greater than the Rayleigh limit $\dot{\bar{c}}_{max}$, then (16.46) is inverted by first setting $\dot{\bar{c}} = \dot{\bar{c}}_{max}$ and then solving for $\dot{\omega}$ which is then used in (16.45):

$$\dot{\mathbf{e}}^c = \dot{\bar{c}}_{max} \frac{3\bar{c}^2}{c_r^3} \mathbf{m} \tag{16.51}$$

The result is equivalent to an overstress approach in which case it is possible for Ψ_D to be greater than zero. It follows from (16.47) and (16.48) that for paths involving large enough strain rate, the dissipation rate becomes

$$D_s = \dot{\bar{c}}_{max} \frac{3\bar{c}^2}{c_r^3} [\boldsymbol{\sigma} : \mathbf{m} - \bar{c}bm] = \dot{\bar{c}}_{max}(\Psi_D + \Psi_{DO}) \tag{16.52}$$

16.5. Tensile Crack Strains

16.5.1. Case 1

Because of its simplicity, the first case considered is the one where all eigenvalues of the stress are positive. It follows from (16.32) that

$$\mathbf{m} = A\left[\frac{3}{2}\boldsymbol{\sigma} + \frac{3}{4}\mathbf{i}(\mathrm{tr}\ \boldsymbol{\sigma})\right] \qquad A = \frac{4}{3} \cdot \frac{64\pi N_0(1 - \nu)}{15G} \tag{16.53}$$

The dissipation rate of (16.47) is

$$\begin{aligned} D_s &= \dot{\omega}_t A\left[\frac{3}{2}\boldsymbol{\sigma} : \boldsymbol{\sigma} + \frac{3}{4}(\mathrm{tr}\ \boldsymbol{\sigma})^2\right] - \dot{\omega}_t b \frac{c_r^3}{3\bar{c}} \\ &= \dot{\omega}_t A\left[\frac{3}{2}\boldsymbol{\sigma}^d : \boldsymbol{\sigma}^d + \frac{5}{4}(\mathrm{tr}\ \boldsymbol{\sigma})^2\right] - \dot{\omega}_t b \frac{c_t^3}{3\bar{c}} \end{aligned} \tag{16.54}$$

in which the subscript "t" is used with ω to emphasise the point that here the evolution parameter is associated with tensile cracking strains. The use of (16.48) yields

$$\Psi_{D1} = A\left[\frac{3}{2}\boldsymbol{\sigma}^d : \boldsymbol{\sigma}^d + \frac{5}{4}(\mathrm{tr}\ \boldsymbol{\sigma})^2\right] - b\frac{c_r^3}{3\bar{c}} - \Psi_{01} \tag{16.55}$$

The subscript "1" indicates that this particular function is associated with Case 1.

16.5.2. Cases 2 and 3

Different damage functions for each of the other tensile cases are constructed in an analogous manner. Based on the expressions for the total crack strain in (16.35) and (16.37), it follows that

$$\mathbf{m} = A\lambda_3\mathbf{S} \qquad A_2 = \frac{64\pi N_0(1-\nu)}{5G} \qquad A_3 = \frac{-64\pi N_0(1-\nu)}{15G} \tag{16.56}$$

With $A = A_2$ or $A = A_3$, the dissipation rate becomes

$$D_s = \dot{\omega}_t A\lambda_3 \boldsymbol{\sigma} : \mathbf{S} - \dot{\omega}_t b\frac{c_r^3}{3\bar{c}} \tag{16.57}$$

where

$$\boldsymbol{\sigma} : \mathbf{S} = \lambda_1 S_1 + \lambda_2 S_2 + \lambda_3 S_3 \tag{16.58}$$

because $\boldsymbol{\sigma}$ and $\mathbf{S}$ have the same principal basis. Since the function, $\mathbf{S}$, assumes different forms for these two cases the dissipation expression actually yields two expressions for the damage function:

$$\Psi_{D2,3} = A\lambda_3 \boldsymbol{\sigma} : \mathbf{S} - b\frac{c_r^3}{3\bar{c}} - \Psi_{02,3} \tag{16.59}$$

The subscript "2, 3" indicates the particular function associated with Cases 2 and 3, respectively, for tensile cracking. The value of Ψ_{02} is picked to ensure a smooth continuation of the surface associated with Ψ_{D1}, and similarly for Ψ_{03}.

Although there are three different cases governing tensile cracking, only one case is active for a given stress state so it is appropriate to utilize only one evolution parameter, ω_t, with all three cases.

16.5.3. Relationship to Existing Approaches

The classical concept of a stress intensity factor is often utilized in the construction of a suitable damage function. For example, Addessio and Johnson (1989) assume a crack will extend when $F^t \geq 0$ where

$$F^t = \mathbf{t}^n \cdot \mathbf{t}^n - \frac{\nu}{2} t_n^n - \frac{K}{c} \tag{16.60}$$

K is related to the critical stress intensity factor and c is again the crack length. If the distribution of all crack lengths and orientations is taken into account, then the following integration must be performed

$$F^{ti} \equiv \int_{\forall|\mathbf{n}|=1} \int_0^\infty F^t(\mathbf{n}, c) p(\mathbf{n}, c)\, dc\, d\mathbf{n} \tag{16.61}$$

The result is

$$F^{ti} = 2\pi N_0 \left[\frac{2(5-2v)}{15} \boldsymbol{\sigma}^d : \boldsymbol{\sigma}^d + (2-v)\frac{1}{9}(\mathrm{tr}\,\boldsymbol{\sigma})^2 - \frac{2K}{\bar{c}} \Gamma(0) \right] \tag{16.62}$$

in which $\bar{c}$ is the average crack length, as before, and

$$\Gamma(s) \equiv \int_0^\infty t^{(s-1)} e^{-t}\, dt \tag{16.63}$$

is the gamma function. When evaluated at $s = 0$, the result $\Gamma(0)$ is infinite. Addessio and Johnson avoided this undesirable feature by multiplying F^t by c prior to performing the integration. With a suitable resolution of the singularity, the condition for propagation of distributed cracks reduces to the criterion $F^* \geq 0$ where

$$F^* = \frac{3}{2} \boldsymbol{\sigma}^d : \boldsymbol{\sigma}^d + \frac{5}{4} \frac{(2-v)}{(5-2v)} (\mathrm{tr}\,\boldsymbol{\sigma})^2 - \frac{B}{\bar{c}} \tag{16.64}$$

where B is a constant. Again, the initial value of $\bar{c}$ is chosen to be some positive value.

Note that if the coefficient $(2-v)/(5-2v)$ is replaced with one, then (16.64) is of a form similar to that given in the damage function of (16.55). In light of the assumptions already introduced in connection with the averaging process, the simplification of (16.55) relative to (16.64) is not considered to be significant.

16.6. Shear Crack Strains

For the situation where shear crack strains are possible, it follows from (3–13) that

$$\mathbf{m} = A_s \boldsymbol{\sigma}^d \qquad A_s = \frac{128\pi N_0 (1-v)}{5G(3-v)} \tag{16.65}$$

With the use of another evolution parameter, ω_s, associated strictly with shear crack strains (modes II and III), the dissipation rate is

$$D_s = \dot{\omega}_s A_s \boldsymbol{\sigma}^d : \boldsymbol{\sigma}^d - \dot{\omega}_s b \frac{c_r^3}{3\bar{c}} \tag{16.66}$$

from which the damage function becomes:

$$\Psi_{Ds} = A_s \boldsymbol{\sigma}^d : \boldsymbol{\sigma}^d - b \frac{c_r^3}{3\bar{c}} - \Psi_{0s} \tag{16.67}$$

The subscript "s" indicates that this particular function is associated with shear. The formulation is reminiscent of the von Mises criterion for plasticity.

In the development of the crack shear strain, the term involving the Coulomb term was ignored. For simplicity we incorporate the mean pressure into the damage function to account for the Coulomb effect present in (16.23) in an approximate but smooth fashion:

$$\Psi_{Dsc} = A_s[\boldsymbol{\sigma}^d : \boldsymbol{\sigma}^d - \alpha^2 P^2] - b\frac{c_r^3}{3\bar{c}} - \Psi_{0sc} \tag{16.68}$$

in which $P = \frac{1}{3}\,\mathrm{tr}\,\boldsymbol{\sigma}$ is the mean pressure and α is a material parameter corresponding to the friction coefficient, μ. The result is analogous to Prager-Drucker plasticity with a nonassociated flow rule.

It is probable that either tensile cracking or shear cracking will be activated at any instant. However, it is possible for both to be activated simultaneously in which case the consistency conditions for both surfaces must be satisfied simultaneously to obtain values for $\dot{\omega}_t$ and $\dot{\omega}_s$. This is a classical procedure in plasticity (Koiter, 1953) although numerical algorithms sometimes use a simpler approach in which only the surface is activated for which the initial trial stress provides the maximum error.

16.7. Composite Damage Surface

There are four damage functions so the actual damage surface is the greatest lower bound of the union. To provide a graphical representation of the composite surface, a particular stress path was chosen in which the material is first loaded hydrostatically to a pressure P (positive in compression) followed by a superposition of shear stress τ until a damage surface is reached. The principal values of stress are $\lambda_1 = -P - \tau$, $\lambda_2 = -P$, $\lambda_3 = -P + 1$. All terms were divided by A_s and Poisson's ratio was picked to be 0.18. Then $A_1/A_s = 0.627$, $A_2/A_s = 1.41$, and $A_3/A_s = 0.47$. A value for Ψ_{01} was chosen to provide a "reasonable" value for damage to first occur along the hydrostatic tensile axis (negative P). Then Ψ_{02} and Ψ_{03} were chosen to ensure continuity of the composite damage surface. Normally experimental data would be available for selecting these parameters.

For the choice of material parameters given as follows:

$$\frac{b}{A_s}\frac{c_r^3}{3\bar{c}} = 794\ (\mathrm{MPa})^2$$

$$\frac{\Psi_{01}}{A_s} = 210\ (\mathrm{MPa})^2$$

$$\frac{\Psi_{02}}{A_s} = 226\ (\mathrm{MPa})^2 \tag{16.69}$$

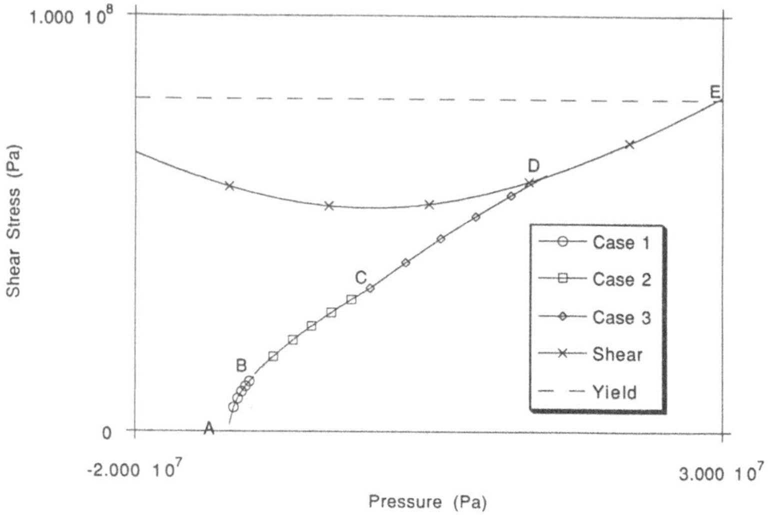

Figure 16.3. Composite damage surface formed from the union of shear damage surface and the damage surfaces obtained for Cases 1, 2, and 3.

$$\frac{\Psi_{03}}{A_s} = 0$$

$$\frac{\Psi_{0sc}}{A_s} = 5020\ (\text{MPa})^2$$

$$\alpha = 2.83$$

the composite surface is shown in Fig. 16.3 where the maximum shear predicted by the various surfaces is plotted as a function of pressure. The segment that starts on the P-axis, identified with circles (A–B), is the contribution when all principal values of the stress tensor are positive (Case 1). The following segments denoted with boxes (B–C) and diamonds (C–D) are formed from Cases 2 and 3, respectively. Note that the union of these segments is smooth but may not be convex. The line labeled with x's is the contribution from shear damage and is not convex, a feature readily seen from the analytical representation of the damage surface. For large P, this segment of the surface asymptotically approaches a linear relation between τ and P, a feature purposely imposed in the formulation. Only the segment D–E of the shear-crack surface will be activated because segments of other surfaces are closer to the origin.

The form of the damage surface shown in Fig. 16.3 is representative of yield surfaces often seen in the literature in which a smooth "tensile cap" is superposed on a conventional Mises surface. The purpose for such a cap is to provide a constitutive model that reflects, in at least an approximate manner, the fact that damage occurs when tensile components of stress are sufficiently large. Here, an approximate microstructural analysis has been used to obtain such a surface but with an interpretation associated with damage as reflected by crack strains rather than with plasticity.

The composite damage surface is meant to be appropriate for primarily tensile states of stress which could be defined loosely as states of stress for which P is close to zero. For large values of P, most materials respond plastically. A plasticity formulation could be combined with the current damage model to provide a constitutive relation with a wide range of applicability. For example, if a Mises model for plasticity is considered appropriate, the damage surface shown in Fig. 16.3 could be supplemented with a plasticity surface such as one illustrated with a horizontal (dashed) line at a specified yield stress. The yield surface would be activated if the stress is associated with a large value of P (to the right of point E). It is also possible to activate the plasticity part for low values of mean pressure if the effective damage surface expands and allows large values of shear because of a large strain rate. Therefore, there is considerable generality with the proposed model even though the formulation is not complicated.

16.8. Conclusion

A formulation for describing the development of microcracks has been constructed utilizing the following assumptions and concepts:

(1) the total strain can be considered a sum of elastic and crack strains,
(2) the Helmholtz free energy is a sum of elastic and crack-surface energies,
(3) damage functions are derived from expressions for dissipation,
(4) evolution equations for crack strength and average crack size are postulated from theoretical derivations available in the literature for crack strains, and
(5) all possible combinations of principal components of stress have been included in the formulation for tensile cracking.

The result is a damage evolution theory consistent with the restrictions of thermodynamics but simple enough to be amenable to numerical analyses. The actual damage surface is continuous but not necessarily convex.

Numerous issues have not been addressed. One of these is deactivation of damage which occurs when cracks close and the stress state is compressive. An approach involving projection operators has been developed by Hansen and Schreyer (1993) but the details for this particular formulation involving crack strains have not been formulated. Therefore, the current results can only be expected to hold for paths that are monotonic and approximately "radial." Other issues involve (i) specific expressions for the maximum allowable average crack speed, (ii) correlation with experimental data, (iii) the addition of plasticity, and (iv) the ease of use in computational codes.

Another important topic is failure as exhibited by loss of ellipticity. As shown by Neilsen and Schreyer (1993), loss of ellipticity can be greatly affected by the stress path. Since the basic formulation implies softening, and since the tangent stiffness is not symmetric, there is a strong possibility for strain localization; more so for some paths than for others. In this context, localization can be thought of as the development of a macrocrack. One path for which this is always the case is that of uniaxial stress in tension. However, for other paths, macrocracking may not occur until a significant amount of softening has occurred. Since a combination of rate dependence and dynamic loading impedes localization, no attempt to investigate localization has been attempted with this model. Also, the problem of ill-posedness in the presence of rate effects remains, for the most part, an open issue.

Nevertheless, the balance of microstructural detail with the simplicity inherent in the current formulation should make it useful for numerical analyses involving the mechanical response of brittle materials.

16.9. References

Addessio, F.L. and Johnson, J.N. A Constitutive Model for the Dynamic Response of Brittle Materials, *J. Appl. Physics*, **67**, 3275–3286 (1989).

Budiansky, B. and O'Connell, R.J. Elastic Moduli of a Cracked Solid, Int. J. Solids and Structures **12**, 81–97 (1976).

Costin, L.S. A Microcrack Model for the Deformation and Failure of Brittle Rock, *J. Geophys. Res.* **88**, 9485–9492 (1983).

Dienes, J.K. Statistical Crack Mechanics, *Proceedings of the Colloque Internationale du CNRS no. 351: Failure Criteria of Structured Media*, Villard-de-Lans, France (1983).

Hansen, N.R. and Schreyer, H.L. A Thermodynamically Consistent Framework for Theories of Elastoplasticity Coupled with Damage, *Int. J. Solids and Structures* (In press)

Johnson, G.R. and Holmquist, T.J. A Computational Model for Brittle Materials Subjected to Large Strains, High Strain Rates, and High Pressures, *Shock-Wave and High-Strain-Rate Phenomena in Materials*, edited by M.A. Meyers, L.E. Hurr and K.P. Staudhammer, Marcel Dekker, Inc., New York, 1075–1081 (1992).

Koiter, W.T. Stress-Strain Relations, Uniqueness and Variational Theorems for Elastic-Plastic Materials with a Singular Yield Surface, *Q. Appl. Math.*, **11**, 350–354 (1953).

Kipp, M.E. and Grady, D.E. *Numerical Studies of Rock Fragmentations*, Sandia Report SAND79-1582, Sandia National Laboratories, Albuquerque, NM (1979).

Kipp, M.E., Grady, D.E., and Wise, J.L. Planar-Shock and Penetration Response of Ceramics, *Shock-Wave and High-Strain-Rate Phenomena in Materials*, edited by M.A. Meyers, L.E. Murr, and K.P. Staudhammer, Marcel Dekker, New York, 1083–1091 (1992).

Lewis, M.W. *Investigation of a Thermodynamically Consistent, Dynamic Brittle Damage Constitutive Model for Ceramics*, Masters Thesis, Department of Mechanical Engineering, The University of New Mexico, Albuquerque, NM (1991).

Margolin, L.G. Elastic Moduli of a Cracked Body, *Int. J. Fracture*, **22**, 65–79 (1983).
Margolin, L.G. Microphysical Models for Inelastic Material Response, *Int. J. Engng. Sci.*, **22**, 1171–1179 (1984).
Neilsen, M.K. and Schreyer, H.L. Bifurcations in Elastic-Plastic Materials, *Int. J. Solids Structures*, **30**, 521–544, (1993).
Rajendran, A.M. *High Strain Rate Behavior of Metals, Ceramics, and Concrete*, WL-TR-92-4006, Final report for period September 1986—September 1991, Materials Directorate, Wright Laboratory, Air Force Systems Command, Wright-Patterson Air Force Base, Ohio, 45433-6533 (1992).
Shockey, D.A., Marchand, A.H., Skaggs, S.R., Cort, G.E., Burkett, M.W., and Parker, R. Failure Phenomenology of Confined Ceramic Targets and Impacting Rods, *Int. J. Impact Engng.* **9**, 263–275 (1990).
Stumvoll, M. and Swoboda, G. Deformation Behavior of Ductile Solids Containing Anisotropic Damage, *J. Eng. Mechanics*, **119**, 1331–1352 (1993).
Taylor, L.M., Chen, E., and Kuszmaul, J.S. Microcrack-Induced Damage Accumulation in Brittle Rock under Dynamic Loading, *Computer Methods in Applied Mechanics and Engineering*, **55**, 301–320 (1986).
Thorne, B.J. *A Damage Model for Rock Fragmentation and Comparison with Blasting Experiments in Granite*, Sandia Report SAND90-1389, Sandia National Laboratories, Albuquerque, NM (1990).

Author Index

Subject Index

GPSR Compliance

The European Union's (EU) General Product Safety Regulation (GPSR) is a set of rules that requires consumer products to be safe and our obligations to ensure this.

If you have any concerns about our products, you can contact us on ProductSafety@springernature.com

In case Publisher is established outside the EU, the EU authorized representative is:

Springer Nature Customer Service Center GmbH
Europaplatz 3
69115 Heidelberg, Germany

Batch number: 10837688

Printed by Printforce, the Netherlands